MICHIGAN ACADEMY OF SCIENCE ARTS AND LETTERS

VOLUME XXIX (1943)

CONTAINING PAPERS SUBMITTED AT THE ANNUAL MEETING IN 1943

VOLUME XXIX IS AVAILABLE IN FOUR PARTS:

PART I: BOTANY AND FORESTRY

PART II: ZOOLOGY

PART III: GEOGRAPHY AND GEOLOGY

PART IV: GENERAL SECTION

Anthropology, Economics, Folklore
History and Political Science
Language and Literature
Philosophy, Sociology

PAPERS OF THE MICHIGAN ACADEMY OF SCIENCE ARTS AND LETTERS

EDITORS

EUGENE S. McCARTNEY

HENRY VAN DER SCHALIE

VOLUME XXIX (1943)

"Pusilla res mundus est nisi in illo quod quaerat omnis mundus habeat."

— SENECA, *Naturales Quaestiones*

ANN ARBOR: THE UNIVERSITY OF MICHIGAN PRESS

LONDON: HUMPHREY MILFORD, OXFORD UNIVERSITY PRESS

1944

Set up and printed,
April, 1944
Published May, 1944

Printed and bound by CPI Group (UK) Ltd, Croydon, CR0 4YY

PRO PATRIA

GRACE E. POTTER, Assistant Editor, Ensign, U. S. N. R.

AUTHORS

PIERCE BRODKORB, First Lieutenant, Army Air Forces

FRED J. ERICSON, Second Lieutenant, Army Air Forces

FRED W. FOSTER, Lieutenant (j.g.), U. S. N.

HOWARD A. MACCORD, Captain, United States Army

CARL P. MANSON, Special Liaison Security Officer, War Production Board

JOHN M. ROBERTS, Supervisor of Operations, Biologic Products Laboratory, United States Army, Medical Department, Lansing, Michigan

JOSEPH A. RUSSELL, Major, A. U. S.; Chief Research Section, Geographic Branch, G–2

OFFICERS FOR 1943

PRESIDENT, H. R. Hunt, Michigan State College
VICE-PRESIDENT, Malcolm H. Soule, University of Michigan
SECRETARY, Harry W. Hann, University of Michigan
TREASURER, Mischa Titiev, University of Michigan
EDITOR, Henry van der Schalie, University of Michigan
LIBRARIAN, Warner G. Rice, University of Michigan

SECTION CHAIRMEN

ANTHROPOLOGY, Robert T. Hatt, Cranbrook Institute of Science
BOTANY, Herbert C. Beeskow, Michigan State College
ECONOMICS, Walter A. Terpenning, Albion College
FINE ARTS, Avard Fairbanks, University of Michigan
FOLKLORE, Emelyn E. Gardner, Wayne University
FORESTRY, A. B. Bowman, Michigan State College
GEOGRAPHY, H. Thompson Straw, Western Michigan College
GEOLOGY AND MINERALOGY, G. M. Stanley, University of Michigan
HISTORY AND POLITICAL SCIENCE, J. A. Perkins, University of Michigan
LANDSCAPE ARCHITECTURE, Charles F. Boehler, Lansing
LANGUAGE AND LITERATURE, Julio del Toro, University of Michigan
PHILOSOPHY, Raymond Hoekstra, Wayne University
PSYCHOLOGY, Heinz Werner, Wayne County Training School
SANITARY AND MEDICAL SCIENCE, John W. Bean, University of Michigan
SOCIOLOGY, Arthur E. Wood, University of Michigan
ZOOLOGY, Edward T. Boardman, Cranbrook Institute of Science

CONTENTS

BOTANY

FORESTRY

ZOOLOGY

LANGUAGE AND LITERATURE

PHILOSOPHY

SOCIOLOGY

ILLUSTRATIONS

PLATES

BOTANY

FORESTRY

ZOOLOGY

ANTHROPOLOGY

FIGURES IN THE TEXT

BOTANY

PAGE

ZOOLOGY

GEOLOGY

ANTHROPOLOGY

BOTANY

THE DISCRIMINANT FUNCTION APPLIED TO SPORE MEASUREMENTS *

WILLIAM DOWELL BATEN

INTRODUCTION

LENGTH and width measurements of spores of *Galera crispa* and *Galera lactea,* which grow side by side on the campus of Michigan State College, were made by Bessey (1) in order to determine whether or not there were significant differences in arithmetic averages of spore dimensions pertaining to these species of fungi. Measurements of 140 spores from four specimens of the former and of 160 spores from four specimens of the latter were made. An analysis of variance was carried out on the length and width measurements of each species to determine homogeneity among length and width averages pertaining to specimens; these analyses revealed no evidence of heterogeneity.

There are several methods of finding significant differences between the averages of these measurements. One is to test for significance between means of length measurements and then to make a similar test between means of width measurements.

99 PER CENT ELLIPSES

One of the methods used in this paper is that of inquiring into the significance between the points $(\bar{x}, \bar{y})$ and $(\bar{x}', \bar{y}')$ in a plane (where $\bar{x}$ and $\bar{y}$ represent respectively the means of length and width measurements of *Galera crispa* spores and $\bar{x}'$ and $\bar{y}'$ represent respectively means of similar spore measurements of *Galera lactea*). These points are (12.26, 8.21) and (12.15, 7.49); they are plotted in Figure 1, together with their 99 percentage ellipses. Rietz (6) shows how these ellipses can be obtained. Since they do not overlap, the centers are significantly separated, and hence the arithmetic averages of lengths and widths taken together for one species are significantly different

* Authorized as journal article No. 632, N. S., of the Michigan Agricultural Experiment Station.

from those of the other species. This indicates that spores of *Galera crispa* are definitely larger than those of *Galera lactea*. The interpretation of this test is that, if many samples of 140 spores of *Galera crispa* are taken at random from the parent population, on the average

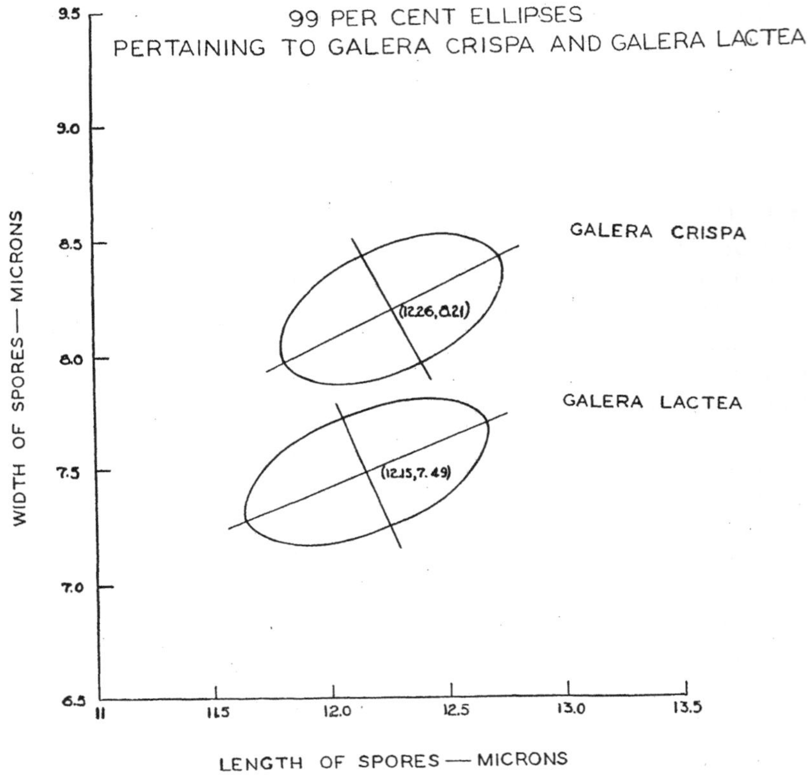

FIG. 1. Ninety-nine per cent ellipses pertaining to averages of length and width measurements of spores of *Galera crispa* and *Galera lactea*

about 99 per cent of the points (one point determined from each sample) will fall within this ellipse; and since these points fall about the point representing the true means of length and width measurements, the odds are 99 to 1 that the point representing the true means will fall in this ellipse, or 1 to 99 that it will fall on the outside. The same interpretation can be given to the point representing the true averages pertaining to the other species. If this is true, it is clear that points representing the true parent universes (*Galera crispa* and *Galera lactea*) are statistically separated or that these two species differ in averages pertaining to spore measurements.

THE DISCRIMINANT FUNCTION

Let a linear compound made up of spore length and spore width be $X = ax + by$, where x and y represent respectively length and width of *Galera crispa* spores and a and b are constants to be found. Let the linear compound pertaining to the other species be $X' = ax' + by'$. The two functions X and X' have the same a and b constants. Let the difference between the means of these be represented by

$$D = ad_1 + bd_2,$$

where $d_1 = \bar{x} - \bar{x}'$ and $d_2 = \bar{y} - \bar{y}'$. Let

$$\begin{aligned} S = a^2[\Sigma(x - \bar{x})^2 + \Sigma(x' - \bar{x}')^2] & \\ & + 2ab[\Sigma(x - \bar{x})(y - \bar{y}) + \Sigma(x' - \bar{x}')(y' - \bar{y}')] \\ & + b^2[\Sigma(y - \bar{y})^2 - \Sigma(y' - \bar{y}')^2], \end{aligned}$$

or the sum of squares of the compounds *within species*. A linear function composed of length and width measurements is desired which will discriminate the most between these species or a linear function among all possible linear functions that will show the most difference between these species. This one compound or discriminant function is found by maximizing the ratio D^2/S. On maximizing this ratio the following simultaneous equations are obtained,

$$[\Sigma(x-\bar{x})^2+\Sigma(x'-\bar{x}')^2]a+[\Sigma(x-\bar{x})(y-\bar{y})+\Sigma(x'-\bar{x}')(y'-\bar{y}')]b=d_1,$$
$$[\Sigma(x-\bar{x})(y-\bar{y})+\Sigma(x'-\bar{x}')(y'-\bar{y}')]a+[\Sigma(y-\bar{y})^2+\Sigma(y'-\bar{y}')^2]b=d_2,$$

from which the values of a and b are derived. From the data these equations are

$$789.06a + 246.59b = 0.11,$$
$$246.59a + 290.59b = 0.72,$$

from which $a = -\ 0.000863$ and $b = 0.003210$. The compound X is

$$X = -\ 0.000863x + 0.003210y.$$

This discriminant function enables one to evaluate a combination of length and width measurements of spores. It furnishes a measure for comparing compounds made up of length and width measurements of spores from the two species. The difference between the means of the compounds pertaining to the two species is

$$(1)\quad D = -\,0.000863(0.11) + 0.003210(0.72) = -\,0.000949 + 0.002311 = 0.002216.$$

The following table, containing an analysis of variance of the two discriminant functions, shows that there is a significant difference between the arithmetic averages of these linear compounds. This indicates that the discriminant function pertaining to one species is significantly different from the discriminant function made up of length and width of spores pertaining to the other.

TABLE I

ANALYSIS OF VARIANCE PERTAINING TO THE DISCRIMINANT FUNCTIONS

Source of variation	Degrees of freedom	Sum of squares	Mean square
Total	299	$D\left[1 + \frac{N_1 \cdot N_2}{N_1 + N_2} D\right] = 0.002583$	
Between species means	2	$\frac{N_1 \cdot N_2}{N_1 + N_2} D^2 = 0.000367$	0.000184*
Within species	297	$D = 0.002216$	0.000007

* Significant at the 1 per cent point.

On examining the numerical value of D in formula (1) it is seen that the difference between the averages of widths multiplied by b is larger than the difference between the averages of length measurements multiplied by a; hence width measurements are more important than length measurements for determining significance between these species.

The discriminant function can be used in comparing two linear combinations of several variables. Its coefficients furnish weights (if written in standard units) to the quantities measured. Literature pertaining to this function can be found in references 2–5.

SUMMARY

Three methods have been presented for testing significance between means pertaining to length and width measurements of spores from two species of fungi. Special attention was given to the 99 percentage ellipses and to R. A. Fisher's discriminant function (5). The discriminant function enables one to test for a difference between two linear compounds of measurements under consideration.

I am grateful to Dean E. A. Bessey for supplying the data for this study.

MICHIGAN STATE COLLEGE
EAST LANSING, MICHIGAN

LITERATURE CITED

1. BESSEY, E. A. Unpublished data.
2. BRIER, G. W., SCHOTT, R. G., AND SIMMONS, V. L. 1940. The Discriminant Function Applied to Quality Rating in Sheep. Proc. Am. Soc. Animal Prod., 1: 153–160.
3. COX, GERTRUDE, AND MARTIN, W. P. 1937. Use of a Discriminant Function for Differentiating Soils with Different Azotobacter Populations. Iowa State Coll. Journ. Sci., 11: 323–332.
4. DAY, BESSE B., AND SANDOMIRE, MARION M. 1942. Use of the Discriminant Function for More Than Two Groups. Journ. Am. Stat. Assn., 37: 461–472.
5. FISHER, R. A. 1936. The Use of Multiple Measurements in Taxonomic Problems. Ann. Eugenics, 7: 179–188.
6. RIETZ, H. L. 1927. Mathematical Statistics. The Open Court Publishing Company, Chicago, Illinois.

STUDIES UPON *GALERA CRISPA*

ERNST A. BESSEY

IN 1899 Burton O. Longyear (5) described as new, under the name *Galera crispa*, a small mushroom that had been collected in the vicinity of Michigan Agricultural College (now Michigan State College) for a number of years. It is still abundant in the same locality and is reported by Kauffman (2) at Ann Arbor and other points in the Lower Peninsula of Michigan. According to Hard (1), it was observed by Longyear in Denver, Colorado, in 1905. It is a sod-inhabiting fungus, mostly growing in lawns or pastures of Kentucky bluegrass (*Poa pratensis* L.) and of Bermuda grass (*Cynodon dactylon* (L.) Pers.) where there is little or no shade. The little mushrooms appear from mid-July to early September two to four days after a heavy rain or a period of gentle showers. They push up during the night and last but a few hours if the day is sunny. The elements of the gills, like those of the closely related species *Bolbitius tener* Berk., soon begin to dissolve and disintegrate, but they do so much more slowly. Between ten and twelve o'clock on a warm day the details of the structure of the trama, basidia, cystidia, and other parts cannot be made out.

Often intermingled with *G. crispa* in Michigan, and usually more abundant, is a species that has been confused by Kauffman (2), Hard (1), and other American botanists as well as by some of the European authors with *G. tenera* (Fr.) Karst. Specimens collected at Las Cruces, New Mexico, early in September, 1942, and sent to me by Dr. Horace L. Barnett agree in all essential particulars with this fungus. In the Herbarium of the Department of Botany of Michigan State College some collections made prior to 1899 are labeled "*G. lateritia* (Fr.) Karst" and others "*G. tenera.*" That there is a whole series of closely similar forms that perhaps constitute a *G. tenera* complex is likely. There is need for an intensive study of all the species, varieties, and forms of this assemblage that have been distinguished in part and given names by Kühner (3), Lange (4), and others. Until this has been done it seems best to call this common species of central Michigan, not *G. tenera* or *G. lateritia*, but *G. lactea* Lange. The chief distinction between the last two lies in the color. *Galera lateritia*, according to Lange and Kühner, is more or less tile red (as indicated by the specific name), and the bulbous enlargement at the base of the stipe is 6 to 10 mm. in diameter, whereas in *G. lactea* the color is cream white and the bulb is small and rounded. In both of these the tenpin-shaped cheilocystidia are abundant along the edges of the gills, and similar cells are absent or almost so from the stipe. In *G. tenera* the stipe is covered with very numerous structures of this type, and the pileus and stipe are ocher to cinnamon-colored when fresh. The common Michigan species, which I tentatively call *G. lactea*, has a smooth pileus with only a very few long hairs similar to those occurring on the stipe. Kühner says that hairs of this type are abundant on the pileus of *G. lateritia* as he interprets it. The latter, moreover, is mostly larger than the Michigan fungus.

Galera crispa and *G. lactea* are alike in the color of the freshly picked specimens, which is as follows: pileus ivory yellow,[1] with a central chamois spot that varies in size; stipe pure white or almost so for the greater part of its length, becoming ivory yellow near the base. The bulbous enlargement is usually buried in the ground and is more often pure white. On drying all parts become darker, the pileus and stipe uniformly chamois or honey yellow. Fresh specimens show striations from the margin halfway or two thirds the distance to the center of the pileus. On dry specimens these are more accentuated and extend nearly to the center. The fresh stipe is usually minutely striate, but sometimes this condition is not apparent. The surface of the stipe is nearly smooth, but dotted with numerous patches of various size, from 30 to 100 μ in diameter or even larger, which consist of more or less crowded lanceolate hairs 5 to 8 μ thick and about 30 μ long. A varying number of these hairs extend as slender, flexuous hyphae 2.2 to 3.3 μ thick and 36 to 108 μ long. Sometimes these slender hairs lack the lanceolate base. In many of the patches only the lanceolate hairs are present. In only one out of 225 such patches examined were found capitate hairs resembling the tenpin-shaped cheilocystidia of the gills, and it had only a few. The narrowly hollow stipe is slender and nearly or quite straight, enlarging very gradually from the apex to the base, where it forms

[1] Color based on R. Ridgway, *Color Standards and Color Nomenclature*. Washington, 1912.

TABLE I

COMPARATIVE DATA ON DIMENSIONS OF *GALERA CRISPA* AND *G. LACTEA*

	Galera crispa	*Galera lactea*
Height, overall	4.3 – 8.0 cm.	4.0 – 8.4 cm.
Pileus		
Height	8.0 – 12.0 mm.	8.0 – 16.0 mm.
Breadth	10.0 – 21.0 mm.	10.0 – 25.0 mm.
Stipe diameter		
Near top	1.0 – 1.7 mm.	1.0 – 2.0 mm.
Bulb	2.0 – 3.5 mm.	2.5 – 5.0 mm.
Basidia	19.0–30.0 × 7.0–12.0 μ	19.0–29.0 × 9.6–12.0
Basidiospores		
Range	7.2–20.2 × 5.8–12.7 μ	8.6–17.5 × 5.8–10.3 μ
Mean	12.65 × 8.8 μ	12.2 × 7.7 μ
Standard deviation		
Length	2.01 μ	1.98 μ
Breadth	1.33 μ	0.94 μ
Coefficient of correlation of length and breadth	0.62	0.554
Cheilocystidia		
Height	20.0 – 30.0 μ	19.0 – 29.0 μ
Diameter of body	7.0 – 11.0 μ	9.6 – 12.0 μ
Diameter of head	3.0 – 4.6 μ	2.9 – 4.6 μ
Percentage of basidia		
4-spored	54	80–95
3-spored	31 }	5–20
2-spored	15 }	

a bulbous swelling. This bulb may be a gradual swelling or may be abrupt, with a distinct margin.

The surface of the pileus is "hymeniform," that is, covered with a palisade of pyriform or clavate cells. This structure is characteristic of the genus Conocybe as delimited by Kühner (treated as a subgenus by Lange). The gills are light-colored, but very soon the maturing basidiospores give them a dark-brown color. The basidia are somewhat pyriform and densely and finely granular at first, scattered among larger sterile hyaline cells two to three times the diameter of the basidia. At the edge of the gill are numerous capitate cheilocystidia shaped like tenpins. They are not evenly distributed but occur in groups of from two or three up to six or eight. The gills are narrow and only slightly attached at the summit of the stipe. The typical spore, which is dark ocher, is more rounded at one end

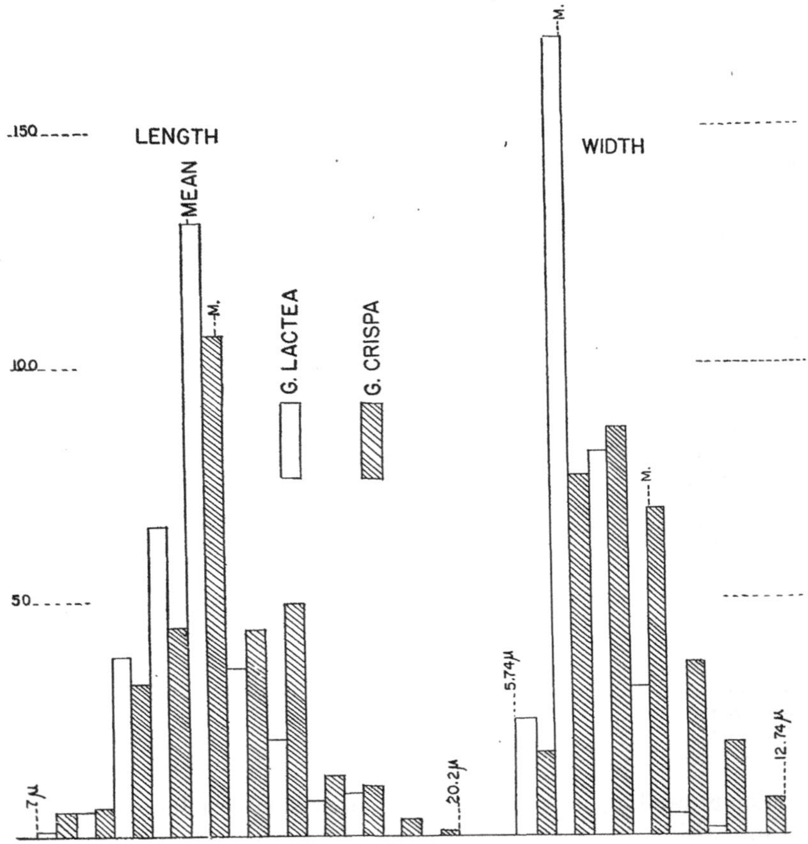

FIG. 1. Distribution of 310 basidiospores in each of the two species, *Galera lactea* and *G. crispa*, by length and width classes. The length classes have a range of 1.2 μ; for the width classes the range is 1.0 μ. In each class the rectangles representing the number of spores of the two species are placed side by side.

and obliquely apiculate at the other end, by which it is attached to the sterigma.

The differences between the two species lie in the shape of the mature pileus, the form of the gills, the number of basidiospores to a basidium, and the variation in the size of the spores. In *G. crispa* the line of attachment of each gill to the pileus is wavy, whereas in *G. lactea* it is straight; in the former the free edges of the gills are more or less crisped, in the latter not at all except occasionally in the more mature specimens. The main gills of *G. crispa* are connected by numerous transverse and oblique narrow anastomoses, but these are lacking in *G. lactea.* The crisping of the gills indicates the occurrence of considerable elongation of their free edges, the pilear tissue failing to elongate correspondingly, so that an outward curvature is brought about resulting in a campanulate shape in *G. crispa* in contrast to the oval form in *G. lactea.*

In *G. crispa* over 40 per cent of the basidia bear but two or three spores, but 80 to 95 per cent (varying in different hymenophores) are four-spored in *G. lactea.* Frequently basidia occur in the former with four sterigmata developed, one bearing a spore of maximum size, to either side of which the spores appear normal but small, whereas opposite the large spore is a colorless or only slightly colored spore, obviously incapable of development. Attempts were made to study the nuclear behavior of the basidia in both species, but apparently the material was fixed too late in the morning, so that disintegration of the tissues had already begun.

The spores of *G. crispa* are more variable in size and of greater range of dimensions than those of *G. lactea.* Through the kindness of Dr. W. D. Baten, of Michigan State College, a comparative statistical study was made of the spore sizes. The results will be presented in a separate paper by Dr. Baten (see pp. 3–7).

In Table I are given the comparative dimensions of the various structures of the two species. In Figure 1 is shown in the form of a histogram the distribution of the basidiospores by size classes.

Drawing conclusions from the foregoing data I am inclined to believe that *Galera crispa* represents a mutation from *G. lactea,* perhaps one that has occurred at various times and places. Attempts were made, unsuccessfully, by one of my graduate students, John B. Routien, to bring both of these species into culture with the intention of making crosses between them for the purpose of genetic studies that might throw light on the problem.

MICHIGAN STATE COLLEGE
EAST LANSING, MICHIGAN

LITERATURE CITED

1. Hard, M. E. 1908. The Mushroom, Edible and Otherwise, Its Habitat and Its Time of Growth, xii + 609, 505 figs. Ohio Library Co., Columbus, Ohio.

2. Kauffman, C. H. 1918. The Agaricaceae of Michigan, 2 vols. Michigan Geological and Biological Survey, Publication 26, Biological Series 5. Lansing, Michigan.

3. Kühner, Robert. 1935. Le Genre Galera (Fries) Quélet. Encyclopédie Mycologique, Vol. 7, 240 pp., 75 figs. Paul Lechevalier, Éditeur, Paris.

4. Lange, Jakob E. 1939. Flora Agaricina Danica, 4: 1–119, pls. 121–160. Society for the Advancement of Mycology. Copenhagen.

5. Longyear, B. O. 1899. Two New Michigan Fungi. Bot. Gaz., 28: 272–273.

THE INFLUENCE OF SOIL DEPTH UPON DISTRIBUTION OF ACTINOMYCETES *

KENNETH LESTER JONES

DURING the past five years the author has made from time to time plate counts of actinomycetes from soils in the vicinity of Ann Arbor, Michigan. Most samples were taken from the surface, but the numerous excavations for "defense houses" made convenient the collection of others from freshly exposed subsoil.

The samples were collected and either plated at once or stored at 5° C. and plated within a few days. In preparing the dilutions a fifty-gram sample was suspended in 500 c.c. sterile tap water, and the dilution series were made from 10 c.c. portions of suspension in 90 c.c. water blanks. Ordinarily the 1 : 10,000 or 1 : 100,000 dilutions were employed as inocula, although in the subsoil the organisms were so few that even a 1 : 10 dilution was too high. A medium used by Jensen (1930) proved very successful. Ten replicate plates of each dilution were poured. The cultures were incubated at 28° C. for two weeks, and counts were then tabulated from observations under a wide field binocular (20 ×). Doubtful colonies were observed under a magnification of 44 diameters.

Table I gives the "counts" of actinomycetes in various surface soils stated in millions per gram of dry soil. It also gives the pH of the soil, the water content, and the percentage of actinomycetes compared with the total number of bacteria and actinomycetes.

The "number" of actinomycetes in a gram of dry soil ranged from 100,000 to 25,400,000, and the percentage from 2 to 70. The grassland soils had a count of 5.6–25.4 millions, with 21–57 per cent actinomycetes; garden soils, 4–12.5 millions, with 14–62 per cent actinomycetes; woodland soils, 2.5–16.9 millions, with 19–67 per cent actinomycetes; sphagnum raw humus, 0.2–7.0 millions, with 3–29 per cent actinomycetes; and barren soils, 0.1–3.5 millions, with 5–46 per cent actinomycetes.

* Papers from the Department of Botany of the University of Michigan, No. 730.

The samples were not of a composite nature. Had they been, they would have been more significant from the agronomist's point of view, as Waksman and Fred have shown (1922). The variables introduced by the season of the year when the samples were collected,

TABLE I

NUMBERS OF ACTINOMYCETES IN THE SURFACE LAYER OF VARIOUS SOILS

No.	Soil type	Water pct.	pH	Actinomycetes		Month of collection
				Millions per gm.	Pct. of total	
	Garden soils					
1	Loam	18	6.0	12.5	62	October
2	Loam	9	6.0	4.0	25	October
3	Clay	18	6.0	5.4	32	October
4	Clay loam	15	7.0	5.3	14	October
5	Loam	13	6.0	10.5	55	October
6	Loam	24	7.0	5.7	34	October
7	Loam	11	5.5	5.7	50	October
	Grassland soils					
8	Clay loam	15	7.0	25.4	19	June
9	Loam	19	5.0	5.6	25	October
10	Loam	19	7.0	13.1	34	October
11	Loam	12	6.0	10.4	57	October
12	Loam	18	4.5	7.5	36	October
13	Clay loam	20	7.0	18.7	42	March
14	Clay loam	19	7.0	18.6	32	April
15	Clay	18	6.5	10.2	36	April
16	Clay loam	15	6.0	13.7	28	June
17	Loam	14	6.0	11.8	21	October
	Woodland soils					
18	Loam	19	7.0	16.9	40	October
19	Clay	10	7.5	4.8	19	October
20	Loam	22	7.0	8.5	41	October
21	Clay	25	7.0	11.0	52	October
22	Clay	16	6.0	2.5	24	October
23	Loam	28	6.0	9.1	39	October
24	Loam	28	5.5	5.4	67	October
25	Loam	24	5.0	3.2	19	October
26	Clay	23	5.5	5.6	41	October
27	Clay	33	6.0	6.4	19	October

TABLE I (*Concluded*)

No.	Soil type	Water pct.	pH	Actinomycetes: Millions per gm.	Actinomycetes: Pct. of total	Month of collection
	Miscellaneous wet soils					
28	Low meadow, loam....	50	7.0	12.4	31	October
29	Grassland, riverbank, loam...............	48	7.5	6.8	16	October
30	Edge of pond, exposed, clay...............	41	6.5	15.8	30	October
31	Edge of pond, exposed, gravel.............	20	7.0	7.0	70	August
32	Riverbank, exposed....	57	7.5	2.0	6	October
	Bog soils					
33	Sphagnum peat.......	81	5.5	3.9	52	October
34	Sphagnum peat.......	80	5.0	5.0	58	October
35	Sphagnum raw humus.	87	− 3.0	0.6	3	October
36	Sphagnum raw humus.	90	− 3.0	1.0	15	October
37	Sphagnum raw humus.	93	− 3.0	1.5	12	October
38	Sphagnum raw humus.	96	6.0	6.2	10	October
39	Sphagnum raw humus.	95	− 3.0	0.2	4	October
40	Sphagnum raw humus.	98	− 3.0	7.0	29	October
41	Hardwood bog, raw humus	91	6.5	9.8	11	August
	Barren soils					
42	Sand.................	6	6.0	3.5	46	October
43	Clay with limestone...	13	8.0	0.1	5	October
44	Sand.................	10	7.0	1.8	8	August
	Special soils					
45	Pine seedling bed......	8	6.0	0.3	2	October
46	Railroad embankment, gravel..............	14	7.0	1.6	24	October

by temperature, by mineral content of the soil, to mention but a few, do not warrant a detailed analysis of the data. The following facts stand out, however: (1) Soils with the same vegetation showed great variance in numbers of actinomycetes; (2) The numbers were high compared with those reported by many other investigators; (3) Organic matter together with a favorable reaction promoted growth. These aspects will be commented on briefly, in turn.

In regard to the great variance in numbers it must be remembered that plate counts are indeed crude estimates of the actual number of actinomycetes in a gram of dry soil. In the soil one individual may overgrow many times the volume of a single gram of soil, and yet in this volume millions of conidia may develop. Conidia, in the author's experience with single-cell isolations, are very likely to grow, whereas short filaments torn from a mycelium are not. Species cultured by the author in sterilized soil differ markedly in their sporulation. Some cover the soil with masses of conidia; in others there is no macroscopic evidence of spores. The plate method for actinomycetes gives primarily the number of conidiospores (and "fragmentation spores") capable of developing on the medium employed. It has been stated repeatedly that environmental factors are responsible for the great variance in actinomycetes in different soils, but the differences in sporulation, perhaps genetical in origin, have been overlooked.

The numbers were high compared with those reported by many investigators; 14 soils out of 46 tested showed a count above 10 millions per gram. There follows a partial but representative record of the findings of other workers. Conn (1913) reported that on his plates about 40 per cent of the soil organisms, or 12–14 millions, were actinomycetes. Krainsky (1914), who first emphasized the use of synthetic media, secured only 20,800, which was, however, about 30 per cent of the total number of organisms that grew. Waksman and Curtis (1918) reported 7,500–2,400,000 in 25 soils from widely different regions. Waksman (1922) found in 51 samples from one plot 8–25 millions per gram. Löhnis (1926) counted 5.5–9.7 millions in field soils used in legume experiments. Subrahmanyan and Norris (1929) studied 50 Indian soils in which the numbers of actinomycetes ranged from 1 to 3 millions, and they were believed to occur only as conidia. Jensen (1930) made counts on 56 Danish soils and found the numbers to be 0–12.7 per gram — only three soils had more than 10 million. Jensen (1934) published counts of 0.1–35.7 millions per gram on 50 Australian soils; eight soils showed over 10 millions. He further made about 50 periodic counts of one soil at weekly or biweekly intervals and secured 3.6–12.4 millions per gram, with four counts above 10 millions. A part of the variance in results must be due to differences in method, e.g. some investigators used single soil samples and others composite ones; there was no uniformity in media, time of incubation, observation of colonies, or allowance for moisture in the soil.

The importance of organic matter together with a favorable soil reaction is the third item for which Table I gives support. In grasslands, for example, the numbers were uniformly higher than under other plant cover. This vegetation furnished readily decomposable organic matter and a favorable pH. Conn (1916) advanced the idea that actinomycetes flourish in sod and are active in the decomposition of grass roots. The lowness of the numbers in barren soil with a favorable reaction indicates the necessity for organic matter, and the paucity of organisms in raw humus demonstrates the importance of the reaction. The lowness of the numbers in raw humus is better realized when one considers that the water content is about 90 per cent and that the actual numbers in a gram of moist soil are approximately as low as 10,000. In fact, Jensen (1930) found some acid soils with no actinomycetes and proved experimentally the effect of soil reaction on growth of certain organisms. He later (1934) made the interesting observation that there is a correlation with the numbers of bacteria. "Whether this is due to an actual stimulation of the actinomycetes by the bacteria, or to some unknown factor that stimulates both groups of organisms simultaneously, remains an open question, but it is worth noticing that actinomycetes play an important role in the decomposition of dead microbial matter in the soil, so that we might expect an increase in actinomycetes at periods when bacterial numbers run high and much bacterial protoplasm is being produced."

Table II gives the distribution of actinomycetes in different depths of soil from six sites where fresh profiles of Miami clay loam were exposed. None of the soils had recently been disturbed by man, and all supported a vegetation that was essentially grass.

There was a sharp and irregular decline in actinomycetes in progressively lower depths of soil. This finding agrees in general with those of Waksman and Curtis (1916) and of Brown and Benton, as reported by Waksman and Starkey (1931 : 46). The decline in numbers is presumably due to lack of oxygen and organic material. The differences in pH and in water content in various levels were not

TABLE II

COMPARATIVE "COUNTS" OF ACTINOMYCETES FROM DIFFERENT DEPTHS OF SOIL, BASED ON THE NUMBER IN A GRAM OF DRY SOIL

Site	Date 1942	Surface soil			Two feet		
		pH	Moisture pct.	Actinomycetes	pH	Moisture pct.	Actinomycetes
1	March 22	7	20	18,700,000	7.3	17	3,493,000
2	April 14	7	19	18,600,000	5.5	15	600,000
3	April 15	6.5	18	10,200,000	7	15	970,000
4	May 21	6	15	13,700,000	7	14	1,620,000
5	May 21	7	15	25,400,000	7.5	12	1,900,000
6	June 2	7	14	15,100,000	5.5	13	2,091,000
Site	Date 1942	Four feet			Eight feet		
		pH	Moisture pct.	Actinomycetes	pH	Moisture pct.	Actinomycetes
1	March 22	7.5	11	16,850	..	11	67
2	April 14	5.0	14	402,000	7	7	2
3	April 15	7.5	13	214,000	7.3	12	680
4	May 21	7.5	9	31,200	7.5	11	1,770
5	May 21	7.5	10	27,100	7.5	9	10
6	June 2	7.5	12	52,000	7	11	6,740

TABLE III

THE PERCENTAGE OF ACTINOMYCETES, COMPARED WITH THE TOTAL NUMBER OF BACTERIA AND ACTINOMYCETES, IN DIFFERENT DEPTHS OF SOIL

Site	Percentage of actinomycetes			
	Surface	2 feet	4 feet	8 feet
1	42	59	*94*	29
2	32	*55*	51	2.5
3	*36*	30	24	1.4
4	28	31	*48*	25
5	19	*25*	20	3
6	10	19	*25*	1.7

extreme enough to have a marked effect. Failure to develop spores in the lower layers of the soil is also a likely cause of the decrease in plate counts. The author's cultures in sterilized soil (maintained as stock cultures for studies in variation) show that conidial production is mostly at the surface of the soil. It is also singular that the actinomycetes which developed on the plates from soil at eight feet were invariably asporous on agar media.

The falling off in numbers of actinomycetes at two and four feet is usually proportionately less than that of bacteria, as is seen in Table III. This is in keeping with the findings of others — Waksman and Curtis (1916), for example. In some soils these authors found a higher number of actinomycetes at eight inches than at the surface. They can continue to develop on a lower quantity of nutrient materials than is true of most soil bacteria. At eight feet the bacteria may again be relatively more abundant because of a greater tolerance for low oxygen concentration; the actual numbers are too low, however, for all organisms at this depth to permit close comparisons.

SUMMARY

1. The "number" of actinomycetes found in the surface layer of 46 soils in the vicinity of Ann Arbor, Michigan, ranged from 100,000 to 25,400,000 per gram of dry soil. Fourteen soils showed "counts" of over 10 millions per gram. The percentage in the total bacteria and actinomycete flora was 2–70.

2. The combination of favorable pH and organic matter was usually associated with high counts of actinomycetes.

3. Stock cultures of actinomycetes in sterilized soil have shown extreme differences in the quantity of spores produced. The author believes that plate counts indicate primarily the number of spores in the soil and that specific differences in spore production is a major factor in the great variance in counts.

4. Actinomycetes were studied in six uncultivated soils at the surface and at depths of two, four, and eight feet. The numbers fell off sharply and irregularly as the depth increased. The decline at two and four feet was proportionately less than that for bacteria, but at eight feet the bacteria were again preponderant.

University of Michigan

LITERATURE CITED

CONN, H. J. 1913. Distribution of Bacteria in Various Soil Types. Journ. Am. Soc. Agron., 5 : 218–221.

—— 1916. A Possible Function of Actinomycetes in Soil. Journ. Bacteriol., 1 : 197–207.

JENSEN, H. L. 1930. Actinomycetes in Danish Soils. Soil Sci., 30 : 59–77.

—— 1934. Contributions to the Microbiology of Australian Soils. I. Numbers of Microorganisms in Soil, and Their Relation to Certain External Factors. Proc. Linn. Soc. New South Wales, 59 : 101–117.

KRAINSKY, A. 1914. Die Actinomyceten und ihre Bedeutung in der Natur. Centralbl. Bakt., II, 41 : 649–688.

LÖHNIS, F. 1926. The Effect of Growing Legumes upon Succeeding Crops. Soil Sci., 22 : 355–389.

SUBRAHMANYAN, V., AND NORRIS, R. V. 1929. Studies on Soil Actinomycetes. I. Introduction. II. Their Mode of Occurrence in the Soil. Journ. Ind. Inst. Sci. Bangalore, 12 : 53–68 (abstract in Centralbl. Bakt., II, 79 : 457, 1929).

WAKSMAN, S. A. 1922. Microbiological Analysis of Soil as an Index of Soil Fertility: I. The Mathematical Interpretation of Numbers of Microorganisms in the Soil. Soil Sci., 14 : 81–101.

—— AND CURTIS, R. E. 1916. The Actinomycetes of the Soil. *Ibid.*, 1 : 99–134.

—— —— 1918. The Occurrence of Actinomycetes in the Soil. *Ibid.*, 6 : 309–321.

—— AND FRED, E. B. 1922. A Tentative Outline of the Plating Method for Determining the Number of Micro-organisms in the Soil. *Ibid.*, 14 : 27–29.

—— AND STARKEY, R. L. 1931. The Soil and the Microbe, 260 pp. John Wiley & Sons, New York.

PRECIPITATION PROSPECTS, 1943–47, FOR OHIO AND NEAR-BY STATES

EDWIN LINCOLN MOSELEY

SOME degree of success in forecasting excessive and deficient rainfall has been obtained by using a cycle of 90.4 years, which is four times the magnetic sun-spot cycle. In New York State early records of rainfall are sufficiently numerous to make possible the forecasting of precipitation in this way. In Michigan and other states of about the same longitude there are few meteorological records which are old enough to be of use, but until quite recently these states contained some very old trees whose stumps still afford a record of rainfall during a period of several times 90.4 years.

When conditions are favorable for tree growth the layer of new wood called the annual ring, formed just beneath the bark, is wider than in unfavorable years. In an area large enough to include counties with various soils and topography, if a majority of the trees whose rings are studied are found to have made a good growth in a certain year, we may be reasonably sure that climatic conditions favored growth at that time.

In the regions where I have studied tree rings I have found a good correlation between ring width and the amount of rain falling in the growing season; this holds true as far back as there are rainfall records for the region. Temperature also affects tree growth, but it varies so little, except for brief periods, that it does not often obscure the effect of rainfall, which is subject to much wider variation at the time when vigorous growth requires a large amount of water.

The effect of plentiful rain or the lack of it may be easily seen by anyone who will examine the end of a log or the top of a sound stump. Knowing that the outermost ring of wood represents the last growing season before the tree was cut down, that the next ring inside it grew the year immediately preceding, and so on, the observer can readily find rings corresponding to any years which he can recall as having been wet or dry.

In Ohio and states near by the summers of 1934 and 1936 were dry;

those of 1935 and 1937 were wet. The rings which record the growth of the tree in these summers will show the effect of these differences in rainfall. The summer of 1925 was dry; in 1927 and 1928 there was abundant rain. This also is plainly revealed by the corresponding tree rings.

The earliest of the very dry summers in my recollection was that of 1871, when forest fires in Michigan and Wisconsin drove many families from their homes and when there occurred at Chicago one of the most destructive fires in human history. Weather records and tree rings show that both 1870 and 1871 were dry.

If one counts the rings from that of 1870 back toward the center of a stump until he has reached ninety, he finds that the ring formed in that year, 1780, is outstanding as a narrow one. That was a dry year in Washtenaw, Lenawee, Cass, and Berrien counties, Michigan, in all of which I have cut sections from stumps of old trees. It was dry also in the Allegheny State Forest in western Pennsylvania and in the neighborhood of Easton in eastern Pennsylvania, in different parts of Ohio, and in Indiana, Kentucky, and northeast Louisiana. We know that earlier still, at intervals of 90 or 91 years, rain was not sufficient to enable the trees to make a good growth.

The years 1872 and 1874 were also dry; 1873 and 1875 had more than average rain in parts of some states near Ohio, but none of the six years 1870–75 yielded as much rain as the average of the nine following years. Measurements of the radial growth made by a large number of trees during these six years and the next nine years, 1876–84, show that in a majority of them it was more than one and one-half times as great in the longer period. The same results hold for six- and nine-year periods antedating these by 90 years, and also in corresponding periods 91 years earlier still. Tree growth in other short periods of years has been measured with similar results — after 90 or 91 years wet periods are followed by wet periods, dry periods by dry periods.

The early periods in which we are now most interested are those that correspond to the present and the immediate future. I have measured the growth in 1846–52 of more than 250 trees and the growth of nearly as many in the preceding six or seven years. My measurements of the years following 1852 began somewhat later than those preceding, and for a time I was uncertain how many rings I should measure, but soon found that the five years after 1852 gave a strong contrast in growth rate with the seven years ending with 1852, so that I have used 1853–57 in most of the measurements which I have made since the summer of 1937. For the period preceding 1846 all recent measurements have included the seven years 1839–45. In almost every area where the trees were studied, from southern Michigan and northern Pennsylvania to southern Tennessee, a majority made more growth in 1846–52 than in the preceding seven years. In the five-year period after 1852 a majority grew more slowly.

Ninety years earlier we again find a strong contrast in growth rate between periods corresponding to those just mentioned. In 1756–62 trees grew decidedly faster than in the preceding seven years or in the ensuing five years. Likewise, in the previous century they grew faster in 1665–71 than in the seven years preceding or in the five years following. This is true also of the corresponding periods in the sixteenth century.

Extending this series of dry and wet periods to the twentieth century, we find that our recollection as well as weather records and tree rings show there was a shortage of rain in the period of nearly seven years beginning with 1930. Since late summer in 1936 rainfall has averaged greater in most places, as was predicted. The twelve-month period ending with May, 1942, was an exception; this also was predicted. After that there was plenty of rain until October. From then until the present, March, 1943, the average rainfall for the whole period has been near normal. It remains to be seen whether this seven-year period which, taken as a whole, was expected to afford more than average rain, will end with a grand finale in the coming spring as I expect it to.

It is probable that in the summer and fall of 1943 we shall have much less rain than we had in these seasons in 1942; at many places there is likely to be somewhat less than usual, but I have no reason to expect a drought anywhere east of the Mississippi, then or later, until about the middle of 1946. I expect the rainfall to be much above average in the latter part of 1945 and the early part of 1946.

A severe drought affecting many states should be anticipated for the latter half of 1946 and the greater part of 1947. After that for many years this part of our country will probably not experience so serious a drought.

The probability of this drought comes from the fact that there has been a drought at every corresponding period as far back as we have any tree record; the earliest that we know about was two years after Columbus's first voyage to America. Only four of our trees, two hemlocks in Pennsylvania and two white oaks in Ohio, were growing in 1494, but each of these formed a very narrow ring that year. Ninety years later there was another severe drought. We have sections from twenty-two trees that were growing then in this part of the United States. The next droughts in this series occurred in 1675, 1765, and 1856.

The severity of the drought of 1856 is attested by weather reports as well as by tree rings. Records of precipitation for that year have been preserved for thirteen places in Ohio, of which four are in the northern part of the state; at three in central and western Pennsylvania; at two in Kentucky; and at Richmond, Indiana. At every one of these nineteen places rainfall in 1856 was below normal; in most of them it was much below. The great state of Michigan has one place whose record includes 1856 — Fort Brady, Chippewa County; it is incomplete, but in the first eight months of that year the total rainfall was only 7.26 inches.

In Illinois six places have records for 1856; at five of them rain was below normal, but at only two was it very far below. In that year St. Louis had three inches of rain more than average, and Oregon in the same state a fraction of an inch more than average. In eastern Iowa there was some shortage at three places, but a larger surplus at three others.

Leavenworth, Kansas, had plenty of rain in 1856, but not in either 1854 or 1855. At St. Paul, Minnesota, less than six inches of rain fell in the whole growing season of 1856, but the year's total was 27.04 inches. Milwaukee got a little less than usual that year; it was much less than in either of the two years preceding.

One fact which has been impressed upon me recently will be of particular interest to people living in and near southern Michigan — the region near the Great Lakes is less affected by droughts or floods than are states farther south, as well as less than those farther west. The drought of 1930–36 was more serious in the latitude of Kentucky than in that of Michigan, and farther west it was more serious in other latitudes also. In the summer of 1936 I went to the Pacific Coast by way of North Dakota and Montana, returning by way of southern California and Kansas. Almost everywhere along the way, except where the land was irrigated, crops were suffering for lack of rain, and in many places they had failed in several preceding years. In northern Indiana they were better, and in northern Ohio they were rather good. In Wood County and doubtless in some others in northern Ohio there has been no general crop failure for a great many years. Throughout most of the West such counties are scarce or entirely wanting unless the land is irrigated.

In the Kentucky area, too, devastating floods are more frequent than they are farther north. In January, 1937, nearly as much rain fell on a large part of Tennessee and Kentucky as at Ann Arbor in the entire year of 1934.

This difference between the two latitudes was true in the nineteenth century likewise and before that. In Jefferson County, Kentucky, in which Louisville is located, the average annual rainfall in 1846–52 was twelve inches more than in the five following years, and at Cincinnati it was 12.44 inches more. My tree-ring measurements show greater differences in growth, owing to greater differences in water supply, in that region than farther north. This is true also of the corresponding periods ninety years earlier still (see Table I).

Table II shows that at all stations where records were made rainfall was above average in 1850 and 1855 and at all except Portsmouth in 1852. In 1856 it was below average at all stations. In such cases there is much probability that similar weather will occur over a great part of the same area about 90.4 years later. In the years 1853 and 1854 there was no such agreement among the stations. This causes difficulty in making forecasts for the two years beginning in the summer of 1943. The tree rings in the eighteenth century show a greater preponderance of dry weather during the corresponding period than in the nineteenth. On the whole, it seems probable that in the two years following the spring of this year (1943) somewhat less than the usual amount of rain will fall on the area I have been studying, but that in much of it rain will be about as usual and in some of it more than usual. Compared with recent years, there should be less danger from floods, less trouble in harvesting crops and in working land that is too wet.

Further data either from tree rings or from unpublished records of rainfall in the middle of the last century may make possible more precise statements regarding probable future rainfall in different areas.

Bowling Green State University
Bowling Green, Ohio

TABLE I

TREE GROWTH IN THE PARTS OF THE FIVE PREVIOUS CENTURIES WHICH CORRESPOND TO THE PRESENT PART OF THE TWENTIETH

Area	Counties	1839–45	1846–52	1846–52	1853–57	1749–55	1756–62	1756–62	1763–67	1658–64	1665–71	1665–71	1672–76	1568–74	1575–81	1575–81	1582–86	1477–83	1484–90	1484–90	1491–95
Southern Michigan and adjacent parts of Indiana and Ohio	7	6	33	22	20	11	21	21	12	6	9	7	12	1	6	3	2	0	1	1	0
Northern Ohio, vicinity of Bowling Green	4	4	22	15	5	9	13	20	0	2	7	3	1	0	1	..	..	..	..	..	..
Central Ohio, vicinity of Marion	2	4	14	15	4	8	10	15	6	4	8	7	3	2	2	2	1	1	0	..	..
Southwestern Ohio	5	2	13	11	3	6	9	7	4	1	4	1	4	..	..	..	..	..	..	..	..
Southern Kentucky near Burnside... } Northern Tennessee near Oneida .... }	6	20	28	22	0	7	22	12	1	0	1	0	1	0	1	1	0	..	..	..	..
Western half of Tennessee, logs at mills	12	9	14	6	2	4	10	3	1	..	..	..	..	..	..	..	..	..	..	..	..
Northeastern Louisiana, West Carroll Parish	2	7	11	12	3	3	6	2	4	..	..	..	..	..	..	..	..	..	..	..	..
Southern Indiana and adjacent Kentucky	7	2	6	1	0	0	14	1	0	..	..	..	..	..	..	..	..	..	..	..	..
Northwestern Pennsylvania, Allegheny State Forest	1	2	4	5	1	1	6	4	1	4	0	3	2	0	3	1	0	2	0	2	0
Northeastern Pennsylvania near Easton	3	4	3	4	2	0	2	2	0	..	..	..	..	..	..	..	..	..	..	..	..

Numbers record how many trees grew faster in the period indicated at the top of the column than in the period at the top of the adjacent column. Measurements showing approximately equal rates are not included.

TABLE II

PRECIPITATION IN THE PART OF THE NINETEENTH CENTURY WHICH CORRESPONDS TO THE PRESENT PART OF THE TWENTIETH *

Area	Normal	1839–45	Departure	1846–52	Departure	1853–57	Departure	1850	1851	1852	1853	1854	1855	1856	1857
OHIO															
Cincinnati	40.38	43.00	+ 2.62	51.82	+ 11.44	39.38	− 1.00	54.76	31.70	54.06	41.23	50.77	47.19	22.85	34.88
Marietta	42.25	38.52	− 3.73	45.49	+ 3.24	38.94	− 3.31	52.36	39.94	46.50	37.04	38.80	45.75	32.46	40.64
Steubenville	41.15	36.6–	− 4.55	47.42	+ 6.27	40.22	− 0.93	46.98	28.59	49.30	35.50	30.11	47.93	32.35	54.98
Portsmouth	41.06	40.57	− 0.49	43.7	+ 2.64	. .	. .	57.20	30.97	39.62	28.74	31.33	48.51	. .	. .
Dayton, 1846–49	39.76	. .	. .	56.21	+ 16.45	. .	. .	. .	. .	. .	. .	. .	. .	. .	. .
Germantown,† 1853–56	38.51	. .	. .	. .	. .	36.1	− 2.41	. .	36.28	41.18	28.54	44 72	47.00	24.15	. .
Urbana	39.84	. .	. .	. .	. .	42.54	+ 2.70	. .	. .	58.84	45.20	41.35	57.47	30.85	39.82
North Lewisburg	40.19	. .	. .	. .	. .	. .	. .	. .	. .	58.35	45.10	41.25	57.50	31.90	38.85
Hudson	37.92	33.84	− 4.08	. .	. .	. .	. .	. .	. .	. .	. .	. .	. .	. .	. .
OTHER STATES															
Richmond, Ind.	40.42	. .	. .	. .	. .	39.38	− 1.04	. .	. .	46.36	31.97	45.06	56.77	23.32	39.30
Springdale, Ky.‡	48.40	43.17	− 5.23	51.84	+ 3.44	39.84	− 8.56	67.10	42.34	51.91	35.56	46.20	. .	30.91	46.70
Pittsburgh, Pa.	36.07	33.2	− 2.87	38.77	+ 2.7	33.83	− 2.24	37.41	29.64	41.36	33.63	26.67	43.33	26.59	38.96
Lebanon, Pa.	43.63	36.41	− 7.22	44.02	+ 0.39	43.26	− 0.37	64.12	35.50	43.90	43.11	37.58	53.63	31.55	50.61
Gettysburg, Pa.	40.79	37.61	− 3.18	41.17	+ 0.38	36.08	− 4.71	45.63	30.99	43.31	32.08	33.86	47.62	28.63	38.23

* Data were missing where leaders occur.

† Dayton and Germantown are in Montgomery County; Urbana and North Lewisburg, in Champaign County.

‡ Data for 1839, 1840, and 1855 are lacking.

ECOLOGICAL STUDIES OF PIERPORT, BEAR LAKE, AND EDGEWATER BOGS IN MICHIGAN

IRVING E. W. OLSON

DURING the summer of 1935 I became interested in the large number of bogs in the west-central portion of the Southern Peninsula of Michigan. I observed that they differed somewhat in floristics and structure from the generally accepted concept of bog succession as set forth by the earlier investigators of bogs in the Huron River valley, and, more recently, by others in the region of Douglas Lake, where the University of Michigan Biological Station is located. Furthermore, I noted that there was a scarcity of literature, especially on floristics and succession in bogs.

Up to the present time ecological studies of bogs in Michigan have been limited almost entirely to two widely separated regions: the Huron River valley, in the southeastern part of the Southern Peninsula, where the earlier investigations were conducted, and the Douglas Lake region far to the north, where more recent studies have been carried out by students and staff of the Biological Station. The only other places where bogs have been studied are: North Manitou Island (Transeau, 1903); on Isle Royale, in Lake Superior (Cooper, 1912); in the extreme southwestern part of the state, at Baroda, in Berrien County (Kurz, 1928) and at Hartford, in Van Buren County (Osvald, 1935); and numerous bogs occurring throughout the Southern Peninsula as listed by Davis (1907), who has done much work on peat deposits in Michigan. According to Davis, bogs are frequent throughout the state; however, a number listed by him in 1907 have since changed considerably because of drainage of lands for agricultural purposes. For example, Four Mile Lake, west of Dexter, in Washtenaw County, which used to have a bog, shows the effect of drainage in that the characteristic bog plants have disappeared, and this once large lake is now little more than a grassy lowland and mud flat with a small stream draining it.

During the summers of 1935, 1936, 1939, and 1941 I visited twenty bogs in the west third of the central region of the Southern Peninsula of Michigan. This area is bounded roughly by a line extending from Grand Rapids, in Kent County, northward about 250 miles to Kalkaska, in Kalkaska County, and westward to Lake Michigan. Since it was not practicable to visit every bog within this area, those chosen will serve, for the purposes of this paper, as examples of the types that occur in it. That many more than those listed might be found in this section is quite possible, for in driving through it one often sees in the distance the characteristic trees — tamarack and spruce — that grow on bogs.

From the studies made to date of the bogs in this area I suggest that many of them have followed different courses in their development from those given in the generally accepted concept of plant succession on bogs. Since field investigations of succession thus far made are not conclusive, further studies are planned.

Plant succession on bogs as outlined by Weaver and Clements (1929) is reviewed briefly as follows:

1. Submerged stage: plants in open water up to 20-foot depths, prominent among which are Potamogeton, Myriophyllum, and Elodea.
2. Floating stage: plants in water 6 to 8 feet deep, largely Nymphaea and Polygonum.
3. Floating mat stage: composed essentially of Carex species; occasionally Decodon is the mat former.
4. Moss — low-shrub stage: Sphagnum, Menyanthes, Drosera, Sarracenia, and others, followed by shrubs such as Andromeda, Ledum, and Chamaedaphne.
5. Bog-tree stage: principal trees, Larix and *Picea mariana.*
6. Climax forest: *Pinus Strobus* in the Great Lakes forest.

This report will be restricted to three bogs in a region in which no bogs have been described, with notes on their floristics and physiognomy. Two lie near Bear Lake, in Manistee County; one is in the Platte Plains, in Benzie County.

Bogs Visited

The bogs studied are grouped according to stability of substratum and physiognomy of vegetation:

Group A. Bogs with stable substratum and open water; mainly low plants;

Group B. Bogs with stable substratum and no open water;

Group C. Bogs with stable substratum and more or less tree cover.

GROUP A. BOGS WITH STABLE SUBSTRATUM AND OPEN WATER; MAINLY LOW PLANTS

Water is present; mat is quaking or partly quaking; Chamaedaphne scrub with few to many tamaracks; spruce may or may not occur; when it does, it becomes well established; pine, when found, is usually on the firmer areas of the mat.

1. Pierport Bog. Along Highway M–22, in Manistee County, midway between Frankfort and Manistee, and one mile east of the village of Pierport.
2. Bog at Big Bay of Bear Lake. In Manistee County, northwest of the village of Bear Lake, along Highway U.S.–31.
3. Mud Lake Bog. In Newaygo County, along Highway M–37 just south of Bitely. The lake was formerly known as Placid Lake.
4. Greenan Bog. Just south of Mud Lake Bog. Named for the owner of the property on which it is located.
5. Newaygo Bog. In Newaygo County, about two miles north of Newaygo, to the west of Highway M–37.
6. Aman Lake Bog. At Aman Park, in Ottawa County, eight miles west of Grand Rapids, along Highway M–50.
7. Bog about four miles east of Grand Rapids, in Kent County, along Highway M–50.
8. Bog about three miles north of Grand Rapids, in Kent County, along a county road.
9. Dead River Bog. In Benzie County, lying between the two Platte lakes.
10. Sugar Loaf Bog. In Kalamazoo County, just south of Portage, along Highway U.S.–131.
11. Spring Run Bog. In St. Joseph County, at Spring Run, about three miles north of Moore Park, along Highway U.S.–131.
12. Prairie River Bog. In St. Joseph County, about two miles south of Three Rivers, at Prairie River, along Highway U.S.–131.

GROUP B. BOGS WITH STABLE SUBSTRATUM AND NO OPEN WATER

Chamaedaphne meadow with tamarack or spruce or both; open water not present; mat not quaking.

13. Bog west of Blue Lake. In Newaygo County, about one and one-half miles south of Brohman and east of Highway M–37.
14. Mud Pond Bog. In Benzie County, near the mouth of Platte River, along Highway M–22.
15. Walkerville Bog. In Oceana County, about two miles north of the village of Walkerville, along either side of county road.
16. Gordon Lake Bog. In Manistee County, about four and one-half miles south and west of the village of Bear Lake.

17. Bog near Grass Lake. In St. Joseph County, three miles west of White Pigeon, along Highway U.S.–112.
18. Bog along county road. In Kalkaska County, about five miles west of the village of Kalkaska.

GROUP C. BOGS WITH STABLE SUBSTRATUM AMD MORE OR LESS TREE COVER

Chamaedaphne meadow being shaded out by spruce and pine; mat is becoming quite firm as the forest advances on it.

19. Edgewater Bog. In Benzie County, about thirteen miles north of Frankfort and one-eighth mile east of Benzie County State Park, along Highway M–22.
20. Little Bay Bog of Bear Lake. In Manistee County, north and west of the village of Bear Lake, along Highway M–51, at Little Bay.

GEOLOGY OF THE REGION

The Southern Peninsula of Michigan lies in the eastern lake section of the central lowland province east of the Mississippi River (Fenneman, 1938). In this part of the state at least five distinct glacial periods can be traced back by moraines that are the result of numerous advances and recessions of continental glaciers. Although it is not the purpose of this study to include a detailed discussion of the history and development of these glaciers, a brief account of some of the outstanding physiographic features of the area about each bog will be given.

The general region in which the bogs studied in detail are located is roughly that lying between the terminal moraine of the Lake Michigan glacial lobe (Chamberlin, 1883) and the present shoreline of Lake Michigan in Manistee and Benzie counties.

The texture and composition of the soils in this area vary greatly and include rich humous soils formed from peat beds, loamy and gravelly soils of outwash plains, and rather extensive areas of sandy soils, both ice- and wind-borne. These sandy soils are acid in reaction and were often covered with nearly pure stands of pine, commonly referred to as "pineries" in the prelumbering days.

The Southern Peninsula is dotted with lakes, large and small, which resulted chiefly from glacial action. Some are pits in outwash plains where stagnant ice blocks melted; others have been formed by wind-blown sand deposited across outlets of large shallow basins or in depressions between ridges. In many of these lakes in places where drainage was retarded or completely shut off conditions were favorable for the development of bogs, which are quite common throughout the Southern Peninsula.

Description of Three Bogs in West-Central Michigan

The three bogs in west-central Michigan that were studied in detail lie in the lowlands between ridges in the Manistee moraine, which extends northward from Manistee along the shores of Lake Michigan through Benzie County and into Leelanau County. This moraine curves around the ends of transverse sand ridges that come out as headlands along this part of the shore of Lake Michigan. Between these ridges the ice pushed into the lowlands for several miles from the shore of the lake, so that the moraine makes a series of loops in crossing the lowlands between the prominent ridges.

Bear Lake was formed as a pit on an outwash plain between two spurs of this moraine. One of the largest bogs in the region has its origin on Big Bay of the lake.

The first bog to be described lies in a small depression on the moraine itself. The second occurs on Bear Lake. The third lies between two sand ridges in the extensive sand plains known as Platte Plains.

1. Pierport Bog

Location

Owing to its proximity to the village of Pierport, just one mile to the west on the shores of Lake Michigan, the bog surrounding a small lake known locally as Kuenzer's Lake will be designated "Pierport Bog" (Fig. 1). It is in the extreme northwestern section of Onekama Township, Manistee County, Section 3, R. 24 N., T. 16 W. It is readily accessible and begins approximately 300 feet to the west of Highway M–22, about midway between Frankfort and Manistee.

The bog has formed completely around a somewhat circular lake in a depression in the morainal deposits along Lake Michigan. It covers slightly over seven acres, of which about three are open water. The level of the lake at the present time is approximately fifty feet below the general level of the upland, which rises rather abruptly in all directions from the edge of the bog except to the

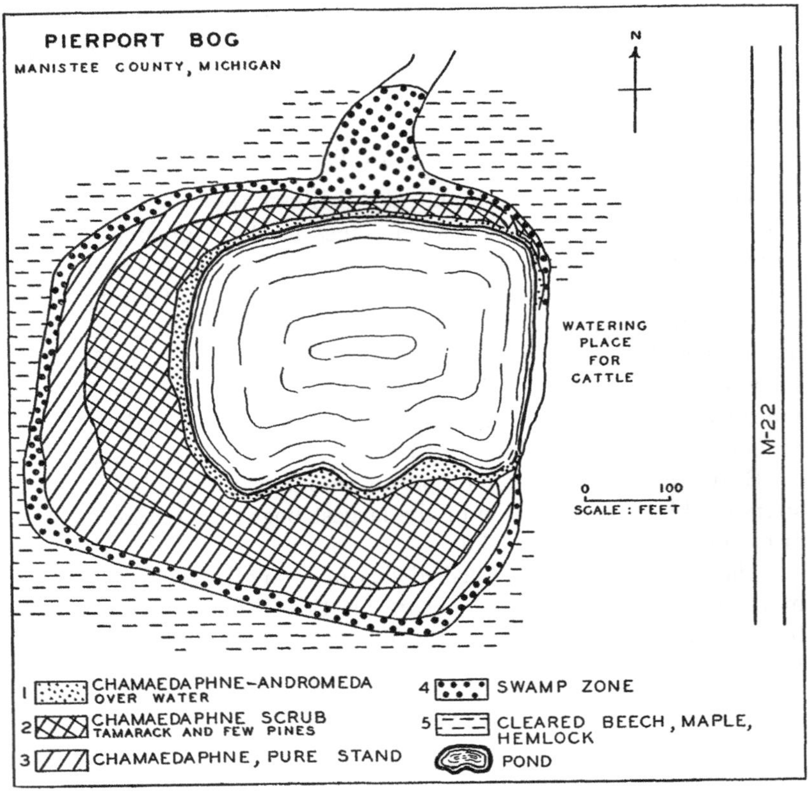

FIG. 1

(In the text the various areas of Figures 1–3 are called "zones.")

north, where a narrow ravine extends gradually to the upland for a distance of about 500 feet.

The lake and surrounding bog are a part of a 160-acre farm owned in 1936 by Mrs. R. W. Todd, of Bangor, Michigan. Nearly all the farm has been under cultivation for many years, most of which is at present in pasture land. There is no apparent drainage, and the lake is probably still fed by hidden springs. A large one at the head of the ravine to the north formerly fed the lake, but ceased flowing some thirty-five years ago. The entire bog is accessible to cattle, which graze on the surrounding upland. The swamp zone at the east end of the lake is devoid of vegetation.

Early History

For data on the early history of this bog and the surrounding area I am indebted to Mr. O. C. Kuenzer, R. F. D. 1, Bear Lake, Michigan. In an interview in August, 1936, he gave me a vivid description of conditions as they existed in 1875, the year in which the general tract containing his father's farm, on which the bog is located, was homesteaded. He was nine years old when he came to this farm, on which he worked until the forest fires of 1885 forced the family to move to a near-by farm. He was in a position to describe many changes in the vegetation resulting from the clearing of the forests as well as to give information on which the historical account of this paper is based.

When this tract was homesteaded, the farm and the land around it were completely forested with beech, maple, and hemlock — the beech and maple predominating to the south, west, and north. To the east, across the present Highway M–22, there was a nearly pure stand of hemlock covering several square miles of land. On the upland to the north were two spruces, the only ones known in the vicinity. These were cut for Christmas trees in 1895. A small number of pines were scattered throughout the region, and a few balsam firs were seen as recently as 1900.

In 1885 a local fire that devastated several square miles of forest land began to the southeast of the bog and spread to a three-mile front. It burned all the vegetation to the ground and reached the shores of the lake on the west and the south. On the upland to the east of the lake two white pines with diameters of 22 and 30 inches were not damaged by the flames and are still standing. A pine and two hemlocks on the south shore survived the fire, as did a pine and a hemlock on the north shore. All of these trees were large. The pines were cut for lumber about 1900; the hemlocks, in 1910.

The water level of the lake, which, as has been stated, was fed by a large spring coming from the upland to the north, has been lowered about six feet. The flow of water from it gradually decreased after the fire and the clearing of the subsequent second growth of trees until it ceased in 1910. Before the lake dropped to the present level a moat with an average depth of six feet of water and a width of from ten to twelve feet had to be crossed to get on the mat. The mat was accessible by foot in only one or two places; a boat was generally used to reach the mat for the gathering of cranberries. A number of the large logs, all of which were under water until 1905, can still be seen at the south edge of the present mat.

Witch-hazel and lady-slipper orchids used to be found here in abundance, the former in the swamp zone along the wooded upland and the latter on the bog mat. In this area there have been no oaks since the fire of 1885. Three species of cranberry, one red, one red and white, and one grayish, were plentiful on the mat. The grayish berry may have been the creeping snowberry now sparsely distributed over the mat. Before the lowering of the level of the lake, limeweed, a local name for pondweed, covered the water along much more of the shoreline than it does now.

During the years 1917–18 a considerable amount of peat was removed from the south border of the present mat and hauled to the uplands to the south as a fertilizer. In this disturbed area some grasses have come in from the surrounding upland, which has been in pasture for many years.

About 1930, when the larch sawfly killed many tamaracks in the general area, the tamaracks on the mat began to show a decided lack of vitality and growth, and many of the larger trees have since died, probably because of the ravages of the fly.

Floristics and Zonation

In the lake are three distinct areas of *Nymphaea advena* Ait.,[1] the largest of which is along the east shore. The other two lie along the east and west shores.

In Zone 1 (Fig. 1) rooted in a dense mat of Sphagnum is a heavy stand of *Chamaedaphne calyculata* (L.) and *Andromeda glaucophylla* Link. The branches of many plants of both species extend out over the water from two to four feet, and where the tips of these branches touch the water, roots develop at the nodes. This suggests the possibility of these shrubs appearing as pioneer plants or mat formers in the development of bogs in this general area. This zone is well defined and reaches completely around the lake, except

[1] With the exception of the genus name Polytrichum, the names used in the identification of plants listed in this study are those of *Gray's New Manual of Botany.*

for a small area along the east shore, where vegetation has been trampled out by cattle coming for water. On the south are two protrusions where the Chamaedaphne–Andromeda mat appears to be spreading over the water at a rate faster than that on the rest of the mat. To the east and northeast such protrusions are smaller and less obvious, although readily recognized. According to local testimony, the mat has advanced noticeably and rather uniformly over the water about ten feet since 1905.

In Zone 2 Chamaedaphne is the dominant shrub, with Andromeda less frequent. This portion of the mat is also quaking and contains a heavy stand of Sphagnum, with which the following plants are commonly associated: *Vaccinium Oxycoccus* L., *Drosera rotundifolia* L., *Kalmia polifolia* Wang., *Sarracenia purpurea* L., *Eriophorum angustifolium* Roth., and *Eriophorum viridi-carinatum* (Engelm.) Fernald. Less common plants are *Menyanthes trifoliata* L., *Chiogenes hispidula* (L.) T. & G., *Eriocaulon articulatum* (Huds.) Morong., and *Hypericum spp.* Only two plants of *Habenaria blephariglottis* (Willd.) Torr. have been observed on the mat, both of which blossomed in July, 1935. *Larix laricina* (DuRoi) Koch. is very abundant in this zone and also in Zones 1 and 3. In Zone 1 a number of the trees grow along the water's edge; in Zone 2 an exceptionally heavy stand of tamarack is becoming established south of the pond at the east and at the west and northwest portions of the mat. As previously stated by Mr. Kuenzer, many of the larger trees have died since 1930. Examination based on a count of growth rings in several large dead trees indicated growing periods ranging from 90 to 120 years. Some of these trees may have been killed by the sawfly, which is known to have destroyed many tamaracks in this part of the state.

In Zone 3 the mat is partly quaking, with a practically pure stand of Chamaedaphne. In some sections Chamaedaphne appears to be thinning out, especially in the southwest, where the large trees on the edges of the upland partly shade the mat. Tamarack is common in the zone, but white birch (*Betula papyrifera* Marsh.) is invading the mat and reaching into Zone 3 and even to the edge of the water in Zone 1. Ten or twelve trees of *Pinus Strobus* L., together with a few seedlings, are on the mat. These have probably developed from cones from the two pines on the upland to the east that survived the fire of 1885. Two or three small hemlocks (*Tsuga canadensis* (L.) Carr.) and several beeches (*Fagus grandifolia* Ehrh.) have also invaded the quaking portions of the mat to the south of the pond.

In Zone 4 the swamp zone is relatively dry. Since about 1900, when the large spring feeding the lake from the uplands began to fail, the level of the lake has dropped about six feet, as has already been noted. In 1905 the logs of hemlock, which are still found in the moat, began to show above the water, which at one time was six feet deep. In this zone the following plants occur: *Cephalanthus occidentalis* L., *Sambucus canadensis* L., *Lonicera caerulea* L. var. *villosa* (Michx.) T. & G., *Salix sericea* Marsh., *Ilex verticillata* (L.) Gray, *Rubus idaeus* L. var. *aculeatissimus* (C. A. Mey.) Regel & Tiling, *Potentilla palustris* (L.) Scop., and *Alnus incana* (L.) Moench. In the portions of this zone to the south, where peat was dug in 1918, and to the east and north of the watering place for cattle *Scirpus cyperinus* (L.) Kunth var. *pelius* Fernald, *Carex trisperma* Dewey, and *Juncus effusus* L. are found. Other plants growing locally in the swamp zone are: *Polytrichum spp.*, *Maianthemum canadense* Desf., seedlings of wild cherry (*Prunus virginiana* L.), *Ribes vulgare* Lam., *Fragaria virginiana* Duchesne, *Rosa carolina* L., *Solanum Dulcamara* L., and *Woodwardia virginica* (L.) Sm.

On the upland slope to the east there are a number of large hemlocks, red maples (*Acer rubrum* L.), hard maples (*Acer saccharum* Marsh.), white birches (*Betula papyrifera*), and sweet birches (*Betula lenta* L.). Under these trees are also bracken (*Pteris aquilina* L.) and wintergreen (*Gaultheria procumbens* L.). A species of blueberry (*Vaccinium sp.*) is invading the mat over the more stable portions. Along the southwest edge of the mat is a clump of wild cherry (*Prunus virginiana*), with white birch (*Betula papyrifera*). On the edge of the mat winterberry (*Ilex verticillata*) and serviceberry (*Amelanchier canadensis* (L.) Medic.) are thriving locally to the north and northwest.

On the upland to the north and extending eastward around the swamp zone and south to the watering place for cattle one finds the following plants: white birch (*Betula papyrifera*), sweet birch (*Betula lenta*), hard maple (*Acer saccharum*), large-toothed poplar (*Populus grandidentata* Michx.), hemlock (*Tsuga canadensis*), white pine (*Pinus Strobus*), and serviceberry (*Amelanchier canadensis*). Ground plants among these trees include bracken (*Pteris aquilina*), blueberry (*Vaccinium spp.*), *Polytrichum spp.*, and cinnamon fern (*Osmunda cinnamomea* L.).

In July, 1941, it was noted that the level of the lake had dropped about one foot since 1925, exposing the mud of the Nymphaea zones above the surface of the water, but the reason for the lowering is not known.

Borings were made in a number of places on the mat to the south of the lake. The greatest depth, 27 feet 4 inches, was reached in Zone 2, about 100 feet south of the edge of the mat, where the peat was found to be deposited on sand.

According to Mr. Kuenzer, the mat was very narrow in front of a pine tree which stands near the edge of the mat along the northwest shore of the lake, for there was just room for him to stand on the mat between the tree and the water when he fished in the lake for the last time in 1905. In July, 1941, it extended out over the water from the base of the pine tree exactly fifteen feet. The mat has apparently grown out over the water at this point at the rate of about one foot every three years. The diameter of the pine tree was nine and one-half inches four feet above the mat.

2. BEAR LAKE BOG

Location

Bear Lake Bog (Fig. 2) lies in Sections 25 and 36 of Arcadia Township and Sections 30 and 31 of Pleasanton Township, Manistee County, Ranges 15–16 and 23–24 N.

The floating mat is most easily reached by following the improved road south and west around the south shore of Bear Lake to the lakeside cottage camp and by hiking from this point northward to the area of Big Bay along the shore of the lake. The more mature portion of the bog, indicated on the map as Zone 3, is readily reached from any place along the beach of Bear Lake in the area of Big Bay.

The bog itself has formed over Big Bay along the shores of Bear Lake to the uplands and extends northwestwardly for nearly a mile.

Geological Features

Bear Lake is in the northwestern part of Manistee County about midway between Manistee and Beulah. At the east end of the lake is the village of Bear Lake.

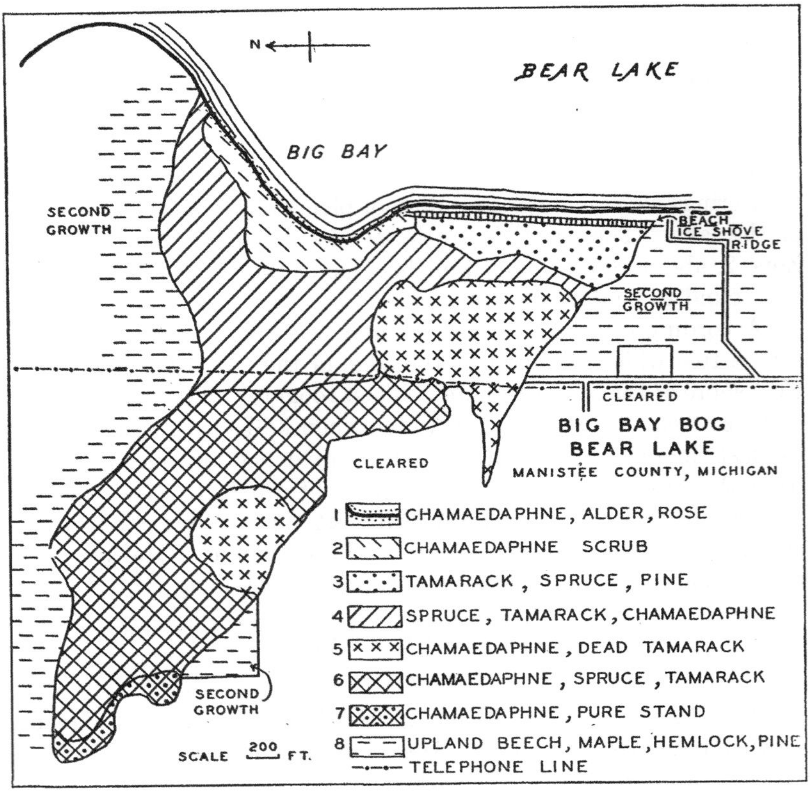

FIG. 2

This lake, an open expanse of water of very regular outline and with dimensions of about two and one-fourth and one and one-fourth miles (Scott, 1921), covers an area of 1,744 acres. The longer axis lies in an east–west direction. The western part of the basin hardly exceeds fifteen feet in depth, but depths of fifty to sixty feet are known for the eastern part. The original outlet, which flowed from the east end, was shifted somewhat to the south when the road was built. The present lack of a definite channel and, possibly, the obstructing road may account for occasional higher water levels of the lake. However, the cause of high water has not been definitely determined.

The glacial formations may readily be made out as one approaches the lake on the road from the south. At first the route is through a rugged moraine, from the crest of which a broad outwash plain below reaches northward to another moraine some six miles away. Beyond the crest the moraine slopes sharply to the north, and the entire lake appears to view; it extends from the foot of the slope to the outwash. The basin is bounded on the east by upland, but stretches eastward as a low heavily wooded swamp to Bear Creek several miles distant. In brief, the lake basin is part of a depression in the outwash at its junction with the moraine. The soil map for the county shows that the soil for about a mile north of the lake and west about two and one-half miles is classed as Rubicon loamy sand. The entire area of the several bogs around the lake, together with that of Pierport Bog, is included in the region of this sand.

Early History

Information regarding the vegetation of the area before the days of lumbering was obtained from Mr. O. C. Kuenzer. According to him, the area for many square miles between Lake Michigan and Bear Lake was heavily forested with beech, maple, and hemlock, and a pure stand of hemlock grew just east of the Pierport Bog, about two miles distant from the shores of Bear Lake. Along the south and west shores of Bear Lake was a rather extensive area with a nearly pure stand of white pine. Local fires did not occur in this immediate vicinity; the nearest one stopped at the edge of the lake on which Pierport Bog is located. The heavy timber was removed during the lumbering operations from 1880 to 1900. With the exception of a small patch of mixed second-growth timber along the north half of the west shore of Bear Lake, now used as a resort, all the upland along this shore is now under cultivation. To the north the cultivated areas extend to the edge of the bog at Big Bay.

Floristics and Zonation

The sandy ridge indicated on the map (Fig. 2) as "ice-shove ridge" is about five feet high and from five to ten feet wide. According to report, this sandy ridge was built up during the lumbering days by the action of lake ice. On it the following plants occur: *Acer rubrum*, *Acer saccharum*, *Alnus incana*, *Amelanchier canadensis*, *Pyrus arbutifolia* (L.) L. f., *Betula alba* (L.), *Betula papyrifera*,

Gaylussacia spp., *Ilex verticillata*, *Nemopanthus mucronata* (L.) Trel., *Pinus Strobus*, blueberries (*Vaccinium spp.*), *Viburnum cassanoides* L., *Viburnum acerifolium* L., *Viburnum Opulus* L. var. *americanum* (Mill.) Ait., and *Woodwardia virginica.*

Along the beach to the east of the sand ridge the following plants appear: *Scirpus americanus* Pers., *Scirpus validus* Vahl., *Eupatorium perfoliatum* L., *Scutellaria galericulata* L., *Mentha piperita* L., *Utricularia resupinata* B. D. Greene, *Hypericum spp.*, *Gerardia spp.*, *Aster spp.*, and *Solidago spp.*

The bog at Big Bay is the largest of the many which I visited. It extends in a southwesterly direction from the shore of the lake more than 4,000 feet. Its width is about 1,400 feet in the region of the telephone line, which crosses the mat at a line approximately 1,000 feet west of the shore of the lake. All along the edge of the floating mat it reaches around Big Bay northward some 2,400 feet from the north end of the "ice-shove ridge."

Zonation on this bog is probably not so readily observed as the map would indicate. Along the edge of the floating mat is a narrow fringe (Zone 1) in which *Chamaedaphne calyculata*, *Andromeda glaucophylla*, *Alnus incana*, and *Rosa carolina* are the dominant plants. Other plants are: *Sarracenia purpurea*, *Kalmia polifolia*, *Sphagnum spp.*, *Menyanthes trifoliata*, *Larix laricina*, *Calopogon pulchellus* (Sw.) R. Br., and *Picea mariana* (Mill.) BSP. At the southern tip of this zone a large bed of *Nymphaea advena* borders the mat. Just back of the Nymphaea, *Decodon verticillatus* (L.) Ell. is the principal plant at the edge of the mat. Throughout this zone, in a mat made up largely of Sphagnum, branches of Chamaedaphne, Andromeda, and Rosa extend out over the water and roots appear to develop at nodes on the submerged portion of the stems. Tamarack and spruce are both found at the water's edge.

Zone 2 is indicated on the map as Chamaedaphne scrub, which covers the remainder of the quaking mat. In this zone numerous plants grow. Tamarack and spruce are abundant, but are so scattered that they offer little if any shade. The chief plants in this zone are: *Chamaedaphne calyculata*, *Andromeda glaucophylla*, *Kalmia polifolia*, *Calopogon pulchellus*, *Pogonia ophioglossoides* (L.) Ker., *Habenaria blephariglottis*, *Eriophorum spp.*, *Vaccinium Oxycoccus*, *Drosera rotundifolia*, *Menyanthes trifoliata*, *Sarracenia purpurea*, *Sphagnum spp.*, *Xyris caroliniana* Walt., *Aspidium spinulosum*

(O. F. Müller) Sw., *Chiogenes hispidula, Eriocaulon articulatum* (Huds.), *Rynchospora alba* (L.) Vahl., *Habenaria clavellata* (Michx.) Spreng., *Eleocharis spp., Larix laricina, Picea mariana,* and *Picea canadensis* (Mill.) BSP.

The portion of the bog which may be classed as mature is indicated as Zone 3. It lies just west of the "ice-shove ridge," a sandy ridge that apparently prevents flooding during the high water stages of the lake. It is firmly grounded and is so heavily shaded that ground cover of herbs and shrubs is sparse as compared with the vegetation in Zone 4.

On this mature mat there is a rather heavy stand of beech, hard maple, red maple, and serviceberry, and, occasionally, white oak and white pine. The ground plants include blueberries (*Vaccinium spp.*), *Coptis trifolia* (L.) Salisb., bracken, wintergreen, *Corallorrhiza trifida* Chatelain, *Cornus canadensis* L., *Cypripedium acaule, Woodwardia virginica, Monotropa uniflora* L., Sphagnum locally, *Melampyrum lineare* Lam., *Osmunda cinnamomea,* and *Aspidium Thelypteris* (L.) Sw.

In Zone 4 there is a heavy stand of tamarack and spruce. These trees are on the firmer portions of the mat and have produced such dense growth that Sphagnum has been almost completely shaded out, as have been many of the other plants associated with Sphagnum. In this entire zone Chamaedaphne persists as the dominant mat covering. Locally small hummocks of Sphagnum may occur with *Vaccinium Oxycoccus, Chiogenes hispidula,* and a dwarf blueberry (*Vaccinium sp.*). *Cypripedium acaule* Ait. is common in the dense shade under the large spruces.

The northern dwarf mistletoe (*Arceuthobium pusillum* Peck) is widely distributed on spruce in Zones 1, 2, and 4, but not on spruce that occurs on the drier portions of the mat in Zone 6. Although the host trees were heavily infected, no noticeable disturbance to growth was apparent other than occasional "witches'-brooms" on some of the taller trees.

Beyond the zones of trees, toward the cultivated uplands, is a region of pure stand Chamaedaphne. In this zone all the smaller ground plants appear to have been replaced by Chamaedaphne which thrives wherever it occurs on the mat.

At the southern tip of the larger area in Zone 5 a small waste section is covered with weeds and swamp grass. This has resulted from cutting during the harvesting of hay on the uplands. In the uplands to the north of the bog the forest of second-growth hardwoods is largely made up of hard maple and beech, with an occasional *Populus grandidentata* and *Prunus virginiana*. *Amelanchier canadensis* appears nearer the edge of the mat. On the edge there is a shrub zone in which *Ilex verticillata, Nemopanthus mucronata, Pyrus arbutifolia,* and *Salix spp.* are the principal shrubs.

Two regions on the mat with no spruce are indicated in Zone 6. In them Chamaedaphne occurs in almost pure stand. Tamaracks were once widely distributed, as is shown by the large number of dead trees still standing. Many were killed by the larch sawfly in 1930.

Throughout Zone 6 the heavy stand of Chamaedaphne is broken only by rather widely scattered tamaracks and spruce. The trees are small and do not show the luxuriant growth manifested by spruce in Zones 1, 2, and 4.

In July, 1941, a number of borings were made on the mat. The maximum depth of peat was 25 feet 4 inches at a point about 800 feet to the southwest of the north end of the "ice-shove ridge" and about 175 feet in from the edge of the lake. About midway between this boring and the center of Zone 4 the greatest depth of peat was 10 feet. A boring some 150 feet farther to the southwest showed the depth of peat to be only two and one-half feet.

All borings except one indicated that peat was deposited on sand. At the point where peat was bored to its greatest depth the upper 12 to 16 feet of the mat was rather firm and yielded cores of peat, but the lower 8 to 10 feet produced only an unsolidified mass, with no deposition of peat on the sand bottom.

3. EDGEWATER BOG

Location

The Edgewater Bog (Fig. 3) is located along Highway M–22 about one-eighth mile east of the Platte River bridge in Benzie County. It lies in almost the center of Section 27 of Lake Township, R. 86 W. and 15 N. It is situated between two sand ridges, one of which separates the bog from the Platte River to the south. Lake Michigan lies just a mile to the north, and the intervening area is made up of a series of sand ridges that parallel the shores of the lake.

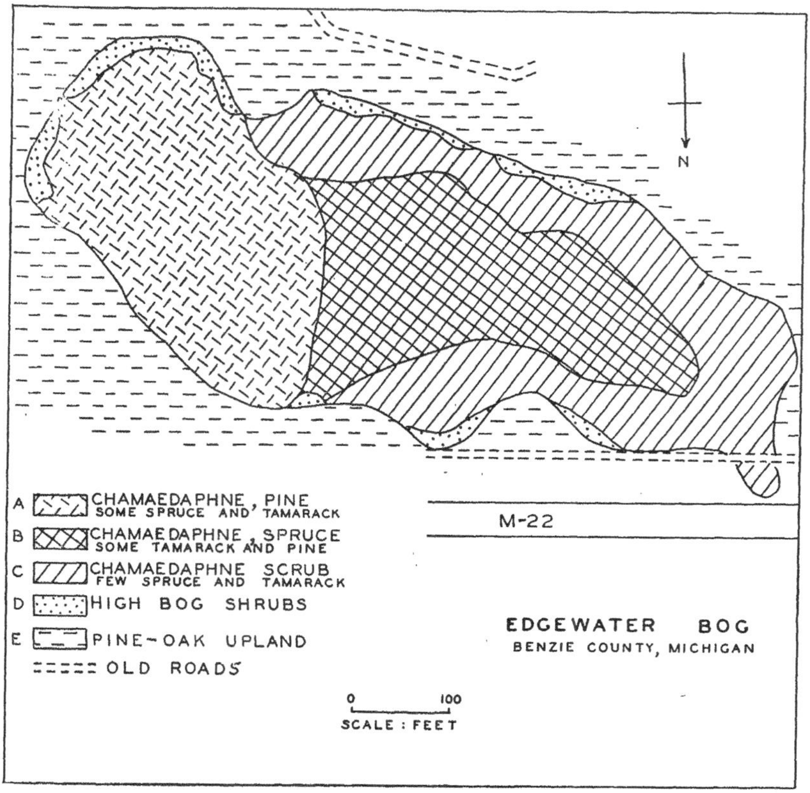

FIG. 3

The bog is roughly 850 feet in length, with an average width of about 300 feet. The longest axis is in an east–west direction. It covers an area of between five and six acres and is easily accessible at any point of its periphery.

Geological Features

The region in which the bog is located was once a great bay in Lake Algonquin, the site of which is now occupied by a series of sand ridges with depressions between, known as the Platte Plains (Waterman, 1917, 1922a, 1922b). This area may be described as a right-angled triangle, the base of which extends east and west eight miles and the east side about the same distance north and south. The hypotenuse is formed by the shoreline of Lake Michigan from the moraine south of Empire to a point almost three miles northwest of Point Betzie. Between the sand ridges and the Algonquin bluff is a wide crescent-shaped trough of relatively slight depth containing a series of lakes or ponds more or less connected with and draining into the Platte River, which meanders through the sand ridges and reaches Lake Michigan about the center of the hypotenuse of the right-angled triangle. From east to west these lakes are Long Lake, Rush Lake, Platte Lake, and Little Platte Lake. On the east three small lakes are connected by Otter Creek.

There are two small lakes, Loon Lake and Mud Lake, on the lower stretches of the Platte River. When the glaciers were finally melting and opening the Straits of Mackinac, the waters of Lake Algonquin gradually receded and formed a series of sand bars with lagoons between them. On these bars the winds have built up low dunes, which now constitute the sand ridges of the Platte Plains. The larger depressions remain as lakes, but the smaller ones have been more or less filled by the growth of aquatic vegetation.

Early History

The bog is appropriately named "Edgewater Bog," for during the lumbering days a logging community was established along the Platte River about where the present Highway M–22 crosses the river. When lumbering operations ceased, the community disappeared, although the name "Edgewater" still appears on the maps of the region.

In many areas of the Platte Plains the vegetation has been influenced by local fires in prelumbering days (Waterman, 1917). The region in which the bog is located appears to have been disturbed only by logging operations. Whether or not any large trees were removed from the bog is not known.

Floristics and Zonation

The zonation of the vegetation on this bog is less distinct than that on any of the other bogs studied. Zones A, B, and C, as shown in Figure 3, grade into one another almost imperceptibly, the most noticeable difference in the zones being perhaps the somewhat less frequent occurrence of spruce and the smaller size of trees in Zones B and C than in Zone A.

In Zone A there is a heavier stand of conifers, white pine, red pine, tamarack, and spruce. The dense stand of Chamaedaphne appears to be giving way to trees that are now producing heavy shade. Sphagnum is found locally in small hummocks, accompanied by *Vaccinium Oxycoccus*, *Sarracenia purpurea*, and *Drosera rotundifolia*. It appears to be dying out wherever it occurs on the mat.

In the more heavily shaded portion of this zone, to the east, the principal ground plants that are becoming established in areas where Chamaedaphne has been killed are: *Pteris aquilina*, *Gaultheria procumbens*, *Eriophorum spp.*, Polystichum and Polytrichum moss, and blueberries (*Vaccinium spp.*). *Cypripedium acaule* occurs under the larger pines.

The east end of the mat also shows invasion of trees from the upland such as *Quercus alba* L., *Acer rubrum*, *Populus grandidentata*, and *Betula papyrifera*.

The vegetation in Zone B differs only slightly from that in Zone A. The conifers are not so densely distributed since spruce is slightly less abundant, and it is smaller. Chamaedaphne, however, makes a heavy and luxuriant growth in this zone, with fewer areas in which it has been killed, apparently by shade. Sphagnum is the only plant found locally with *Vaccinium Oxycoccus* and *Chiogenes hispidula*.

Zone C is a narrow zone in which the conifers are slightly less abundant than in either Zone A or Zone B. In it Chamaedaphne is vigorous and occurs in almost pure stand, and Andromeda appears only occasionally here and also on other parts of the mat. Beside the highway and along the west margin of the mat the upland trees are invading the mat, the principal species being white oak and large-toothed aspen.

Zone D is the zone in which the high-bog shrubs grow on the mat. It is narrow but quite extensive along the south and southwest margins of the mat. At the north margin these shrubs occur only in three small clumps. The principal shrubs in this zone are: *Nemopanthus mucronata*, *Pyrus arbutifolia*, *Ilex verticillata*, *Hamamelis virginiana* L., and *Salix spp.* Arbor vitae and spreading junipers grow in clumps along the north margin.

The mat has become quite firm. The Chamaedaphne stage is apparently giving way to the bog forest, which at present is made up chiefly of red pine, white pine, tamarack, and spruce. In the more mature Zone A upland trees are rapidly invading the mat, as they are in Zone C at the west end. *Picea canadensis* is occasionally found in Zones A and B.

The surrounding upland has a vegetation typical of the sand ridges of the Platte Plains. The dominant plants of the ground cover are: *Pteris aquilina, Gaultheria procumbens, Epigaea repens* L., *Campanula rotundifolia* L., *Arctostaphylos Uva-ursi* (L.) Spreng., and blueberries (*Vaccinium spp.*). Among the trees characteristic of this area the most important are: *Quercus alba, Acer rubrum, Populus grandidentata, Prunus virginiana,* and *Amelanchier canadensis.*

Westminster College
New Wilmington, Pennsylvania

LITERATURE CITED

Chamberlin, T. C. 1883. Terminal Moraine of the Second Glacial Epoch. U. S. Geol. Surv., Third Ann. Rep., 1881–1882, p. 295.

Cooper, Wm. S. 1912. The Ecological Succession of Mosses as Illustrated upon Isle Royale in Lake Superior. Plant World, 15: 197–213.

Davis, C. A. 1907. Peat: Essays on Origin, Uses, and Distribution in Michigan. Rep. Mich. Geol. Surv., 1907: 94–395.

Fenneman, N. M. 1938. Physiography of the Eastern United States. McGraw-Hill Book Co., New York.

Kurz, Herman. 1928. Influence of Sphagnum and Other Mosses on Bog Reactions. Ecology, 9: 56–69.

Osvald, Hugo. 1935. A Bog at Hartford, Michigan. Ecology, 16: 520–528.

Scott, I. D. 1921. The Inland Lakes of Michigan. Mich. Geol. and Biol. Surv., Publ. 30, Geol. Ser. 25: 370–371.

Transeau, E. N. 1903. On the Geographic Distribution and Ecological Relations of the Plant Bog Societies of North America. Bot. Gaz., 36: 401–420.

Waterman, W. G. 1917. Ecology of Northern Michigan Dunes: Crystal Lake Bar Region. Rep. Mich. Acad. Sci., 19: 197–208.

—— 1922a. Development of Plant Communities of a Sand Ridge Region in Michigan. Bot. Gaz., 74: 1–31.

—— 1922b. Forest and Dunes from Point Betzie to Sleeping Bear. Northwestern University, Evanston, Ill.

Weaver, John Ernest, and Clements, Frederic E. 1929. Plant Ecology, pp. 55–74. McGraw-Hill Book Co., New York.

ADDITIONS TO THE MYCOLOGICAL FLORA OF THE CHICAGO REGION

JOHN M. ROBERTS

THREE authors have previously reported on the agarics to be found in and about Chicago, Illinois. In 1909 Moffatt (5) listed 210 species; in 1927–33 Graham (1–3) added 46; and in 1936–40 Routien (6–7) increased the number by 41. In the summer of 1940 the present author collected 115 species of agarics in a small part of this same area. Of these species 43 had not hitherto been published as being from this region. The total of agarics for the Chicago region is now 48 genera, 339 species, and 18 varieties.

The genera with the greatest number of species are: Hygrophorus, 24; Russula, 22; Collybia, 21; Marasmius, 20; Lactarius, 19; and Clitocybe, 17. There are, in addition, 6 varieties of Hygrophorus and 2 of Clitocybe. From the author's observations, however, the genera most dominant in the number of fruiting bodies appear to be Hypholoma and Coprinus.

The increase in the number of species is evidently due to each author having limited his collecting to places readily accessible to him and not studied by the others. In this way each worker collected under different conditions and found different species. To suppose that the new species were introduced by natural agencies seems illogical because the areas that are favorable for the growth of agarics are isolated from one another by fields and urban districts. Most of the author's collections were made in a wood along the Des Plaines River, an area of not more than twelve acres.

In his paper Moffatt (5) wrote: "The region at the head of Lake Michigan is by no means an ideal one for the collector." Graham (3) estimated that there were at least 700 species of mushrooms (using the term in a broad sense). As already stated, there are known to be at least 339 species of agarics alone in this limited area. The fact that there have been found at least one third as many agarics as Kauffman (4) listed for Michigan, in addition to a great diversity of other fungi in the larger wooded localities, does not uphold Moffatt's assumption.

In general, the species from the Chicago region correspond with the descriptions given by Kauffman. However, thirteen of these species were larger, on the average, in size of pileus, stipe, or spores, and eighteen were smaller in some structure. In spite of these minor, though constant, differences, the Chicago species were evidently the same.

The present additions to the agaric flora of the Chicago region are: *Agaricus subrufescens* Pk., *Amanitopsis vaginata* (Fries) Karst. var. *alba* Sacc., *Clitocybe candida* Bres., *C. pulcherrima* Pk., *Collybia cirrata* (Fr.) Quél., *C. colorea* Pk., *C. hariolorum* (Fr.) Quél., *C. lentinoides* Pk., *C. longipes* (Fr.) Sacc., *C. myriadophylla* (Pk.) Sacc., *Coprinus arenatus* Pk., *C. sterquilinus* (Fr.) Quél., *Crepidotus haerens* (Pk.) Sacc., *C. versutus* (Pk.) Sacc., *Entoloma indigoferum* (Ellis) Sacc., *Galera lateritia* (Fr.) Quél., *Hebeloma capnoides* (Fr.) Quél., *H. saccharinophilum* Pk., *Inocybe longicystis* Atk., *Lentinus vulpinus* Fr., *Lepiota rugulosa* Pk., *Marasmius foetidus* Fr., *M. graminum* (Libert.) Berk. & Br., *Mycena lasiosperma* Bres., *M. polygramma* (Fr.) Quél., *Naucoria lignicola* (Pk.) Sacc., *Omphalia luteola* Pk., *Panaeolus campanulatus* (Fr.) Quél., *Panus angustatus* Berk., *Pholiota discolor* (Pk.) Sacc., *P. duroides* Pk., *P. squarrosoides* (Pk.) Sacc., *Pleurotus petaloides* (Fr.) Quél., *Pluteus longistriatus* (Pk.) Sacc., *Psilocybe agariella* Atk., *P. atrorufa* (Fr.) Quél., *P. murcida* (Fr.) Karst., *Russula albida* Pk., *R. lepida* Fr., *R. subdepallens* Pk., *R. uncialis* Pk., and *Tricholoma resplendens* Fr.

ACKNOWLEDGMENTS

The author wishes to thank Dr. A. E. Edgecombe, of Northwestern University, for his assistance and guidance in the collection and identification of the fungi, and to express his gratitude to Dr. E. A. Bessey, of Michigan State College, for his aid in determining the correct authorship of the different species and for his advice in the preparation of this manuscript.

LITERATURE CITED

1. GRAHAM, V. O. 1927. Ecology of the Fungi in the Chicago Region. Bot. Gaz., 83: 267–287.

2. —— 1932. The Genus *Hygrophorus* in the Chicago Region. Proceedings of the Illinois Academy of Science, 23: 160–168.

3. —— 1933. Mushrooms of the Chicago Region. Chicago Academy of Sciences, Program of Activities, 4 (3): 42–63.

4. KAUFFMAN, C. H. 1918. The Agaricaceae of Michigan. Lansing, Michigan.

5. MOFFATT, W. S. 1909. The Higher Fungi of the Chicago Region. Chicago Academy of Sciences, Bulletin VII.

6. ROUTIEN, J. B. 1936. Mushrooms of Evanston and Vicinity. M. A. thesis. Northwestern University, Evanston, Illinois.

7. —— 1940. New Agaricaceae of the Chicago Region. Proceedings of the Indiana Academy of Science, 49: 84–85.

MICHIGAN STATE COLLEGE
EAST LANSING, MICHIGAN

MARINE ALGAE OF HONG KONG. V
THE GENUS HERPOSIPHONIA *

CHENG KWEI TSENG

THE genus Herposiphonia was proposed by Nägeli in 1846 in a paper dealing with the morphology of two plants of this group. Curiously, no specific names were mentioned in the entire article. Nevertheless the illustrations and the very detailed descriptions show without the least doubt that the species Nägeli had in mind in proposing the genus were *Hutchinsia tenella* C. Ag. (*Polysiphonia tenella* (C. Ag.) J. Ag.) and *Hutchinsia secunda* C. Ag. (*Polysiphonia secunda* (C. Ag.) Zan.), the first of which has since been practically accepted as the type species of the genus. It should be noted here that in 1845 Kützing had already used the same name, Herposiphonia, for sections of the genus Polysiphonia, and that J. Agardh employed it later, in 1863. The section Herposiphonia of Kützing is different from that of J. Agardh, however, as well as from the genus Herposiphonia of Nägeli. The two species mentioned above also belong to the section Herposiphonia in the Kützingian classification, but are not included in the Herposiphonia section of J. Agardh. The latter phycologist seemed to disagree with Nägeli in placing them in a special genus separate from Polysiphonia, and he preferred to group them together with several others that we now recognize as members of Herposiphonia Näg., in the subsection Pectinatae of the section Ptilosiphonia of the genus Polysiphonia.

The combinations *Herposiphonia tenella* and *H. secunda* were first used by Ambronn in 1880. Several years later Schmitz transferred to Herposiphonia *Polysiphonia prorepens* Harv. and Okamura *Polyzonia fissidentoides* Holmes. In his monograph of the family Rhodomelaceae Falkenberg (1901) transferred eleven other Polysiphonia species to this genus, including a doubtful one that was subsequently removed. De Toni transferred three more species to it,

* Papers from the Department of Botany of the University of Michigan, No. 824.

all doubtfully. Later Pilger (1911, 1920) added two new species, and Okamura, Howe, and Gardner each added one. This is a total of twenty-two species of Herposiphonia, including three whose affinity with the genus is not definitely fixed. Two more species will be described in this article.

Plants of this genus are characterized by a creeping filamentous dorsiventral axis, sending stout vigorous rhizoids downward, long shoots laterally, alternately and horizontally, and erect subdorsal dwarf shoots in two rows, which are seldom strictly dorsal. A typical member, for instance, *H. tenella* (C. Ag.) Ambr., has the arrangement of its shoots repeated on the following pattern: a long shoot at the left, lateral, then a dwarf shoot on the second segment at the right subdorsal, another on the third at the left subdorsal, still another on the fourth at the right subdorsal, and then, on the fifth segment, another long shoot laterally at the right side. The long shoots are the branches; eventually they come to resemble the main axis and are indefinite in growth. The dwarf shoots are the branchlets and are definite in growth, finally giving rise to the reproductive organs. The apices of the axis and the horizontal branches are always more or less inwardly curved, and the growth is exogenous, the apical cell dividing transversely or slightly obliquely. The filaments are distinctly segmented, and the pericentral cells are numerous, ranging from 8 to 18, and naked, not covered by any corticating cells or filaments. The trichoblasts are deciduous or persistent, being quite simple or repeatedly subdichotomously branched. The tetrasporangia are formed in long single series in the middle part of the dwarf shoots. The spermatangial clusters are subcylindrical, replacing the trichoblasts on the dwarf shoots, and are spirally disposed at the distal ends. The subspherical, ovate, or urceolate cystocarps are solitary or in pairs, lateral near the apices, or subterminal on the dwarf shoots.

Members of this genus are inhabitants of the warmer waters, being especially abundant in the Australian and Indo-Pacific regions. None has so far been recorded from the Hong Kong area, although two have been reported from near-by Amoy (Tseng, 1936). Briefly, the characteristics that are employed in the identification and classification of the species of this small genus are: the habit and size of the plant and its filaments, the branching, the number of the pericentral cells, the length of the segments, the number of segments in the dwarf shoots, the spermatangial clusters and their disposition, and the shape, size, and position of the cystocarps.

The four species now found in this region may be distinguished from one another by the following key:

A. Dwarf shoots branched above; number of pericentral cells 12–18 .... 4. *H. ramosa*
A. Dwarf shoots simple; number of pericentral cells 8–12
  B. Frond more rigid, with certain segments bare; long and dwarf shoots rather irregularly disposed .... 3. *H. insidiosa*
  B. Frond more flaccid, all segments with branches or branchlets that are very regularly arranged
    C. Frond beautiful pinkish red, habit loose, epiphytic on other algae; dwarf shoots with 12–14 segments .... 1. *H. Pecten-Veneris*
    C. Frond dark purplish, forming dense pulvinate tufts, generally on rocks; dwarf shoots with 20–30 segments .... 2. *H. caespitosa*

1. *Herposiphonia Pecten-Veneris* (Harv.) Falk.

Rhodomelaceen, 1901, p. 315; Howe, Bahama Alg., 1920, p. 573.
Polysiphonia Pecten-Veneris Harv., var. a Harv., Nereis Boreali-Americana, II, 1853, p. 46, pl. 16 C.

Frond small, delicate, beautiful pinkish red, loosely epiphytic on various algae; about 3 cm. long, patently and alternately decompoundly branched, with a branch or branchlet from every segment. Primary filament giving rise to both the dwarf and the long shoots, with three of the former between every two of the latter in the typical Herposiphonia pattern. These are arranged in four rows, two lateral, alternate rows of long shoots and two dorsal ones of the dwarf shoots. Main axis 120–150 μ in diameter, with segments 1.5–2.0 times as long as broad, attached to the host by long rhizoids of about 30 μ in diameter, each separated by a transverse wall from the mother pericentral cell, more or less flexuous, with its apex strongly and gracefully recurved, which results in its dorsal convexity. The long shoot, when fully mature, similar to the main axis in size, in the strong incurving of the tip, in the creeping habit, and in the indefinite growth. The dwarf shoot definite in growth and erect, 50–80 μ in diameter, and up to 1.0 mm. high, obtuse at the tip and rather strongly curved toward the axis. Each branchlet with 12–14 segments, which are slightly longer than broad below and shorter above. Trichoblasts present at the tips of the younger dwarf shoots, deciduous. Pericentral cells ranging from 8 to 12 in each segment. No reproducing specimens secured.

Habitat. — On other algae in the littoral region, Shaukiwan in May (*Tseng 345*) and Shek-O in March (*Tseng 649*), both on Hong Kong Island.

Distribution. — Key West, Florida (type locality), and the West Indies.

The present species, so far as its records show, has not been formally reported outside the Florida–West Indies region. Nevertheless, the writer has reasons to believe that it is quite extensively distributed. The difficulty in defining its distribution lies in the controversy whether it should be treated merely as a form of the more widely distributed *H. tenella* (Ag.) Ambr. or kept as a separate species. Falkenberg (1901) feels rather uncertain in regard to this question. American phycologists, Collins, Howe, and Taylor, have separated the two, whereas Børgesen seems to prefer to keep the plant under *H. tenella,* although he never did formally put Harvey's species as a synonym of the former. To judge from Okamura's description and illustrations (1930, p. 23, pl. 264, figs. 1–9) he has evidently adopted the same broad view of *H. tenella* that Børgesen did and included this species under the former. In the past the present writer also shared the view of Børgesen and Okamura and identified a similar alga from Amoy with *H. tenella* (Tseng, 1936, p. 59, pl. 6, fig. 36). Now, having studied many specimens of these two species from various places, he has come to agree with the American phycologists in keeping them separate and in differentiating the present species from *H. tenella* on the basis of its strongly incurved tips with dorsal convexity, its much shorter dwarf shoots with, as a rule, not more than fourteen segments each, and, consequently, its fewer sporangia, generally 4–7 in a series. Unfortunately, the cystocarps and spermatangial clusters, which would facilitate further comparison, are not known for both species. When found, they will surely help in solving the problem.

2. **Herposiphonia caespitosa,** sp. nov.

(Plate I)

Frons fuscipurpurea, mollis, dense pulvinata, caespitosa, repens, ca. 3 mm. alta, lateraliter decomposite ramosa; filamentis omnino sine cortice, 8–12-siphoniis, apice ascendentibus, articulis omnibus ramiferis, dispositione ramorum 4-seriata, ramis longioribus lateraliter, ramis nanis subdorsaliter dispositis; filamentis primariis et ramis longioribus decumbentibus dense radicantibus, dorsiventralibus, ad 200 μ (generaliter 120–150 μ) latis, articulis omnibus isodiametientibus vel brevioribus quam latioribus; ramis nanis subteretibus, erectis, ca. 3 mm. altis, 90 μ latis, apice obtusis vel rotundatis, articulis 20–30, isodiametris vel duplo longioribus quam brevioribus; trichoblastis 5- vel 6-plo dichotome ramosis, copiose ad apicem ramorum nanorum spiraliter ordinatis; tetrasporangiis ca. 35 μ latis, in medio ramorum nanorum spiraliter et uniseriatim ortis; cystocarpiis maturis urceolatis usque ad 460 μ latis, 520 μ longis, ut videtur in ramis nanis terminalibus, sed vero ab initio quasiterminalibus; acervulis spermatangialibus numerosis (10), subcylindricis usque ad 45 μ latis, 170 μ longis, superne cellula sterili praeditis, sed ad apicem ramorum nanorum spiraliter (uno in quoque articulo) dispositis. Species *Herposiphoniae prorepenti* (Harv.) Schmitz et *H. densae* Pilg. proxima.

Specimen typicum: *Tseng 2752* (in herbario auctoris), ad rupes zonae inferioris littoralis, prope Putoi Island, 22 Apr., 1940.

Frond flaccid, dark purplish, forming densely pulvinate tufts to about 3 mm. tall, laterally fastigiately decompoundly branched, with ecorticate filaments consisting of 8–12 rows of pericentral cells. Long and dwarf shoots disposed in four rows, as typical in the genus, with two lateral rows of long shoots and two subdorsal ones of dwarf shoots, and a branch or branchlet on each segment, hence always three dwarf shoots between the two alternating long shoots (Pl. I, Fig. 1). Primary filament and long shoots decumbent, dorsiventral, with their apices upwardly incurved, densely beset with stout unicellular rhizoids, which are generally in pairs, sometimes one or three to each segment, and up to 200 μ, generally 120–150 μ in diameter, with segments shorter than broad or subequal. Dwarf shoots subcylindrical, erect, up to 3 mm. tall, 90 μ in diameter, broader below, attenuated upward, sometimes to as much as 1.8 mm., long when mature, obtuse-rounded at the tips, and composed of 20–30 segments, which are longer below, to about twice longer than broad, and shorter above. Trichoblasts well developed, abundantly spirally disposed at the distal ends of the dwarf shoots, 5–6 times subdichotomously branched with a large basal cell about 20 μ in diameter. Chromatophores of the cells short and linear, transversely zonately arranged, giving the cells a striate appearance. Young tetrasporangia, about

35 μ in diameter, found only once, in the middle portion of a dwarf shoot in a single spiral series of nine (Pl. I, Fig. 4). Spermatangial clusters abundantly formed at the upper ends of dwarf shoots, subcylindrical, each terminating in a large sterile cell, about 45 μ broad and 170 μ long (Pl. I, Fig. 3), spirally arranged, replacing the trichoblasts and, because of the short segments there, very much crowded. When young, they are only about 200 μ long, but grow progressively with the branchlets and increase in number to 10–12, the male shoots maturing to a length of about 1 mm., about half or two thirds the length of the surrounding sterile shoots. Cystocarps subglobular to ovate at first, becoming more or less urceolate, to 460 μ in diameter, 520 μ long, including a neck of about 60 μ (Pl. I, Figs. 1–2), apparently terminal and generally solitary on each fertile shoot, the cystocarps occasionally occurring in terminal pairs, but one always larger and the other smaller and forming later (Pl. I, Fig. 2). Female branchlets are generally 0.75–1.0 mm. They are often only about one third to one half the length of the neighboring sterile filaments. Their segments, of the same number as those of the sterile filaments, are necessarily very much shortened.

Habitat. — On surf-beaten rocks in the lower littoral region, Shek-O, Hong Kong Island, in March and July (*Tseng 2661*) intermingled with *Taenioma* and other algae, and on Putoi Island, in April (*Tseng 2752*, TYPE).

The present species is unique among species of Herposiphonia in the at least apparently terminal position of the cystocarp. The cystocarps seem to be lateral in origin, just as they are in other species of the genus. This is inferred from the frequent occurrence of a small, inconspicuous protrusion near the base of some of them, which may represent the original apices of the female shoots, whose growth has been inhibited by the growth of the cystocarps (Pl. I, Fig. 2). In this respect there seems to be a gradual reduction of the active growth of the apex of the fertile dwarf shoot, from typical species like *H. secunda* through smaller species such as *H. prorepens* and *H. densa* to the present species. In the first species the apex of the shoot remains very active after the formation of the cystocarp, and sometimes even grows out to produce other dwarf shoots just like ordinary long shoots (Falkenberg, 1901, pl. 3, fig. 2, and Børgesen, 1920, p. 474, fig. 431 a, b). In the second one the growth of the apex is not so vigorous, but it still outgrows the cystocarp (cf. Harvey, 1862, pl. 185 B, figs. 2–3). In the third the activity of the apex has been reduced, and the cystocarp outgrows the apex of the fertile shoot, which, however, still remains distinct although short (cf. Pilger, 1911, fig. 17). In the present species the growth of the cystocarp is so rapid that the growth of the apex of the fertile shoot is entirely stopped; the apex is reduced to a very small, inconspicuous pseudolateral protrusion or, more frequently, is entirely aborted (Pl. I, Figs. 1–2).

This species is undoubtedly most closely related to *H. densa* from Africa which, as has been mentioned, has the fertile apex of the fertile branchlet very much reduced. They resemble each other in the dwarf, dense caespitose habit, the arrangement of the branches, the number of the pericentral cells, and the short segments. The main differences between the two lie chiefly in the position of the cystocarps, as noted above, and the much taller erect branchlets of the African plant (7–8 mm. as against 2–3 mm. in the present species). The striated closely transversely parallel arrangement of the short linear chromatophores seems to be quite characteristic of the Hong Kong plant, but is not reported for the African alga. *H. prorepens* is very closely related to these two species, but it is a much more delicate plant. The diameter of the axis and the branches is less, as are the height of the dwarf shoots and the number of segments. The sterile and fertile shoots are of about the same height.

3. *Herposiphonia insidiosa* (Grev.) Falk.

Rhodomelaceen, 1901, p. 317; Okamura, Icones of Japanese Algae, VI (3), 1930, p. 35, pl. 264, figs. 10–16; Tseng, Mar. Alg. Amoy, 1936, p. 60, pl. 6, fig. 34; Børgesen, Contrib. South Indian Marine Algal Flora, II, 1937, p. 352, fig. 18.

Polysiphonia insidiosa Grev. in J. Agardh, Species algarum, II (3), 1863, p. 926.

Frond fragmentary, only about 1 cm. tall, up to 120 μ in diameter in the main filaments and 80–90 μ in diameter in the branchlets, with well-developed rhizoids and segments much shorter than broad, to subequal; pericentral cells 8–12, sometimes slightly spirally twisted; branches and branchlets irregularly disposed, by no means like such typical species as *H. tenella* and *H. Pecten-Veneris*. The apices of the main filament and the long shoots upwardly curved, irregularly produced, often with variable numbers of dwarf shoots between them, and the axis sometimes with occasional bare segments. Dwarf shoots are more or less alternately disposed, very much curved toward the apex of the plant, and strongly inrolled in a spiral, with 20–26 segments, and obtuse apices. Trichoblasts generally absent or, if present, simple and unbranched. The tetrasporangia are up to 60 μ in diameter, in a single longitudinal series in the central part of the dwarf shoot. Other reproductive organs were not found.

Habitat. — Intermingled with Laurencia on littoral rocks, Big Wave Bay, Hong Kong Island, in May (*Tseng 2577*).

Distribution. — East Indies (type region); Ceylon; Amoy, China; Japan.

4. **Herposiphonia ramosa,** sp. nov.

(Plate II)

Frons nigrescentipurpurea, aliquantum rigida, ca. 7 mm. alta, lateraliter decomposite ramosa, repens, ramulis rhizoideis bi- vel tricellularibus ad algas alias adfixa; filamentis omnino sine cortice, 12–18-siphoniis, apice ascendentibus, ramis aliis longioribus lateraliter ordinatis, aliis nanis subdorsaliter ordinatis, articulis isodiametientibus vel brevioribus quam latioribus, normaliter articulo ramo nano praedito sequente articulum ferentem ramum longiorem et articulis plerumque 2 (sed interdum 3–4) nudis sequentibus eundem ramum nanum; filamentis primariis et ramis longioribus decumbentibus, radicantibus, dorsiventralibus usque ad 240 μ (generaliter 180–200 μ) latis; ramis nanis subteretibus, erectis, ramosis, usque ad 7 mm. altis, 120–150 μ latis, apice obtusis vel rotundatis, articulatis 30–40; trichoblastis 4- vel 5-plo dichotome ramosis, apice ramorum nanorum spiraliter ordinatis; partes aliae desunt. Species *Herposiphoniae secundae* (C. Ag.) Ambr. proxima.

Specimen typicum: *Tseng 2647a* (in herbario auctoris) ad Corallinam, cum Falkenbergia epiphyticum, in zonae inferioris litoralis, prope Stanley Bay, Hong Kong Island, 19 Jun., 1939.

Frond somewhat rigid, blackish purple, about 7 mm. high, decompoundly laterally branched. Main filaments and long shoots decumbent, dorsiventral and attaching to other, especially coralline, algae, with long 2- to 3-celled rhizoidal branchlets, about 30 μ in diameter, with 12–18 pericentral cells (Pl. II, Figs. 1, 3), the segments all shorter than broad, or subequal. Long shoots horizontal and lateral dwarf shoots erect and dorsal, a dwarf shoot always following a long shoot, generally followed in turn by two, sometimes even three or four, segments that are naked, after which the long shoot again occurs, on the other side (Pl. II, Fig. 1). A peculiarity observed in this plant is the possession of branches on the dwarf shoots (Pl. II, Fig. 2), which branches issue exogenously and are irregularly subdichotomous or subpinnate. The dwarf shoot, supposedly definite in growth, thus resumes branching, resembling somewhat the main axis and the long shoots. The main filament and the long shoots are generally 180–200 μ, sometimes to 240 μ, in diameter. They have their apices upwardly incurved and segments shorter than, or subequal to, the diameter. The dwarf shoots are up to 7 mm. high, 120–150 μ, sometimes 180 μ, in diameter. They are obtuse-rounded at the tips and are composed of 25–50 segments, which are about as long as broad below, becoming longer, to 1.5 times longer than broad in the middle, and shortening again above. Trichoblasts are 4–5 times dichotomously branched, with basal cells to 20 μ in diameter and are spirally disposed on the distal ends of the dwarf shoots. Specimens collected are all sterile.

Habitat. — Epiphytic on corallines, together with Falkenbergia, in the lower littoral region, Stanley Bay, Hong Kong Island, in June (*Tseng 2647a*, TYPE) and Clear Water Bay, Port Shelter, Kowloon in May (*Tseng 2803a*).

The present species is unique among the members of the genus Herposiphonia in the possession of branches by the dwarf shoots. This characteristic and the large number of pericentral cells (12–18) as well as its brownish to blackish purple rather rigid frond remind one very much of *Lophosiphonia obscura* Auct. ex Howe. However, it cannot be referred to that species, or even to the genus Lophosiphonia, because of the exogenous growth and the regularity exhibited by the long and dwarf shoot sequence. Among other species of Herposiphonia probably the most closely related is *H. secunda*, which also has certain segments naked without any branch or branchlet and in which the number of segments between the two alternating long shoots sometimes varies between 3 and 5. From that species *H. ramosa* differs, not only in the branching of dwarf shoots, but also in the greater number of pericentral cells (12–18 against 8–12) and of segments in the dwarf shoots (25–30 against 7–20), and in the more robust frond (240 μ in diameter, against 150 μ).

SUMMARY

Of the four species of Herposiphonia reported in this article two are described here for the first time. *Herposiphonia caespitosa* Tseng, sp. nov., a species closely related to *H. densa* Pilg. from Africa, is unique in having apparently terminal cystocarps. *Herposiphonia ramosa* Tseng, sp. nov., a species more nearly related to the widely distributed *H. secunda* (C. Ag.) Ambr. than any others, is unique in having branched dwarf shoots. Of the two other known species, *H. Pecten-Veneris* (Harv.) Falk. of the Florida–West Indian region, generally confused with the widely distributed *H. tenella* (C. Ag.) Ambr., is here formally credited to the China coast for the first time, although it had been reported previously under the latter name from Amoy. *H. insidiosa* (Grev.) Falk., originally described from the East Indies, probably widely distributed in the Indo-Pacific region, has been reported from Amoy on the China coast.

The writer wishes to thank Professor Wm. Randolph Taylor, of the University of Michigan, for his advice during this study.

UNIVERSITY OF MICHIGAN

LITERATURE CITED

AGARDH, J. G. 1863. Species, genera et ordines algarum, II, Part 3, 701–1291. Lund.

BØRGESEN, F. 1920. The Marine Algae of the Danish West Indies. Vol. II. Rhodophyceae. Dansk Bot. Arkiv, 3: 1–504.

—— 1937. Contributions to a South Indian Marine Algal Flora, II. Journ. Ind. Bot. Soc., 16 (6): 311–357.

FALKENBERG, P. 1901. Die Rhodomelaceen des Golfes von Neapel und der abgrenzenden Meeresabschnitte. Fauna und Flora des Golfes von Neapel. 26. Monographie. 754 pp., 24 pls. Berlin.

HARVEY, W. H. 1853. Nereis Boreali-Americana. Part II. Rhodospermeae. 258 pp., pls. 13–36. Washington.

—— 1862. Phycologia Australica; or, a History of Australian Seaweeds, etc., IV: i–viii, pls. 181–240 (with descriptions). London.

HOWE, M. A. 1920. Algae. *In* N. L. Britton and S. F. Millspaugh, The Bahama Flora, pp. 553–631. New York.

NÄGELI, C. 1846. *Herposiphonia.* *In* Schleiden, M. J., und Nägeli, C., Zeitschrift für wissensch. Bot., 1 (3–4): 238–256, pl. 8.

OKAMURA, K. 1930. Icones of Japanese Algae. VI, No. 3: 19–27, pls. 261–265. Tokyo.

PILGER, R. 1911. Die Meeresalgen von Kamerum. Engler's Bot. Jahrb., 46: 294–323.

—— 1920. Algae Mildbraediannae Annabonenses. Bot. Jahrb., 57: 1–14. Figs. 1–34.

TSENG, C. K. 1936. Notes on the Marine Algae from Amoy. Amoy Mar. Biol. Bull., 1 (1): 1–86, pls. 1–4.

EXPLANATION OF PLATE I

FIG. 1. Habit sketch of portion of cystocarpic plant. × 83

FIG. 2. Cystocarps in different stages of development, with two in pair. × 44

FIG. 3. Upper part of male branchlet, showing gametangial cluster. × 83

FIG. 4. Upper part of young tetrasporic branchlet, showing young trichoblasts. × 166

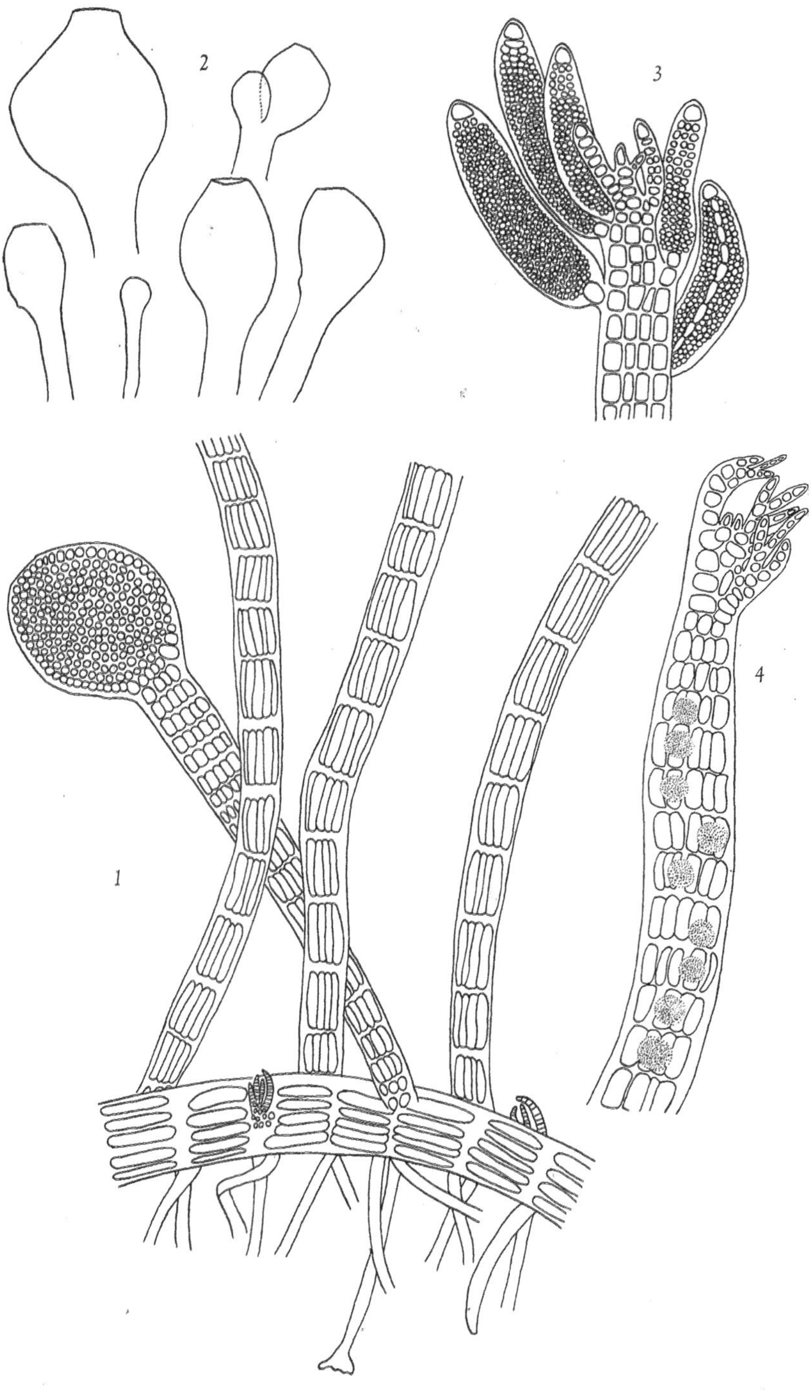

Herposiphonia caespitosa Tseng, sp. nov.

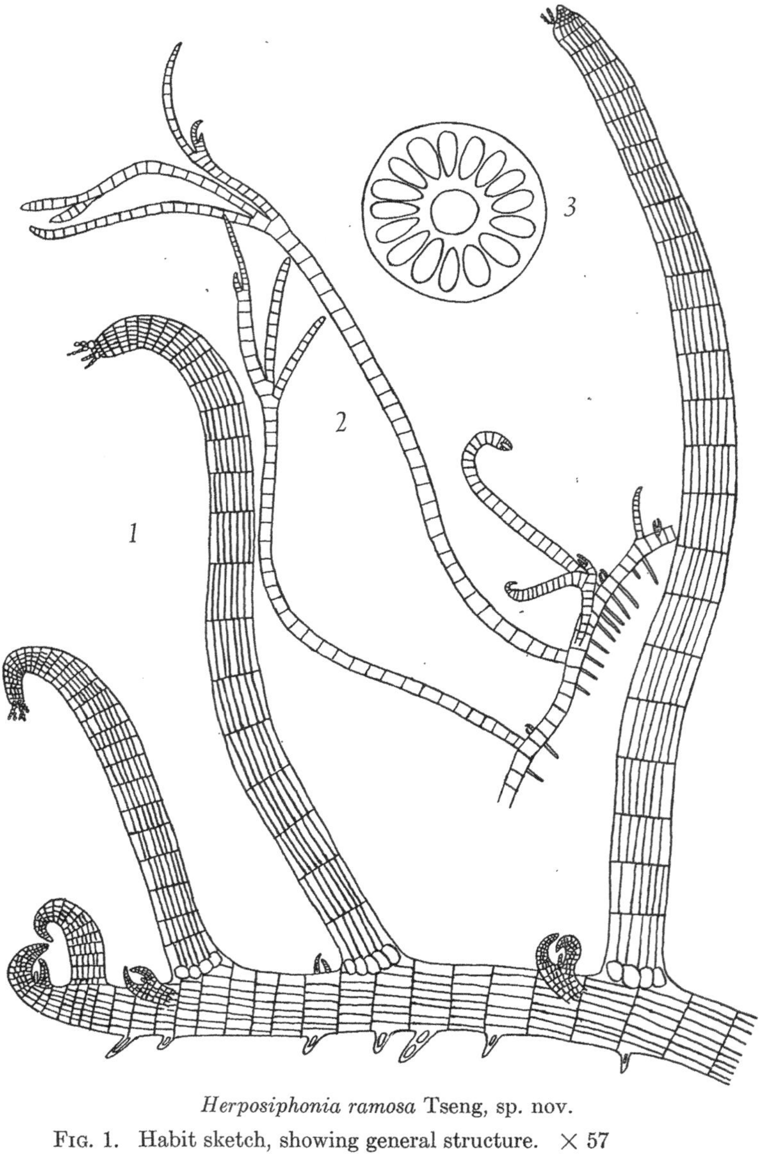

Herposiphonia ramosa Tseng, sp. nov.

FIG. 1. Habit sketch, showing general structure. × 57
FIG. 2. Habit sketch, showing branching of dwarf shoots. × 16
FIG. 3. Transverse section of branchlet. × 108

MARINE ALGAE OF HONG KONG. VI
THE GENUS POLYSIPHONIA *

CHENG KWEI TSENG

THE generic name Polysiphonia was proposed by Greville in 1824 as a substitute for Hutchinsia C. Agardh (1817) since the latter name was already preoccupied by a group of cruciferous plants, namely, Hutchinsia R. Brown (1812). Bonnemaison (1822) had already proposed the name Grammita for the ineligible Hutchinsia C. Ag., but Greville objected to it because of the possibilities of confusion with the generic name Grammitis, which had been used for one group of fungi and for another of ferns. Greville also overlooked the earlier published Vertebrata S. F. Gray (1821) referring to the same group of algae. Theoretically, therefore, Polysiphonia Grev. is a superfluous name and would be rejected if it had not been legalized in the status of a *nomen conservandum* by the International Botanical Congress (cf. Briquet, 1935, p. 88).

Plants of this genus are either entirely erect, rising from a more or less discoid holdfast, or with decumbent basal filaments, which attach to the substratum by a number of stout rhizoids originating from the pericentral cells, or, in corticated species, from the cortical cells. As a rule, the frond is abundantly dichotomously or laterally branched. The branches are filamentous, finely capillary to very coarse. Structurally the main axis and the branches of all orders are similar and polysiphonous, being composed of a series of axial cells surrounded by four to twenty-four rows of pericentral cells of similar length but generally of larger diameters. The pericentral cells are parallel to the central siphon; a few, however, are spirally twisted. In some species most of the frond, except the ultimate branchlets, is lightly or densely corticated by small cells or rhizoidal downgrowths that originate from the pericentral cells and may build up a pseudoparen-

* Papers from the Department of Botany of the University of Michigan, No. 825.

chymatous cortex. Trichoblasts, which usually are present, are deciduous or persistent, occurring as delicate colorless dichotomously branched filamentous hairs, near the tips of the branchlets; these are regarded by some as equivalent to "leaves."

Tetrasporangia are formed in the ultimate branchlets, generally near the tips, and are slightly or greatly swollen to distorted. They are solitary, one in each segment, and occur in straight or spirally twisted series. In the formation of these spores the fertile pericentral cell divides lengthwise, and the inner daughter cell divides again transversely, the upper one becoming the sporangium mother cell, which, by meiosis and tetrahedral cytokinesis, gives rise to the four tetrahedrally grouped tetraspores. The spermatangial clusters are colorless and ovoid to subcylindrical; each is transformed from an entire trichoblast, or a part of it, at the distal end of the branchlet. The fertile portion is polysiphonous, bearing on the pericentral cells the numerous small colorless oval spermatangia. Procarps are also developed from trichoblast rudiments. Cystocarps are typically urn-shaped, each having at the top an ostiole through which the numerous pear-shaped, or sometimes subglobose to ovate, carpospores are discharged.

Polysiphonia is one of the largest genera among the marine algae and, although many of its former members have been transferred to various other genera lately proposed, a few hundred species are still recognized in the genus. A great percentage of these presumably "valid" species may eventually be found to be synonymous with others when a critical revision of the genus can be made. Geographically the genus is a cosmopolitan one that is found in practically all the oceans and seas from the Antarctic to the Arctic. Generally speaking, the members in the warmer waters are more delicate ecorticate or lightly corticated tetrasiphonous species, whereas those of the colder waters are more robust, often corticated, and tetrasiphonous as well as polysiphonous.

Plants of this genus vary greatly in color, size, and habitats. They are found on rocks, sand, and mud, and in sheltered as well as exposed places. Some inhabit deep seas; others may occur only near high-water mark or far up the estuaries of tidal rivers. There are many epiphytic forms growing on various coarser algae.

The genus is divided into two sections, Oligosiphonia, with four pericentral cells, and Polysiphonia, with more than four. Further division of the sections is based on the presence or the absence of the cortication. Other characteristics employed in the differentiation of the various species are: the method of branching, the nature of the attachment organs, the relative length of the articulation, the origin of the trichoblasts and of the branches, the nature of the tetrasporangial branchlet, the position of the spermatangial clusters, and the shape and size of the cystocarps.

The only record of Polysiphonia in the Hong Kong region is that of *P. Harlandii* Harv., described as occurring on materials collected from Hong Kong Island by "Charles Wright in the North Pacific Exploring Expedition under Captain John Rodgers." From other parts of the North China coast three more species have been reported, namely, *P. japonica* Harv., *P. urceolata* (Lightf.) Grev., and *P. Morrowii* Harv. Collins' record of "*P. ferulacea*" has been shown by Howe (1924, p. 141) to be based on a plant of *P. japonica* and should be taken from the floristic list of China. It is certainly incredible that there would be only four species of this large widely distributed genus on so extensive and varied a coast as that of China. The difficulty of delimiting the Polysiphoniae has probably discouraged reports of their occurrence. In fact, in the writer's own collections there are about two dozen species still waiting to be reported.

At least six species are now known to be represented in the flora of Hong Kong. Except for *P. Harlandii* Harv., originally described from this region, they are all recorded here for the first time from any part of China. The following key serves to distinguish them:

A. Frond corticated in the lower portion . . . . 6. *P. Harlandii*
A. Frond ecorticate
  B. Filaments slender, not over 90 μ in diameter; rhizoids continuous with the mother pericentral cells; trichoblasts generally absent or, if present, poorly developed and deciduous . . . . 1. *P. subtilissima*
  B. Filaments coarser, to 180 μ in diameter or more; rhizoids separated from the mother pericentral cells by cross walls; trichoblasts mostly present and well developed
    C. Frond solitary, epiphytic, small, to 1 cm. high . . . . . 3. *P. Savatieri*
    C. Frond in tufts on rocks, larger, 2 cm. or more high
      D. Frond regularly sparsely and distantly subdichotomously branched, to 180 μ in diameter . . . . 2. *P. coacta*
      D. Frond profusely, closely alternately branched, to 250 μ or more in diameter
        E. Frond reddish purple, soft, segments generally as long as broad, pericentral cells elongated-rectangular, tetrasporangia forming elongated spiral series in the otherwise ordinary branchlets . . . . 4. *P. gracilis*

E. Frond reddish brown (blackish when dried), rigid, segments generally half as long as broad, pericentral cells subquadrate, tetrasporangia in short series in special dichotomous branchlets, the fertile portion more or less swollen and distorted .... 5. *P. ferulacea*

1. *Polysiphonia subtilissima* Mont.

(Plate I)

Pl. cell. cent., II, 1840, p. 199; Harvey, Ner. Bor.-Am., II, 1853, p. 34; Kützing, Tab. Phyc., XIII, 1863, pl. 28, figs. a–e; Farlow, Mar. Alg. New Engl., 1881, p. 170; Taylor, Alg. Northeastern Am., 1937, p. 365.

The present species forms dense purplish brown soft and flaccid, but nongelatinous tufts, about 2–3 cm. high. The frond is tetrasiphonous (Pl. I, Fig. 4) and ecorticate throughout, and consists of a horizontally running stoloniferous basal filament attaching to the substratum by stout rhizoids and giving rise to the erect branches upward at irregular intervals, almost perpendicularly (Pl. I, Figs. 1–2). The creeping filaments have thick walls and are sinuose, 60–75 μ in diameter, with segments about as long as broad. The rhizoidal filaments are about 45–60 μ in diameter and are merely modified pericentral cells; in other words, in the formation of the rhizoid the pericentral cell simply extends by lateral growth without forming another rhizoid-initiating cell, as it does in most other species. The cell contents of the rhizoids are therefore continuous with those of the pericentral cells. These rhizoids generally end in broad lobed discs (Pl. I, Fig. 2).

The erect branches are repeatedly subdichotomously branched below and alternately or sometimes unilaterally branched above (Pl. I, Fig. 1). The basal parts of these branches are of sizes similar to those of the stolon, 60–75 μ in diameter, with segments about as long as broad. Upward the diameter of the segments gradually decreases, but the length increases until they are about one and one-half times longer than broad. At the terminal part the segments again shorten a great deal, and are generally only about one third of the diameter long; the filaments there measure 20–30 μ in diameter. A large apical cell terminates the branchlet (Pl. I, Fig. 3). The trichoblasts are generally absent, although a few poorly developed ones are found. Most probably they are deciduous, and that may be why they are generally reported to be absent. The branches are lateral in origin, not subtended by the trichoblasts.

Only young tetrasporic plants were collected. The tetrasporangia, about 60 μ in diameter when mature, are in a straight, longitudinal series in the central portions of the ultimate branchlets, which are slightly swollen (Pl. I, Fig. 5).

Habitat. — On mud-covered littoral rocks in sheltered places, Shatin, Tide Cove, Kowloon; in May (*Tseng 2786*), intermingled with Caloglossa in April (*Tseng 2727a*) and Lophosiphonia in April (*Tseng 2754a*) and July (*Taam A144a*).

Distribution. — Cayenne, French Guiana, on northern South America (type locality), northward to Massachusetts; Africa.

The writer has hesitated to refer his plants to the rather imperfectly known *P. subtilissima*. The difficulty lies in the fact that the specimens collected from various places in the Americas and assigned by various phycologists to this species are probably heterogenous, consisting of more than one species. Consequently the writer has to select certain specimens out of the mixture and give to them the present specific name, basing his judgment, unfortunately, not on the type specimen, which is not available to him, but, rather, on the original description of Montagne, *loc. cit.*, supplemented by that of Harvey, *loc. cit.*, who claimed that he had studied one of the original specimens, and the little note by Howe (1920, p. 569). With these American specimens thus regarded as approaching typical *P. subtilissima* Mont., those from Hong Kong, cited above, agree very well.

2. **Polysiphonia coacta,** sp. nov.

(Plate II)

Frons fulvipurpurea, pulvinata, caespitosa, ca. 2 cm. alta; filamentis primariis decumbentibus radicantibus; ramis principalibus erectis, inferne hinc illinc inter se per tenacula rhizoidea adhaerentibus, parce distanterque subdichotome ramosis; filamentis subtetragonis, tetrasiphoniis, sine cortice, omnino articulatis, usque ad 180 μ latis, sursum attenuatis, ad apicem 30–40 μ latis; articulis omnibus isodiametientibus vel diametro brevioribus quam latioribus; ramis trichoblastorum locum tenentibus; tetrasporangiis ca. 45 μ latis, in superiore parte ramorum spiraliter ortis et in seriebus brevibus praesentibus; cystocarpiis (immaturis) ovatis, subsessilibus. Species *Polysiphoniae incomptae* Harv. proxima videtur.

Specimen typicum: *Taam A 62* (in herbario auctoris) ad rupes arenosas zonae inferioris litoralis, prope Panglongwan, Port Shelter, 26 Mar., 1941

Plants of the present species form brownish purple pulvinate tufts about 2 cm. high. The primary filaments are decumbent, attaching to the substratum by producing downward short, stout rhizoids. The latter, to 70 μ in diameter, grow out from the pericentral cells, separating by cross walls, and end in broad adhesive discs, many of which usually adhere to one another and thus form a very effective attachment organ (Pl. II, Fig. 1). The erect filaments are sparsely and distantly subdichotomously branched, occasionally with some alternate or unilateral branchlets. The two arms of the dichotomy are unequally developed (Pl. II, Fig. 2). The lower parts of the erect filaments are so intricately and firmly united with one another by the rhizoids, acting as tenacula at places of contact, that it is impossible to isolate the individual filaments without breaking certain parts of them. The filaments are about 180 μ in diameter in the decumbent basal portions, gradually attenuating upward to 30–40 μ in diameter near the tips. In transverse section they appear somewhat tetragonal and are composed of four large pericentral siphons surrounding a small central one (Pl. II, Fig. 3). They are clearly articulate and ecorticate throughout. The segments are always shorter than broad, generally about half the diameter long in the lower parts to about as long as broad in the middle, and then shorter again to equal one third the diameter near the tips. The trichoblasts are well developed near the apices in a one-quarter divergence. The branches are nonaxillary, originating without reference to these trichoblasts. The tetrasporangia are about 45 μ in diameter and are spirally arranged in short series near the tips of the branchlets; the sporogenous portions are only slightly swollen. Some young cystocarps, about 120 μ in diameter, are found. They are ovate and subsessile (Pl. II, Fig. 4).

Habitat. — Abundant on sand-covered rocks in the lower littoral region, Panglongwan, Port Shelter, in March (*Taam A62*, TYPE).

The present species is probably more closely related to the Cape species, *P. incompta* Harv., than to others. That species is rather imperfectly known; it was described on sterile materials and is familiar to the writer only through descriptions. It is said to have the decumbent filaments with the longest segments to 1.5 times the diameter, gradually becoming smaller upward. The present species, on the contrary, has much shorter segments in the basal filaments, longer ones in the middle and shorter ones again near the tips. In describing the African plant no mention was made of rhizoidal tenacula, which are characteristic of the Hong Kong alga. From other pulvinate-caespitose species of the ecorticate Oligosiphonia the present species differs in the characteristics mentioned above and also in the sparsely and distantly subdichotomously branched frond.

3. *Polysiphonia Savatieri* Hariot

(Plate III, Figure 1)

Alg. Mar. Yokoska, 1891, p. 226.

To this species is referred a smaller reddish epiphytic Polysiphonia. The frond is solitary, being only about 1 cm. high and is repeatedly alternately branched; the branches are subfastigiate and the branching more or less flabelliform. Owing to the vigorous growth of the side branches it sometimes appears to be dichotomous. The attachment to the host is by means of unicellular rhizoids, produced by and separated from the pericentral cells by cross walls. The filament is ecorticate and tetrasiphonous throughout, to 180 μ in diameter below, attenuating to 60–90 μ in diameter near the tips. The segments are distinct, shorter than broad near the base, soon becoming longer, to about 1.3 times longer than broad, and then gradually becoming shorter, to about half a diameter, long. The trichoblasts are 3–4 times dichotomously branched, spirally arranged near the tips in a one-quarter divergence. Young branches originate by replacement of the trichoblasts. The tetrasporangia, to 60 μ in diameter, form long single spiral series, to ten each, at the ultimate branchlets, extending from the tips downward to the third internodes (Pl. III, Fig. 1).

Habitat. — On Acanthophora, from Shek-O, Hong Kong Island, in May (*Tseng 2581a*); on Caulerpa, washed ashore, Clear Water Bay, Port Shelter, in February (*Taam A33d*).

Distribution. — Yokoska, Japan (type locality).

Since *P. Savatieri* Hariot was first described it has never been reported, not even from its "home," Japan, whose phycologists have been extensively investigating their algal flora for the last fifty years. The discovery of this little plant so far away from the type locality without its having been previously reported from places in between is certainly of some interest. Possibly it is a common alga in the intervening area and has been overlooked by the collectors because of its small, rather inconspicuous frond. To be sure, the writer has not had the opportunity to study the type or any authentic specimens for comparison, and the species has never been illustrated. The identification of this alga must, therefore, be merely tentative. The type specimen, according to the original description, is cystocarpic. The tetrasporangia are here reported for the first time, if the determination be correct. Among the common species of Oligosiphonia *P. Savatieri* seems to be near to *P. ferulacea*, as Hariot, *loc. cit.*, pointed out. That species, however, is a much more robust plant in all respects, being entirely different in substance and color and having much shorter segments.

4. **Polysiphonia gracilis,** sp. nov.

(Plate III, Figures 2–3)

Frons rubripurpurea, ca. 6 cm. alta, mollis, gracilis, e filamentis decumbentibus assurgens; filamentis cylindricis, sine cortice, tetrasiphoniis, articulis; filamentis basalibus usque ad 450 μ latis, articulis diametro longitudineque subaequalibus, ad geniculas leviter incrassatis, rhizoideis substrato adfixis; filamentis erectis irregulariter lateraliter decompositis, articulis diametro sesquilongioribus vel subaequalibus, raro duplo longioribus; ramis ramulisque subfastigiatis, extra-axillaribus trichoblastorum locum tenentibus; tetrasporangiis usque ad 90 μ diam., in superiore parte ramorum ramulorumque ortis, in serie longa spiraliter ordinatis. Cystocarpia et acervuli spermatangiales ignota.

Specimen typicum: *Tseng 2723* (in herbario auctoris), ad rupes argillosas litoreas, prope Tsuenwan, Kowloon, 15 Apr., 1940.

This species has a reddish purple frond about 6 cm. high. It is soft and slender, arising from the decumbent filaments. The filaments are cylindrical, ecorticate, tetrasiphonous, and distinctly articulate throughout. The basal filaments, up to 450 μ in diameter, have segments about as long as broad, slightly thickened at the nodes, and attaching to the substratum by means of rhizoids, which are separated from the mother pericentral cells by cross walls (Pl. III, Fig. 3). The erect filaments are irregularly laterally decompound, with segments about as long as broad, or one and one-half diameters long, occasionally to about twice as long as broad. The branches of all orders are generally subfastigiate and directed upward (Pl. III, Fig. 2). Occasionally, however, some of the young branchlets are rather patent and may even be slightly curved downward. Trichoblasts are in general very poorly developed at the tips of a few young branchlets; apparently they are deciduous since they are usually not observed. The branches and branchlets are evidently lateral in origin since they arise without apparent reference to the trichoblasts. All specimens collected are tetrasporic. The tetrasporangia are about 90 μ in diameter, forming an elongated spiral series in the upper portions of the branches and branchlets (Pl. II, Fig. 2). The fertile parts are not swollen or in any way distorted; they usually extend through several internodes, sometimes continuously, sometimes in several interrupted series. The youngest sporangia are generally found about a dozen cells below the apices of the fertile branches, which continue to grow and branch normally without being in the least distorted by the formation of the sporangia.

Habitat. — On mud-covered littoral rocks in sheltered places, Tsuenwan, Kowloon, in April (*Tseng 2723*, TYPE).

The present species is undoubtedly most closely related to *P. platycarpa* Børg. from India, differing from it chiefly in the much coarser filaments (about twice as large), the shorter segments (about half as long), and the absence or the poor development of the trichoblasts (very well developed in the Indian plant). Børgesen later strongly emphasized the close relationship of his plant with the Australian *P. mollis* Hook. f. et Harv., which has also been reported from Japan by Yendo and from the Malay Archipelago by Weber-van Bosse, in both cases doubtfully. The Australian species, according to descriptions and illustrations by Kützing, appears to be a much more robust plant than both the Indian and the Hong Kong algae and, moreover, has a discoid holdfast that is not found in these two species.

5. *Polysiphonia ferulacea* Suhr ex J. Agardh

(Plate III, Figures 4–7)

Species algarum, II (3), 1863, p. 980; De Toni, Sylloge algarum, IV (3), 1903, p. 892; Collins and Hervey, The Algae of Bermuda, 1917, p. 124; Yendo, Alg. Jap., VIII, 1918, p. 75; Børgesen, Mar. Alg. Dan. W. Ind., II, 1920, p. 277, figs. 277–280, Ind. Rhodophyc., 1931, p. 16, figs. 10–11.

P. breviarticulata Harv., Ner. Bor.-Am., II, 1853, p. 36, pl. 16, B (non *P. breviarticulata* (C. Ag.) Zanard.).
? *P. fracta* Harv., *loc. cit.*, p. 38.
Forma **implicata**, f. nov.

Planta a specie differt frondibus tenuioribus, usque ad 300 μ diam., intricatis, hinc illinc inter se per tenacula rhizoidea adhaerentibus. Specimen typicum: *Tseng 2682* (in herbario auctoris) prope Clear Water Bay, Port Shelter, 22 Mar., 1940.

This new form of *P. ferulacea* has a densely intricate frond with the filaments here and there united to one another by rhizoidal filaments acting as tenacula. These rhizoids, about 45 μ in diameter, are either unicellular or septate, and all are separate from the mother pericentral cells by cross walls (Pl. III, Fig. 4). The branching is irregularly alternate, appearing subdichotomous at places, with some apices more or less forked like those of the members of Ceramium (Pl. III, Fig. 5). The filaments are rigid and brittle, easily broken, often with stunted tips, frequently proliferating new, regenerating filaments. They are 200–300 μ in diameter, branches of all orders being of approximately the same thickness, except near the tips, where they may taper to 90 μ in diameter. They are tetrasiphonous (Pl. III, Fig. 7), ecorticate, and distinctly articulate throughout. The pericentral cells are subquadrate in surface view, one half to one diameter long or slightly longer. The segments are very short, always shorter than broad, frequently only about one half to one third the diameter in length (Pl. III, Fig. 4). The trichoblasts are well developed, and are arranged on a one-quarter divergence (Pl. III, Fig. 4). The branches are lateral in origin, replacing the trichoblasts. The tetrasporangia are 60–80 μ in diameter when mature; they are spirally arranged in densely forking terminal branchlets, more or less distorted and swollen, the fertile segments about 150 μ in diameter (Pl. III, Fig. 6).

Habitat. — Stanley Beach, Hong Kong Island, where they are washed ashore; in May (*McClure A 81*) and Clear Water Bay, Port Shelter, in May (*Tseng 2682*, TYPE of f. *implicata*).

Distribution. — West Indies (the probable type region), Mexico, Guadeloupe Island, Sandwich Islands, Australia, South Japan, and, India.

Since *Polysiphonia ferulacea* is a very widely distributed plant of the tropical and subtropical waters, there has naturally been some controversy, fortunately very little, as to the interpretation and delimitation of it. With regard to the origin of the branches, Collins and Hervey, *loc. cit.*, mention that "Leaves are abundant, branches being produced from their axils." Having examined numerous West Indian specimens of the species, the writer has come to agree with the views of Howe, Børgesen, and Taylor that the branches are independent of the trichoblasts and are lateral in origin. In the matter of length of the segments, Yendo, *loc. cit.*, made the following remark: "The length of the articuli was also greatly variable, sometimes as long as 6–8 times the diameter as Harvey noted, but frequently as short as only 1.5 times." It must be stated that Harvey, *loc. cit.*, mentioned "articulations in all parts of the frond much shorter than their diameter" rather than "6–8 times the diameter," as Yendo wrote. It seems to the writer that Yendo had a very broad view of this species, apparently including a heterogenous group, since the shortness of the segments has been generally accepted by phycologists working more intensively on tropical forms as one of its more constant and conspicuous characteristics.

Collins and Hervey, *loc. cit.*, remarked that they found plants agreeing closely with *P. fracta* Harv., which seemed to them "to be an old and battered form of *P. ferulacea.*" They added, further: "Authentic specimens of the latter, in the Farlow Herbarium, confirm this view." It seems, therefore, that a rechecking of the type specimens of *P. fracta* is very necessary and, if it proves to be the same as *P. ferulacea*, as Collins and Hervey suspected, the specific epithet *fracta* of Harvey (1853) has to be used instead of the now current "*ferulacea*" of Suhr, ex J. Agardh (1863).

The forma *implicata* differs from the typical form of the species in its intricate frond, with abundant tenacula produced everywhere, attaching the filaments to the substratum as well as to one another, making it almost impossible to separate the branches from one another without breaking some of them. The Hong Kong form is also more slender and shorter than the typical one, and has smaller filaments. Its apices are also generally forcipate, like those of the ceramiums. In its external characteristics, therefore, it is very different from the typical erect, tufted, rather regularly and abundantly subdichotomously pinnate form of the West Indian region. In other important characteristics, for instance, the cells, the segments, the tetrasporangial stichidia, there is no fundamental difference that the writer can find to distinguish it from the typical form. It is therefore regarded merely as an environmental form of the West Indian plant. Since the typical form has been reported from Japan on the east of this region and from India on the west, it seems less probable that the Hong Kong plant is a geographic variety. As noted above, McClure's No. 81 also belongs to the present form. This specimen, by the way, bears the identification of Professor Setchell as "*Polysiphonia sp.* cf. *P. ferulacea*, probably undescribed."

6. *Polysiphonia Harlandii* Harv.

(Plate IV)

Charact. New Alg., 1860, p. 330; J. Agardh, Species algarum, II (3), 1863, p. 1063; De Toni, Sylloge algarum, IV (3), 1903, p. 913; Yamada, Notes on Japanese Algae, V, 1933, p. 283, pl. 13, fig. 3.

The present species was described originally from Hong Kong by Harvey, *loc. cit.*, on the basis of the material collected by Charles Wright in the United States North Pacific Expedition under Captain John Rodgers. It was listed under "Species Inquirendae" by J. Agardh, *loc. cit.*, who, as well as De Toni, *loc. cit.*, merely transferred Harvey's description to his work. The species was not again heard of until quite recently, when Yamada, *loc. cit.*, reported it from Formosa, claiming that he "was fortunate enough to have been able to compare these specimens with the type specimens of Harvey at Dublin and came to the conclusion that the Formosan specimens were to be referred to the present species." He published a photograph of two specimens, but, unfortunately for fellow phycologists, did not describe the plants at all. Plants that appear to be of this species, to judge from the photograph published by Yamada and from Harvey's original description, are very common from December to May in exposed places on the southeastern part of Hong Kong Island. In the past few years the writer had paid special attention to the collection of specimens showing the various stages of development and the fruiting materials. In view of the fact that the species is still so imperfectly known a rather detailed description of this plant will be given in the following pages which, the writer hopes, may help to define it more clearly.

Plants of this species attain a height of 6.5 cm., and are dark reddish to brownish purple, rather rigid below and very soft above, tufted from a basal disclike holdfast, with a dozen or more erect filaments in the same group. The holdfast is a large aggregation of rhizoids that grow out from the small cortical cells of the basal parts. The main axis is percurrent and laterally branched in all directions. This and its main branches are more or less naked in the lower portion, and are only sparsely beset with small branchlets, but very densely clothed with branchlets in all directions above. The branchlets are densely and repeatedly forked, with tufts of well-developed trichoblasts at the tips; they are lateral, originating without reference to these "leaves."

The filaments are tetrasiphonous, with rather light cortication in the main axis and the principal branches, where the articulations are not clear (Pl. IV, Figs. 3–4). In the younger parts and the branchlets they are ecorticate and distinctly articulate. There the segments are short, generally one half to one diameter long. The main axis reaches a diameter of 1.2 mm. The dichotomously decompound branchlets are 160–200 μ in diameter below, decreasing gradually to about 30–45 μ in diameter at the apices. The trichoblasts are repeatedly dichotomous, about 10–12 μ in diameter at the base and 4–6 μ near the tips.

The tetrasporangia are found near the tips and, in mature fronds, extend from the apices downward to two, rarely three, orders of the last few dichotomies (Pl. IV, Fig. 4). They occur in rather long spiral series; as many as twelve have been counted in a single consecutive series, with the oldest ones lowermost and the youngest uppermost. The older, mature, ones are, as is to be expected, more prominent, but the younger ones are generally rather inconspicuous. This might have accounted for the following description of the sporangia by Harvey, "sub apice ramulorum solitariis," since the small, inconspicuous sporangia, especially in dried specimens, are rather difficult to detect, whereas the one or two large, mature, conspicuous sporangia in the lowermost part of the series could easily attract one's attention. The fertile branchlets are greatly or slightly swollen and distorted. Mature sporangia attain a diameter of 75 μ.

The pericarps are short-pedicellate, with a stalk about 60 μ long on the longer, dorsal, side and less than 10 μ on the shorter, ventral, one. They are abundantly found on the dichotomously decompound ultimate branchlets. They are obliquely and fastigiately placed, long-ovate when young, becoming broadly subspherical when mature, generally to about 0.4 mm. (sometimes 0.45 mm.) broad, and 0.36 mm. (sometimes 0.40 mm.) long, each without a protracted neck and with a very broad ventrally situated terminal ostiole (Pl. IV, Fig. 2). The spermatangial clusters are developed on the trichoblasts, each replacing an arm of the first dichotomy (Pl. IV, Fig. 1). They are subcylindrical, about 45 μ in diameter and 165 μ long, each terminating on a sterile cell.

Habitat. — Abundant on surf-beaten rocks in the lower littoral region, Hong Kong Island; Shek-O, in March (*Tseng 655, 2660*) and December (*Tseng 2601* and *Taam A 12*); and Stanley Beach, in March (*Tseng 576*).

Distribution. — Hong Kong (type locality) and Formosa.

As mentioned above, in identifying these specimens with *P. Harlandii* Harv. the writer has not been able to compare them with authentic specimens. No illustrations have ever been published to show the important detailed microscopic characteristics of the species. Nevertheless, in spite of these difficulties, he has no hesitation in referring them as indicated above, because these specimens represent the most common species of Polysiphonia on the island and are the only ones that match perfectly with the original description of the Harveyan species.

The description given above refers to the typical form of *P. Harlandii* Harv., according to the interpretation of the writer, which is based on topotype material. This he proposes to name forma **typica,** nom. nov., based necessarily on Harvey's type specimen. In the course of his study of other algae the writer found several small corticated epiphytic Oligosiphonia with full loads of tetrasporangia in some and cystocarps in others. A study of them shows that they are not fundamentally different from *P. Harlandii,* so far the only corticated species discovered in this region, in spite of the fact that these plants are only one tenth to one twentieth as large as the normal mature ones of the typical form. This the writer proposes to call forma **minutissima,** f. nov.

Planta a forma typica differt frondibus et partibus omnibus multo tenuioribus; frondibus ca. 4 mm. altis, filamentis usque ad 0.5 mm. diam., cystocarpiis 280 μ latis, 300 μ longis; ramulis ultimis minus ramosis. Specimen typicum: *Tseng 2698a,* in *Laurencia undulata,* prope Cape d'Aguilar, Hong Kong Island, 24 Mar., 1940.

Plants of this new form are epiphytic on *Laurencia undulata* and are about 2–4 mm. high, yet fully mature asexually and sexually. They attach to the host by means of basally issued rhizoids. The main filaments are generally 250–350 μ in diameter, but in some as much as 0.5 mm. in diameter. Like that in the typical form, the cortication is light, and the articulated portions have segments much shorter than broad. The tetrasporangia are about 60 μ in diameter, and are disposed as in the species. The fertile branchlets are also swollen and distorted, but are much less branched, with only few dichotomies. The pericarps are ovate to subspherical, short-pedicellate, to 280 μ broad and 300 μ long.

SUMMARY

Of the six species of Polysiphonia two are described here for the first time. These are: *P. coacta* Tseng, sp. nov., probably closely related to the Cape species, *P. incompta* Harv.; and *P. gracilis* Tseng, sp. nov., most closely related to the Indian species, *P. platycarpa* Børg. Of the remaining species, *P. subtilissima* Mont., *P. Savatieri* Hariot, and *P. ferulacea* Suhr ex J. Agardh are recorded here for the first time for the Hong Kong region as well as for the entire China coast. A new form, f. *implicata* Tseng, f. nov., is now first described for the West Indian species, *P. ferulacea*. The last member, *P. Harlandii* Harv., technically a "native" of this region since it was based on Hong Kong material, is described in some detail, based on topotype materials. Besides the typical form (f. *typica* Tseng, nom. nov.) a new form, f. *minutissima* Tseng, f. nov., was found as a small epiphyte on other algae. All these species are four-siphoned, belong to the section Oligosiphonia, and, with the exception of *P. Harlandii*, are all ecorticate.

The writer wishes to thank Professor Wm. Randolph Taylor, of the University of Michigan, for his advice during this study.

UNIVERSITY OF MICHIGAN

LITERATURE CITED

AGARDH, J. G. 1863. Species, genera, et ordines algarum, II, Part 3, 701–1291. Lund.

BØRGESEN, F. 1915–20. The Marine Algae of the Danish West Indies, Vol. II. Rhodophyceae. Dansk Bot. Arkiv, 3 : 1–504.

Børgesen, F. 1931. Some Indian Rhodophyceae, Especially from the Shores of the Presidency of Bombay. Kew Bull. Misc. Inform., 1931 (1) : 1–24.

Briquet, J. 1935. International Rules of Botanical Nomenclature. Dritte Ausgabe, Jena.

Collins, F. S., and Hervey, A. B. 1917. The Algae of Bermuda. Proc. Am. Acad. Arts and Sci., 53 (1) : 1–195, pls. 1–6.

De Toni, G. B. 1903. Sylloge algarum omnium hucusque cognitarum, IV (Sylloge Floridearum), Part 3, 775–1525. Patavium.

Farlow, W. G. 1881. The Marine Algae of New England. Rep. U. S. Comm. Fish. for 1879, Appendix A–I : 1–210.

Hariot, P. 1891. Liste des algues marines rapportées de Yokoska (Japon) par M. le Dr. Savatier. Soc. Sci. Nat. Cherb. Mém., 27 : 211–230.

Harvey, W. H. 1853. Nereis Boreali-Americana. Part II. Rhodospermae. 258 pp., pls. 13–36. Washington.

—— 1860. Characters of New Algae, Chiefly from Japan and Adjacent Regions, Collected by Charles Wright in the North Pacific Exploring Expedition under Captain John Rodgers. Proc. Am. Acad. Arts and Sci., 4 : 327–335.

Howe, M. A. 1920. Algae. *In* N. L. Britton and C. F. Millspaugh, The Bahama Flora, pp. 553–631. New York.

—— 1924. Chinese Marine Algae. Bull. Torr. Bot. Club, 51 : 133–144, pls. 1–2.

Kützing, F. T. 1863. Tabulae phycologicae, XIII : 31, 100 pls.

Montagne, C. 1840. Seconde centurie de plantes cellulaires exotiques nouvelles. Décades I et II. Ann. Sci. Nat., Sér. 2, Bot., 13 : 193–207.

Taylor, W. R. 1937. Marine Algae of the Northeastern Coast of North America, ix + 427 pp., 60 pls. Ann Arbor.

Yamada, Y. 1933. Notes on Some Japanese Algae V. Journ. Fac. Sci. Hokkaido Imp. Univ., Ser. V, 2 (3) : 277–285, pls. 10–13.

Yendo, K. 1918. Notes on Algae New to Japan, VIII. Bot. Mag. Tokyo, 32 (376) : 66–81.

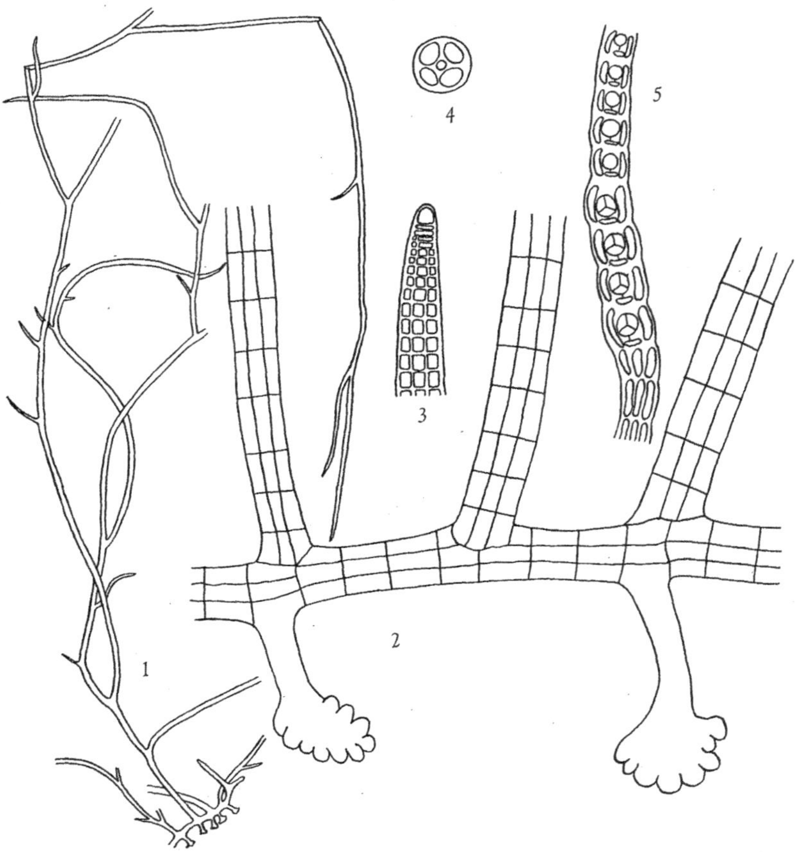

Polysiphonia subtilissima Mont.

FIG. 1. Habit sketch. × 11

FIG. 2. Part of basal stoloniferous filament, showing origin of rhizoids. × 100

FIG. 3. Tip of filament, showing apical cell. × 206

FIG. 4. Transverse section of filament. × 206

FIG. 5. Part of tetrasporiferous filament. × 100

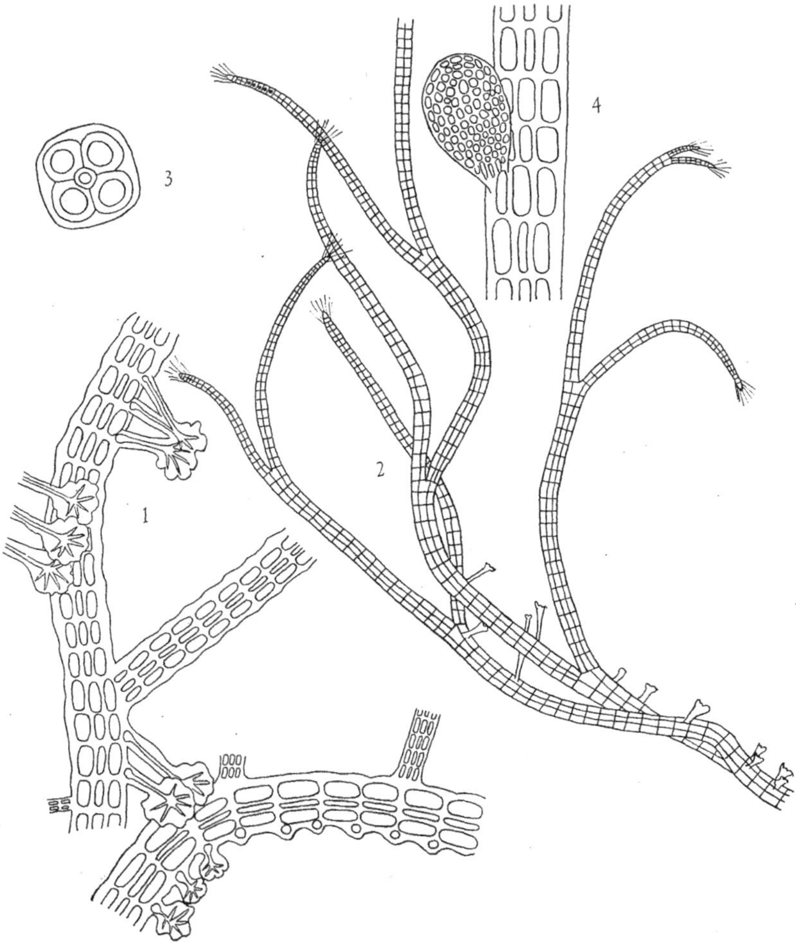

Polysiphonia coacta Tseng, sp. nov.

FIG. 1. Basal filaments, showing rhizoids. × 46
FIG. 2. Habit sketch of tetrasporiangiate plant. × 19
FIG. 3. Transverse section of filament. × 100
FIG. 4. Part of filament, with young cystocarp. × 100

EXPLANATION OF PLATE III

FIG. 1. *P. Savatieri* Hariot. Tetrasporangiate branchlet. × 76

P. gracilis Tseng, sp. nov.

FIG. 2. Part of tetrasporic plant, showing branching and elongated tetrasporangial series. × 17

FIG. 3. Part of branchlet, showing rhizoid. × 76

P. ferulacea Suhr ex J. Ag., f. *implicata,* f. nov.

FIG. 4. Part of filament, showing pericentral cells, trichoblast rudiments, and rhizoids. × 76

FIG. 5. Habit sketch of part of plant, with broken filaments of others of same plant, and intersegmental rhizoidal tenacula. × 10

FIG. 6. Tetrasporangiate branchlet. × 40

FIG. 7. Transverse section of filament. × 40

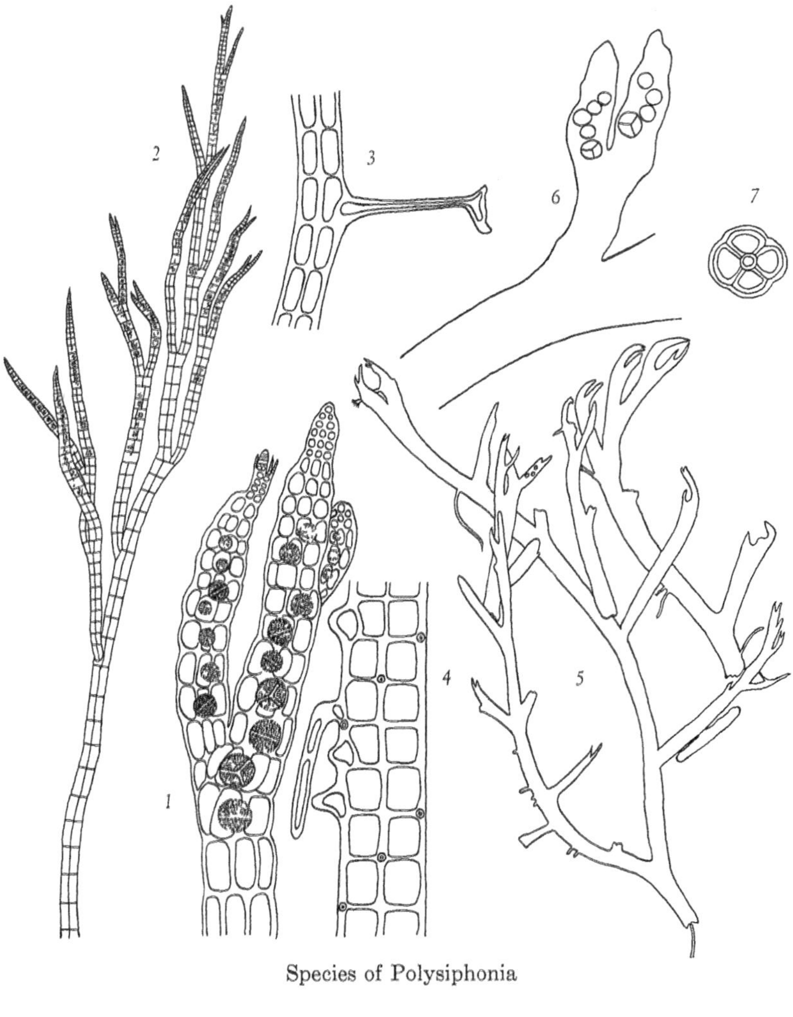

Species of Polysiphonia

EXPLANATION OF PLATE IV

FIG. 1. Part of branch, showing spermatangial cluster. × 212

FIG. 2. Part of branch bearing a cystocarp. × 106

FIG. 3. Transverse section of young branchlet (left) and another of a principal branch. × 56

FIG. 4. Part of branch with tetrasporangiate branchlet, showing cortication. × 30

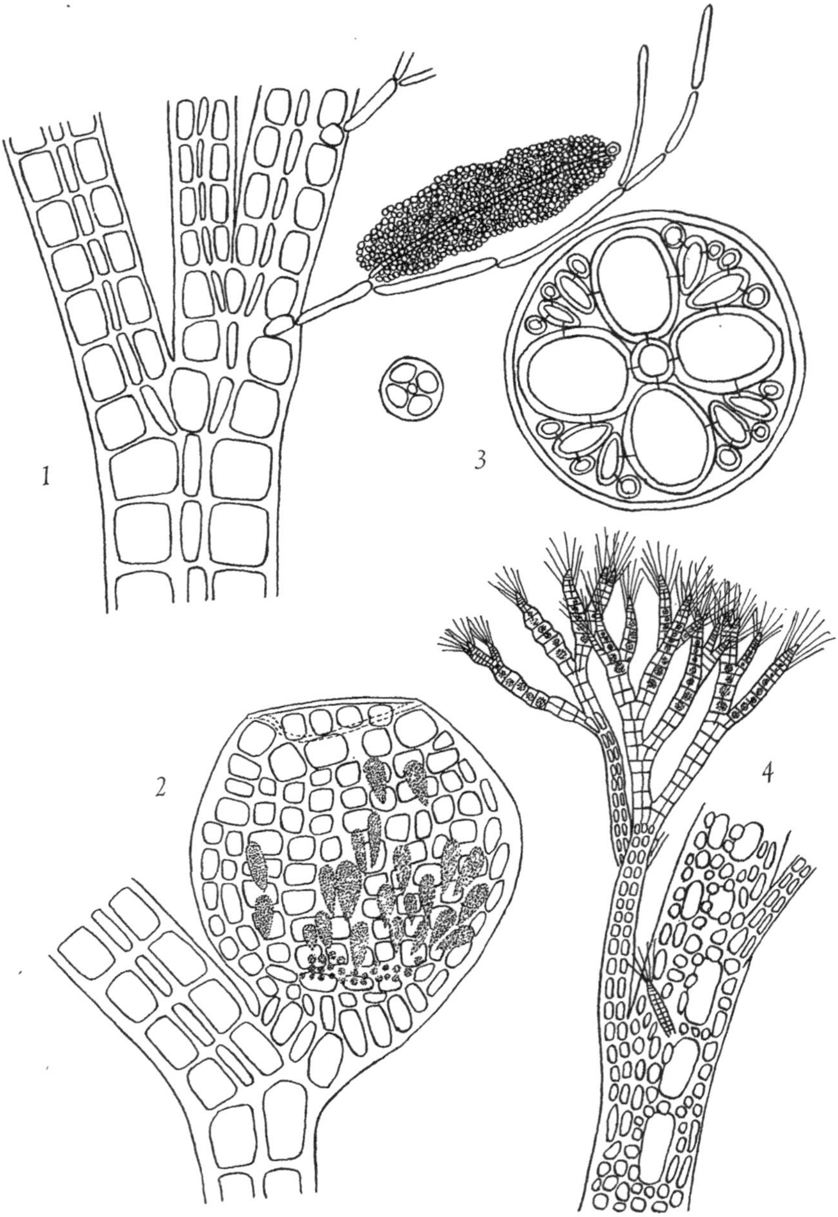

Polysiphonia Harlandii Harv. f. *typica* Tseng, nom. nov.

FORESTRY

SOME RESUPINATE POLYPORES FROM THE REGION OF THE GREAT LAKES. XV *

DOW V. BAXTER

THE genera Poria, Trametes, and Polystictus (which is often resupinate in the tropics as well as elsewhere) include several species of fungi that attack woods used for military purposes. The losses they cause, both directly and indirectly, may be shown by numerous communications from war fronts all over the world. That decay in crating and shipping containers is serious is shown by the following note received from Africa: "Pine boxes, opened up the other day [censored words] were lined with *Poria incrassata* (sp. ?) a quarter inch thick. They nearly fell apart with a few blows of the hammer."

In a communication from "down under" the prevalence of decay in wood used locally by the army is indicated by this excerpt: "I am on an island somewhere in the southwestern Pacific . . .; have been cutting some of the so-called Australian pine . . ., which only grows along the coast, seldom over more than 100 yards away from the ocean . . .; butt rot seems to be present in definite groves. Some trees will be rotten clear through, while others just three or four feet up in the trunk."

An employee of the Board of Economic Warfare in Ecuador writes: "There is a factor here that affects all calculations, and it is extremely hard to figure. Balsa is a delicate wood. If the logs are

* Throughout the work on these monographs I have been indebted to many individuals and institutions for suggestions, help, and privileges. My appreciation is expressed particularly to the men who have accompanied me on my ten expeditions to Alaska, the Yukon Territory, and the Northwest Territories. They aided in the collection and care of specimens, sometimes under rather difficult circumstances. I am under obligation to Professors T. G. Halle and Gunnar Samuelson, of Naturhistoriska Riksmuséet in Stockholm, with whom I have had the pleasure of association. Thanks are due to several American institutions and scholars also: to the authorities at the New York Botanical Garden; to Dr. W. H. Long, Albuquerque, New Mexico; to the staff of the Division of Pathological and Mycological Collections of the United States Department of Agriculture; and to Professor H. H. Bartlett and Reed W. Varner, of the University of Michigan.

left long before milling they are greatly damaged by worms and checks. Decay starts about as soon as the tree is felled. The result is that many logs are absolutely worthless by the time they reach the mill. This degrade is shown by the fact that since the beginning of the year the most efficiently operating mill in the area averaged 68 board feet out of logs averaging 144 board feet per log."

Since long periods of transit are inevitable during a war and since the methods of handling war materials are becoming more complex, there is urgent need of information on the rate of growth of wood-destroying fungi and on the temperatures at which they thrive. It is my plan to discuss these fungi in three groups: (*a*) white, (*b*) brown, and (*c*) those other than white and brown. The present paper, in which the first group is treated, is based largely upon studies of the rate of growth in culture and other features. It is intended to serve as a guide to the principal white resupinate polypores in North America.

THE RESUPINATE POLYPORES IN CULTURE

Identification of the causal organisms of decay is greatly facilitated by a study of variations under field conditions. About one hundred and fifty isolates and species collected from widely separated areas of the continent form the basis for this work. The practicability of a key that separates the different species and isolates is dependent, however, upon an understanding of variable reactions of the fungi themselves (even under controlled environments), method and type of culture, and an awareness that physiological characteristics do not always conform to morphological differences. Two plants that are distinct morphologically may exhibit like features in culture.

The usual initial classification of a fungus, based upon mycelial features, is made by comparing the unknown culture with a stock collection from named plants. Obviously the larger and more representative the reference collection of species and races is, the greater are the possibilities of arriving at the correct identification of a fungus.

An adequate number of the resupinate polypores from such a vast area as the North American continent has yet to be brought together. There is great danger, in work of this nature, of not having an unidentified species, or races of it, represented among the named plants. Furthermore, if pore formation does not occur in an isolate from wood, it may not be a polypore at all, but some other fungus, such as a member of the Thelephoraceae. Five cases of flask cultures are included in the collection of the School of Forestry and Conservation of the University of Michigan (see Pl. VI). This stock collection of the various fungi on red-gum test pieces, which is duplicated on white pine blocks, is most useful in making comparisons much as one does between an unknown and an authentic herbarium specimen when descriptions are difficult to interpret. In fungus cultures many of the possible elements are nearly alike, or they are often variable, or some elements occur in so many different species that they cannot be used as a basis for separation. After a determination of a fungus in a key, a final comparison with a reference collection is highly desirable and often necessary.

Since the concepts of Friesian species are based largely upon morphological features it seems reasonable to consider such characters of fundamental importance in the identification of a polypore. If possible, the determination of a species should be based upon structures of the fruiting body itself. Cultures, on the other hand, are most helpful in determining races. Furthermore, they are often essential in identification work if fruiting bodies are absent or immature. Reactions in culture may be used to separate plants that are closely allied morphologically if the structural features described for the species have only minor distinguishing characteristics, or if they are known to exhibit variations in different environments.

There are certain objections to the selection of almost any element that may be adopted to form the structure of a key to the fungi in culture. Some of these relate to the time, the equipment, and perhaps the difficulty involved in making the tests, whereas others are concerned with the variations shown by the fungi themselves. In some species color, for example, varies with the intensity of the light, and it often changes with the aging of a culture. Temperature may also affect color, as can be observed when tests are made at constant temperatures. Often the color in young cultures, in *Poria ferrea*, for example, is mostly white and not at all like that of the brown fruiting body of the plant in nature.

In general, microscopic structures are less variable than such features as color, texture, and rate and form of growth, and so may often be used to advantage. The presence or the absence of clamp connections, chlamydospores, conidia or oidia, staghorn and other types of branching of the mycelium, the diameter of the hyphae, setal hyphae, and setae are important elements in making up a key. The existence or the lack of clamp connections, however, was of little use in constructing the key presented here since they occur so generally in the white group of resupinate species reported. As a rule, clamp connections do not occur at all in the brown group, which is composed of a number of species, such as *Fuscoporia* (*Poria*) *rufitincta*, *Fomitiporia dryophylla*, *F. Earleae*, *Fomes nigrolimitatus*, *Poria laevigata*, *P. prunicola*, *P. inermis*, and *P. tsugina*. By means of the presence or the absence of clamp connections one might easily separate a group of fungi in which these plants are included from a category containing white species. However, one never experiences any difficulty in distinguishing between isolates of this brown group and the white or whitish species. The two groups can be separated by other means more conveniently than by a microscopic examination of the mycelium, for color alone can usually differentiate them.

Difficulty is encountered when it becomes necessary to distinguish between the fungi within the brown group or within a large group of white plants without resorting to physiological characteristics. Microscopic structures such as the spines that occur on the mycelium of *Poria spiculosa* Campbell and Davidson (4) are important and definite.

The use of certain other microscopic features may be limited by the fact that their incidence varies somewhat with environmental factors, such as the age of the culture and thickness of the medium. It is well known that moisture may influence not only spore production, but also the type of spore formed. In several species of the Polyporaceae, for example, Bose (2) reports that, after the rainy season in India and at intervals between showers, basidia in fruiting bodies are transformed into hyphal elongations with terminal spores. Conversely, if showers occur in the middle of the dry season, the hymenial elongations soon become reconverted into basidia. He states that it is possible to demonstrate such conversion of spore formation in cultures.

Among the physiological reactions of a fungus that may enable the investigator to make certain distinctions between plants is the one determined by the use of the Bavendamm oxidase test (1). This test is made by culturing the fungus on malt agar containing 0.5 per cent tannic or gallic acid. In general, it is possible to separate the fungi that cause brown rot from those that bring about a white decay. Those that induce brown rot react negatively, and so do not darken the medium under the fungus mat. If fungi that cause a white rot are cultured, the agar beneath and surrounding the mat becomes discolored.

Two of the chief disadvantages of using the reaction to the oxidase test as an element of a key are the time and the effort required for the task. One can ordinarily determine whether a white or a brown rot is produced by an examination of the wood from which the isolation is made. If, on the other hand, the isolation is made from a fruiting body, means of identification more direct than cultures are usually available.

Time, effort, equipment, and variation in growth of the fungus are also factors important in making identifications on the basis of rate of growth at constant temperature. Much more time and effort are needed to test the growth of mycelium on coniferous and hardwood blocks. Such tests, however, have a multiple use. Records giving data on rate of growth, mycelial development on wood, and the effect of the fungus on wood not only are useful for purposes of identification in a key, but have other values that are often significant in pathology.

If one does not have wood-block cultures for comparison, the microscopic and other features listed in the key with the isolates are for the most part characteristic of the fungus in two-week-old petri-dish tests, so that comparisons may be made on this basis. Examination of the plates accompanying the text will be helpful if texture and form of growth are elements mentioned.

In cultures of *Fomes annosus* the practically constant occurrence of conidiophores with spiked heads, which can be seen under the microscope, is distinctive for the species and is used as the first element to separate this fungus from the others in the white group. Their presence can be determined without experiencing the difficulty of keeping the fungi at different temperatures, and these structures are constant in the fungus. It is necessary, however, to resort to the use of both physiological and other features as elements of a long key.

The descriptions mentioned in the key presented in this paper are based in general on not fewer than eighteen tests, and for those experiments in which wide variations occurred, upon numerous other tests. It is characteristic of certain species to be relatively constant in their reactions and of others to vary greatly. In most species greatest variation occurs at temperatures critical for their growth. Those which are not constant in their growth reactions are listed in more than one division of the key.

Allowances are necessary, again particularly at critical temperatures, for growth rates for those isolates which have been cultured constantly on malt agar over a number of years. Although various species are affected differently, this is likely to be particularly true if in the meantime the fungus has not been grown on wood or on a medium other than malt-extract agar. Several cultures of the resupinate polypores when not made on wood and reisolated have shown at least some indication of growing more slowly at critical temperatures within two-week-old tests than they did twelve years ago. For this reason, too, certain species are included in more than one section of the key.

Only cultures that are made from actively growing mycelium should be compared with those mentioned in this paper. Cultures obtained from stock that has been stored at cold temperatures for long periods or cultures prepared directly from dishes in which the agar has become desiccated may be slow in starting their growth. The petri-dish records reported in this paper are based on inoculum obtained from two-week-old transfers.

Some species cannot be traced beyond a general grouping in a key based solely on features exhibited in culture. For example, *Poria aneirina*, *P. mollusca*, *P. xantha crassa*, and *Trametes serialis* are distinct species, yet some isolates of these fungi show characteristics when cultured which cause them to be brought together in sections of the key.

Submerged mycelium is not always visible unless the petri dishes are held up to the light. Consequently the total amount of growth indicated in the key may not be shown clearly by the plates accompanying this paper. Furthermore, when wide variation has been the pattern of growth, photographs that indicate the amount of mycelium most usually obtained in the tests are presented.

Young cultures of several of the brown resupinates and those of a number of other colors are white at first. Likewise certain porias that have truly white fruiting bodies become discolored in one-year-old flask cultures. Some porias, for example, *Poria unita pulchella*, are white or yellow, depending upon light conditions. The temperature ovens used in these studies are dark. However, the principal resupinates of the American continent that fall into the categories mentioned are included in the key.

KEY TO THE PRINCIPAL WHITE RESUPINATE POLYPORES IN CULTURE

The white or whitish mycelium shown in both petri-dish tests and one-year-old wood-block cultures characterizes the resupinate polypores listed in this key. Certain species having fruiting bodies that are brown or of some other color are included in this group, however, since they either exhibit a whitish mycelium for a number of weeks or show white as well as colored mycelium in cultures. *Poria ferrea*, *P. rufa*, and *P. nigra* are examples. A key to the colored resupinate polypores will be presented in Number XVI of this series.

All growth measurements are averages of the two longest radii emanating from the inoculum, which was not measured.

The color terms enclosed in quotation marks are those of R. Ridgway, *Color Standards and Color Nomenclature* (1912).

I. Conidiophores, simple or branched (and suggestive of those of an Aspergillus), produced on broad hyphae; the swollen apex of the structure left with a number of pointed sterigmata after the spores have fallen
Fomes annosus on coniferous timber

I. Conidiophores not present or, if secondary spores are formed, not produced on the swollen apex of a structure with a number of pointed sterigmata II

II. Superficial mycelium with staghorn branches
F. lobatus isolate reported by Campbell (3) (mycelium forming a dense compacted yellow to drab central zone, containing cuticular cells)
F. applanatus isolate reported by Campbell (3) (cannot always be separated from *F. lobatus*, which usually has a yellow to drab central zone. Both have cuticular cells)
F. fraxineus isolate reported by Campbell (3) (production pores and chlamydospores distinguish this fungus from *F. applanatus* and *F. lobatus*)

II. Superficial mycelium without staghorn branches 1

1. Mycelial growth filling petri dishes in two-week-old malt-agar cultures at 35° C. 2
1. Mycelial growth not filling petri dishes in two-week-old malt-agar cultures at 35° C. 5

2. Mycelium yellow, i.e. "Naples yellow," at least on inoculum, as well as white

   A. Mycelium in one-year-old red-gum block cultures "avellaneous" as well as white; no pores formed; mycelium not much branched and not particularly narrow
      T. rigida on *Q. Shumardii,* Florida (Pl. I)
      T. subserpens sensu Murr. on *Nyssa biflora,* Florida. (This isolate may prove to be *T. rigida* when *T. subserpens* is better known.)

   A. Mycelium in one-year-old red-gum block cultures white or yellow,[1] but not "avellaneous"; pore formation common; mycelium usually much branched and narrow
      P. unita on *Quercus sp.,* Michigan (Pl. I)
      P. unita pulchella on *Populus balsamifera,* Great Slave Lake, Northwest Territories

2. Mycelium white or whitish, but not at all "Naples yellow" 3

3. Cultures at 35° C. appearing water-soaked or arachnoid, i.e. not leathery or like chamois skin, but agar visible from beneath mycelium; culture therefore not distinctly snow white, mycelium never so luxuriant that it grows out of the petri dish

   A. Mycelium in one-year-old red-gum block cultures growing so luxuriantly that the shape of the test pieces is masked from view, i.e. blocks completely enveloped and edges not visible
      F. pinicola on *Pinus ponderosa,* Oregon

   A. Mycelium in one-year-old red-gum block cultures not growing so luxuriantly that the shape of the test pieces is masked from view, i.e. edges of blocks visible . . . . B

   B. Mycelium in one-year-old red-gum block cultures nodular, often forming irpiciform pore surfaces
      P. albobrunnea on *Pinus caribaea,* Georgia (Pl. II)

   B. Mycelium in one-year-old red-gum block cultures not nodular C

   C. Mycelium in one-year-old red-gum block cultures distinctly rhizomorphic, mycelium often becoming "avellaneous" or light brown in old cultures . . . . E
   C. Mycelium in one-year-old red-gum block cultures not distinctly rhizomorphic . . . . D

---

[1] The color of the mycelium in *Poria unita* and *P. unita pulchella* is influenced by light conditions.

D. Mycelium in one-year-old red-gum block cultures forming a white mat that covers the wood and remains white in old cultures
P. albipellucida on *Tsuga heterophylla*, Washington (Pl. I)[2]

D. Mycelium in one-year-old red-gum block cultures not forming a white mat that completely covers the wood and remains white in old cultures; but often powdery and all or most of the wood visible from beneath the fungous growth
P. tacamahacae on *Populus balsamifera*, Waterways, Alberta (Pl. I)

E. Mycelium in one-year-old red-gum block cultures pinkish brown, rhizomorphs often salmon-colored
P. semitincta on *Q. marilandica*, North Carolina

E. Mycelium in one-year-old red-gum block cultures mostly snow white, not pinkish brown, rhizomorphs whitish or "avellaneous"
P. subfusco-flavida on *Pseudotsuga taxifolia*, locality unknown (Pl. II)

3. Cultures not appearing water-soaked or arachnoid at 35° C., i.e. agar not visible from beneath the snow-white mycelium, or, if water-soaked in spots, the mycelium growing at such a rapid rate that it extends out between the lid and the dish 4

4. Mycelium appearing distinctly flaky, i.e. not exhibiting a uniform growth in cultures at 30° C., mycelium developing so luxuriantly that it grows out of the petri dishes
Pol. anceps on *Picea glauca*, Fort Wrigley, Northwest Territories (Pl. II)

4. Mycelium not distinctly flaky, but exhibiting a generally uniform type of growth over the agar surface at 30° C. and in some species developing so luxuriantly that it extends out from beneath the lid of the petri dish; often like chamois skin 22

5. Mycelium not filling petri dish at 35° C., but growing more than 10 mm.[3] in two-week-old cultures 6

5. Mycelium showing no growth or growing less than 10 mm. at 35° C. . . 9

6. Culture appearing water-soaked or mycelium appressed at 35° C.

A. Mycelium in one-year-old red-gum block cultures pink or pinkish or pinkish brown C

A. Mycelium in one-year-old red-gum block cultures not pink or pinkish . . . . B

B. Mycelium completely enveloping red-gum blocks in one-year-old cultures so that their shape is entirely masked, i.e. no sharp corners or edges visible
F. pinicola on *Populus balsamifera*, Michigan

---

[2] Spores are not allantoid, but globose or subglobose, (3) $3.5 - 4.5 \times 4 - 4.5$ (5) μ.

[3] Fungi exhibiting a growth between 5 and 10 mm. are included in both divisions throughout the key.

B. Mycelium not growing so luxuriantly that the wood blocks become masked from view; corners and edges remain distinct

P. albipellucida on *Tsuga heterophylla*, Washington (no pore formation; irregular growth mat in petri dishes at 30° C.) (Pl. I)

P. tacamahacae on *Populus balsamifera*, Waterways, Alberta (no pore formation; irregular growth mat in petri dishes) (Pl. I)

P. xantha crassa U. 28 on *Salix interior*, Providence, Northwest Territories (no pore formation)

P. xantha on *Pinus palustris*, Florida (no pore formation; smooth growth mat in petri dishes)

C. Mycelium in one-year-old red-gum block cultures pink or pinkish, nodular fructifications exhibited, mycelium not at all rhizomorphic

F. subroseus on *Larix laricina*, Michigan

C. Mycelium in one-year-old red-gum block cultures pinkish brown, mycelium rhizomorphic

P. semitincta on *Quercus marilandica*, North Carolina

6. Cultures not appearing water-soaked at 35° C. 7

7. Mycelial growth thin, arachnoid at 25° C. and somewhat water-soaked at this temperature, even though not appearing water-soaked at 35° C.

P. xantha crassa, on greenhouse bench timber, England 381

7. Mycelial growth thick, like chamois skin or cottony at 25° C. 8

8. Mycelial growth "avellaneous" or pinkish at 25° C., at least on inoculum, as well as white in two-week-old cultures (in *Poria nigra* 71099 as "avellaneous" as white, whereas in *F. fraxinophilus* more of the mycelium white than "avellaneous")

A. Mycelium in one-year-old red-gum block cultures dark brown, i.e. "bister," "Verona brown," or "Mars brown," as well as whitish

P. nigra ex Division of Forest Pathology, Washington, 71099 on *Quercus velutina*, Illinois (some hyphae with brown walls)

Unknown Poria on *Taxodium distichum*, railroad ties, Georgia. (Six-month-old red-gum block cultures are identical in color with *Poria nigra* 71099. Both fungi grow luxuriantly, but growth of *P. nigra* does not completely mask general shape of test pieces from view. Hyphae hyaline)

A. Mycelium in one-year-old red-gum block cultures white to "avellaneous," but not dark brown, i.e. "bister"

F. fraxinophilus on *Fraxinus americana*, Michigan

8. Mycelial growth not "avellaneous" but snow white at 25° C. in two-week-old cultures

A. Mycelium in one-year-old red-gum block cultures somewhat cottony, at least not smooth as in a glassy surface

T. Morganii on *Quercus sp.*, Virginia

A. Mycelium in one-year-old red-gum block cultures not at all cottony, but smooth as in a painted surface

T. serpens on *Carpinus sp.*, sensu Lloyd Cat. No. 35980, Florida (not *T. serpens* Fr. sensu Romell and contemporary Swedish scholars. This Florida isolate is most nearly related to *T. subserpens* Murr. or *T. rigida* Fr.)

9. Mycelium essentially filling petri dishes at 30° C. in two-week-old cultures 10
9. Mycelium not filling petri dishes at 30° C. in two-week-old cultures . . . 17

10. Cultures whitish yellow [4] at 30° C., not snow white, i.e. yellow or yellow cream as well as whitish; mycelium mostly thin on agar surface or forming coarse strands 11
10. Cultures white at 30° C. or "avellaneous" or slightly pinkish or water-soaked, i.e. agar color showing from beneath mycelium, but culture not cream or yellowish, mycelium thin or thick 12

11. Mycelium appearing in coarse strands (i.e. not fine arachnoid threads) which radiate out from inoculum or form a distinct fanlike or radiate type of growth

P. incrassata ex U. S. Forest Products Laboratory, Madison (Pl. II)

11. Mycelium forming uniform growth over medium, not in coarse strands and not exhibiting a radiate type of growth

P. xantha crassa ex Canadian Forest Products Laboratory, 349–3–Q, Canada

P. xantha crassa, "A," Mt. McKinley, Alaska

P. xantha crassa on *Salix interior*, U. 28. Providence, Northwest Territories

P. xantha on *Pinus palustris*, Florida

P. xantha crassa U. 34 on *Picea glauca*, Fort Wrigley, Northwest Territories

---

[4] Throughout the key, where the separation of species is based upon relative features, the plants are placed in both sections. Since it may be difficult to determine, for example, whether a culture is whitish or "avellaneous," it becomes necessary to list the species in more than one group.

P. xantha crassa, on greenhouse bench timber, England 381

12. Culture appearing water-soaked, i.e. mycelium not completely masking agar from view at 30° C. 13
12. Culture not appearing water-soaked, i.e. mycelium masking agar from view at 30 ° C. 14
13. Mycelium more arachnoid (i.e. in fine threadlike strands that extend outward) than granular or powdery; culture may or may not always exhibit aërial mycelium in two-week-old cultures (often trace of cream color in various isolates of *Poria xantha,* the variety of *P. xantha crassa,* and often a trace of pink or "avellaneous" in area about inoculum in *P. semitincta*)

A. Mycelium in one-year-old cultures "vinaceous-fawn" as well as white; blocks collapsed from decay so that they no longer are square in cross section

P. semitincta on *Quercus marilandica,* North Carolina

A. Mycelium in one-year-old cultures not "vinaceous-fawn" as well as white, blocks collapsed or not collapsed .... B

B. Mycelium forming a thick mat over the surface of red-gum test pieces in one-year-old cultures, so that the edges of the blocks are rounded, i.e. not sharp, and often the shape of the test piece completely concealed C

B. Mycelium not forming a thick mat over the surface of red-gum test pieces in one-year-old cultures, so that the edges of the blocks are hidden from view .... D

C. Mycelium in one-year-old red-gum block cultures forming large globular masses of fungous tissue; mycelium appearing as fleecelike coverings of the test pieces which can be separated from the wood blocks by shaking the flasks

F. pinicola on *Populus balsamifera,* Michigan (cultures more "avellaneous" than pink)

C. Mycelium in one-year-old red-gum block cultures not forming large globular masses of fungous tissue, although sometimes nodular; mycelium not separable from the wood blocks as in C 1

P. versipora on *Alnus oregona,* British Columbia (cultures "pinkish buff"; cultures impossible to distinguish by mat features or microscopically from various isolates of *P. subacida*)

P. subacida[5] on *Castanea dentata* ex Division of Forest Pathology, Washington

P. subacida on *Acer circinatum,* British Columbia (cultures "pinkish buff")

---

[5] All cultures of *Poria subacida* are on the border line with reference to their classification as "water-soaked," and are therefore included in this section.

P. subacida on *Thuja plicata*, British Columbia (cultures "pinkish buff")
P. subacida on *Tsuga heterophylla*, Washington (cultures "pinkish buff") (Pl. III)
P. xantha crassa, "A," Mt. McKinley National Park, Alaska (cultures "pinkish buff"; petri-dish cultures usually more yellowish when young than are those of *P. subacida*)
Unknown Poria ex University of Idaho V. I. F. P. 721 IX 14.40 (culture white)

D. Mycelium in one-year-old red-gum block cultures forming a smooth mat (as in a painted surface), uniform in texture and color, that completely masks the wood from view

P. eupora on *Quercus borealis maxima*, Michigan

D. Mycelium in one-year-old red-gum block cultures not forming a smooth uniform mat (as in a painted surface) .... E

E. Mycelium in one-year-old red-gum block cultures forming nodules that produce regular tubes and the nodules and tubes appearing flesh-colored

P. vaporaria (?) W. D. 28, ex Canadian Forest Products Laboratory (identification not certain since fungus has not yet been compared with Swedish cultures)

E. Mycelium in one-year-old red-gum block cultures not forming flesh-colored nodules that produce pores; or, if nodules and pores are formed, the nodules are brownish or some other color typical of mycelium in cultures of this age

P. xantha crassa ex Canadian Forest Products Laboratory, 349–3–Q, Canada (pore formation common; mycelium in petri-dish tests for all isolates appearing as a smooth mat over agar at 30° C.)
P. xantha crassa on *Salix interior*, Providence, Northwest Territories (no pore formation)
P. xantha on *Pinus palustris*, Florida (no pore formation)
P. xantha crassa U. 34 on *Picea glauca*, Fort Wrigley, Northwest Territories (no pore formation)
P. xantha crassa "A," Mt. McKinley National Park, Alaska (pore formation)
T. serialis on *Tsuga heterophylla*, Sitka, Alaska (no pore formation;

culture often displaying some brown as a result of drying on wood in cultures of this age; mycelium in petri-dish tests appearing somewhat flaky at 30° C.)

Unknown Poria, ex University of Idaho V. I. F. P. 721 IX 14.40

P. aneirina U. 32 on *Populus tremuloides*, Fort Wrigley, Northwest Territories (Pl. III) (blocks covered by thin mat of mycelium; most of wood visible from beneath mycelium, rhizomorphic; mycelium in petri-dish tests appearing arachnoid in all isolates at 30° C.)

P. aneirina U. 33 on *Populus balsamifera*, Burnt Island, Great Slave Lake, Northwest Territories (Pl. III) (blocks covered by thin mat of mycelium; most of wood visible from beneath mycelium, rhizomorphic)

P. aneirina U. 41 on *Populus balsamifera*, Russian River, Alaska (Pl. III) (blocks covered by only thin mat of mycelium)

P. mollusca on *Acer floridanum*, Florida (Pl. IV) (conspicuous rhizomorphs present on wood; mycelium in petri-dish tests appearing somewhat powdery, but oidia present at 30° C.)

P. corticola on *Populus tremuloides* (thin smooth mat of mycelium covering the blocks in cultures of this age; mycelium in petri-dish tests appearing powdery at 30° C.) (Pl. V)

P. vaporaria sensu Burt non Swedish concept (culture exhibiting practically no aërial mycelium in two-week-old petri-dish cultures, but submerged and appressed mycelium fills dish at 25° and 30° C.)

13. Mycelium more granular or powdery in appearance than arachnoid, i.e. fine threadlike strands not extending out from inoculum at 30° C.; culture always showing aërial mycelium 25

14. Mycelium snow white at 30° C. 15
14. Mycelium "light buff" or "avellaneous" or pinkish as well as white at 30° C. in two-week-old cultures 16

15. Mycelium exhibiting a radiating type of growth at 30° or 25° C. in two-week-old cultures

A. Mycelium in one-year-old red-gum block cultures growing so luxuriantly that the shape of the blocks is completely masked; mycelium as much brown (i.e. "snuff brown") in culture as white and giving the entire culture a brown appearance; black-line formation present
P. sitchensis on *Picea sitchensis*, Sitka, Alaska

A. Mycelium in one-year-old red-gum block cultures not growing so luxuriantly that the shape of the test pieces is masked from view; mycelium white or "avellaneous," but not brown to the extent that much of the culture appears "snuff brown"; no black-line formation . . . . B

B. Mycelium in one-year-old red-gum block cultures forming a smooth mat (as in a glossy surface) on the test pieces
T. serpens on *Carpinus sp.*, Florida (see 8 A in key)

B. Mycelium in one-year-old red-gum block cultures flocculent or rhizomorphic, not forming a mat that is smooth as a glossy surface
P. aneirina U. 32 on *Populus tremuloides*, Fort Wrigley, Northwest Territories (Pl. III)
P. aneirina U. 33 on *Populus balsamifera*, Burnt Island, Great Slave Lake, Northwest Territories (Pl. III)
P. aneirina U. 41 on *Populus balsamifera*, Russian River, Alaska (Pl. III)

15. Mycelium having a uniform type of growth, i.e. not radiating at 30° or 25° C. in two-week-old cultures

A. Mycelium in one-year-old red-gum block cultures showing pink fruiting structures, often nodular
F. subroseus on *Larix laricina*, Michigan

A. Mycelium in one-year-old red-gum block cultures not showing pink fruiting nodules; if present, not pink . . . . B

B. Mycelium in one-year-old red-gum block cultures showing much dark brown, i.e. "bister," "Verona brown," or "Mars brown"
P. nigra ex Division of Forest Pathology, Washington, 71099 [6] on *Quercus velutina*, Illinois
P. nigra ex Division of Forest Pathology, Washington, 71073 [6]

---

[6] *Poria nigra* does not properly belong to the white group of porias, and even old cultures may appear distinctly brown. Often, however, cultures are light, so that there is a possibility of confusing them with other fungi that are "avellaneous," and hence *Poria nigra* is listed in this key. One isolate of *P. nigra* 71118, however, does not fall in this section of the key because of color distinctions.

B. Mycelium in one-year-old red-gum block cultures white or "avellaneous," but not showing dark brown C

C. Mycelium in one-year-old red-gum cultures separating from the wood in fleecelike masses when the flask is shaken
F. pinicola on *Populus balsamifera*, Michigan (culture more "avellaneous" than pinkish)

C. Mycelium in one-year-old red-gum cultures not separating from the wood when the flask is shaken . . . . D

D. Mycelium in one-year-old red-gum cultures growing so luxuriantly that three fourths or more of the entire flask is filled with fungous tissue; mycelium conspicuously nodular
F. fraxinophilus on *Fraxinus americana*, Michigan

D. Mycelium in one-year-old red-gum cultures not growing so luxuriantly that most of the flask is filled with fungous tissue; mycelium not nodular or, if nodular at all, inconspicuously so . . . . E

E. Mycelium in one-year-old red-gum cultures completely covering the blocks with a thick woolly mat that masks the edges, i.e. edges of wood rounded with mycelial growth . . . . F

E. Mycelium in one-year-old red-gum cultures covering or not covering the wood, but never enveloping the block in a thick woolly mat; the edges of the test piece remaining distinct . . . . G

F. Mycelium "pinkish buff" in one-year-old red-gum block cultures
P. subacida on *Thuja plicata*, British Columbia (hyphae hyaline in all isolates)
P. subacida on *Tsuga heterophylla*, Washington
P. subacida on *Castanea dentata* ex Division of Forest Pathology, Washington
P. subacida on *Acer circinatum*, British Columbia
P. versipora on *Alnus oregona*, British Columbia (hyphae hyaline)
P. nigra ex Division of Forest Pathology, Washington, 71118 (hyphae with both hyaline and brown walls)

F. Mycelium white in one-year-old red-gum block cultures
P. subacida (?), "white cedar fungus" on *Thuja occidentalis*, Michigan

G. Mycelium in one-year-old red-gum block cultures uniformly "pinkish buff"
P. nigra ex Division of Forest Pathology, Washington, 71118

G. Mycelium in one-year-old red-gum block cultures not uniformly "pinkish buff"; if "avellaneous" or some other related color, color not uniform throughout culture on wood and agar . . . . H

H. Mycelium in one-year-old red-gum block cultures forming a uniformly smooth mat over the wood, the mat not flocculent, not granular, and not conspicuously rhizomorphic, but smooth as in a glossy surface

P. undata on *Fraxinus nigra*, Michigan (culture mostly snow white; secondary spores produced) (Pl. IV)

T. serpens on *Carpinus sp.*, Florida (culture white or turning somewhat brown in patches. The mycelium in cultures of this age is so appressed that it may appear skin-like. Secondary spores not produced. See 8 A in key.)

H. Mycelium in one-year-old red-gum block cultures not forming a uniformly smooth mat as in a glossy surface

P. aneirina U. 32 on *Populus tremuloides*, Fort Wrigley, Northwest Territories (all isolates characteristically showing no growth whatever at 35° C. in petri-dish tests) (Pl. III)

P. aneirina U. 33 on *Populus balsamifera*, Burnt Island, Great Slave Lake (Pl. III)

P. aneirina U. 41 on *Populus balsamifera*, Russian River, Alaska (Pl. III)

P. tacamahacae on *Populus balsamifera*, Waterways, Alberta (growth apparent at 35° C. in petri-dish tests)

P. ambigua (?) on *Nyssa biflora*, Florida

P. mollusca on *Acer floridanum*, Florida (culture at first rhizomorphic but not so later; oidia present; mat thin or flocculent; not chamois-like in petri-dish tests) (Pl. IV)

P. Morganii on *Quercus sp.*, Virginia (culture avellaneous; granular or flocculent, not distinctly rhizomorphic; no secondary spores; mat chamois-like in petri-dish tests) (Pl. IV)

P. vaporaria (?), Switzerland; doubtfully the *P. vaporaria* of Sweden (culture avellaneous, mycelium somewhat powdery; no secondary spores; mat somewhat flocculent in petri-dish tests)

P. undata on *Fraxinus nigra*, Michigan (blocks covered with thin mat of mycelium; not rhizomorphic or not distinctly so; secondary spores; mat thin at 25° C. in petri-dish tests) (Pl. IV)

16. Mycelium more silky [7] than cottony 26
16. Mycelium more cottony than silky 24

17. Mycelium at 25° C. in two-week-old cultures exhibiting a growth of at least 15 mm. 18
17. Mycelium at 25° C. and not exhibiting a growth of as much as 15 mm. .. 20

18. Agar turning yellowish about inoculum at 30° C. and turning yellowish throughout dish at 25° C.; mycelium faintly brown

Pol. alboluteus on stringers of *Picea sitchensis*, dockyard, Juneau, Alaska

Pol. alboluteus U. 6 on *Picea glauca*, Lower Russian Lake, Alaska

18. Agar not turning yellowish about the inoculum at 30° C. and not changing color at 25° C.; mycelium not faintly brown 19

19. Mycelium appressed, silky, granular, powdery, not cottony at 25° C. in two-week-old petri-dish cultures

A. Mycelium in one-year-old red-gum block cultures mostly uniform "cream color" or "cream buff" .... B

A. Mycelium in one-year-old red-gum block cultures "avellaneous" or some other color, but not uniformly "cream color" or "cream buff" C

B. Mycelium in one-year-old red-gum block cultures forming a mat that completely masks the wood from view

Poria species B (unknown), Mt. McKinley National Park, Alaska

B. Mycelium in one-year-old red-gum block cultures not masking the wood, i.e. wood not at all covered by the fungus

P. mollusca, substratum unknown, Florida (Pl. IV)

C. One-year-old red-gum block cultures exhibiting pore formation

T. alaskana on *Tsuga heterophylla* Lost Lake Trail, Seward, Alaska

T. alaskana U. 43 on *Picea sitchensis*, Kodiak, Alaska

T. alaskana U. 47 on No. 3 common Sitka spruce stringers in mill yard, Juneau, Alaska (this isolate defi-

---

[7] Where comparative features are described the plants are listed in two different sections of the key.

nitely more cottony in petri-dish tests than all the other isolates of *T. alaskana*)

C. One-year-old red-gum block cultures not exhibiting pore formation .. D

D. Mycelium not covering red-gum blocks in one-year-old cultures

T. heteromorpha U. 10 on *Picea sitchensis*, Juneau, Alaska

T. heteromorpha U. 44 on *Tsuga Mertensiana*, Hope, Alaska

Unknown Poria 25 W. on *Tsuga heterophylla* (check name for *T. alaskana*, *T. sitchensis*, and *T. serialis*), Wrangel, Alaska (oidia present)

Unknown Poria 28 W. on *Tsuga heterophylla*, Wrangel, Alaska

P. mollusca on *Acer floridanum*, Florida (oidia present) (Pl. IV)

P. xantha crassa (?) on *Pseudotsuga taxifolia*, Oregon. (This fungus, unlike other isolates, produces a somewhat uneven white granular to arachnoid mat in petri dishes.)

D. Mycelium covering red-gum blocks in one-year-old cultures .... E

E. Culture "clay color" as well as white; red-gum blocks collapsing as a result of decay so that they are no longer square in cross section

T. isabellina 6 W. on *Picea glauca*, Selkirk, Yukon Territory

T. isabellina 10 W. on *Picea glauca*, Selkirk, Yukon Territory

E. Culture not "clay color"; red-gum blocks not collapsing as a result of decay so that they are no longer square in cross section

P. xantha crassa (?) on *Pseudotsuga taxifolia*, Oregon (see note for isolate under D)

P. mollusca on *Acer floridanum*, Florida (oidia present) (Pl. IV)

P. corticola on *Populus tremuloides*, Michigan (mycelial mat in petri-dish cultures granular) (Pl. V)

P. crustulina on *Picea glauca*, Tanana, Alaska (mycelium in petri-dish cultures cottony)

T. serialis, *Tsuga heterophylla*, Sitka, Alaska (mycelium in petri-dish cultures flaky)

19. Mycelium not appressed, not silky, not granular, but cottony 23

20. Submerged mycelium showing hyphal swellings at 30° C. in two-week-old cultures

A. Cultures exhibiting "clay color" as well as white in one-year-old red-gum cultures; blocks collapsing within period because of decay so that they are no longer square in cross section

T. isabellina 6 W. on *Picea glauca*, Selkirk, Yukon Territory

T. isabellina 10 W. on *Picea glauca*, Selkirk, Yukon Territory

A. Cultures not exhibiting "clay color" as well as white in one-year-old red-gum cultures; blocks not collapsing within period because of decay so that they are no longer square in cross section

T. variiformis on *Picea glauca*, Michigan (mats of all isolates more or less cerebriform in petri-dish cultures)

T. variiformis on *Picea sp.*, Toklat, Alaska

T. alaskana U. 43 on *Picea sitchensis*, Kodiak, Alaska (mats not cerebriform in petri-dish cultures)

20. Submerged mycelium not exhibiting hyphal swellings at 30° C. in two-week-old cultures 21

21. Mycelium showing a growth of at least 5 mm. at 25° C.

A. Cultures exhibiting "clay color" as well as white in one-year-old red-gum cultures; blocks collapsing within period because of decay so that they are no longer square in cross section

T. isabellina 6 W. on *Picea glauca*, Selkirk, Yukon Territory

T. isabellina 10 W. on *Picea glauca*, Selkirk, Yukon Territory

A. Cultures not exhibiting "clay color" as well as white in one-year-old red-gum cultures; blocks not collapsing within period because of decay so that they are no longer square in cross section . . . . B

B. Mycelium in one-year-old red-gum cultures forming a thin white mat over the surface of the wood C

B. Mycelium in one-year-old red-gum cultures not covering the surface of the wood blocks

T. heteromorpha U. 10 on *Picea sitchensis*, Juneau, Alaska

T. heteromorpha U. 44 on *Tsuga Mertensiana*, Hope, Alaska

Unknown Poria 25 W. on *Tsuga heterophylla*, Wrangel, Alaska (oidia present)

Unknown Poria 28 W. on *Tsuga heterophylla*, Wrangel, Alaska

C. Mycelium in one-year-old red-gum cultures distinctly rhizomorphic and not uniformly smooth over the entire surface of the wood

P. crustulina on *Picea glauca*, Tanana, Alaska

C. Mycelium in one-year-old red-gum block cultures not distinctly rhizomorphic, but forming a uniformly smooth mat over the surface of the wood

Unknown Poria F. P. L. 2503 on Cupressus

21. Mycelium not showing a growth of at least 10 mm. at 25° C. in two-week-old cultures in petri dishes

Polystictus abietinus on *Picea glauca*, Healy, Alaska

22. Mycelium "avellaneous" or pinkish on inoculum in two-week-old cultures and frequently growing on the surface of the agar at 35°, 30°, or 25° C. as well

A. Mycelium in six-month-old [8] red-gum block cultures filling flask so that the test pieces are completely masked from view

P. rufa on *Picea glauca*, Fort Wrigley, Northwest Territories

Unknown Poria on *Pinus ponderosa*, locality unknown

Unknown Poria on *Taxodium distichum* railroad ties, Georgia

A. Mycelium in six-month-old red-gum block cultures not filling flask so that the test pieces are completely masked from view

T. rigida on *Quercus Shumardii*, Florida

22. Mycelium not "avellaneous" but white on the inoculum in two-week-old petri-dish cultures, as well as on the surface of the agar at 35°, 30°, or 20° C.

A. Mycelium in one-year-old red-gum block cultures growing so luxuriantly that the shape of the test pieces is masked from view .... B

A. Mycelium in one-year-old red-gum block cultures not growing so luxuriantly that the shape of the test pieces is masked from view

P. lenis on *Juniperus virginiana*, Maryland (Pl. V)

Unknown Poria U. 23 on Gowan's cypress, California

B. Mycelium in one-year-old red-gum block cultures exhibiting "fawn color" and tendency to red line formation

P. tinctipora (?) on *Liquidambar styraciflua*, Georgia (Pl. V)

B. Mycelium in one-year-old red-gum block cultures white, whitish, or exhibiting patches of red brown or some color other than "fawn color" C

C. Mycelium in one-year-old red-gum block cultures exhibiting red-brown patches

---

[8] The fungus *T. rigida* Fries can be distinguished readily from the *Poria rufa* group by its growth in six-month-old red-gum block cultures, but not by the standard one-year-old cultures used generally throughout the key.

T. serialis from mill-roof timber, Rhode Island. (This isolate has been classified (5) under the name *Polyporus palustris* B. & C.)

C. Mycelium in one-year-old red-gum block cultures white to "avellaneous," but not exhibiting red-brown patches

P. lenis on *Juniperus virginiana*, Maryland

23. Mycelium exhibiting a growth of only approximately 10 mm., often less, at 30° C.

A. Mycelium in one-year-old red-gum block cultures "clay brown" as well as white; blocks collapsing because of decay so that their shape is no longer square in cross section

T. isabellina 10 W. on *Picea glauca*, Selkirk, Yukon Territory

T. isabellina 6 W. on *Picea glauca*, Selkirk, Yukon Territory

A. Mycelium in one-year-old red-gum block cultures not "clay brown" as well as white; blocks not collapsing because of decay .... B

B. One-year-old red-gum block cultures exhibiting pore formation

T. alaskana U. 43 on *Picea sitchensis*, Kodiak, Alaska (in petri-dish tests fungus growing approximately 10 mm. at 30° C. in two weeks)

T. alaskana U. 47 on stringers and caps, Juneau, Alaska (in petri-dish tests fungus growing approximately 10 mm. in two weeks)

B. One-year-old red-gum block cultures not exhibiting pore formation

T. heteromorpha U. 10 on *Picea sitchensis*, Juneau, Alaska (growing more slowly than *T. alaskana* in two-week-old petri-dish cultures; i.e. it exhibits only 1 mm. growth at 30° C.)

T. heteromorpha U. 44 on *Tsuga Mertensiana*, Hope, Alaska (growing somewhat more slowly than *T. alaskana* in two-week-old petri-dish cultures; i.e. it grows approximately 4 mm. at 30° C.)

23. Mycelium exhibiting a growth of approximately 20 mm., often more, at 30° C. in two-week-old cultures

A. Red-gum wood blocks collapsing in one-year-old cultures so that they are no longer square in cross section; mycelium forming a white network of strands over the wood

P. ferruginosa on *Thuja occidentalis*, Minnesota

A. Red-gum blocks not collapsing in one-year-old cultures so that they no longer appear square in cross section; mycelium not forming a white network of strands over the wood .... B

B. Mycelium distinctly more "avellaneous" or cream color than white in one-year-old red-gum block cultures and forming a smooth and uniform mat over the surface of the wood

T. malicola on *Populus grandidentata*, Michigan (mat somewhat chamois-like in petri-dish tests) (Pl. V)

P. corticola on *Populus tremuloides*, Michigan (mat granular or powdery in petri-dish tests) (Pl. V)

P. mollusca (substratum unknown), Florida (mat flocculent in petri-dish tests) (Pl. IV)

B. Mycelium white or at least more whitish than "avellaneous" in one-year-old red-gum block cultures

P. mollusca on *Acer floridanum*, Florida (oidia present) (Pl. IV)

P. alaskana on *Tsuga heterophylla*, Seward, Alaska (growing faster than other isolates of the species; i.e. it grows only 10 mm. or less at 30° C. in two weeks)

24. Mycelium pink as well as white in two-week-old petri-dish cultures ... 27
24. Mycelium white, not pink in two-week-old petri-dish cultures

A. One-year-old red-gum block cultures exhibiting pink nodular fruiting structures; mycelium also showing pink

F. subroseus on *Larix laricina*, Michigan

A. One-year-old red-gum block cultures not exhibiting pink nodular fruiting structures; mycelium not showing pink .... B

B. Mycelium chocolate brown, i.e. "bister," as well as white

P. nigra ex Division of Forest Pathology, Washington, 71099 on *Quercus velutina*, Illinois

B. Mycelium not chocolate brown at all, but white or "avellaneous"

P. mollusca on *Acer floridanum*, Florida (Pl. IV)

25. Mycelium pinkish at 30° C. in two-week-old petri-dish cultures

P. purpurea on *Pinus contorta*, Jasper, Alberta

25. Mycelium not pinkish at 30° C. in two-week-old petri-dish cultures

A. Mycelium in one-year-old red-gum block cultures distinctly "avellaneous," i.e. more "avellaneous" than white

P. corticola on *Populus tremuloides*, Michigan (Pl. V)

A. Mycelium in one-year-old red-gum block cultures; more of it white than "avellaneous"

P. mollusca on *Acer floridanum*, Florida (oidia present) (Pl. IV)

26. Mycelium pinkish in two-week-old petri-dish cultures
    A. Blocks in one-year-old red-gum cultures collapsing so that the test pieces are no longer square in cross section
    P. semitincta on *Quercus marilandica*, North Carolina
    A. Blocks in one-year-old red-gum cultures not collapsing
    P. purpurea on *Pinus contorta*, Jasper, Alberta

26. Mycelium not pinkish in two-week-old petri-dish cultures
    A. Mycelium in one-year-old red-gum block cultures pinkish, i.e. "fawn color"
    P. semitincta on *Quercus marilandica*, North Carolina
    A. Mycelium in one-year-old red-gum block cultures not pinkish
    P. aneirina U. 32 on *Populus tremuloides*, Fort Wrigley, Northwest Territories (Pl. III)
    P. aneirina U. 33 on *Populus balsamifera*, Burnt Island, Great Slave Lake, Northwest Territories (Pl. III)
    P. aneirina U. 41 on *Populus balsamifera*, Russian River, Alaska (Pl. III)

27. Mycelium exhibiting a cerebriform type of growth at 25° C. in two-week-old petri-dish cultures
    P. purpurea on *Pinus contorta*, Jasper, Alberta

27. Mycelium not exhibiting a cerebriform type of growth at 25° C.
    A. Pink nodular fructifications appearing in one-year-old red-gum block cultures
    F. subroseus on *Larix laricina*, Michigan
    A. Pink nodular fructifications not appearing in one-year-old red-gum block cultures
    Unknown Poria on western yellow pine, locality unknown
    Unknown Poria on *Taxodium distichum* railroad ties, Georgia

University of Michigan

LITERATURE CITED

1. Bavendamm, W. 1928. Über das Vorkommen und den Nachweis von Oxydasen bei holzzerstörenden Pilzen. Zeitschr. f. Pflanzenkrank., 38 : 257–276.

2. Bose, S. R. 1943. Moisture-relation as a Determinant Factor in the Transformation of Basidia of Certain Polyporaceae. Mycologia, 35 : 33–46.

3. Campbell, W. A. 1938. The Cultural Characteristics of the Species of Fomes. Bull. Torr. Bot. Club, 65 : 31–69.

4. Campbell, W. A., and Davidson, Ross W. 1942. A Species of Poria Causing Rot and Cankers of Hickory and Oak. Mycologia, 34 : 17–26.

5. Nobles, M. K. 1943. A Contribution toward a Classification of the *Trametes serialis* Complex. Can. Journ. Res., 21 : 211–234.

PLATES I–VI

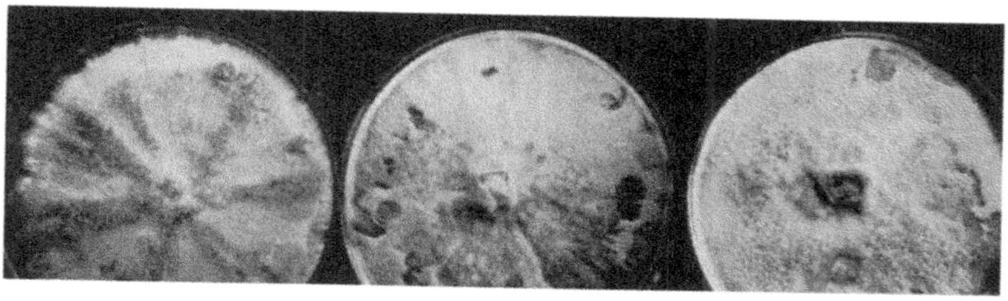

Trametes rigida on *Quercus Shumardii*, from Florida

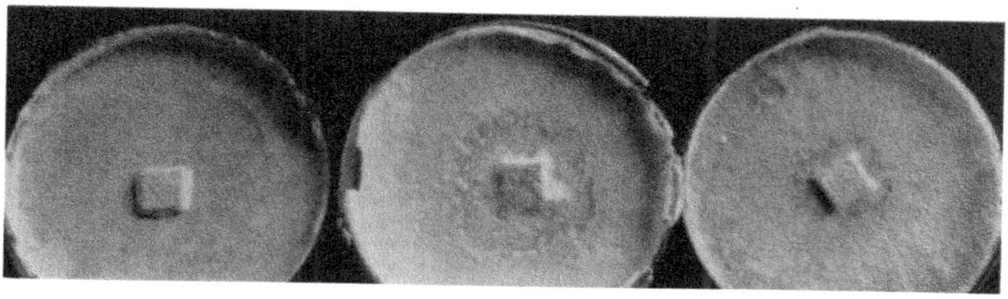

Poria unita on *Quercus sp.*, from Michigan

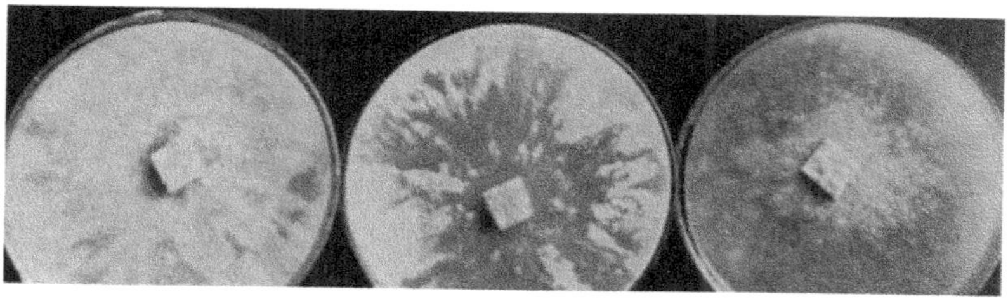

Poria albipellucida on *Tsuga heterophylla*, from Washington

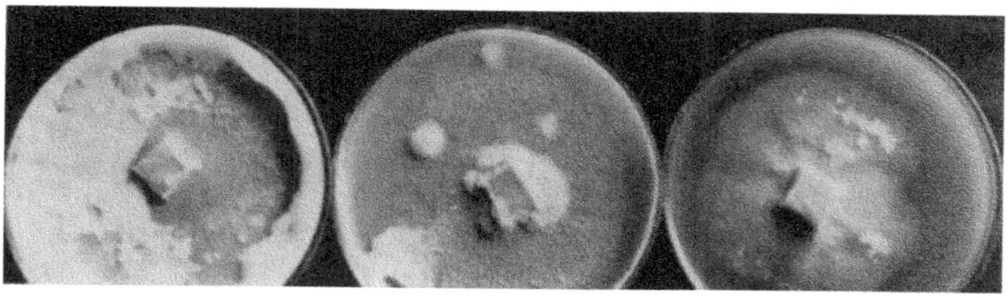

Poria tacamahacae on *Populus balsamifera*, from Waterways, Alberta

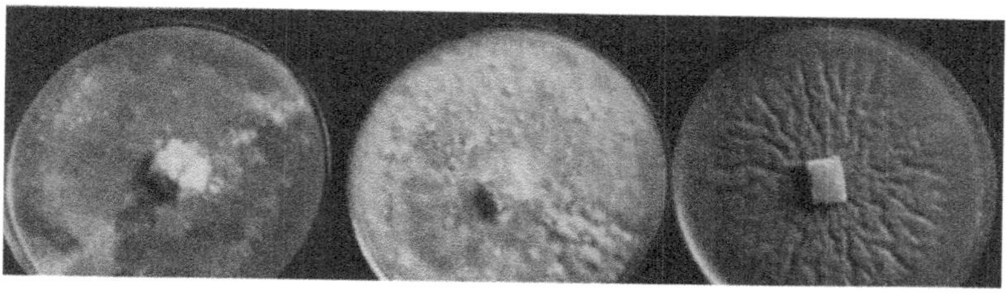

Poria subfusco-flavida on *Pseudotsuga taxifolia*; locality unknown

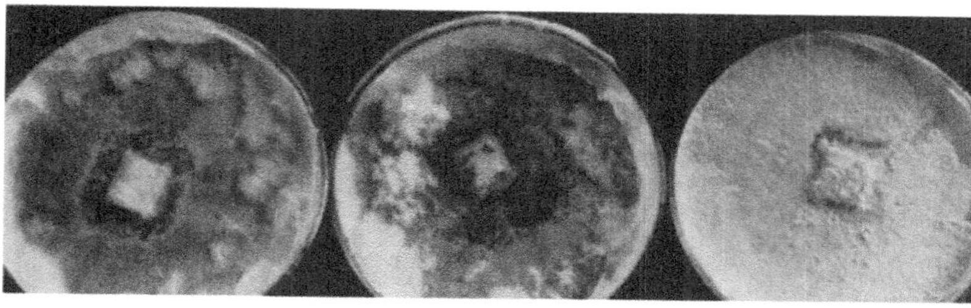

Polyporus anceps on *Picea glauca*, from Fort Wrigley, Northwest Territories

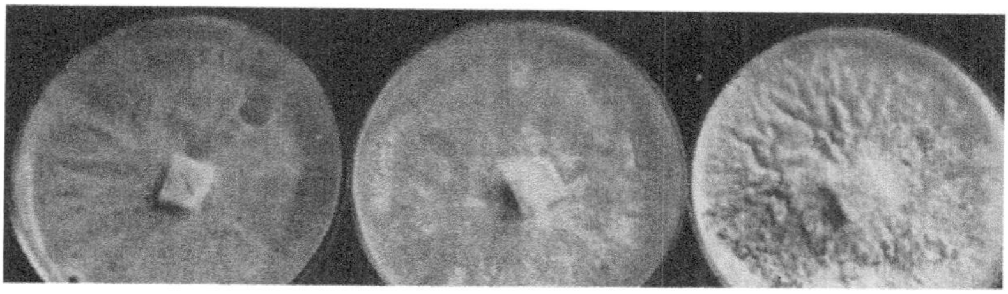

Poria albobrunnea on *Pinus caribaea*, from Georgia

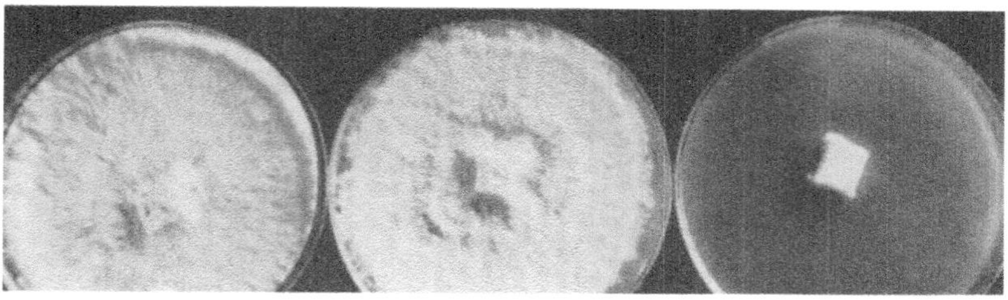

Poria incrassata, ex United States Forest Products Laboratory, Madison, Wisconsin

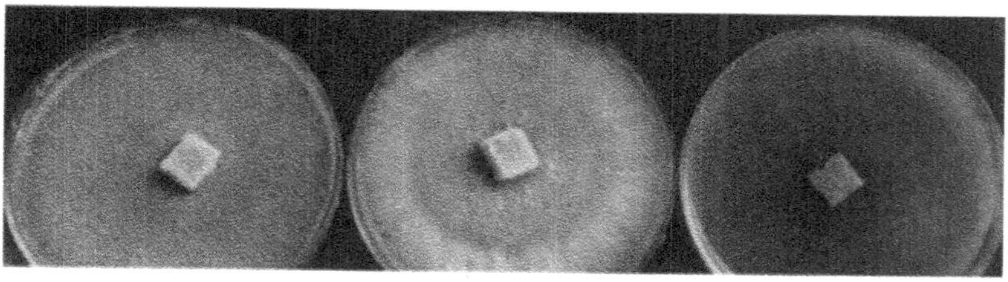

Poria subacida on *Tsuga heterophylla*, from Washington

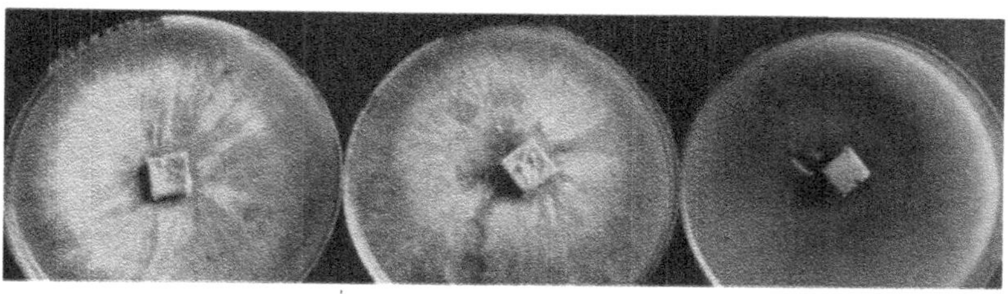

Poria aneirina U. 32 on *Populus tremuloides*, from Fort Wrigley, Northwest Territories

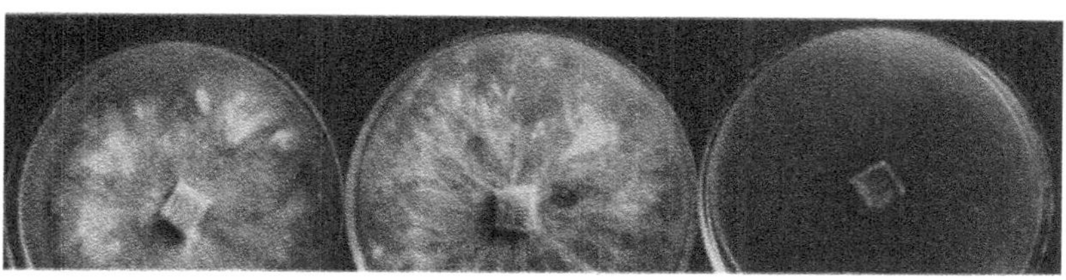

Poria aneirina U. 33 on *Populus balsamifera*, from Fort Wrigley, Northwest Territories

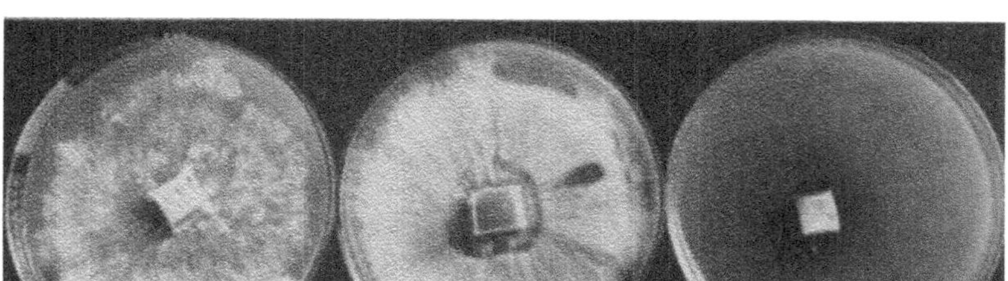

Poria aneirina U. 41 on *Populus balsamifera*, from Russian River, Alaska

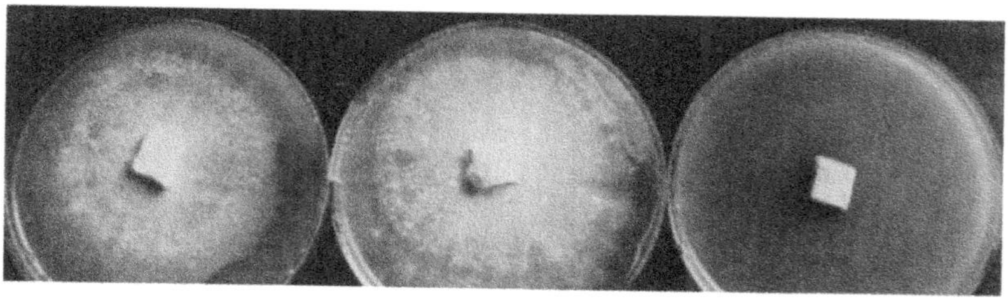

Poria mollusca on *Acer floridanum*, from Florida

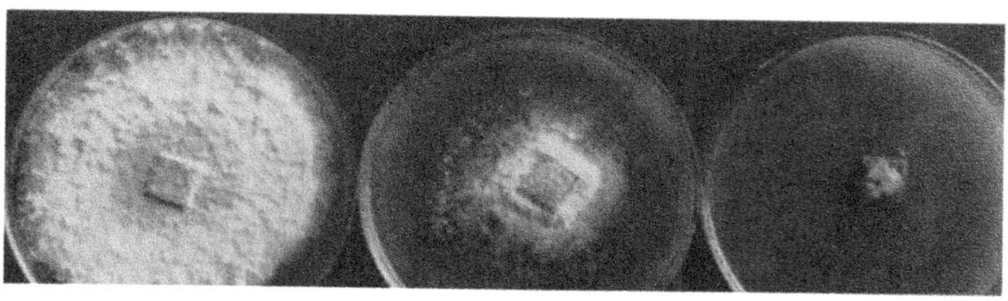

Poria mollusca, substratum unknown, from Florida

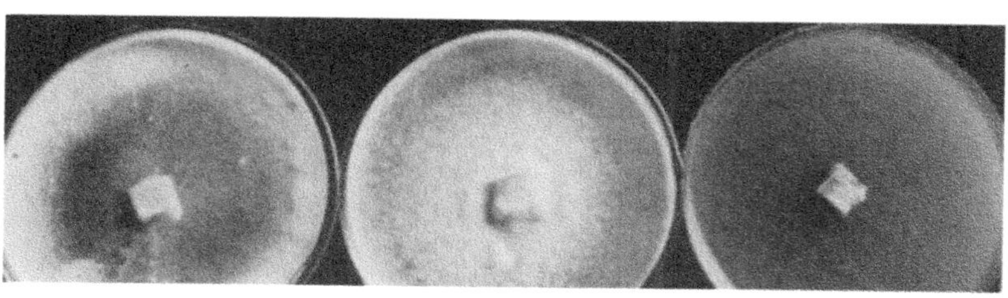

Poria undata on *Fraxinus nigra*, from Michigan

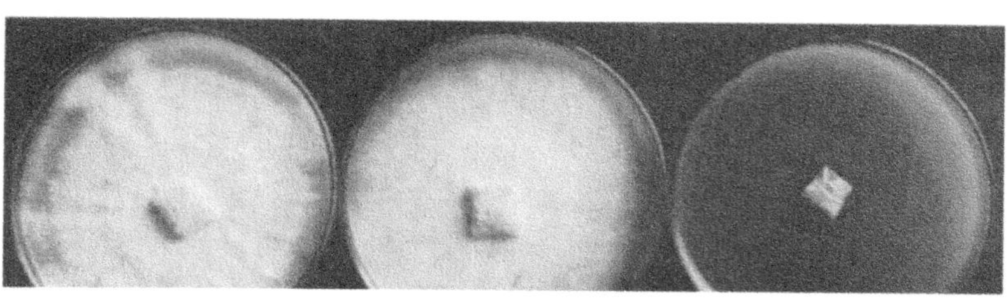

Trametes Morganii on *Quercus sp.*, from Virginia

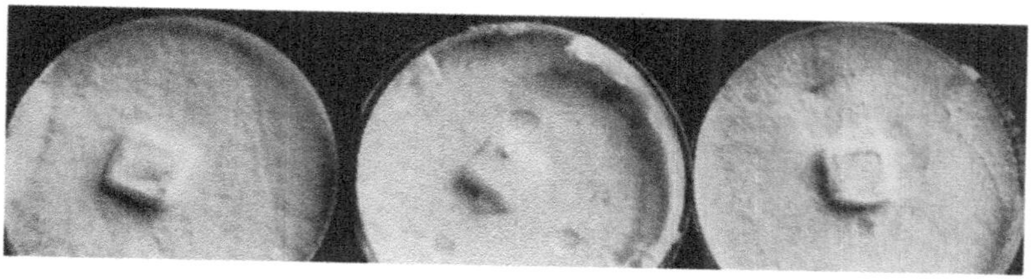

Poria lenis on *Juniperus virginiana*, from Maryland

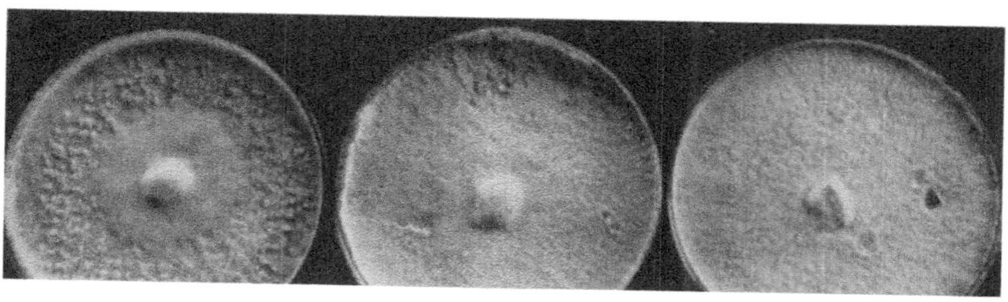

Poria tinctipora (?) on *Liquidambar styraciflua*, from Georgia

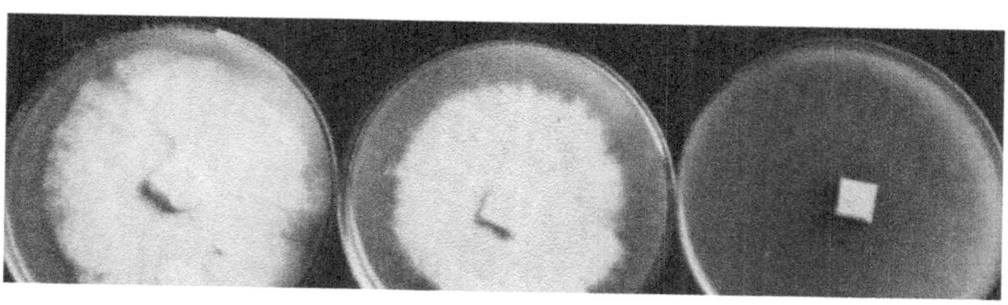

Trametes malicola on *Populus grandidentata*, from Michigan

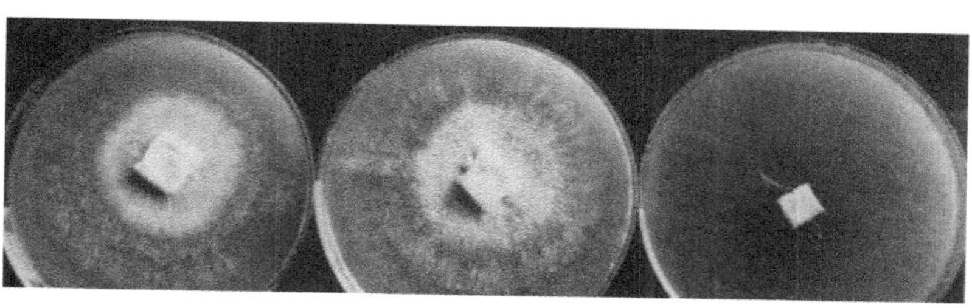

Poria corticola on *Populus tremuloides*, from Michigan

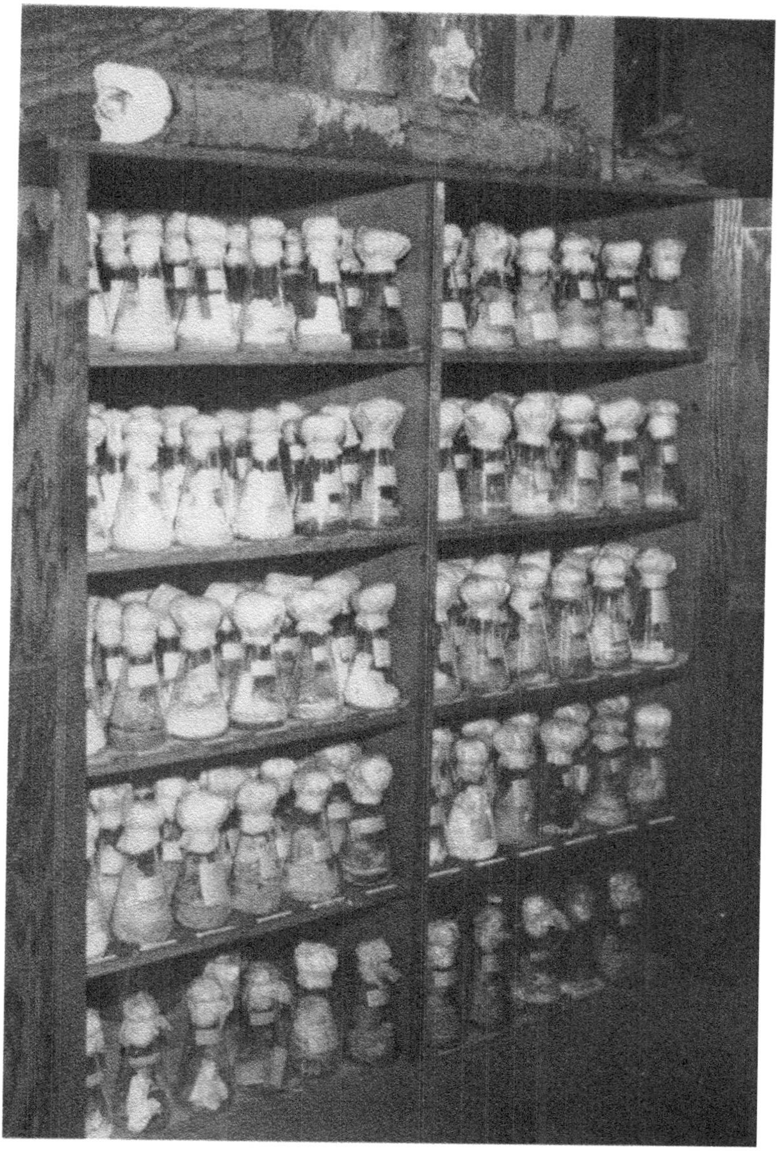

One of five cases of cultures that constitute a part of the reference collection of North American polypores at the School of Forestry and Conservation, University of Michigan. The fungi, which were collected on many types of substrata in various parts of the continent, are cultured on red-gum test pieces and duplicated on white-pine blocks

SHADE METHOD OF DIRECT CAMBIUM-TO-BARK DEVELOPMENT

CARL GEORGE FENNER

THE healing of bark-deep wounds upon which the cambium is undisturbed has always been a matter of great interest to tree scientists, park men, tree workers, and owners of fruit and shade trees. Many are the stories (some true and some false) concerning the success of the tree "doctor" in healing and saving completely girdled trees by the use of cow dung. Some twenty years ago I saw an old gardener treat and bring to complete recovery a fine but tender ten-inch white birch by means of this material. This observation inspired a search for a wound covering that would be easier and more pleasant to handle, and perhaps as efficient. Over a period of ten years many coverings of soft sterile materials were tried without encouraging results. Failure appeared to be due to the impossibility of preventing various parasitic fungi from destroying the sensitive exposed cambium cells. All coverings tested actually protected and favored the growth of these fungi rather than that of the cambium.

During the treatment and subsequent observation of hundreds of wounds caused by motor vehicles, horses, dogs, children, and other agencies it was often found that small bark-deep injuries on the north or shaded side of the trunk developed healthy bark directly from cambium when no covering or dressing was applied. Subsequently it was noted that the cambium on areas on the sunny side of tree trunks dried out and died when left exposed. Apparently the drying effect of direct sun rays was the cause of death. Attempts to prove this theory gave birth to the idea of shading by artificial means all bark-deep wounds. It was found that the cambium cells could be protected and saved from desiccation by the erection of a burlap barrier or shade. The shading material is held on steel or wooden frames, which usually form an enclosure about the trunk. Wrapping fails to allow sufficient air circulation. In fact, it became necessary to provide aëration by keeping the top of the enclosure entirely open, and also by leaving a four-inch space below the lower edge of the shading material. Each shade is erected in such a manner as to hold the burlap rigidly in position six inches from the cambium to be protected. All direct contact with the open wound is avoided since even slight abrasion destroys patches of the extremely tender cambium.

The method described has proved itself so thoroughly that it is now routine in our organization to shade all new trunk wounds of valuable trees as soon as a service man can reach the location. Several girdled maple, elm, hackberry, and mountain-ash trees have been successfully treated; one Norway maple girdled for a vertical distance of five feet was saved without foliage wilting or appreciable retardation of the rate of growth. Four to eight weeks of shading is usually sufficient time to secure survival of the cambium and its development into thick healthy bark.

A complete koda-chrome slide record was made of the cambium development.

LANSING, MICHIGAN

ZOOLOGY

GEOGRAPHICAL VARIATION IN THE BLACK VULTURE

PIERCE BRODKORB

THE black vulture, *Coragyps atratus* (Bechstein), has a wide distribution, from the southeastern United States south to Argentina and Chile. Within this immense area size is the only character that is known to vary geographically. The most significant measurement is that of the wing. Measurements of tail and bill show the same trends as the wing, but to a less degree.

HISTORY

J. A. Allen (1905, p. 275) was the first author to suggest the geographic variability of the black vulture. He stated that South American birds were smaller than those of North America. Almost simultaneously Hellmayr (1906, p. 567) reached the conclusion that South American birds were separable subspecifically from those of North America. Unfortunately neither author published measurements to substantiate the claim to distinctness of the South American bird.

In recent years there has been little agreement among ornithologists on the advisability of recognizing a southern subspecies of this vulture. Some, notably Todd and Carriker (1922, p. 142) and Friedmann (1933, p. 187), have held that the differences are too inconstant for taxonomic separation. Others (Swann, 1924–1930, pp. 7–10; Chapman, 1926, p. 217; Dickey and van Rossem, 1938, p. 97) have accepted the division into a northern and a southern race. It has even been suggested (Wetmore, 1926, p. 91; Peters, 1929, p. 415; Aldrich, 1937, p. 38) that three subspecies instead of two may be involved.

MATERIAL

There have been used in the present study forty-four specimens from various parts of the range of the black vulture. This number, though greater than that available to previous investigators, was thought to be not entirely adequate, since individuals in certain stages of molt had to be discarded. I have therefore incorporated published measurements of individual birds when these could be segregated by locality. Thus wing measurements of five were taken from Aldrich, four from Wetmore, three from Chapman, three from Swann, and one each from Dickey and van Rossem and from Salvin and Godman (1897–1904, p. 133).

SEXUAL VARIATION

According to Aldrich, sexual variation in size is insignificant in the black vulture. My material supports this statement; the wings of birds from the United States, for example, measure 418–448 mm. in six males, 417–443 mm. in five females. In samples from other localities sometimes one sex, sometimes the other, happens to be the larger. I have therefore combined the measurements of both sexes, since this permits the inclusion of specimens whose sex was not recorded.

GEOGRAPHIC VARIATION

When the wing measurements of vultures from North America were compared with those of South American birds, both lots were found to show about the same limits of variation. In birds from Panama northward the wing measured 381–454 mm. Among South American specimens the variation was 391–453 mm. Friedmann found 80 per cent of his specimens to fall within the range in wing length shown by birds of both continents. Obviously no separation can be made between North American birds as a whole on the one hand and South American birds as a whole on the other.

A different division was then made. The first group contained United States birds. The second lot consisted of specimens from tropical America, from Mexico south to about the Tropic of Capricorn. The third category was composed of individuals from the Temperate Zone of South America. This grouping brought out the fact that the vultures from the tropical area are small, while those from the temperate regions to north and south are large (Table I).

In order to test the significance of the geographic differences in wing length the figures were treated statistically by the graphic method recently devised by Hubbs and Perlmutter (1942). The

TABLE I

VARIATION IN WING LENGTH IN THE BLACK VULTURE

Locality	No.	Wing length in millimeters																				
		Mean	Min.	Max.	380–383	384–387	388–391	392–395	396–399	400–403	404–407	408–411	412–415	416–419	420–423	424–427	428–431	432–435	436–439	440–443	444–447	448–451	452–455
United States...	18	431.4	415	454	0	0	0	0	0	0	0	0	2	2	1	2	2	3	1	1	2	2	1
Tropical America	20	403.5	381	415	1	0	2	2	1	2	4	5	3	0	0	0	0	0	0	0	0	0	0
Temperate South America.....	16	430.9	418	453	0	0	0	0	0	0	0	0	0	2	2	1	4	5	0	0	0	0	2

thin vertical line (Fig. 1) shows the range of variation in the sample. The solid vertical bar gives one standard deviation on either side of the mean, which is indicated by the cross bar. The hollow rectangle represents twice the standard error on either side of the mean.

This method seems particularly adapted to taxonomic studies. In a normal curve 84 per cent of the specimens will fall within the range indicated by the solid vertical bar, a percentage that approximates the degree of differentiation usually accepted as the minimum for subspecific separation.

A significant difference between two populations is regarded as demonstrated if the gap between two rectangles exceeds 10 per cent of the length of the shorter rectangle. In Figure 1 the gap between the rectangle representing the Tropical Zone population and the rectangles representing both the North Temperate and the South Temperate Zone populations amounts to over 200 per cent of the length of the short rectangle. The difference in wing length between Tropical and Temperate Zone birds is thus taxonomically significant.

The length of the rectangle representing the South Temperate Zone birds is contained wholly within the height of the rectangle representing the North Temperate Zone birds. Thus no significance is to be attributed to a difference between the means of the two samples.

NOMENCLATURE OF THE TEMPERATE ZONE RACE

Although it is contrary to the usual theory for the range of one subspecies to consist of two separate parts interrupted by the range of another subspecies, I can see no justification for applying different names to the two Temperate Zone populations of black vultures. I therefore propose to treat them as the same subspecies. The proper

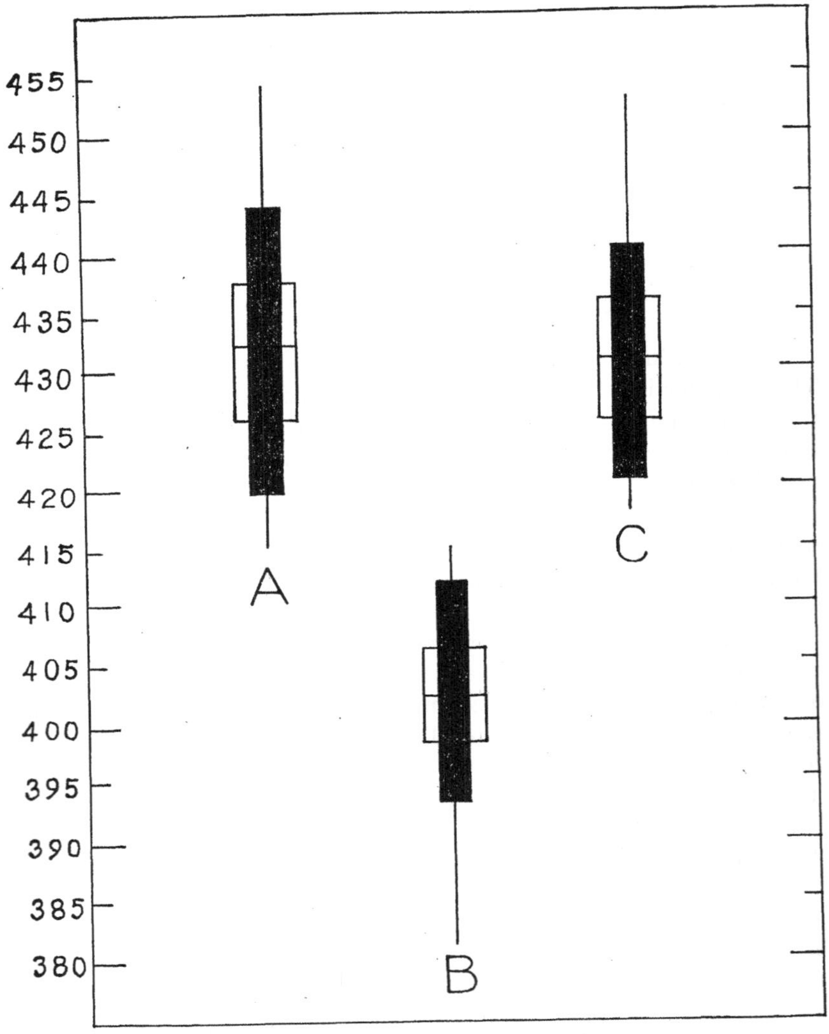

FIG. 1. Variation in wing length (mm.) in the black vulture, *Coragyps atratus*. *A*, United States (18 specimens); *B*, tropical America (20 specimens); *C*, temperate South America (16 specimens)

name for this form is *Coragyps atratus atratus* (Bechstein), based upon a bird from Florida.

The name *Cathartes foetens* Lichtenstein is a synonym of *C. atratus*. Lichtenstein's name, currently employed for a South American race, was founded on Azara's description of a bird from Paraguay. The small tropical race occurs only in the northwestern corner of the Chaco, beyond the area worked by Azara. Thirteen specimens from the remaining parts of the republic have wing measurements of 418–453 mm., and thus belong with *C. a. atratus*. In order to avoid any future confusion I restrict the type locality of *Cathartes foetens* to Asunción, Paraguay.

The possibility remains, of course, that future studies may disclose some overlooked difference between the northern and the southern colonies of Temperate Zone birds. In that case the name *foetens* would be available for the southern birds.

NOMENCLATURE OF THE TROPICAL ZONE RACE

As has already been demonstrated, the name *foetens* cannot be used for this subspecies. The first available name is *Coragyps atratus brasiliensis* (Bonaparte). Bonaparte described his bird as being smaller than the North American form, so that the applicability of his name can scarcely be questioned. He gave its distribution as South America and the Antilles. The Antillean locality is of course erroneous, and Swann restricted the type locality to "S. Brazil." A further restriction may be necessary, for although a specimen from Fazenda Miranda, Matto Grosso, belongs to the race *C. a. brasiliensis*, it seems likely that *C. a. atratus* will prove to be the subspecies of the southernmost states of Brazil.

POSSIBLE MIGRATION OF TEMPERATE ZONE BIRDS

There is some evidence which suggests that the Temperate Zone birds of both hemispheres are partially migratory. Five Mexican specimens, from localities from Sonora to Chiapas, including all the birds taken during the breeding season, are of the subspecies *C. a. brasiliensis*, with wings measuring 388–409 mm. Four specimens shot during September and March at Acacoyagua and Unión Juárez, Chiapas, are as big as others from the United States. Their wings measure 418–430 mm. These large nonbreeding black vultures from Mexico are interpreted as migrant individuals of *C. a. atratus*.

Nine black vultures from the Chaco west of Puerto Casado, Paraguay, some of which were collected during the breeding season, likewise belong to *C. a. brasiliensis*, since their wing measurements are 391–414 mm. Two others, taken in the winter months of February and May, measure 422 and 424 mm. They are considered migrants of *C. a. atratus* from temperate South America.

The turkey vulture, *Cathartes aura*, a species with which the present one often associates, is known to be partially migratory in the northern part of its range. It will be desirable to investigate further the possibility that the black vulture is also migratory.

EVOLUTION

It is interesting, though perhaps not very profitable, to speculate whether subspeciation in the black vulture took place in the center of the range of the species or whether the character of large size arose as an independent parallel mutation at the northern and southern ends of its range. Subspeciation is generally understood to occur at the periphery of the range of another form. There are, however, a few known exceptions, and a case in point is that of the cinnamon teal, *Anas cyanoptera*. This duck occurs from the western United States to Argentina. Within this area its characters are constant, except at Lake Titicaca, in the Andes of Peru. A local race, differentiated by larger size, inhabits that lake. Its distribution is so restricted that there can be little doubt that it originated there as a local mutation. Here is an example of subspeciation occurring near the center of the range of another form.

Argument that the small central form of black vulture was derived from the large end form could be made on paleontological grounds. Besides the modern bird in its two subspecies, the genus *Coragyps* contains a Pleistocene form from California and nearby areas. The Pleistocene bird, *Coragyps occidentalis* (L. H. Miller), had a wing even longer than that of *C. a. atratus*. Its known habitat lay largely beyond that of the modern bird. There is thus perhaps a trend, progressive both geographically and in point of time, from larger to smaller size.

University of Michigan

LITERATURE CITED

ALDRICH, JOHN WARREN. 1937. Annotated List of Birds. *In* Aldrich, John Warren, and Bole, Benjamin Patterson, Jr., The Birds and Mammals of the Western Slope of the Azuero Peninsula. Sci. Publ. Cleveland Mus. Nat. Hist., 7: 1–196, map.

ALLEN, J. A. 1905. Supplementary Notes on Birds Collected in the Santa Marta District, Colombia, by Herbert H. Smith, with Descriptions of Nests and Eggs. Bull. Am. Mus. Nat. Hist., 21: 275–295.

CHAPMAN, FRANK M. 1926. The Distribution of Bird-Life in Ecuador. Bull. Am. Mus. Nat. Hist., 55: xiii–784, 11 pls., map.

DICKEY, DONALD R., AND VAN ROSSEM, A. J. 1938. The Birds of El Salvador. Zool. Ser. Field Mus. Nat. Hist., 23: 1–609, 24 pls.

FRIEDMANN, HERBERT. 1933. Critical Notes on American Vultures. Proc. Biol. Soc. Wash., 46: 187–189.

HELLMAYR, C. E. 1906. Revision der Spix'schen Typen brasilianischer Vögel. Abhandl. K. Bayer. Akad. Wiss., Math.-phys. Kl., 21, Bd. 3: 563-726.

HUBBS, CARL L., AND PERLMUTTER, ALFRED. 1942. Biometric Comparison of Several Samples, with Particular Reference to Racial Investigations. Am. Nat., 76: 582–592.

PETERS, JAMES L. 1929. An Ornithological Survey in the Caribbean Lowlands of Honduras. Bull. Mus. Comp. Zoöl., 69, No. 12: 397–478.

SALVIN, OSBERT, AND GODMAN, FREDERICK DUCANE. 1897–1904. Biologia Centrali-Americana. Aves, III, iv + 510. Taylor and Francis, London.

SWANN, H. KIRKE. 1924–1930. A Monograph of the Birds of Prey (Order Acciptres), xi + 487, pls. Wheldon and Wesley, London.

TODD, W. E. CLYDE, AND CARRIKER, M. A., JR. 1922. The Birds of the Santa Marta Region of Colombia: A Study in Altitudinal Distribution. Ann. Carnegie Mus., 14: 3–582.

WETMORE, ALEXANDER. 1926. Observations on the Birds of Argentina, Paraguay, Uruguay, and Chile. Bull. U. S. Nat. Mus., 133: 1–448.

EGG PRODUCTION OF THE NORTHERN PIKE, *ESOX LUCIUS* L., AND THE PERCENTAGE SURVIVAL OF EGGS AND YOUNG ON THE SPAWNING GROUNDS*

WILLIAM F. CARBINE

IN 1937 the writer started an investigation of the life history of the northern pike. The aim was to devise procedures for the management of this important game fish. Particular attention was paid to egg production and to the survival of the young from the time of hatching until the time when they leave the spawning grounds since little information was available on these phases of pike life history.

The work was begun in southern Michigan, but most of the field observations were made from 1939 to 1942 at Houghton Lake, the largest inland lake in Michigan and one of the most productive. Some of the results have already been published (Carbine, 1942).

EGG PRODUCTION

Materials and Methods

During 1939, 1940, and 1942 the writer and various staff members of the Institute for Fisheries Research secured at Houghton Lake thirty specimens of northern pike (Table I) for the making of egg counts. Some were caught during January and February by ice fishermen, but the data were gathered by the fishery investigators. A larger number were taken at random during the spawning run, from the weirs that were installed in ditches connected with the North Bay of Houghton Lake at Peterson's resort (Carbine, 1942). With the exception of two fish that were stripped at the time of capture only fish that were "green" (with egg sacs intact) were saved for samples.

Shortly after its capture each fish was weighed to the nearest gram or ounce, depending upon the size of the fish. Standard and

* Contribution from the Institute for Fisheries Research of the Michigan Department of Conservation.

TABLE I

DATA ON COLLECTION, SIZE, CONDITION, AND NUMBER OF EGGS FOR THIRTY NORTHERN PIKE COLLECTED AT HOUGHTON LAKE *

Counts made by C. B. Obrecht, W. F. Carbine, George N. Washburn and John T. Greenbank

Date collected	Standard length (millimeters)	Total length (inches)	Weight (ounces)	K	Number of eggs	Counted by
4–2–42	338	15.7	11.0 †	0.808	7,691	Carbine
3–29–42	370	17.0	16.0 †	0.896	13,309	"
4–15–39	443	20.3	34.0	1.109	28,871 ‡	Obrecht
2–25–39	446	20.6	26.0	0.831	16,898	"
4–15–39	455	20.7	33.0	0.994	22,966 ‡	"
2–25–39	450	20.7	34.0	1.058	24,030	"
4–22–39	452	20.7	33.5	1.029	25,982	"
4–18–39	458	20.9	31.5	0.930	25,760	"
4–18–39	460	21.0	33.0	0.962	21,086	Carbine
4–21–39	468	21.4	39.0	1.079	27,700	Obrecht
3–27–42	460	21.6	44.0	1.281	33,168	Carbine
4–21–39	475	21.7	33.5	0.886	19,418	Obrecht
4–21–39	490	22.3	39.0	0.940	12,599	"
4–21–39	485	22.3	39.5	0.982	27,550	Washburn
4–16–39	510	23.6	44.0	0.940	23,920	"
4–23–39	512	23.6	48.0	1.014	33,342	Obrecht
2–25–39	525	23.6	50.0	0.980	36,484	"
2–23–39	540	24.5	52.0	0.936	41,116	"
2–4–39	540	25.0	44.0	0.792	19,808	Carbine
2–26–39	559	25.4	52.0	0.844	47,289	Obrecht
4–17–39	570	25.9	56.0	0.857	38,756	"
2–25–39	566	26.0	54.0	0.844	29,945	"
4–13–39	571	26.0	60.0	0.914	37,332	Carbine
4–17–42 §	584	26.1	64.0 †	0.911	43,170	Greenbank and Carbine
2–25–39	578	26.2	60.0	0.881	38,989	Obrecht
4–18–39	578	26.2	70.0	1.027	44,902	"
1–28–43 ‖	600	27.1	67.0	0.879	36,812	Carbine
1–28–39	600	27.2	68.0	0.893	54,085	Greenbank
2–23–39	638	28.4	88.0	0.961	35,754	Obrecht
4–23–39	775	35.0	170.5	1.038	97,273	"
Averages	516.5	23.6	49.8	0.950	32,200	

* See fifth footnote for exception.
† Average weights of other female northern pike from Houghton Lake.
‡ Eggs stripped from fish.
§ Collected by Dr. P. I. Tack from Houghton Lake.
‖ Collected by Dr. L. N. Allison from Fletcher Pond, Thunder Bay River.

total lengths were measured in millimeters, and the ovaries were removed and preserved in 10 per cent formalin.

Actual total counts of the eggs were made on all thirty specimens. In nineteen of the counts Carl B. Obrecht used a counting board, patterned after boards used in the East. He constructed it from a piece of black hard-rubber radio panel, $\frac{3}{8}$ inch thick and measuring 3 by 5 inches. He drilled in it 250 holes $\frac{1}{8}$ inch in diameter and $\frac{1}{8}$ inch deep. Each hole was lightly countersunk. Another board with slightly smaller holes was employed in counting the smaller eggs of winter-caught fish. In operating this board the well-separated eggs are merely spread over the surface so that they will lodge in the holes. After surplus eggs are removed, inspection reveals any empty holes, and the count of 250, minus the number of empty holes, is obtained.

To trace the growth of the eggs and the relationship between the various sizes present in an ovary diameter measurements of ova were made by means of an ocular micrometer in a compound binocular microscope. The magnifications used gave a value of 0.04 mm. for each micrometer unit for the small eggs, and 0.2 mm. for the large eggs. The diameter was determined by placing the micrometer in a horizontal position with respect to the field of the microscope and reading the vertical diameter of each egg, without regard to its shape or position in the field of the microscope. Owing to distortion in the process of preservation ova are seldom perfectly spherical, and this method obviated any selection of the longest or shortest diameter. It gave measurements of the longest diameter of some eggs, the shortest of others, and intermediate ones for still others. Clark (1925), who made a careful test of this method, found it to be reliable and the most satisfactory for constant use.

In order to show the growth history of the ova before the spawning season a random sample of eggs was obtained after all eggs in the ovary had been teased apart. Measurements of the diameter of the large ova were made at the lower magnification until 500 eggs from each fish had been recorded, and the process was then repeated at the higher magnification to obtain the measurements of 500 small ova. To present the relationship between the actual number of all eggs of each size in the ovary a section approximately $\frac{3}{8}$ inch wide and about an inch long was carefully teased from the center of the ovary of one fish, and measurements of all ova (totaling 3,688) were taken in the sample.

To correlate the egg production with the condition of the fish the coefficient of condition (K) was calculated by using the formula $K = \frac{W \times 100}{L^3}$, in which W is the weight in grams and L the standard length in centimeters.

Because of the difficulty and uncertainty ever present in determining the age of northern pike from examination of the scales age determinations are not included in this paper.

Number of Eggs per Female

The actual number of eggs contained in the thirty females (Table I) ranged from 7,691 (for a fish having a total length of 15.7 inches and weighing 11 ounces) to 97,273 (for a 35-inch pike weighing 170.5 ounces). The following are average figures: number of eggs

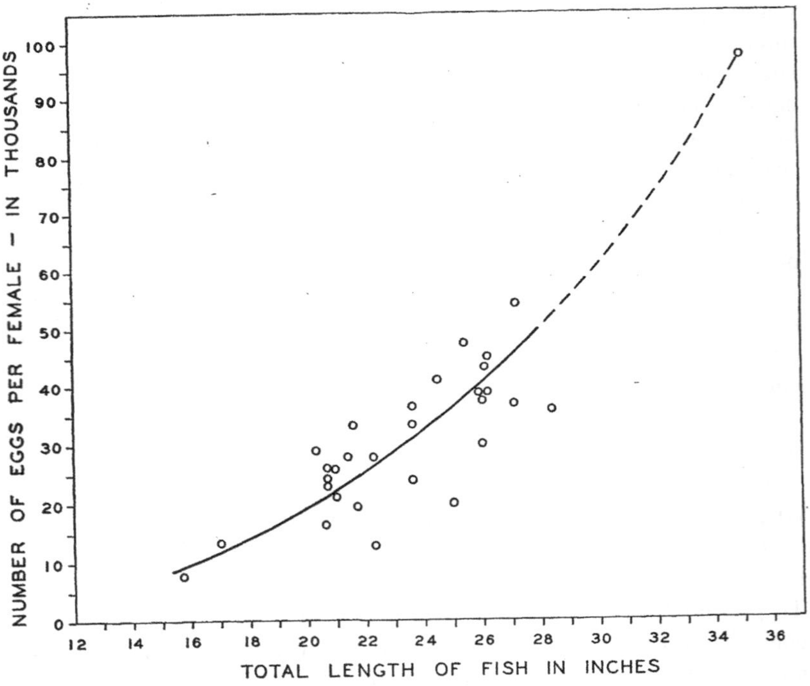

FIG. 1. Relation between the number of eggs and the total length of fish for thirty northern pike

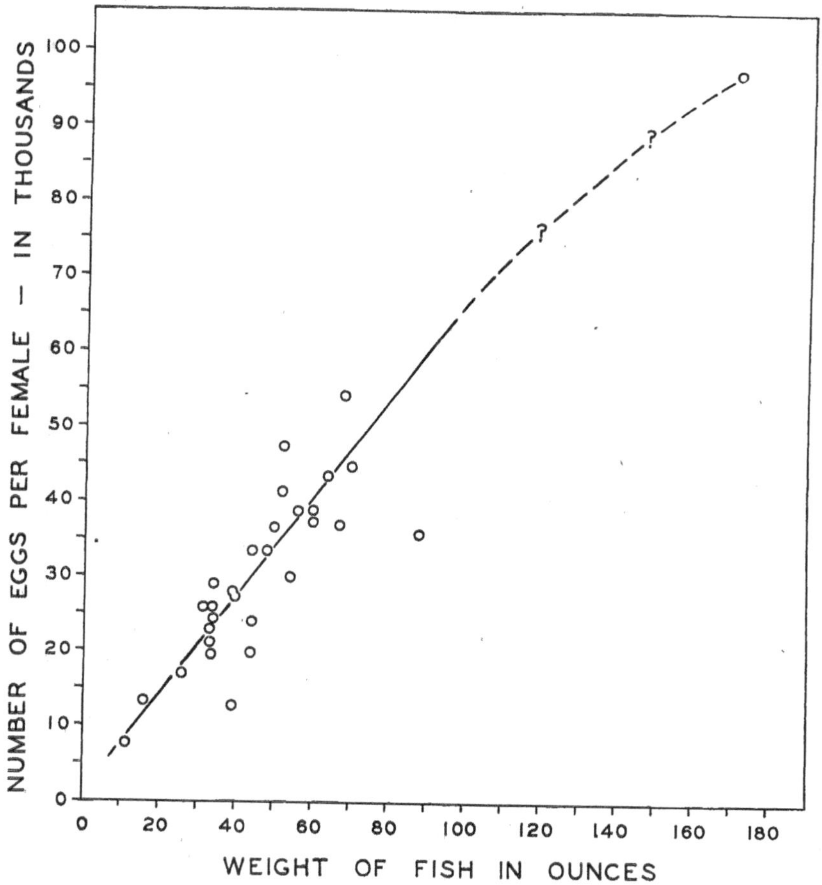

FIG. 2. Relation between the number of eggs and the weight of fish for thirty northern pike

produced, 32,200; total length, 23.6 inches; weight, 49.8 ounces. The average total length of the fish in the sample agrees closely with that of all females which entered the spawning grounds (23.5 inches in 1939; 23.2 inches in 1940; and 22.4 inches in 1942).

The number of eggs increases rapidly with increase in length of fish and is roughly proportionate to the weight (Figs. 1–2). In fish of any particular length or weight, however, it varies greatly.

Despite a large fluctuation there is some positive correlation between the condition of the female and its fecundity (Table I). The values for K fluctuate widely, and do not increase with size of fish; K averages 0.984 for the fourteen fish 15.7 to 22.3 inches long, and 0.919 for the sixteen that measure 23.6 to 35.0 inches. The specimen with the largest K value did, however, produce more eggs than other fish in the same general size range. Likewise, the fish having the lowest K value produced fewer eggs than other specimens in the same general size group. The six fish having the highest number of eggs, if the length of the fish is considered, and the six having the lowest number (as determined by the plus or minus deviation from the average line in Fig. 1) had average K values of 1.007 and 0.893, respectively.

Growth History of the Developing Ova

Because northern pike spawn but once a year and because the spawning season is of short duration the eggs of a female at spawning time are of two markedly distinct sizes: (1) large mature eggs and (2) almost microscopic immature ones. To demonstrate the history of the maturing ova and to check the assumption that all ova that are to be spawned in one year become distinguishable during the seasonal development from the small immature eggs, diameter measurements of ova were made from seven fish collected at various dates before the spawning season (Fig. 3A). The first specimen, a large fish from Drayton Plains hatchery stock known to be in its second year, was preserved on November 4. The fish of January 28 and February 4 and 25 were taken through the ice at Houghton Lake by fishermen. Those of March 27 and April 15 and 26 came from Peterson's Ditches during the spawning run. The one collected on March 27, a green female, and the one taken on April 15, a ripe female, were killed during the upstream migration to the spawning grounds.

The specimen secured on April 26 was one of several that entered the spawning marshes on April 10, at which time they were marked. This particular fish was seen to be ripe on April 10, for some eggs were lost during the process of weighing, measuring, and tagging. After spawning it returned from the marshes to the lake on April 25 with a serious head injury. On April 26 it was found dead, washed up on shore. The ovaries were then removed. This northern pike had therefore spawned some time during the seventeen-day period be-

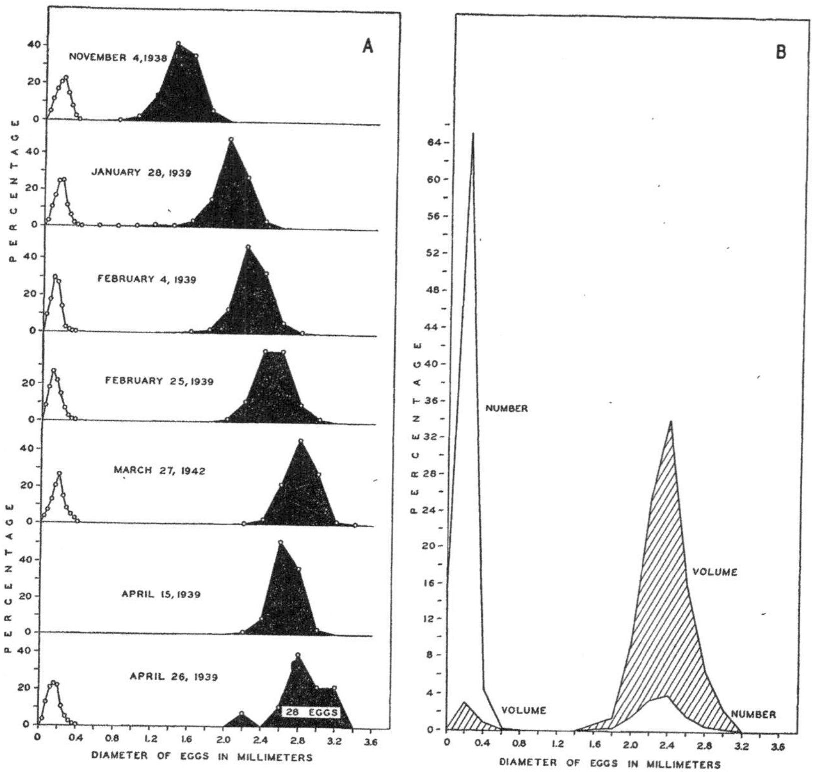

FIGURE 3

A. Frequency polygons showing the growth history of northern pike eggs collected before, during, and after the spawning season. Because each polygon was based on measurements of 500 eggs and because more and different size classes were used in measuring the immature ova, the areas of the polygons are not proportional to the numbers of immature and mature eggs present in the ovary. The fish collected on March 27 (a green female) and on April 15 (a ripe female, which was stripped) were killed on the upstream spawning run. See text (p. 128) for the data concerning the collection of the fish taken on April 26.

B. Frequency polygons showing the proportionate number and proportionate volume of the eggs of each size class in the ovary of a mature northern pike. Based on measurements of the 3,688 ova comprising a random sample of the ovary of a fish collected January 28, 1943, from the Fletcher Pond of the Thunder Bay River. The immature eggs were measured in size classes of 0.04 mm., but these were combined by fives to give size classes of 0.2 mm., to correspond with the measurements of the maturing eggs.

tween April 10 and 26. Because the fish was ripe when it entered the spawning grounds it is probably safe to assume that spawning occurred at least two weeks before the ovaries were removed and preserved. The ovarian wall had shrunk little during this period. Close examination revealed the presence of twenty-eight old, mature eggs held over from spawning. The few large eggs that were well preserved were free from the follicles; those that were still enclosed in the follicles showed varying degrees of resorption. In addition to the twenty-eight mature eggs a tremendous number of extremely small eggs were also present in the ovaries (Fig. 3 A, bottom graph). Some of these eggs were granular in appearance and were apparently degenerating, because they were easily broken after preservation. The remainder of the small eggs still retained the nucleus and were the only ones measured.

In studying the growth of the ova (Fig. 3 A) chief attention should be paid to the eggs with a diameter exceeding 0.4 mm., because these are the maturing eggs. Ova less than 0.4 mm. in diameter obviously remain immature through the late fall and winter; they constitute a distinct group of relatively uniform size in each adult female, during the fall and winter (Fig. 3 A) and no doubt throughout the year. Any increase in size beyond 0.4 mm. indicates the beginning of the growth of the ova toward maturity for the next spawning season. We were unable to obtain a fish with ovaries at this stage. The specimen in the earliest stage in the growth of the ova was taken on November 4, approximately five months before spawning. Differentiation between the two sizes of eggs was already well marked.

The only fish with eggs of intermediate size was collected on January 28. The few eggs of this size had either started to develop a short time before this fish was killed or else had been retarded in growth.

On each succeeding date from November 4, 1938, to April 15, 1939, the maturing ova were found to have attained a constantly greater average diameter. The March specimen had mature eggs of the largest size, either because of individual variation or because the sample was taken in another year (1942).

Mature eggs of the northern pike range in diameter from 2.2 to 3.4 mm.

The immature eggs greatly outnumber the maturing eggs in an ovary of a pike about to spawn, but the volume of the maturing eggs vastly exceeds that of the immature ones (Fig. 3B). To illustrate these relations, measurements were made of the diameter of all the 3,688 eggs comprising a random sample from the ovary of a fish collected on January 28, 1943 (this fish measured 27.1 inches in total length and weighed 67 ounces). The volume of the eggs in each size class was computed mathematically from the average diameter of each class. The number and the volume of the eggs in each size class were computed and expressed as the percentage of the total number and of the total volume, respectively.

Of the 3,688 ova that were measured the maturing eggs constituted 11.7 per cent by number and 95.9 per cent by volume, whereas the immature eggs made up 88.3 per cent by number and 4.1 per cent by volume. Since the total number of maturing eggs in both ovaries of this fish was 36,812 (by count), it is computed that these ovaries contained 278,089 immature ova. Enough small eggs for several spawnings were therefore present in the ovaries of this fish. I do not believe, however, that the 314,901 eggs estimated to be present in this fish would suffice for all spawnings of a pike that would reach a relatively large size and advanced age. A 35-inch pike was found to contain 97,273 mature eggs; in three years such a fish would produce more eggs than the total number of immature eggs estimated to occur in the 27-inch fish. Large pike show no apparent reduction in the relative number of immature ova. It is concluded that new eggs develop during the adult life of the pike. Some evidence was obtained to indicate that a number of the immature ova as well as all unspawned mature eggs are resorbed after the spawning time. Such loss would provide a further need for new egg production in the adults.

The ovary of the fish used in determining the number of eggs in each size group (see Fig. 3B) contained in its anterior, middle, and posterior parts mature ova of the following diameters, as determined by measurements of 500 eggs from each part:

Anterior part, 1.6 to 3.0 mm.; average 2.305 $\pm$ .0103 mm.;
Middle part, 1.6 to 3.0 mm.; average 2.355 $\pm$ .0130 mm.;
Posterior part, 1.6 to 3.0 mm.; average 2.304 $\pm$ .0103 mm.

The average size of the eggs is significantly larger in the middle part of this ovary than in the anterior and the posterior parts (the difference is about three times the standard error of the difference). The difference, however, is only about 0.05 mm. It is therefore concluded that no considerable error was introduced by measuring the eggs (as for Fig. 3) from samples taken from the middle part only of each ovary.

The developmental cycle of the eggs in the northern pike is like that of the salmonoid fishes, lampreys, and other fishes that spawn over a very brief period. Many fishes, for instance, the halibut (Thompson, 1915), the grunion (Clark, 1925), and other species (Hickling and Rutenberg, 1936), deposit eggs more than once in a spawning season, and throughout most of this period have eggs of more than two size groups in the ovary. In addition to the reserve stock of minute ova and the group of maturing ova, one can usually distinguish groups of ova of one or more intermediate sizes. These ova are obviously being prepared for future periods of egg laying.

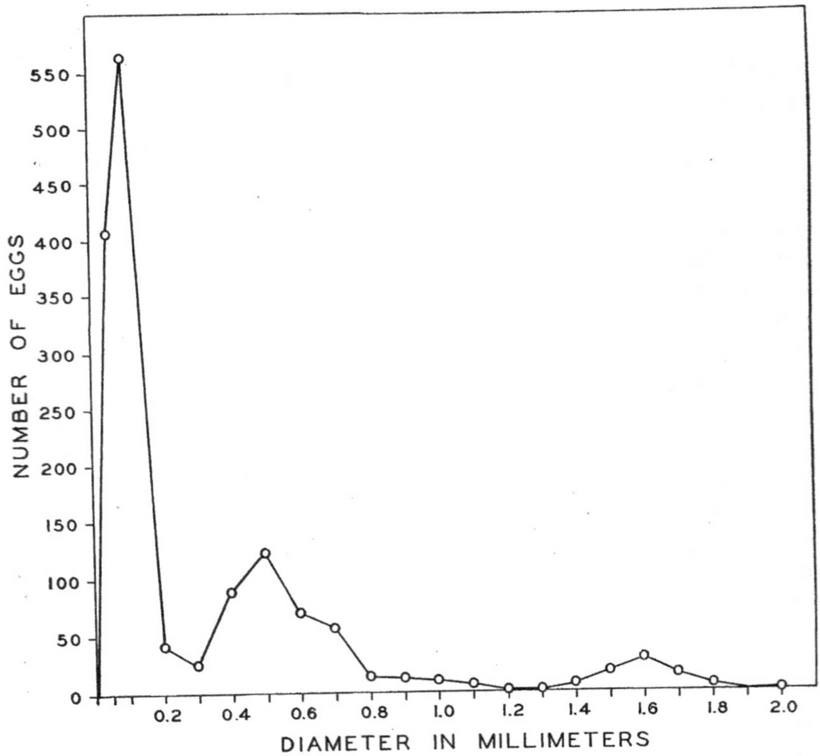

FIG. 4. Frequency polygon showing relationship of the number of eggs of various sizes present in a mud pickerel (*Esox vermiculatus*) taken April 2, 1941, from Whitmore Lake

In general, this type of egg development appears to be characteristic of families or at least of genera. It comes as a surprise, then, to learn that both types of egg development occur in the pike genus *Esox*. The mud pickerel, *Esox vermiculatus*, contains in its ovaries a group of eggs intermediate in size between the immature and the mature eggs. In the one mature female studied the immature eggs averaged about 0.1 mm. in diameter, and the seemingly mature (yellowish and rather free) eggs ranged in diameter from 1.2 to 2.0 mm. (Fig. 4). The mature eggs thus seem to average smaller than those of *Esox lucius*. The female of *Esox vermiculatus* that was studied measured 6.2 inches in total length, and was collected at Whitmore Lake, Washtenaw County, Michigan, on April 2, 1941, during the spawning period of this species, which coincides with or at least overlaps that of the northern pike. Newly hatched mud pickerel may be taken in early April in southern Michigan. The eggs of the intermediate size group ranged in diameter from 0.3 to 1.2 mm. It is estimated that the 6.2-inch mud pickerel contained a total of 15,732 eggs, of which 10,925 were immature, 4,004 of intermediate size, and 803 mature. Several spawnings during the spring are thus indicated. Occasionally mud pickerel spawn in the fall also, according to Lagler and Hubbs (1943).

PERCENTAGE SURVIVAL OF EGGS AND YOUNG ON THE SPAWNING GROUNDS

In the management of any species of fish knowledge of the natural mortality and of the probable yield from any spawning is obviously important. It is desirable to know at what stage of development the mortality rate is highest. In the experiments described in this paper a complete count was obtained of all adult northern pike migrating from Houghton Lake to the spawning grounds under observation, and in 1939 and 1940 complete counts were made of all young migrating from the marshes to the lake (Carbine, 1942). It is possible, therefore, to present figures indicating the survival of the young from the time of egg deposition until the last young pike had migrated from the marshes (a maximum period of 82 days from the time the first fry hatched in 1939 and of 85 days in 1940). Each year a portion of the young were fin-clipped before they were released in Houghton Lake in the hope that an estimate could be obtained of the percentage survival from the fingerling stage to that of legal size. Unfortunately so few captures of these fish were reported (voluntary returns only were available) that this experiment was abandoned.

In 1942 all the migrating adults were trapped during the spawning run, but fry and fingerling weirs were not installed. On May 6, 1942, I stopped at Houghton Lake to check the success of the hatch of pike in the various marshes that had been under examination earlier in the year. It was surprising to find that Mr. John Peterson, who operates a resort near the marshes on which this investigation was being conducted, had installed and maintained weirs for trapping fry and fingerling pike, following the procedure previously used here (Carbine, 1942). During the past few years Mr. Peterson had become interested in our projects and desired to have as many pike as possible placed in Houghton Lake. He also wanted to compare the production obtained in 1942 with that of 1939 and 1940. He had been keeping accurate records of the number of young pike that were taken in the traps from the time of the first migration until about June 1 (when the weirs were removed). According to the records kept by Mr. Peterson up to May 6, slightly over 4,000 young northern pike had been taken in the traps and had been transferred to Houghton Lake. He estimated, from a two-hour tally, that approximately half of the fish went through some small holes in the wings of the weir rather than into the traps. Since his record book was unfortunately lost, it is not possible to give the exact estimate on the total run of young pike for 1942. The figure of 8,000 given for 1942, in Table II, a rough calculation, is considered to be a minimal estimate.

The eggs and young of the northern pike suffered an enormous mortality on the Houghton Lake spawning grounds that were under observation in 1939, 1940, and 1942 (Table II). The approximately four million eggs estimated to have been spawned in 1939 yielded only 7,239 migrating young — about 1,800 young for each million eggs. The mortality was computed to be 99.82 per cent. In 1940 about 700 young resulted from the deposition of a million eggs (mortality — 99.93 per cent) and in 1942, 4,400 young were produced per million eggs (mortality — 99.56 per cent).

The stage at which the greatest loss occurred was not determined. Observations in the marshes and ditches proved that very large numbers survived through hatching and the period of yolk absorption. As many as twenty very young pike were caught in a single dip of a small scap net. Gradually the numbers of young dwindled. Predation, including cannibalism, was observed, but the loss from each of many causes was not estimated.

The survival of eggs and young, 0.44 per cent, was relatively much higher in 1942 than in 1939 or 1940. The cause is thought to lie not in the smaller number of eggs laid, but rather in the higher water level. The marshes that provide the spawning and nursery waters for northern pike were more extensively covered in 1942 than in the other years, and remained covered for a longer period than in any of the last eight to ten years, according to local reports. The water level in 1942 did not become appreciably lower until all the fingerling pike had had a chance to move out to the lake.

TABLE II

ESTIMATES OF THE EGG PRODUCTION IN NORTHERN PIKE AND OF THE SURVIVAL OF EGGS AND YOUNG UP TO THE TIME OF DOWNSTREAM MIGRATION

Data obtained at Peterson's Ditches, Houghton Lake, Michigan

	1939	1940	1942
Number of spawning females	125	65	56
Number of spawning males	280	81	70
Ratio of males to females	224:100	125:100	125:100
Estimated number of eggs deposited *	4,025,000	2,093,000	1,803,000
Number of migrating young †	7,239	1,495	8,000
Percentage survival of eggs and young to time of migration	0.18	0.07	0.44

* Based on average of 32,200 eggs per female; see p. 127.
† See text in regard to estimate for 1942.

A very high mortality during the fry stage was also observed in the pike-propagation experiment conducted in 1937 at the Ortonville Rearing Pond of the Drayton Plains Fish Hatchery of the Michigan Department of Conservation. This three-acre pond was stocked with approximately 150,000 northern pike fry on May 3, 1937, and when it was drained on October 14 (171 days from the date of hatching) the total number of pike recovered was only 362. The mortality from the fry stage was 99.76 per cent, virtually identical with the average mortality of 99.77 per cent estimated to have occurred during the egg and early fingerling stages on the spawning grounds of Houghton Lake.

ACKNOWLEDGMENTS

I should like to express my appreciation to Albert S. Hazzard, director of the Institute for Fisheries Research, for assistance in planning the work and for suggestions and help in the preparation of this report. Thanks are also due to Carl B. Obrecht, who counted most of the eggs; to George N. Washburn and John T. Greenbank, who assisted in some of the egg counts; to Louis A. Krumholz, who drafted the figures; to David S. Shetter, Walter R. Crowe, and other members of the Institute, as well as to various officers of the Department of Conservation who assisted in the collection of material for the egg counts and in other phases of this investigation. I am indebted to Carl L. Hubbs for guidance and help in the analysis of the data and in preparation of the report.

SUMMARY

1. Investigation of the life history of the northern pike, with special emphasis on egg production and survival of eggs and young on the spawning grounds, was begun in 1939 and continued through the spawning season of 1942 at Houghton Lake, Michigan.

2. Most of the migrating young and adult northern pike were captured in weirs, although a few of the adults were caught by ice fishermen.

3. On the basis of actual counts of the eggs contained in the thirty northern pike the average number produced was estimated to be 32,200 (range 7,691 to 97,273).

4. The number of eggs increases rapidly with increase in length of the fish and is roughly proportionate to the weight.

5. Despite a great fluctuation there is some positive correlation between the condition of the female and the number of eggs produced.

6. Diameter measurements of ova made on seven fish collected before, during, and after the spawning season show that all eggs that are to be spawned in one season form a single size group, easily distinguishable throughout the seasonal development from the small immature eggs.

7. Mature eggs of the northern pike range in size from 2.2 to 3.4 mm.

8. The immature eggs present in the ovaries of a pike before the

spawning season constitute 88.3 per cent of the number and 4.1 per cent of the total volume, whereas the maturing eggs made up 11.7 per cent by number and 95.9 per cent by volume.

9. The presence in the ovaries of the mud pickerel *Esox vermiculatus* of eggs of intermediate size indicates that this species spawns more than once during one season and that one type of egg development is not characteristic of a genus.

10. A total of 125 females spawned in the marshes in 1939, 65 in 1940, and 56 in 1942.

11. The estimated egg production, based upon the average of 32,200 eggs per female, was 4,025,000 in 1939, 2,093,000 in 1940, and 1,803,000 in 1942.

12. The approximately four million eggs assumed to have been spawned in 1939 yielded only 7,239 young; in 1940 only 1,495 young hatched from about two million eggs; and in 1942 an estimated 8,000 young from 1,800,000 eggs. The mortality of eggs and young to the time when the young left the marshes is therefore estimated to have been 99.82 per cent in 1939, 99.93 per cent in 1940, and 99.56 per cent in 1942. The increased yield in 1942 seems to be correlated with a higher and more constant water level over the marshes.

Institute for Fisheries Research
Ann Arbor, Michigan

LITERATURE CITED

Carbine, W. F. 1942. Observations on the Life History of the Northern Pike, *Esox lucius* L., in Houghton Lake, Michigan. Trans. Am. Fish. Soc., 71 (1941): 149–164.

Clark, Frances N. 1925. The Life History of *Leuresthes tenuis*, an Atherine Fish with Tide Controlled Spawning Habits. Calif. Fish and Game Comm., Fish Bull., 10: 1–51.

—— 1934. Maturity of the California Sardine (*Sardina caerulea*), Determined by Ova Measurements. *Ibid.*, 42: 1–49.

Hickling, C. F., and Rutenberg, E. 1936. The Ovary as an Indicator of the Spawning Period in Fishes. Journ. Mar. Biol. Assn., N. S., 21 (1): 311–318.

Lagler, Karl F., and Hubbs, Clark. 1943. Fall Spawning of the Mud Pickerel, *Esox vermiculatus* LeSueur. Copeia, 2: 131.

Thompson, William F. 1915. A Preliminary Report on the Life-History of the Halibut. British Columbia. Comm. Fish., Rep. for 1914: N76–N99.

THE SALTICIDAE (JUMPING SPIDERS) OF MICHIGAN

ARTHUR M. CHICKERING

THE Salticidae or jumping spiders are mong the most interesting of our native araneids. They construct no true webs, but they build silken retreats, often cunningly devised, for the protection of the young and the concealment of the parent. They commonly make one or more frail cocoons, which they surround by a nest with one or two openings. Here in the north most of our species appear to lay their eggs early in the summer or late in the spring. The spiderlings are often guarded by the mother for a short time until they disperse to care for themselves. Upon the approach of winter the young probably spin, in some protected spot, a shelter within which they hibernate until the following spring.

The salticid spiders are active hunters, gifted with a keen vision, accurate judgment of distance, and skill in jumping. They spin a dragline wherever they go, and by means of it they can usually regain a position after springing away to capture an insect or to escape an enemy.

Although our northern salticid fauna is drab and somber when compared with the extraordinary richness, brilliance, and beauty of the family in the tropics, the jumping spiders of Michigan are worthy of a closer acquaintance than most people, even professional zoologists, have with them. This paper is presented with the hope that it will be a help to all who wish to become more familiar with the family. As was true of former papers in this series (1939, 1940), an attempt has been made to prepare the descriptions, keys, and drawings so that those with little zoological knowledge may use them to identify local salticids without a great deal of trouble. It must be admitted, however, that it is not easy to provide adequate keys for the separation of the salticid genera and even for certain of the species.

Several figures accompanying this paper were drawn from specimens in the collection of the Museum of Comparative Zoölogy at Harvard College and loaned to me by Miss Elizabeth B. Bryant, of the museum staff. Dr. Willis J. Gertsch, American Museum of Natural History, New York City, kindly presented me with a male and a female *Phidippus altanus* Gertsch for comparison with my Michigan specimen. From the female I have made the first published drawing of the epigynum. All other drawings are from specimens in my private collection. The colors noted in the descriptions are always given as they appear in alcohol unless otherwise stated. In agreement with my usual practice I have included only the most essential synonymy.

Acknowledgments are gratefully extended as follows: to Mr. Nathan Banks and Miss Elizabeth B. Bryant, both members of the staff of the Museum of Comparative Zoölogy at Harvard College, for their continued interest in my work and their never-failing willingness to give me the privileges of the museum and its extensive library; and to that indefatigable collector, Mr. R. R. Dreisbach, Midland, Michigan, for the opportunity to examine his collection of spiders from the region around Midland County.

The thirty-nine species of Salticidae now known from Michigan may be listed as follows:

Admestina tibialis (C. L. Koch)
Evarcha hoyi (Peckham)
Fuentes lineata (C. L. Koch)
Habrocestum pulex (Hentz)
Habronattus agilis (Banks)
H. borealis (Banks)
H. decorus (Blackwall)
H. peregrinus (Peckham)
Hentzia mitrata (Hentz)
H. palmarum (Hentz)
Hyctia bina (Hentz)
Maevia vittata (Hentz)
Marpissa undata (De Geer)
Metaphidippus canadensis (Banks)
M. capitatus (Hentz)
M. flavipedes (Peckham)
M. montanus (Emerton)
Neon nellii Peckham
Paraphidippus marginatus (Walck.)
Peckhamia picata (Hentz)
Peckhamia scorpionia (Hentz)
Phidippus altanus Gertsch
P. audax (Hentz)
P. brunneus Emerton
P. clarus Keyserling
P. insigniarius C. L. Koch
P. pius Scheffer
P. purpuratus Keyserling
P. whitmanni Peckham
Phlegra fasciata (Hahn)
Salticus scenicus (Clerck)
Sitticus palustris (Peckham)
Synemosyna formica Hentz
Tutelina elegans (Hentz)
T. formicarius (Emerton)
T. hartii (Emerton)
T. similis (Banks)
Zygoballus bettini Peckham
Z. nervosus (Peckham)

Because the logical separation of the Salticidae into subfamilies is still very much in doubt I am making no attempt at classification within the family. Hence the simple alphabetical arrangement of all species in this paper.

BRIEF DIAGNOSIS OF THE FAMILY SALTICIDAE IN MICHIGAN

The cribellum, calamistrum, and colulus are all lacking. Six spinnerets with the anterior pair approximated, stouter than the others. Eight eyes, all diurnal, arranged in three rows; anterior medians very large; posterior medians, making up the second row, very small. Margins of the fang groove usually pluridentate, fissidentate, or unidentate. Legs prograde, adapted for jumping; without scopulae but with well-developed claw tufts; two claws, more or less dissimilar, pectinate in a single row. Tracheal spiracle close to base of spinnerets, with respiratory system extending into cephalothorax. (Modified from Petrunkevitch, 1939.)

KEY FOR THE SEPARATION OF THE GENERA OF MICHIGAN SALTICIDAE

1. Pedicel and at least posterior border of carapace overlapped by anterior end of abdomen; not distinctly antlike in form 2
1. Pedicel and posterior border of carapace not overlapped by anterior end of abdomen; distinctly antlike in form 19

2. Usually long and flattened; carapace at most only slightly more than one half as tall as wide at widest place 15
2. Usually not much elongated and not flattened; carapace considerably more than one half as tall as wide at widest place 3

3. Males with well-rounded palpal bulbs, having long curved emboli describing wide arcs from points of origin and long curved apophyses accompanying emboli (Figs. 17–21); females with eipigyna having a more or less conspicuous tubular or conical central part and anterior curved margins (Fig. 24) .... *Habronattus*, p. 151
3. Males and females without combination of characters given above for *Habronattus* 4

4. Very small spiders, when mature only 2 to 3 mm. long; eyes occupy more than one half of length of carapace, posterior eyes almost exactly as far apart as ALE; abdomen yellowish, with many dark narrow bars, oblique lines, and mottles .... *Neon*, p. 178
4. Larger spiders, at least more than 3 mm. when mature; eyes always occupy less than one half of length of carapace except in *Zygoballus;* abdomen never barred, lined, and mottled as in Neon 5

5. Eyes occupy more than one half of length of carapace; carapace slopes steeply to posterior border from posterior eyes; first pair of legs much the stoutest; male chelicerae much enlarged .... *Zygoballus*, p. 218
5. Not with combination of characters given for *Zygoballus* 6

6. Retromargin of fang groove with a fissidentate (compound) tooth; four pairs of anterior ventral tibial spines, third pair displaced laterally; two

pairs of anterior ventral metatarsal spines, more conspicuous in females than in males 7

6. Not with combination of characters given above 8

7. First pair of legs considerably enlarged; male palp large and complicated; epigynum with two oval depressions just in front of posterior margin; abdomen with longitudinal light and dark stripes . . . . . . *Fuentes*, p. 147

7. First pair of legs not much enlarged; male palp not unusually large and not complicated; epigynum with a single small central depression far anterior; abdomen not conspicuously striped, but usually with a series of dorsal chevrons . . . . *Maevia*, p. 165

8. Retromargin of fang groove without teeth; two pairs of anterior ventral tibial spines; carapace with a narrow white marginal line and three narrow dorsal white stripes, more conspicuous in males . . *Sitticus*, p. 206

8. Not with combination of characters given above 9

9. First pair of tibiae without ventral spines; cephalothorax long and narrow, with eye space occupying three sevenths of length of carapace; abdomen with a basal white bar and two pairs of dorsolateral bars *Salticus*, p. 204

9. Not with combination of characters given above 10

10. Males with palpal tibial apophysis a stout blunt extension longer than body of tibia and extending beyond middle of bulb; carapace with a dorsolateral white stripe on each side; abdomen nearly encircled by a dorsolateral light margin. Females with epigyna broader than long, with two large oblique depressions near posterolateral border; color variable but tending to be like that of males . . . . *Evarcha*, p. 145

10. Males and females without combinations of characters given above 11

11. Male palpal tibia with a short, blunt, somewhat membranous ventral process; embolus curled nearly into a circle; female with epigynum having two large, nearly circular contiguous areas . . *Habrocestum*, p. 148

11. Males and females without combination of characters given above 12

12. Large robust spiders, up to 10 mm. or more; frequently hairy; cephalothorax only a little longer than wide, tall and convex laterally; thorax falls gently from posterior eyes and then more steeply to posterior border; ocular quadrangle about one third wider than long. . . *Phidippus*, p. 186

12. Relatively smaller spiders; less robust and without combination of characters given above 13

13. Carapace and legs a bright reddish brown, especially in males; middle row of eyes only slightly nearer to first row than to third; AME more than twice the diameter of ALE; PME less than one third of diameter of PLE; male chelicerae more or less produced 14

13. Carapace and legs a more somber color; body scaly and more or less iridescent; AME about twice the diameter of ALE; carapace not markedly convex, sides nearly parallel, usually about two thirds as wide as long . . . . *Tutelina*, p. 210

14. Male chelicerae produced and excavated along fang groove; a fairly robust genus; first legs not much enlarged; carapace quite convex laterally . . . . *Paraphidippus*, p. 179

14. Male chelicerae only mildly developed and not strongly excavated along fang groove; not a robust genus; first legs often enlarged; carapace less convex laterally . . . . *Metaphidippus*, p. 169

15. First femora and tibiae much swollen; first ventral tibial spines entirely lacking; one pair of ventral metatarsal spines; first metatarsi shorter than first tarsi .... *Admestina*, p. 143
15. Without combination of characters given above 16
16. Four pairs of first ventral tibial spines; carapace widest opposite beginning of last third; male palp with greatly swollen bulb, a robust curved tibial retrolateral apophysis, and a deep excavation along retrolateral side of cymbium; epigynum small, with two small depressions just anterior to marginal notch .... *Hyctia*, p. 162
16. Fewer than four pairs of first ventral tibial spines; not with combination of characters given above 17
17. Cephalothorax as long as or longer than abdomen; carapace with two and abdomen with three narrow light longitudinal dorsal stripes *Phlegra*, p. 202
17. Combination of characters not as given above 18
18. Large species, about 9–10 mm. long; a broad dorsal central serrated light abdominal stripe .... *Marpissa*, p. 167
18. Smaller species, about 4–5 mm. long; no such dorsal serrated stripe as in *Marpissa* .... *Hentzia*, p. 159
19. Fang groove with a retromarginal fissidentate tooth; ocular area occupies more than one half of length of carapace; neither carapace nor abdomen deeply constricted .... *Peckhamia*, p. 182
19. Fang groove with several simple retromarginal teeth; ocular area occupies less than one half of length of carapace; both carapace and abdomen deeply constricted in middle .... *Synemosyna*, p. 208

Genus ADMESTINA Peckham, 1888

Small spiders. First leg with femur and tibia much enlarged. Fang groove with a single simple retromarginal tooth. First tibia without ventral spines; first metatarsus with one pair of stout ventral spines and shorter than first tarsus. Carapace low and flat, nearly twice as long as wide. Ocular region nearly twice as wide as long, occupies about one third of length of carapace. PME (small eyes of second row) nearer to ALE than to PLE. Only one species in Michigan.

Admestina tibialis (C. L. Koch)

(Figures 1–3)

Admestina wheeleri Banks, 1892.
Maevia tibialis Emerton, 1909.
A. tibialis Peckham and Peckham, 1909.

Male. — Length about 3.25 mm. Legs, 4132; first pair much thickened (Fig. 2). Anterior ventral tibial spines lacking; anterior

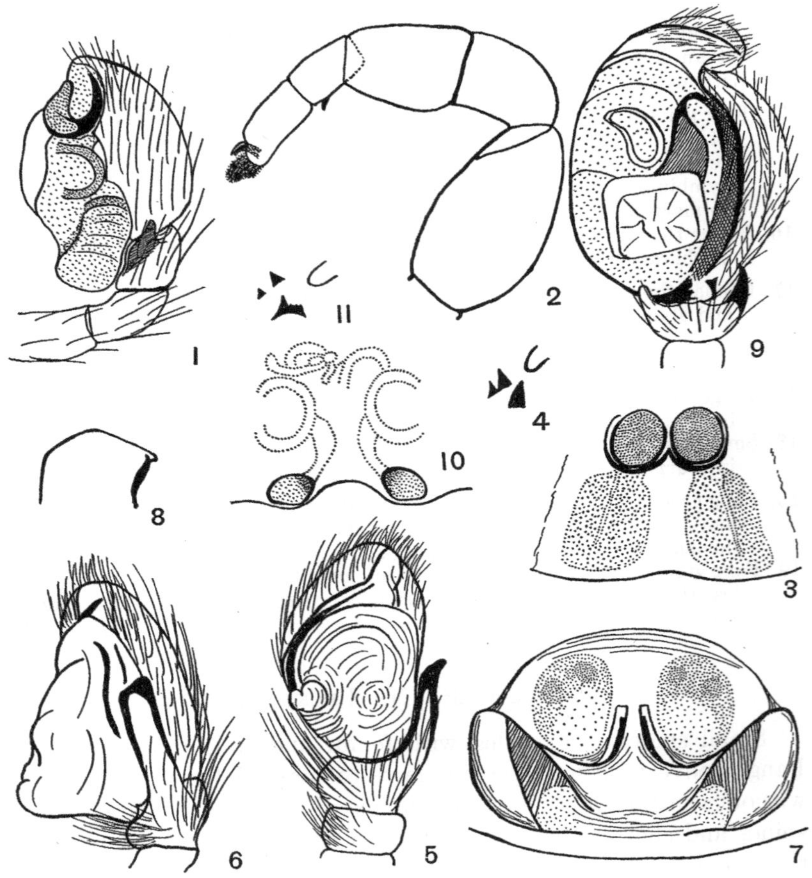

EXTERNAL ANATOMY OF SPIDERS

(Figures 1–11)

1. *Admestina tibialis*, male palpus, retrolateral view
2. *A. tibialis*, first leg of male
3. *A. tibialis*, epigynum
4. *Evarcha hoyi*, left male cheliceral teeth
5. *E. hoyi*, male palpus, ventral view
6. *E. hoyi*, male palpus, retrolateral view
7. *E. hoyi*, epigynum
8. *E. hoyi*, distal end of left maxilla
9. *Fuentes lineata*, male palpus, ventral view
10. *F. lineata*, epigynum
11. *F. lineata*, left male cheliceral teeth

metatarsi with one pair of ventral spines. Cephalothorax longer than wide in ratio of 59 : 37; only slightly swollen opposite second legs; low and flat with very little posterior declivity. Eyes occupy slightly more than one third of length of carapace; PME nearer to ALE than to PLE; width of clypeus nearly equal to one third of diameter of AME, which project considerably. Fang groove with a single small retromarginal tooth and two small promarginal teeth. Palp: femur strongly curved; tibia with a single stout retrolateral apophysis; bulb much swollen, with a short strongly curved embolus at the anterior end; patella slightly longer than short tibia (Fig. 1). Color: cephalothorax generally a dark mahogany brown, almost black in eye region, overlaid with white hairs; first pair of legs and palps brownish, other legs yellowish with dark annulations; abdomen with a broad lanceolate central dorsal brown stripe, with six pairs of more or less conspicuous lateral extensions; laterally the abdomen is yellowish, with many chalky white patches, and ventrally it is whitish, with a broad central dusky stripe and a pair of ventrolateral broken narrow dusky stripes.

Female. — First legs and palps essentially as in male, except that they are lighter, like the other legs. Epigynum with two shallow openings close together at some distance from the posterior margin, which is slightly notched, as shown in Figure 3.

Distribution. — Appears to be widely distributed over New England, southward to North Carolina, westward to Illinois and southwestward to Texas. Taken only once in Michigan, from pines and spruces; at Homer, Calhoun County, August, 1933.

Genus EVARCHA Simon, 1902

PME slightly nearer to ALE than to PLE; diameter of AME about 1.5 times that of ALE; AME separated from one another by about one ninth of their diameter, from ALE by about two ninths; diameter of PLE slightly more than three times that of PME. Anterior metatarsus just a little longer than anterior tarsus, but only about two thirds as long as anterior tibia; third legs a little longer than fourth legs. Carapace moderately high, very gently arched to three quarters of distance from PLE to posterior border and then steeply declivitous. Bulb of male palpus large, cone-shaped near base, blunt distally.

Evarcha hoyi (Peckham)

(Figures 4–8)

Hasarius hoyi Emerton, 1891.
Pellenes hoyi Peckham and Peckham, 1909.

Male. — Length about 5.2 mm. Legs, 1342; first pair only a little the stoutest. Three pairs of anterior ventral tibial spines (basal promarginal spine somewhat smaller and apparently overlooked by some authors); two pairs of anterior ventral metatarsal spines. Cephalothorax longer than wide in ratio of 19 : 14; gently rounded along lateral sides; rather high, with carapace a little more than two thirds as high as wide; gently rising from AME to PLE, then gently sloping to about one fifth from posterior border, then steeply declivitous to posterior border. Eyes occupy slightly less than one half of length of carapace; PME slightly closer to ALE than to PLE; anterior row a little longer than posterior row; PLE occupy about eleven thirteenths of width of carapace at that level. Maxilla with slightly hooked tooth on outer distal corner (Fig. 8). Fang groove with a single stout retromarginal tooth, two promarginal teeth (Fig. 4). Palp: body of tibia short, with a single long stout blunt somewhat flattened apophysis (Figs. 5–6) reaching beyond middle of bulb, which is much swollen into a conical structure invaginated a little at the apex; embolus a curved tube extending more than one fourth of distance around periphery of bulb. Abdomen: about as long as cephalothorax and pointed posteriorly. Color: extremely variable; cephalothorax usually with a dark patch surrounding ocular area, overlaid with considerable white hair; a dorsolateral light stripe begins at the clypeus on each side and extends nearly or quite to posterior border; beneath the stripes carapace is dark to ventral margin. Abdomen dark brown basally and laterally; a narrower yellowish region dorsolaterally almost encircles a darker dorsal area, which bears near posterior end a series of incomplete light and dark chevrons. First femora entirely brownish dorsally, remaining three pairs dark brown over distal halves only; other segments more or less mixed brown and yellowish.

Female. — Length about 6.2 mm. Legs, 3 = 412. Other characters essentially as in male except that color markings are more subdued. Epigynum with two rather large oblique depressions far apart near lateral margins (Fig. 7); no notch in posterior border.

Distribution. — Widely distributed over the eastern part of the United States; also known from southeastern Canada; reported from Brazil. One of our common salticids; taken in many localities in both Northern and Southern peninsulas of Michigan.

GENUS FUENTES PECKHAM, 1894

Carapace rather low, quite flat, with lateral sides only a little swollen, cephalic part very slightly inclined, thoracic part gently declined in first half, then steeply declivitous to posterior border. Ocular region occupies a little less than half the length of carapace, about one fourth wider than long, very slightly wider behind than in front, anterior row of eyes recurved, all close together. Diameter of AME greater than diameter of ALE in ratio of 7 : 4; PME a little nearer to PLE than to ALE; third row occupies about nine tenths of width of carapace at that level. Chelicerae not strongly developed, parallel, vertical with a single broad fissidentate retromarginal tooth on fang groove. First pair of tibiae with four pairs of ventral spines not all in regular order.

Fuentes lineata (C. L. Koch)

(Figures 9–11)

Maevia lineata C. L. Koch.
Icius lineatus Peckham, 1888.
Menemerus lineatus Emerton, 1891.
Onondaga lineata Peckham, 1909.

Male. — Length about 3.3 mm. Legs, 1423; first pair considerably enlarged, slightly fringed. Spines: first pair of tibiae with four pairs of ventral spines, the third pair distinctly offset laterally and all four along promargin displaced so far laterally as to make them appear to some as lateral instead of ventral in position; second pair of tibiae with one pair of ventral terminal spines and two more along retromargin; first two pairs of metatarsi each with two pairs of ventral spines. Outer distal corners of maxillae with a distinct hook. Eyes generic in position and relationships. Fang groove generic in respect to teeth (Fig. 11). Palp: tarsus very large, complicated; tibial retrolateral apophysis a short hooked spur; embolus starts near middle from expanded base, turns retrolaterally and continues to base, turns again and extends along whole prolateral side of tarsus to distal end nearly in a semicircle (Fig. 9). Color: first femora yellowish above and below but reddish brown prolaterally and retrolaterally; first tibiae entirely reddish brown; first patellae darkened proximally but otherwise yellowish; first metatarsi darkened somewhat; first tarsi yellowish; all other legs yellowish-spotted and ringed with reddish brown. Eyes on black areas. Whole dorsal region of carapace except eyes amber color; two dark spots just behind AME showing through chitinous covering such as is often seen in other members of this division; a narrow broken black streak from longitudinal thoracic groove backward to near posterior margin and forward a short distance; a narrow black marginal line and two broad dorsolateral dark brown irregular stripes from posterior eyes to posterior margin. Abdomen yellowish, with four narrow broken dark brown dorsal lines; lateral sides spotted with brown.

Female. — Length about 4.2 mm. Legs, 4123. Spines like those in male except that second tibiae have no terminal ventrals. Color like that of male except that it is much darker; abdomen has a light brown central stripe with yellow margins; on each side of this central stripe is a broad darker brown stripe. Otherwise essentially like the male. Epigynum: a broad but shallow notch in posterior border; at each side of the shallow notch is an oval depression; anteriorly a series of coils of an internal tube (Fig. 10).

Distribution. — Known from Labrador, New England, a few scattered localities along the Atlantic coast to Florida, New York, Michigan, Wisconsin, and westward to Kansas. Collected in Michigan only once: Waterloo Recreational Area, Jackson County, May, 1941.

Genus HABROCESTUM Simon, 1876

The original definition of the genus must be modified for American species. This description is intended only as a working basis for a more critical definition. Cephalothorax high and convex dorsally, highest just in front of PLE; lateral sides nearly parallel; cephalic part inclined forward, thoracic part descends gently from PLE and then very steeply through the last fifth. Ocular region a little wider than long, occupies about two fifths of length of carapace, a little wider in front than behind. AME less than twice as large as ALE; PME nearly equidistant between ALE and PLE. PLE about three times as large as PME.

Habrocestum pulex (Hentz)

(Figures 12–16)

Attus pulex Hentz, 1846.
Saitis pulex Emerton, 1891.
H. pulex Simon, 1901.

Male. — Length about 4.0 mm. Legs, 3 = 42 = 1; first and second pairs a little less than three fourths as long as third and fourth pairs. Anterior ventral tibial spines probably not as recorded by the Peckhams (1909); two along retromargin (none terminal) and only one along promargin (terminal); two pairs of anterior ventral metatarsal spines. Cephalothorax longer than wide in ratio of about 3 : 2; lateral sides nearly parallel, with only little widening opposite second coxae; of moderate height; rises considerably from AME to PLE, then descends very slightly to steep posterior declivity, which is about one fifth from posterior margin. Eyes occupy about three sevenths of length of carapace; PME a little closer to PLE than to ALE; anterior row a little longer than posterior row; PLE occupy about five sixths of width of carapace at that level. Maxillae with a small tooth on outer distal corner. Fang groove with a single fairly stout retromarginal tooth and a compound promarginal tooth (Fig. 14). Palp: body of tibia short with a short blunt somewhat membranous ventral process and a fairly stout well-chitinized retrolateral apophysis about as long as body (Figs. 15–16); bulb swollen retrolaterally and with a distal embolus curled almost into a circle ending in a stout spine. Abdomen: shorter by about one seventh than the cephalothorax, thirteen fifteenths as wide in the middle as long. Color: somewhat variable; a dark brown quadrilateral area, which is scalloped behind, encloses the eyes, and is somewhat overlaid with whitish hairs; remainder of carapace yellowish, with dark radial streaks. Abdomen with dark brown venter and lateral sides, both dotted with yellow; dorsally with a median central light stripe reaching back to a light transverse zigzag bar. The Peckhams' Figure 11, Plate XLII (1909), is a good drawing of the color pattern. Legs: coxae, femora, and patellae light dorsally but dark laterally and on the last two pairs of legs also ventrally; last three segments of first pair of legs all dark; on second pair all but tarsi, which are lighter; other segments mixed brown and yellowish. All segments of palps except tibiae and tarsi light yellowish.

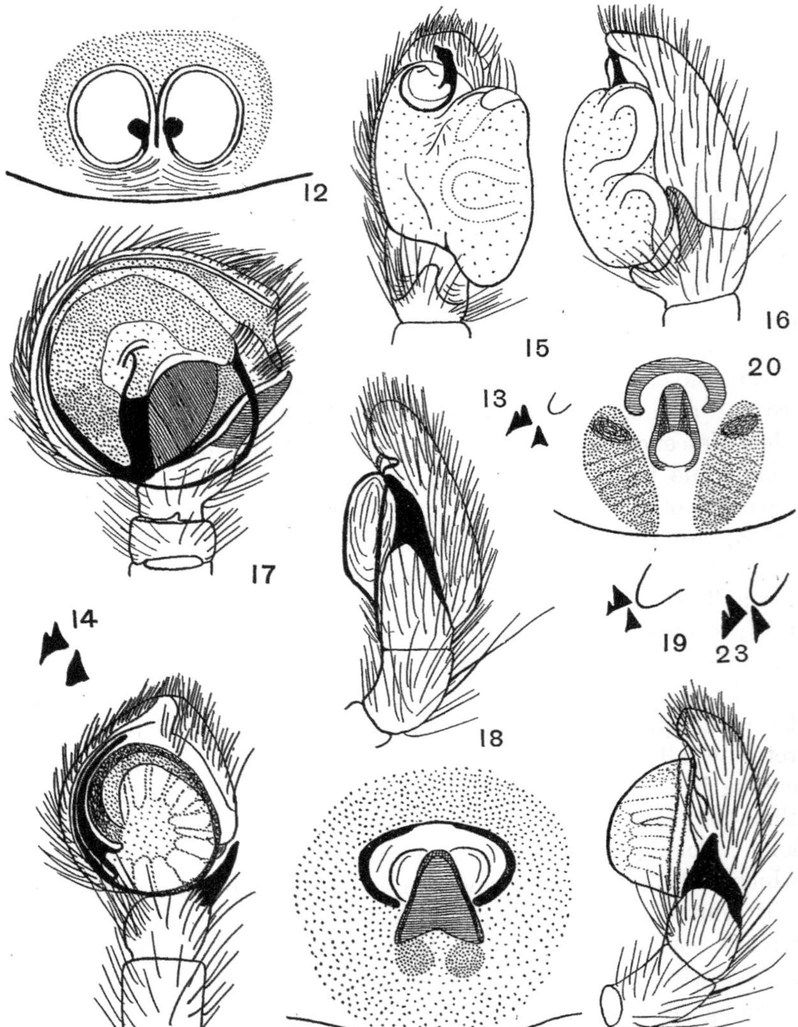

EXTERNAL ANATOMY OF SPIDERS

(For explanation of figures see facing page.)

Female. — Length about 5.25 mm. Legs, 4312. Anterior tibiae have only two retromarginal ventral spines, with none terminal and one promarginal ventral, which is terminal. Color markings similar to those of male except as follows: dark ocular area less definite; a somewhat vague light triangular area behind eye region and lateral sides of cephalothorax are darker; light and dark abdominal areas less definite and more subdued. Epigynum with a pair of somewhat oval contiguous areas each bounded by a definite margin (Fig. 12); no distinct notch in posterior border.

Distribution. — Southeastern Canada; widely distributed over most of the United States east of the Mississippi River and westward into Kansas and Iowa, and probably into surrounding states. Known from both peninsulas of Michigan; appears to be common, especially in the southern part of the state.

Genus HABRONATTUS F. Cambridge, 1901

Also known under the names *Attus, Ephippus, Habrocestum, Hasarius, Evarcha,* and, most commonly, *Pellenes.* Carapace: rather high; considerably swollen laterally and widest opposite second coxae; noticeably wider below than above; only a little more than two thirds as wide as long. Ocular quadrangle about two thirds as long as wide. Eyes occupy about two fifths of length of carapace. Middle row usually midway between first and third rows. Males frequently have some modification of first and third legs which helps to distinguish them. Chelicerae normal in both sexes. Male palps always with expanded, rounded tarsus, with a long slender embolus

---

EXPLANATION OF FIGURES 12–24

12. *Habrocestum pulex,* epigynum
13. *H. pulex,* left female cheliceral teeth
14. *H. pulex,* left male cheliceral teeth
15. *H. pulex,* male palpus, ventral view
16. *H. pulex,* male palpus, retrolateral view
17. *Habronattus agilis,* male palpus, ventral view
18. *H. agilis,* male palpus, retrolateral view
19. *H. agilis,* left male cheliceral teeth
20. *H. agilis,* epigynum
21. *Habronattus borealis,* male palpus, ventral view
22. *H. borealis,* male palpus, retrolateral view
23. *H. borealis,* left male cheliceral teeth
24. *H. borealis,* epigynum

curled around much of the bulbar periphery accompanied by a slender guard or apophysis (Fig. 17). Females are difficult to determine, but they all have epigyna constructed on a standard pattern: a simple curved posterior border with no notch; a somewhat truncated central conical tube more or less surrounded by a well-defined curved margin (Fig. 20).

KEY TO THE SPECIES OF *HABRONATTUS*

MALES

1. Distal end of third patella provided with a stout spur (Fig. 28); tarsal apophysis arising at base of bulb, not at all lateral (Fig. 25) *H. peregrinus*, p. 158
1. No spur at distal end of third patella; tarsal apophysis arising either from center of bulb or from a somewhat lateral position 2

2. Tarsal apophysis arising near center of bulb (Fig.17); embolus arising near tip of tibial apophysis and making a nearly complete circle around bulb *H. agilis*, p. 153
2. Tarsal apophysis arising from periphery of bulb; embolus arising far from tip of tibial apophysis and passing no more than halfway around bulb . . . 3

3. Abdomen a bright iridescent red; first legs without decorative fringes; embolus passes around fully one half of circumference of bulb (Fig. 30) *H. decorus*, p. 155
3. No red on abdomen; first legs with a small and poorly developed fringe on prolateral surface of first femora; embolus passes only about one third of way around bulb (Fig. 21) . . . . *H. borealis*, p. 154

FEMALES

1. Tubelike center of epigynum small and only slightly narrower at anterior end (Fig. 29) . . . . *H. peregrinus*, p. 158
1. Tubelike center of epigynum larger than in *H. peregrinus* and very much narrower at anterior end 2

2. Tubelike center of epigynum with a circular opening; a complete broad anterior rim to epigynum; two large spermathecae obliquely placed and bounding this organ laterally (Fig. 20) . . . . *H. agilis*, p. 153
2. Tubelike center not with a circular opening; epigynum not with a complete broad anterior rim; spermathecae either small or, if large, not oblique in position 3

3. Anterior rim of epigynum definitely divided into two somewhat crescentic ridges; spermathecae large and lateral to tubelike center (Fig. 33) *H. decorus*, p. 155
3. Anterior rim of epigynum not completely divided as in *H. decorus;* spermathecae small and immediately behind central tubelike center (Fig. 24) *H. borealis*, p. 154

Habronattus agilis (Banks)

(Figures 17–20)

Habrocestum agilis Banks, 1893.
H. auratum Emerton, 1902.
Pellenes agilis Peckham and Peckham, 1909.

Male. — Length a little more than 5 mm. Legs, 3412 (the Peckhams, 1909, 3142); first pair only moderately enlarged; with femora, patellae, and tibiae usually moderately fringed (sometimes heavily so, as drawn by the Peckhams, 1909). Anterior tibiae with three ventral retromarginal spines but only one ventral promarginal terminal; anterior metatarsi with two pairs of ventral spines. Cephalothorax quite robust as compared with rather small abdomen. Fang groove with a single retromarginal tooth and two promarginal teeth, the inner of which is much the smaller (Fig. 19). Palp: body of tibia short and broad; retrolaterally drawn out into a long stout somewhat sinuous apophysis (Fig. 18); numerous white clavate scalelike hairs on tarsus, tibia, and patella; prolateral fringe of clavate hairs and dorsal fringe of simple long curved hairs on femur; tarsus large-rounded, somewhat flattened; embolus arises close to apex of tibial apophysis and makes a complete circle ending just distal to its origin; tarsal apophysis arises near middle of bulb, extends proximally to basal end, then turns sharply to prolateral side and curves around the periphery of the bulb about one third of its circumference (Fig. 17). Abdomen: about seven eighths as long as the cephalothorax; about five sevenths as wide as long (including anal tubercle); ovoid. Color: carapace dark brown with well-developed covering of mixed black and light brown hair; between and below ALE and PLE are several long slender black bristles not forming definite tufts; a band of somewhat shorter, stiffer bristles lies above AME; beginning just posterior to ALE are two white stripes about as wide as these eyes and extending backward just above the PLE, converging a little toward the posterior border, which they do not quite reach; the clypeus is white, with a white band reaching back on each side along the margin in a broad stripe to posterolateral corner leaving posterior border dark brown; abdomen with a central dorsal whitish stripe, which does not reach to anal tubercle but anteriorly connects with a basal band that extends dorsolaterally back to base of spinnerets. White stripes composed of clavate hairs. At one place the marginal cephalothoracic stripe is interrupted, and a tuft of long simple white hairs projects over the interval between the first and second coxae.

Female. — Length about 5.75 mm. Legs, 3412. Anterior ventral tibial spines same as in male except that distal retromarginal spine appears to be lacking. Color: in general like that of male with following exceptions: cephalothorax not so dark; a short median dorsal stripe reaching from between AME nearly to posterior row of eyes (mentioned by the Peckhams in males but it seems to be missing in this sex in my specimens). Clypeus is white in center with a dark patch on each side. White marginal cephalothoracic stripe turns upward anteriorly and disappears just lateral to AME. Abdomen also somewhat lighter; whitish stripes more broken. Epigynum: follows usual pattern, with a recurved crescentic anterior margin, tubular center and two large obliquely placed spermathecae (Fig. 20).

Distribution. — Appears to be widely distributed over the eastern part of the United States, westward to Kansas, southwestward to Texas and New Mexico. Collected from a few localities in the Southern Peninsula of Michigan.

Habronattus borealis (Banks)

(Figures 21–24)

Habrocestum borealis Banks, 1895.
Pellenes borealis Peckham and Peckham, 1909.

Male. — Length about 5.5 mm. Legs, 3412; first pair only moderately enlarged; no decorative fringes as in *H. agilis* except a small prolateral one poorly developed on each anterior femur. Anterior tibiae with three retromarginal ventral spines and one (terminal) promarginal spine. Fang groove with a single fairly stout retromarginal tooth and two promarginal teeth, of which the inner is somewhat the smaller (Fig. 23). Palp: tibia short but fairly stout, with a broad apophysis about as long as tibial body; numerous white clavate hairs all over the femur and dorsally on patella and tibia; with no special fringes; tarsus large, rounded; bulb much thickened; embolus arises near bulbar base on prolateral side and passes about two fifths of distance around the bulb accompanied by a fairly stout apophysis, which arises close to base of embolus and curves about one third of distance around bulb (Figs. 21–22). Color (only a single mature male from which to make observations): carapace generally very dark brown, with many whitish clavate hairs, which give a grayish effect (since whitish hairs are easily brushed off in handling, different color patterns are easily produced); many black bristles in eye region not tufted together; first coxae, first trochanters, and mouth parts all dark brown; remaining coxae and trochanters progressively lighter toward posterior end; elsewhere legs yellowish; palpal segments except tarsus dark brown beneath white hairs; dorsally abdomen dark brown, with a white central spot about two thirds of distance from base to spinnerets, and a narrow recurved band which connects with a broad whitish margin encircling whole abdomen.

Female. — Length about 6.0 mm. Legs, 3412. Color: carapace fundamentally dark brown like that of male, but a dirty white hairy covering gives it a grayish appearance in unrubbed specimens; abdomen still more completely covered with yellowish white hairs, giving a glossy yellowish color; in rubbed specimens abdominal color pattern resembles that of male; legs and palps yellowish below, darker above. Epigynum: an ovate marginal rim almost encloses the central tubular structure, which is nearly as broad at posterior end as it is long (Fig. 24). Otherwise essentially similar to male.

Distribution. — A northern species known from New England, New York, Michigan, Wisconsin, and probably other northern states and Canada. My Michigan specimens have all come from three counties in the northern part of the Southern Peninsula.

Habronattus decorus (Blackwall)

(Figures 25–28)

Salticus decorus Blackwall, 1846.
Attus roseus Hentz, 1846.
A. splendens Peckham, 1883.
Pellenes roseus Peckham and Peckham, 1909.
P. splendens Peckham and Peckham, 1909.

Note. — The synonymy is in some doubt, but I cannot improve it until more attention is given to the genus.

Male. — Length about 5.0 mm. Legs, 3412; first pair only a little stouter than others; without fringes. Spines: anterior tibiae with three pairs of ventral spines, but the first and second on the promargin so far prolateral that they could be regarded as prolateral instead of ventral in position; second tibiae with a pair of terminal ventral spines, but only two others along the retromargin, with those on the promargin entirely lacking. Cephalothorax slightly longer than abdomen and fairly stout. Eyes occupy about three eighths

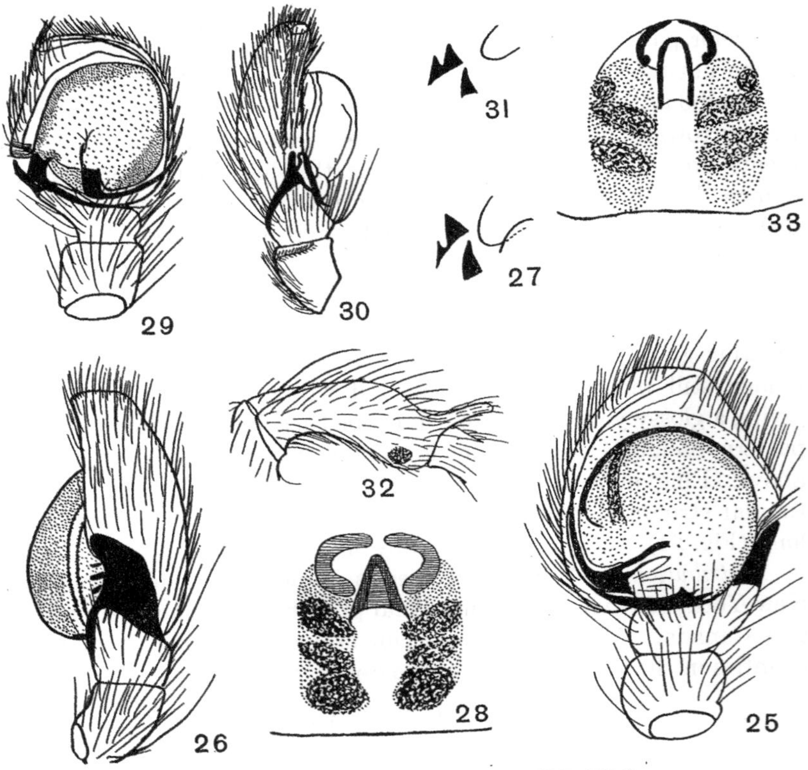

EXTERNAL ANATOMY OF SPIDERS

(Figures 25–33)

25. *Habronattus decorus*, male palpus, ventral view
26. *H. decorus*, male palpus, retrolateral view
27. *H. decorus*, left male cheliceral teeth
28. *H. decorus*, epigynum
29. *Habronattus peregrinus*, male palpus, ventral view
30. *H. peregrinus*, male palpus, retrolateral view
31. *H. peregrinus*, left male cheliceral teeth
32. *H. peregrinus*, third patella of male
33. *H. peregrinus*, epigynum

of length of carapace. Fang groove with a single fairly robust retromarginal tooth and two promarginal teeth close together, with inner one small and slender (Fig. 27). Palp: body of tibia short and broad, drawn out retrolaterally into a stout apophysis, which reaches a little more than two fifths of length of tarsus (Figs. 25–26); embolus originates at base of tarsus and without elbow extends a little more than halfway around circumference of bulb; tarsal apophysis has a sharp elbow, not quite a right angle, and extends a little more than one fourth of circumference of bulb, which is moderately swollen. Color: most brilliant of the Michigan habronattids; whole ocular area almost black, overlaid with a heavy coating of white scales, just behind this area is a light yellowish band somewhat curved posteriorly; elsewhere carapace is brownish, with many fine irregular lines all overlaid with many white scales; a fine black marginal line along ventral margin of carapace; whole clypeus covered by snowy-white hairs and scales. Anterior femora with a black prolateral stripe. A dark patch at base of abdomen dorsally; just behind this a light transverse band; nearly whole dorsum covered by a dark patch overlaid with azure iridescent scales, also with a good deal of reddish pigment; laterally reddish; venter yellowish with no stripes. Peckham's description states that the whole carapace and clypeus are covered with blue scales. Examination of specimens in the collection of the Museum of Comparative Zoölogy, Harvard College, indicates considerable variation in coloration among different individuals.

Female. — Length about 6.4 mm. Generally like the male except in secondary sexual characters and color. Color: differs from male essentially as follows: carapace with fewer white hairs; no prolateral stripes on anterior femora; azure-blue scales, iridescence, and bright red pigment seem to be entirely lacking on the female abdomen. Instead there is a dark-brown basal band, which continues on each side to become narrow ventrolateral stripes; some distance posteriorly there is another dorsal band, which continues laterally on each side as a stripe almost to the spinnerets; also a central light-colored broken sagittate mark and a broken dark-brown stripe on each side of it; background of abdomen is light yellow. Epigynum: follows usual pattern with anterior crescentic margin divided into two smaller lateral crescents facing each other; tubular central part pointed anteriorly and broad posteriorly (Fig. 28).

Distribution. — Reported from southeastern Canada, along the Atlantic seacoast to Florida and through the northern states from New England to Washington and Oregon. Collected only twice in Michigan: Calhoun County, May, 1935; Otsego County, July, 1938.

Habronattus peregrinus (Peckham)

(Figures 29–33)

Attus peregrinus Peckham, 1883.
Habrocestum peregrinum Emerton, 1891.
Pellenes peregrinus Peckham and Peckham, 1909.

Male. — Length about 5.0 mm. Legs, 3412; first pair somewhat fringed. First and second tibiae with a pair of ventral terminal spines and two more along retromargin, with none to match along promargin. First two pairs of metatarsi with two pairs of ventral spines each. Third patellae with a single stout distal spur (Fig. 32). Cephalothorax rather robust, about as long as abdomen, longer than wide in ratio of 41 : 30. Eyes occupy about two fifths of length of carapace. Fang groove with a single slender retromarginal tooth and a pair of promarginal teeth close together, with inner somewhat the smaller (Fig. 31). Palp: body of tibia short and broad; retrolaterally drawn out to a stout, terminally slender apophysis (Fig. 30), which reaches only about one fourth of length of tarsus; tarsus large, with swollen bulb; embolus rises near apex of tibial apophysis and curves about three quarters around bulb, with right angle near its base; tarsal apophysis arises near proximal end, makes a right-angle turn at base, and extends around less than one fourth of circumference of bulb (Fig. 29). Abdomen: about as long as cephalothorax; about three quarters as wide as long; broad at base, widest about two fifths from base and tapered to a point posteriorly. Color: carapace dark brown with a narrow white marginal band of scales (easily rubbed off), which continues as a narrow band across clypeus; ocular area darker, with many nearly white scales (in undisturbed condition); behind this darker area is a lighter one with two whitish bands diverging to posterolateral corners of carapace; sternum dusty yellow. Viewed dorsally, abdomen shows three nearly white longitudinal stripes enclosing two dark stripes; ventrally abdomen has three narrow broken black stripes separated and bounded by narrow yellowish stripes.

Female. — Length about 5.75 mm. Spines essentially like those in male. Color: like that of male, with all contrasts much subdued. Epigynum: the usual general pattern; recurved crescentic anterior margin relatively smaller and reduced in middle; tubular central piece only slightly broader posteriorly than anteriorly (Fig. 33).

Distribution. — Appears to be widely distributed over the eastern half of the United States. Collected in Michigan only from the region around Marquette in the Northern Peninsula.

GENUS HENTZIA MARX, 1883

Usually known as *Wala*; also formerly widely known as *Anoka*. Cephalothorax rather flat, of moderate height, only slightly longer than wide, widest just behind ALE. Ocular region one third wider than long, a little wider behind than in front, occupies two fifths of length of carapace. First legs in male much the longest. Abdomen long and slender. PME about halfway between ALE and PLE; third row somewhat narrower than carapace at that level. First coxae separated by more than width of labium, which is twice as long as wide. Two species occur in our Michigan fauna.

KEY TO THE SPECIES OF *HENTZIA*

MALES

1. First pair of legs reddish brown; chelicerae reddish brown and more or less porrect and flattened; abdomen with a broad central reddish stripe without spots . . . . *H. palmarum*, p. 161
1. First pair of legs yellowish white; chelicerae yellowish white, vertical and not flattened; abdomen with several central dark spots *H. mitrata*, p. 159

FEMALES

1. Abdomen yellowish white with a central reddish light-brown, more or less broken and sometimes obscure stripe composed of spots and dots; no dorsal spines on patellae . . . . *H. palmarum*, p. 161
1. Abdomen yellowish white with three pairs of reddish-brown dorsal spots more or less coalesced; patellae with single dorsal spines *H. mitrata*, p. 159

Hentzia mitrata (Hentz)

(Figures 34–37)

Attus mitratus Hentz, 1845.
Anoka mitrata Peckham, 1894.
Wala mitrata Peckham and Peckham, 1909.

Male. — Length about 4.5 mm. Legs, 1423; first pair much the longest, fringed with many white hairs; other legs nearly equal to one another in length; anterior tibiae longer than anterior patellae. Anterior tibiae with three pairs of ventral spines; second tibiae with a pair of terminal ventral spines but with only one other spine, a retromarginal one near the base. First and second metatarsi with two pairs of ventral spines. Carapace longer than wide in ratio of

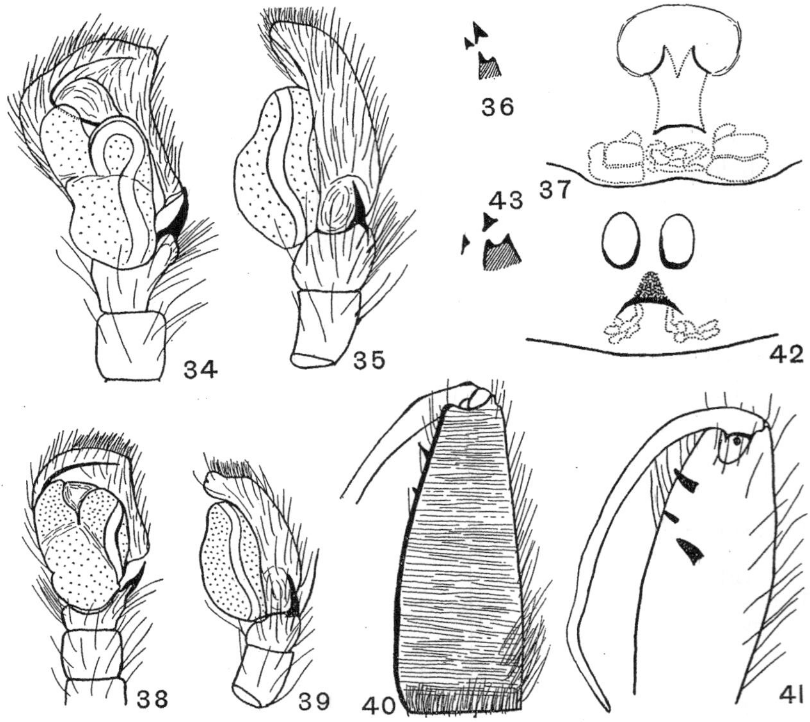

EXTERNAL ANATOMY OF SPIDERS

(Figures 34–43)

34. *Hentzia mitrata,* male palpus, ventral view
35. *H. mitrata,* male palpus, retrolateral view
36. *H. mitrata,* left male cheliceral teeth
37. *H. mitrata,* epigynum
38. *Hentzia palmarum,* male palpus, ventral view
39. *H. palmarum,* male palpus, retrolateral view
40. *H. palmarum,* male chelicera, dorsal view
41. *H. palmarum,* left male chelicera and teeth
42. *H. palmarum,* epigynum
43. *H. palmarum,* left female cheliceral teeth

about 30 : 23, somewhat flattened. Ocular region wider than long in ratio of about 35 : 26, middle eyes closer to ALE than to PLE in ratio of 7 : 10. Eyes occupy about two fifths of length of carapace. Chelicerae vertical and of moderate size; fang groove with two slender promarginal teeth, inner one very small, and a single fissidentate retromarginal tooth (extent of fissidentation variable among different individuals [Fig. 36]). Much long white hair on palps and chelicerae. Palp: patella about equal in length to that of tibia; tarsus much broader anteriorly than at base; with a conspicuous twisted tube in bulb (Figs. 34–35). Abdomen: longer than wide in ratio of about 5 : 2, only slightly flattened, widest about two fifths from base. Color: all appendages and sternum except lip a yellowish white; carapace a reddish brown throughout ocular area and back to posterior margin; lateral sides and clypeus yellowish white; abdomen reddish brown dorsally, with three pairs of brown spots more or less coalesced, while laterally yellowish and reddish narrow stripes alternate, and ventrally there are four narrow light-colored broken stripes surrounded by light reddish areas.

Female. — Length about 5.0 mm. Spines essentially like those of male except that female has a single dorsal spine on each patella whereas the male has none. Other features including color essentially like those of the male. Epigynum: a small opening indicated by a recurved crescentic margin; an inconspicuous curved anterior border, as shown in Figure 37.

Distribution. — Apparently known from southeastern Canada, nearly the whole eastern seaboard and westward through the northern states to Nebraska. Collected in Michigan only at three localities, which are in Calhoun County.

Hentzia palmarum (Hentz)

(Figures 38–43)

Epiblemum palmarum Hentz, 1846.
Anoka palmarum Peckham, 1894.
Icius palmarum Emerton, 1902.
Wala palmarum Peckham and Peckham, 1909.

Male. — Length about 4.2 mm. Legs, 1423; first pair much the longest, scantily fringed with black hairs ventrally on femora and patellae; other legs all of about the same length; with no long white hair on palps and chelicerae. Anterior tibiae with three pairs of ventral spines; first and second pairs of metatarsi with two pairs of ventral spines. Second tibiae apparently with a pair of terminal ventral spines and only one other, on retromargin near the middle, but one or more of these may be lacking. Carapace of moderate height, considerably flattened dorsally, well rounded laterally, longer than wide in ratio of about 73 : 56. Eye region wider than long in ratio of 23 : 15, middle eyes slightly closer to ALE than to PLE. Eyes occupy about two fifths of length of carapace. Chelicerae more or less porrect, developed, and flattened dorsally. Fang groove with a pair of slender promarginal teeth, the inner of which is very small, and a single retromarginal tooth, which sometimes shows signs of fissidentation (Figs. 40–41). Fang long, slender, and sinuous. Palp: patella somewhat longer than body of tibia without the apophysis (Figs. 38–39); not broader distally than farther back; tube of bulb not so sinuous as in *H. mitrata* and less conspicuous; apophysis long and slender and somewhat curved ventrally. Color: first pair of legs, chelicerae, lip, sternum, and all but the cymbium in the palp a reddish brown; palpal cymbium and all legs except first pair a yellowish white; carapace entirely reddish brown, except for a lateral white stripe on each side, which continues across clypeus to unite on opposite side; a broad reddish central abdominal stripe from base to anal tubercle; on each side of reddish stripe is a whitish stripe with reddish dots; elsewhere abdomen is light reddish.

Female. — First pair of legs not so much elongated as in the male; much the stoutest. Spines essentially as in male. Chelicerae not porrect and not developed as in the male. Fang groove with two promarginal teeth, inner one very small; retromarginal tooth distinctly fissidentate (Fig. 43). Epigynum: two small depressions less than a diameter of one of them apart; a recurved crescentic margin bounding the central opening and a coniform tube leading inward (Fig. 42). Color: differs from that of male considerably; all legs and palpi light yellowish; chelicerae, lip, maxillae, and sternum same as in male; carapace lighter in ocular area and without lateral white stripes; otherwise as in male. Abdomen with a central light reddish stripe much spotted with whitish; dorsolateral whitish stripes much dotted with reddish.

Distribution. — Reported from entire eastern seaboard of the United States; among the northern states westward to Oklahoma and Nebraska. Collected in Michigan only a few times, in Calhoun and Jackson counties.

Genus HYCTIA Simon, 1876

Long slender body, with first pair of legs much the stoutest and longest. Cephalothorax long, low, flat, and narrow, nearly twice as long as wide; lateral sides vertical, nearly parallel, widening just a little behind PLE. Quadrangle of eyes one fourth wider than long, equally wide in front and behind, occupies from one third to two fifths of length of carapace. PME about midway between anterior and posterior rows. Posterior eyes nearly as wide as carapace at that level. Labium longer than wide. Chelicerae vertical and each with a single retromarginal tooth. Sternum oval and contracted in front, where anterior coxae are separated by less than width of labium. Only one species so far taken in Michigan, but *H. pikei* Peckham should be here, it would seem; collected from five localities in Calhoun and Branch counties.

Hyctia bina Hentz

(Figures 44–49)

Attus binus Hentz, 1845.
Menemerus binus Emerton, 1891.
H. bina Peckham and Peckham, 1909.

Male. — Length about 5.75 mm. Legs, 1432; without fringes; other features generic. Spines: anterior pair of tibiae with four pairs of ventral spines, third pair offset laterally; second pair of tibiae ordinarily with a pair of terminal ventral spines and two others along retromargin, but none to match them along promargin. First and second metatarsi each with two pairs of ventral spines. Cephalothorax widest some distance posterior to third row of eyes; about two thirds as wide as long. Eyes occupy almost one half of length of carapace. Ocular area about four fifths as long as wide, almost as wide behind as in front. Third row of eyes occupy slightly more than five sixths of width of carapace. Fang groove with a single stout retromarginal tooth and two smaller promarginal teeth, of which the inner is very small (Figs. 47, 49). It is commonly stated that two forms of the male exist, but only one has been found in Michigan. Palp: patella about equal in length to that of tibia, viewed dorsally; bulb much swollen; a deep curved retrolateral excavation along whole length of cymbium; a stout apophysis turned toward cymbial excavation; embolus arises near base and extends prolaterally more than halfway around margin of bulb (Figs. 44–46). Color: first pair of legs much the darkest; all femora with three dorsal stripes, a median dark-brown one between two much lighter ones; anterior tibiae similarly striped; carapace generally a rich mahogany brown,

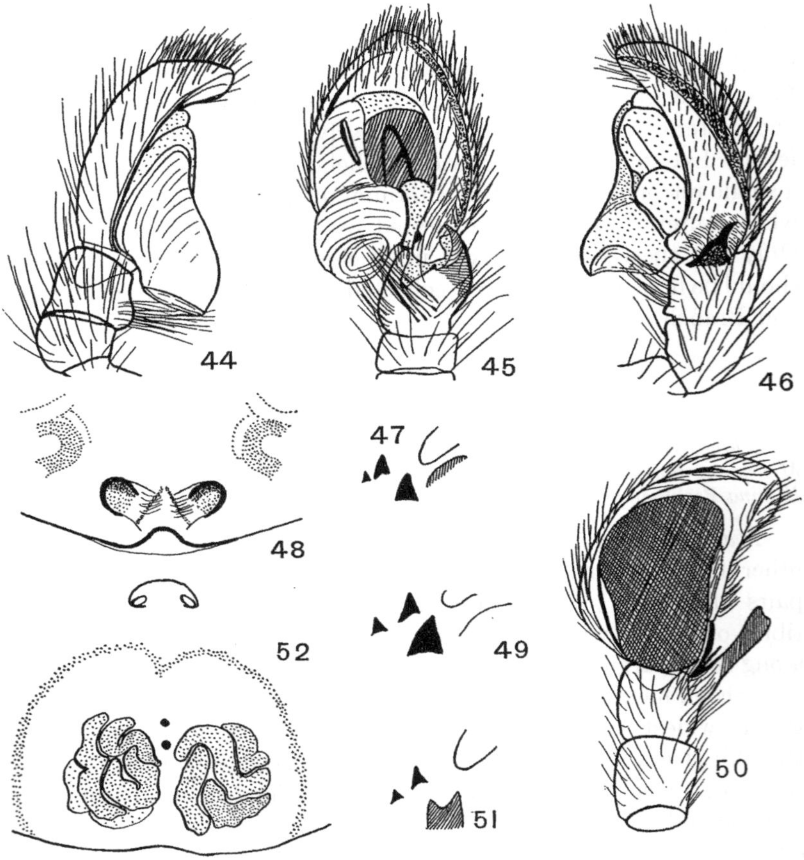

EXTERNAL ANATOMY OF SPIDERS

(Figures 44–52)

44. *Hyctia bina*, male palpus, prolateral view
45. *H. bina*, male palpus, ventral view
46. *H. bina*, male palpus, retrolateral view
47. *H. bina*, left male cheliceral teeth
48. *H. bina*, epigynum
49. *H. bina*, left female cheliceral teeth
50. *Maevia vittata*, male palpus, ventral view
51. *M. vittata*, left male cheliceral teeth
52. *M. vittata*, epigynum

with irregular black pencil marks, ocular area darker; a narrow black marginal line, which extends across the very low clypeus, also a narrow white line just dorsal to the black one, but this is complete only at posterior lateral angles; a spot of white hairs above and between AME, more or less white hair dorsal to and behind ALE; a spot of white on each side just posterior to PLE and another on each side just at beginning of posterior declivity. Abdomen: a dark-brown band at base; a band of white just dorsal to the dark band; behind these are three pairs of whitish spots bounding a brownish central area more or less broken by light-colored recurved chevrons; post-abdomen also nearly white; laterally many narrow light and dark stripes alternating; ventrally a narrow dark central stripe widened at its two ends, bounded by wider yellowish stripes.

Female. — Length about 7.0 mm. Legs as in male. Spines as in male except that on second tibiae there are ordinarily two pairs of matched ventral spines, with one other unmatched on retromargin. Color: considerably different from that of male. First pair of legs dark without stripes; other legs yellowish; carapace much like that of male, but white spots are lacking; much long white hair on low clypeus; considerable bronze iridescence on ocular area; abdomen with an irregular central light stripe and a reddish-brown stripe on each side of the central one; laterally yellowish dotted with dark brown; a central ventral dark-brown stripe and dots. Epigynum: small, with a central notch and two small depressions (Fig. 48).

Distribution. — Along the eastern coastal states from New England to South Carolina; westward through the northern states to Nebraska. Collected in Michigan only in Calhoun County, in the vicinity of Albion.

Genus MAEVIA C. L. Koch

Cephalothorax moderately high, with sides nearly vertical, parallel in front and then widened a little behind; cephalic part a little inclined as far as PLE and then gently arched back to posterior declivity. Ocular region occupies two fifths of length of carapace, slightly wider in front than behind, about one third wider than long. PME about midway between ALE and PLE; third row occupies about nine tenths of width of carapace at that level. Legs rather slender, first pair only slightly the stoutest. Fang groove with a single retromarginal fissidentate tooth.

Maevia vittata (Hentz)

(Figures 50–52)

Attus vittatus Hentz, 1845.
A. niger Hentz, 1845.
M. vittata Emerton, 1891.
M. vittata Peckham and Peckham, 1909.

Male. — Length about 6.0 mm. Legs, 4132; without fringes. In addition to the generic characters: first pair of tibiae with four pairs of ventral spines, third pair displaced laterally; second pair of tibiae with essentially the same spination as the first, but one or more of the ventral spines may be missing; first and second pairs of metatarsi with two pairs of ventral spines. Carapace only a little widened anywhere, but widest between second and third coxae; longer than wide in ratio of 45:28. Fang groove with two small promarginal teeth in addition to the one large retromarginal fissidentate tooth (Fig. 51). Palp: well fringed with black and white hairs, especially on the femur, tibia, and tarsus; patella about as long as body of tibia; tarsus large, broadly curved prolaterally and distally; tibial apophyses much more complex than usually figured; main one is a stout spur, between this and bulb is a thin blade, and from ventral distal surface there extends a low blunt tubercle (Fig. 50). Color: there are two color varieties; the one called *M. niger* (Hentz) has legs of pale yellow; most of the palp is black dorsally except the femur, which is striped white and black; carapace generally a dark irregular brown except for a lighter central thoracic spot divided longitudinally by a dark line; from the area bounded by PME and PLE protrude three large tufts of nearly erect somewhat procurved black hairs; abdomen dark brown with light dots and a series of four or five light dorsal chevrons. The light variety has legs with numerous black dots, rings, and stripes; dorsally the palps are light yellowish, but there are black irregular prolateral stripes; there are no black hair tufts on the head, and the whole carapace is of a lighter color; abdomen like that of dark variety except that it is lighter. In life color more as described by the Peckhams (1909).

Female. — Length about 7.0 mm. Similar to male except for color. Color: appendages light yellowish, with considerable short black hair and a few very small dark spots; carapace nearly black immediately surrounding eyes, but the interocular area much lighter, elsewhere yellowish with much dark hair, a black narrow marginal pencil line extends on each side from posterolateral angle to opposite PME; abdomen with a broad irregular light central stripe, with increasingly darker chevrons toward posterior half; on each side of the light central stripe is a reddish-brown irregular area; laterally abdomen is yellowish, with small dark dots; ventrally the same except that dark dots are lacking in center. Epigynum: a slightly swollen area with a small depression far anterior; just in front of genital furrow body of epigynum shows two irregularly coiled tubules (Fig. 52).

Distribution. — Known from the entire United States; probably also occurs in southern Canada and, perhaps, in Mexico. Very common in Michigan, where I have collected it from many localities in both peninsulas.

Genus MARPISSA C. L. Koch

Cephalothorax well rounded in anterior half; wide, flat, and low, widest a little distance behind posterior row of eyes. Ocular region occupies a little less than one half of total length of carapace, only a little wider than long, equally wide in front and behind, middle row of eyes slightly closer to ALE than to PLE. Anterior row recurved as seen from in front; diameter of AME about twice that of ALE. Posterior eyes occupy about seven tenths of width of carapace at that level. Retromargin of fang groove with a single stout simple tooth. Sternum narrowed in front, where anterior coxae are separated by the width of the labium, which is about as wide as long. Legs usually 1423 in males and 4132 in females. Three pairs of ventral spines on first and second tibiae, two pairs of ventral spines on first and second metatarsi, also an extra spine promarginally offset between second and third ventral on first tibiae (some would consider this a prolateral spine). The genus *Marpissa* is well known from the American tropics, where several species abound. Only one species in our northern fauna.

Marpissa undata (De Geer)

(Figures 53–56)

Aranea undata De Geer, 1778.
Marptusa familiaris Peckham, 1888.
Marpissa familiaris F. Cambridge, 1901.
M. rupicola Banks, 1892.
M. undata Peckham and Peckham, 1909.

Male. — Length about 8.4 mm. Legs, 1432; first pair of legs somewhat fringed on femur and patella. Spines generic. Carapace

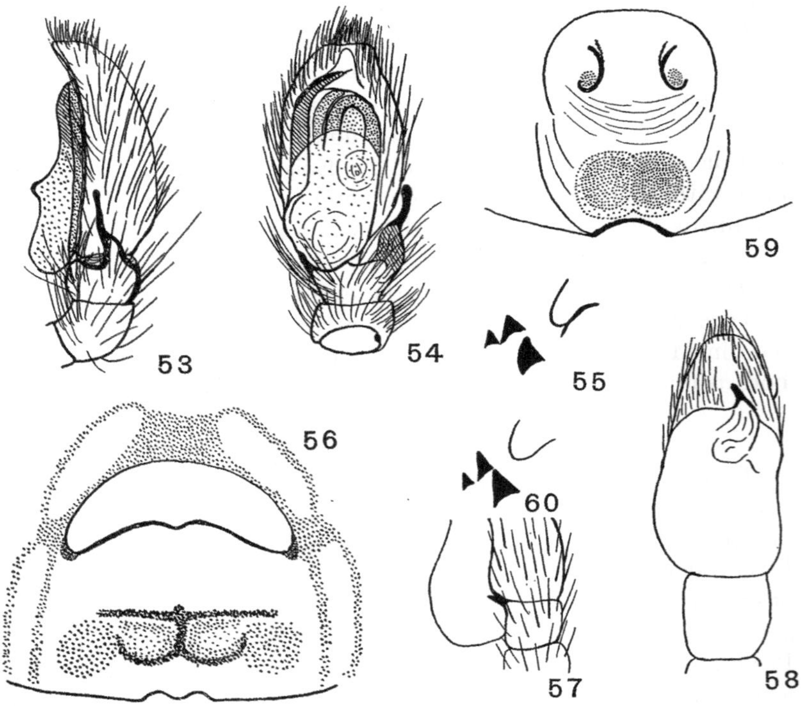

EXTERNAL ANATOMY OF SPIDERS
(Figures 53–60)

53. *Marpissa undata*, male palpus, retrolateral view
54. *M. undata*, male palpus, ventral view
55. *M. undata*, left male cheliceral teeth
56. *M. undata*, epigynum
57. *Metaphidippus canadensis*, part of male palpus, retrolateral view
58. *M. canadensis*, male palpus, ventral view (redrawn from Fig. 2D, Pl. XXXVI, Peckham, 1909).
59. *M. canadensis*, epigynum
60. *M. canadensis*, left female cheliceral teeth

with generic features. Fang groove with two promarginal teeth in addition to large simple retromarginal tooth (Fig. 55). Palp: well fringed with black and white hairs on femur, patella, tibia, and tarsus; femur with a ventral proximal tubercle, strongly curved and somewhat excavated ventrally; patella dorsally about as long as tibia; tibial apophysis a long spur, stout basally, attenuated and sinuous through distal half; tarsal bulb swollen proximally, with a tubercle about midway on prolateral side (Figs. 53–54). Color: all legs brownish, with irregular brown and yellowish spots; carapace with a broad nearly black marginal stripe narrowing to disappearance at clypeus, elsewhere grayish, with a darker patch including ocular area, but this is partly due to fact that gray hairs are easily brushed off. A fringe of erect whitish hair below eyes laterally. Clypeus covered with long whitish hair. Abdomen with a broad central light-colored deeply indented stripe from base to anal tubercle; border of light stripe nearly black; lateral sides dark brown, with venter almost black.

Female. — Length about 11.2 mm. Like male in general except for color. Color: legs much lighter-colored than in male, all legs with a good deal of long whitish hair, especially first pair; palps likewise; carapace with narrower dark marginal stripes and larger and lighter dorsal and lateral areas; only regions immediately around eyes dark; much long whitish hair on clypeus. Abdomen like that of male, but all regions lighter-colored. Epigynum: rather large, with anterior recurved crescentic groove, followed posteriorly by a large slightly swollen area (Fig. 56).

Distribution. — Appears to occur throughout the United States and southward through Mexico and much of Central America. Collected from a few localities only, in Cheboygan, Calhoun, and Jackson counties, in the Southern Peninsula of Michigan.

Genus METAPHIDIPPUS F. Cambridge

This genus was established with no adequate definition on the basis of the author's study of Central American species. Cambridge stated that about the only tangible difference between *Metaphidippus* and *Paraphidippus* is the absence of fringes of hair beneath or at sides of first and second legs in the former. Both genera are rather commonly accepted by araneologists, but no good definitions have been given for either; the status of both here in the north is in doubt. The species included in *Metaphidippus* are generally smaller than those included in *Paraphidippus*. I have a large number of species from Panama which seem to fall into these genera, and it is hoped that some clarification of the confusion surrounding them may be given when I have opportunity to study them adequately. The following features of *Metaphidippus* may be tentatively stated:

Small spiders, or of moderate size; Michigan species seldom exceed 6 mm. Carapace high, moderately convex dorsally and laterally, widest about opposite second coxae, highest at posterior eyes; a little wider below than above, only moderately robust. Anterior row of eyes recurved, AME larger than ALE in ratio of 16 : 7; AME close together, separated by about one eighth of diameter; eyes of second row somewhat nearer to ALE than to PLE. Ocular region occupies about four ninths of length of carapace; wider than long in ratio of 57 : 40; a little wider behind than in front. Body only moderately hairy and legs seldom with enough hair to be regarded as fringed. First pair of legs enlarged only a little. Chelicerae only moderately enlarged in males.

KEY TO THE SPECIES OF *METAPHIDIPPUS*

MALES

1. Anterior end of tarsal bulb terminating in a long stout distally blunt proximally broad spine (Fig. 70); abdomen with five pairs of dorsal black spots (first pair usually obscure) .... *M. montanus*, p. 176
1. Tarsal bulb without long stout distally blunt proximally broad spine; ordinarily without five pairs of dorsal abdominal black spots 2

2. Anterior end of tarsal bulb with short sharply pointed spine; abdomen with three black ventral stripes (median one very narrow), and four nearly white ventral ones .... *M. canadensis*, p. 171
2. Tarsal bulb and abdominal stripes unlike those above 3

3. Tarsal bulb terminating in a pair of long slender spines; carapace without two definite dorsolateral white stripes from ALE to near posterior margin .... *M. flavipedes*, p. 174
3. Tarsal bulb terminating in a single thin broad somewhat twisted spine with black margin; carapace with two definite dorsolateral white stripes from near ALE to close to posterior margin .... *M. capitatus*, p. 172

FEMALES

1. First two pairs of legs with black prolateral stripes; second ventral retromarginal tibial spine lacking on first pair of legs ..... *M. flavipedes*, p. 174
1. First two pairs of legs without black prolateral stripes; almost always with three complete pairs of central tibial spines on first pair of legs 2

2. First pair of legs considerably thickened; first tibia with three pairs of long stout spines, the longest of which are as long as or longer than diameter of segment .... *M. canadensis*, p. 171
2. First pair of legs not unusually thickened; first tibia with shorter and weaker spines, none as long as diameter of segment 3

3. Abdomen with five pairs of definite dorsal dark spots (first pair often obscure); carapace distinctly swollen laterally behind posterior eyes; legs generally yellowish without reddish spots .... *M. montanus*, p. 176
3. Abdomen generally with no paired definite dorsal dark spots; carapace not swollen laterally behind posterior eyes; legs generally with many reddish spots, often annulations .... *M. capitatus*, p. 172

Metaphidippus canadensis (Banks)

(Figures 57–60)

Icius canadensis Banks, 1897.
Dendryphantes canadensis Peckham and Peckham, 1909.

Male. — The male of this species has not yet been collected in Michigan, and I have been unable to borrow a specimen for study. The Peckhams (1909) described the male as very closely resembling the female in all essentials. Palp: tibia and patella both short, with the latter somewhat the longer; tibial retrolateral apophysis a short pointed spur; embolus a short pointed spine (Figs. 57–58).

Female. — Length about 5.5 mm. Legs, 4132; first pair much the stoutest, with coxae, femora, and tibiae considerably thickened. Spines on first legs robust and elongated; first leg with three pairs of ventral tibial spines; second legs with one ventral tibial spine, almost in middle of segment; both first and second legs have two pairs of ventral metatarsal spines. Posterior row of eyes wider than anterior row in ratio of 38 : 31; eye space occupies about seven fifteenths of length of carapace, which is longer than wide in ratio of 74 : 55; posterior eyes occupy nearly full width of carapace at that level; middle eyes small and closer to ALE than to PLE in ratio of 9 : 13. Fang groove with a single fairly stout retromarginal tooth and two smaller promarginal teeth, the inner of which is much the smaller (Fig. 60). Color: first pair of legs and chelicerae a deep amber, other legs and palps a much lighter amber; whole ocular area black, with remainder of carapace a reddish brown with a scanty covering of short white hair. A narrow basal abdominal band extends for some distance along lateral sides; dorsum a golden brown, with a series of irregular paired black spots and chevrons along the middle; near beginning of posterior quarter a pair of small white spots and farther back another pair of smaller white spots, anal tubercle also white; lateral sides obliquely, brokenly, and narrowly striped; venter with two nearly black stripes somewhat converging posteriorly and a very narrow central stripe between the two wider ones (see the Peckhams'

Figs. 2 and 2A, Pl. XXXVI, 1909). Epigynum: notch in posterior border not so deep as in many species; just anterior to notch are two conspicuous spermathecae closely crowded together; still further anterior is a well-defined shallow depression with a distinct curved anterior border (Fig. 59).

Distribution. — Known from Ottawa, Canada, North Carolina, New England, and now Michigan. Apparently nowhere common. Collected only once in Michigan; Albion, Calhoun County, September, 1932.

Metaphidippus capitatus (Hentz)

(Figures 61–65)

Attus capitatus Hentz, 1845.
Dendryphantes capitatus Peckham, 1888; Peckham and Peckham, 1909.
Metaphidippus capitatus F. Cambridge, 1901.

Male. — Length about 3.8 mm., with considerable variation in size of different individuals. Legs, 1423. Spines: first pair of legs with three pairs of ventral tibial spines, but those along promargin occupy only about the distal half, whereas those along retromargin occupy fully two thirds of length of segment; all relatively weak; second pair of legs with one pair of terminal ventral tibial spines and two more farther back on retromargin with none to match them on the promargin. Distal outer corners of maxillae drawn out into distinct hooks. Posterior row of eyes a little wider than anterior row. Eye space occupies two fifths of length of carapace, which is longer than wide in ratio of 25 : 16. Posterior eyes occupy fifty-five sixty-fourths of width of carapace at that level. Fang groove with a single fairly stout retromarginal tooth and two small promarginal teeth (Fig. 63). Palp: patella somewhat longer than body of tibia (dorsal view); tibia with usual small pointed retrolateral apophysis; tarsal bulb swollen basally, with a simple tubular loop, and terminating distally in a simple, thin, somewhat membranous spine with black margins (Figs. 61–62). Color: first coxae usually reddish brown, all others yellowish; all legs generally ringed with reddish brown; palpal femora white dorsally; carapace a rich reddish-brown, with a white stripe on each side extending from ALE beneath posterior eyes to posterior declivity, usually a short white stripe extending from clypeus just above the margin to opposite PLE, more or less white also around clypeus and AME. Abdomen with a reddish-brown (frequently

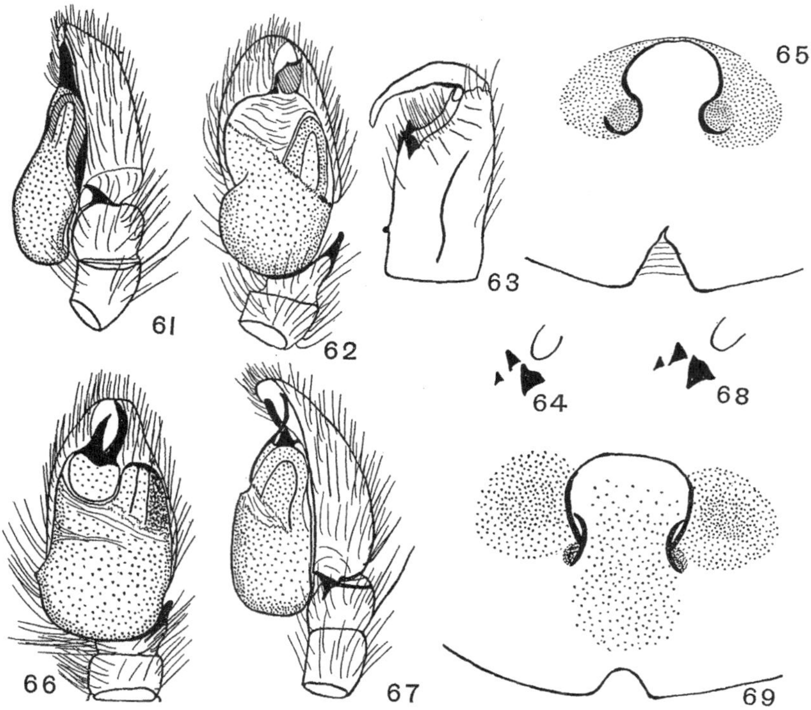

EXTERNAL ANATOMY OF SPIDERS
(Figures 61–69)

61. *Metaphidippus capitatus*, male palpus, retrolateral view
62. *M. capitatus*, male palpus, ventral view
63. *M. capitatus*, left male chelicera, retrolateral view
64. *M. capitatus*, left female cheliceral teeth
65. *M. capitatus*, epigynum
66. *Metaphidippus flavipedes*, male palpus, ventral view
67. *M. flavipedes*, male palpus, retrolateral view
68. *M. flavipedes*, left male cheliceral teeth
69. *M. flavipedes*, epigynum

bronzed) central oval area, within which are often four pairs of more or less obscure round spots; bounding this area is a white margin varying greatly in different individuals; ventrally abdomen usually with a broad central reddish stripe bounded laterally by whitish stripes.

Female. — Length about 5.1 mm., with great variation in size among different individuals. Legs, 4123; first pair somewhat stouter than others. Spines: weaker but essentially like those in male on first legs, except that they occupy a greater part of length of segment; ventral spines of second tibia like those of male except one fewer on retromargin. No hooks on outer distal corners of maxillae. Other features essentially as in male (Fig. 64). Color: all coxae light ventrally, first a little darker; all appendages spotted, ringed, and incompletely striped with reddish brown; carapace nearly black immediately around eyes, elsewhere reddish brown, with many white scales, which are easily removed by handling; abdomen with a dorsal shield-shaped area surrounded by a yellowish area; both areas highly variable in different individuals; shield sometimes more or less broken into a series of paired spots; laterally there is a series of oblique spots narrowly divided by alternate yellowish and reddish-brown stripes; ventrally there is a broad reddish central stripe bounded by yellowish areas; in some individuals this stripe may be broken into two yellowish stripes and three reddish ones. Epigynum: a deep central posterior notch; just anterior to notch is a somewhat raised area; still farther anterior is a recurved chitinous margin enclosing a shallow depression (Fig. 65). In the epigynum great variations have been noted that are due, apparently, to differences in age.

Distribution. — Probably flourishes over much of southern Canada; known from most of the United States and Mexico. One of our most common Michigan Salticidae; collected from many localities in both peninsulas. Females often confused with females of *P. marginatus.*

Metaphidippus flavipedes (Peckham)

(Figures 66–69)

Dendryphantes flavipedes Peckham, 1881; Peckham and Peckham, 1909.

Male. — Length about 4.4 mm. Legs, 1423; first pair somewhat the stoutest. First pair of tibiae with three pairs of ventral spines, retromarginal set occupy about three quarters of length of segment; second pair of tibiae with one pair of terminal spines and usually two more unmatched along retromargin, but either of these may be lacking; first two pairs of metatarsi with two pairs of ventral spines. Outer distal corners of maxillae with small and obscure hooklike extensions. Posterior row of eyes wider than anterior row in ratio of

14 : 13; eye space occupies eight seventeenths of length of carapace, which is longer than wide in ratio of 18 : 13. Posterior eyes occupy twenty-six thirty-firsts of width of carapace at that level. Fang groove with usual single stout retromarginal tooth and two smaller promarginal teeth, the inner of which is very small (Fig. 68). Palp: patella somewhat longer than body of tibia; tibia with usual short pointed ventrally curved apophysis; bulb swollen basally, terminating in an apophysis which is basally broadened and extended into two slender spines curved toward each other (Figs. 66–67). Color: all legs generally light yellowish dorsally, with few light reddish spots and annulations; first two pairs of legs with a single nearly black prolateral stripe on femora, patellae, and tibiae, sometimes on other legs also; a soft white fringe on first legs extending from femora to metatarsi. Carapace with three spots of white scales, two above ALE and extending between ALE and AME, and one between AME; remainder of ocular area and back over posterior declivity to margin a rich reddish brown characteristic of this genus, but also with many white scales; lateral sides with (1) a white marginal stripe from AME to posterior lateral corners, (2) another white stripe from ALE to posterior declivity, and (3) a reddish-brown stripe between the two white ones. Abdomen with a basal white band, which extends dorsolaterally into a stripe on each side to enclose large reddish-brown area, within the posterior half of which in some individuals appear several chevrons and darker spots; venter typically shows three centrally placed reddish stripes separated and bounded by four light ones.

Females. — Length about 4.7 mm. Legs, 4132. Spines weaker but essentially like those in male along ventral surfaces of first two pairs of tibiae and metatarsi. Teeth along fang groove essentially as in male. Color: legs with much more color than in male; prolateral stripes less well defined, but usually occur on all legs; distal ends of all segments more or less reddish brown; carapace generally a rich reddish brown, with white scales irregularly distributed throughout; clypeus white, and the same white cephalic spots occur in the vicinity of eyes as in male except that they are smaller; no definite lateral stripes like those in male; dorsum of abdomen light reddish brown, with many narrow stripes and streaks; on posterior half are several incomplete chevrons and four pairs of darker spots much less distinct than those in *M. montanus.* Venter usually with three reddish stripes and four yellowish ones as in male. Epigynum: with a fairly deep broad posterior notch; anteriorly there is a fairly well defined recurved margin, which is low anteriorly in the middle; a short excurved fold on each side, where the recurved margin ends (Fig. 69).

Distribution. — Known from Labrador; probably occurs widely in Canada. Known from New England north of Connecticut and from the northern Middle Western states. Collected in Michigan from many localities in both peninsulas.

Metaphidippus montanus (Emerton)

(Figures 70–73)

Dendryphantes montanus Emerton, 1891.
D. montanus Peckham, 1900; Peckham and Peckham, 1909.

Male. — Length about 3.8 mm., with considerable variation noted among different individuals. Legs, 1423; first pair somewhat thickened, especially the femora. Spines: first pair of tibiae with three pairs of ventral spines, a long interval between second and third on retromargin; second pair of tibiae apparently with a pair of terminal ventral spines and two others along retromargin, but there may be three others and one or more on promargin and probably other variations in some individuals. Apparently the usual two pairs of ventral spines on first and second metatarsi. Outer distal angles of maxillae extended somewhat but not into hooklike structures. Posterior row of eyes only slightly wider than anterior row. Eye space occupies slightly less than half of length of carapace, which is longer than wide in ratio of 83 : 61. Posterior eyes occupy almost five sixths of width of carapace at that level. Fang groove with usual stout retromarginal tooth and two smaller promarginal teeth (Fig. 72). Palp: patella somewhat longer than body of tibia (viewed dorsally), which has a short slender pointed apophysis; tarsal bulb most swollen near middle, and it terminates in a stout elongated squarely truncated spine (Figs. 70–71). Color: all coxae light-colored, first somewhat darker; all pedal femora light yellowish proximally and distally but dark centrally except on ventral surface, no distinct stripes; other segments somewhat annulated and spotted with reddish brown; palpal femur white, more distal segments all reddish brown. Carapace generally a rich reddish brown, with a distinct white stripe on each side extending from ALE below posterior eyes to posterior declivity; with a little white color dorsal to first row of eyes and especially above

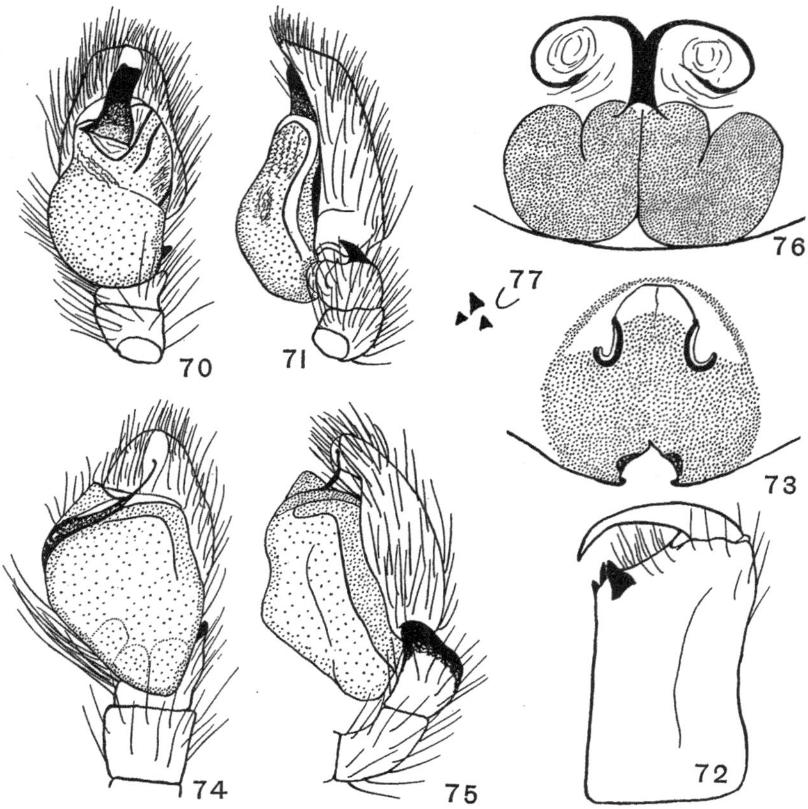

EXTERNAL ANATOMY OF SPIDERS

(Figures 70–77)

70. *Metaphidippus montanus*, male palpus, ventral view
71. *M. montanus*, male palpus, retrolateral view
72. *M. montanus*, left male chelicera
73. *M. montanus*, epigynum
74. *Neon nellii*, male palpus, ventral view
75. *N. nellii*, male palpus, retrolateral view
76. *N. nellii*, epigynum
77. *N. nellii*, left female cheliceral teeth

and between AME; elsewhere carapace is a rich reddish brown; abdomen with a whitish basal band, which is continued dorsolaterally, but there is broken by a series of oblique whitish narrow bands, dorsally a large oval reddish-brown area bearing five pairs of nearly black spots, the first pair obscure and the last pair more or less fused to a single spot; typically the venter has a broad reddish-brown central stripe bounded by two narrow light stripes.

Female. — Length about 4.5 mm. Legs: first pair somewhat stouter than others. Carapace considerably swollen laterally just a little posterior to PLE. Spines: weaker but essentially like those in male, with somewhat more variation indicated. Color: legs and palps nearly uniformly light yellowish, with reddish-brown markings only faintly showing on posterior two pairs of legs. Carapace nearly black immediately around eyes though elsewhere reddish brown, with many white scales but no definite white stripes. Abdomen colored like that of male but less deeply. Epigynum: a deep posterior notch; considerably swollen just anterior to genital furrow; farther anterior a low recurved margin within which occurs a second margin also recurved and terminating on each side in a shallow depression with excurved margins (Fig. 73). I do not see the epigynum as it was figured by the Peckhams (1909).

Distribution. — Known from a few localities in New England, southward to North Carolina; from a few localities in the Midwest. Collected in Michigan only twice, both times by R. R. Dreisbach; Midland County.

Genus NEON Simon, 1876

Very small spiders. Carapace high, a little longer than wide; lateral sides nearly parallel; cephalic part gently arched above, thoracic part at first gently declined for a short distance, and then steeply declivitous to posterior margin. Ocular region a little wider than long, about as wide in front as behind, occupies about three fifths of length of carapace. Anterior row of eyes in a somewhat recurved row (viewed from in front); AME slightly less than twice as large as ALE. Second row of eyes equidistant from ALE and PLE. Eyes of third row large and prominent, occupying full width of carapace at that level. Sternum broad in front, widely separating first coxae by more than width of labium, which is a little wider than long.

Fang groove with a single retromarginal tooth and two promarginal teeth, one larger and one smaller than the retromarginal tooth.

Neon nellii Peckham

(Figures 74–77)

Icius obliquus Banks, 1895.

Male. — Length about 2.5 mm. Legs, 1432; first pair a little the stoutest. Spines: three pairs of very long and slender ventral spines on first tibiae; two pairs of ventral spines on first pair of metatarsi. Since it is very unlikely that this species will be confused with any other in the state, the descriptive material is reduced to the bare minimum. Palp: patella somewhat longer than body of tibia, which has a very stout short retrolateral apophysis; tarsus large, with bulb swollen medially, and an embolus which extends obliquely across near the distal end and then turns distally near the tip (Figs. 74–75). Color: all legs yellowish, with pale gray spots and prolateral and retrolateral broad gray stripes on anterior legs in well-colored individuals; carapace generally an amber color, the ventral margin with a narrow black line; eyes on black spots; posterior declivity and lateral sides somewhat streaked with black; abdomen yellowish, finely mottled with black or gray dots and a series of chevrons dorsally on posterior half.

Female. — Length about 2.9 mm. Legs, 4132. Cheliceral teeth as shown in Figure 77. Other general features so similar to those of male as to deserve no special description. Epigynum: two deep depressions separated by a long narrow rather elevated septum (Fig. 76).

Distribution. — Well known from New England, New York, along the Atlantic seaboard to North Carolina; also from several Middle Western states and Washington State; also southeastern Canada. Collected in Michigan at numerous localities in both peninsulas. Probably fairly common in forest débris.

Genus PARAPHIDIPPUS F. Cambridge, 1901

F. Cambridge found it convenient to establish the genus *Paraphidippus* for numerous species among his Central American spiders which appeared to be sufficiently different from other nearly related genera, such as *Phidippus* and *Metaphidippus*. I regard this as a good genus, and it has been recognized by Petrunkevitch (1928) and others. Cambridge placed *Philaeus militaris* Peckham in the genus *Paraphidippus* on the basis of inadequate figures and description. It now appears clear that the species does not belong in *Paraphidippus*, but nobody seems to be sure where it should go. In view of the uncertainty it is provisionally kept in this genus. There is but one species in our fauna, and this can be identified quite readily by its positive characters of palpi and chelicerae.

Paraphidippus marginatus (Walckenaer)

(Figures 78–82)

Attus marginatus Walckenaer, 1837.
A. militaris Hentz, 1845.
Dendryphantes militaris Emerton, 1891.
Paraphidippus militaris F. Cambridge, 1901.
Dendryphantes militaris Peckham and Peckham, 1909.
Phidippus molinor Chamberlin, 1925.

Male. — Length about 5.0 mm., but much variation in size has been noted among individuals. Legs, 1432; first legs much the longest but not much thickened. Spines: first pair of tibiae with three pairs of ventral spines, those on promargin occupy less than one half of length of the segment, those on retromargin occupy about two thirds of length of segment, interval between second and third retromarginal nearly twice as great as interval between first and second; second pair of tibiae with one pair of terminal ventral spines and two more along retromargin. First and second metatarsi with two pairs of ventral spines. Distal outer corners of maxillae produced into very small hooklike extensions. Posterior row of eyes almost exactly as wide as anterior row. Eye space occupies only a little less than one half of length of carapace, which is longer than wide in ratio of 79 : 62. Posterior eyes occupy about five sixths of width of carapace at that level. Chelicerae considerably enlarged (Fig. 80); fang groove deeply excavated; with a very large retromarginal tooth and two promarginal teeth, one of which is large and the other very small. Palp: slender, patella and tibia about equal in length; tibia with a single simple retrolateral, basally stout, terminally slender apophysis; tarsal bulb ends distally in a slender procurved spine (Figs. 78–79). Color: all coxae usually light-colored, with first somewhat the darkest; first femora usually rich reddish brown with white scales, all other femora usually yellowish in proximal halves

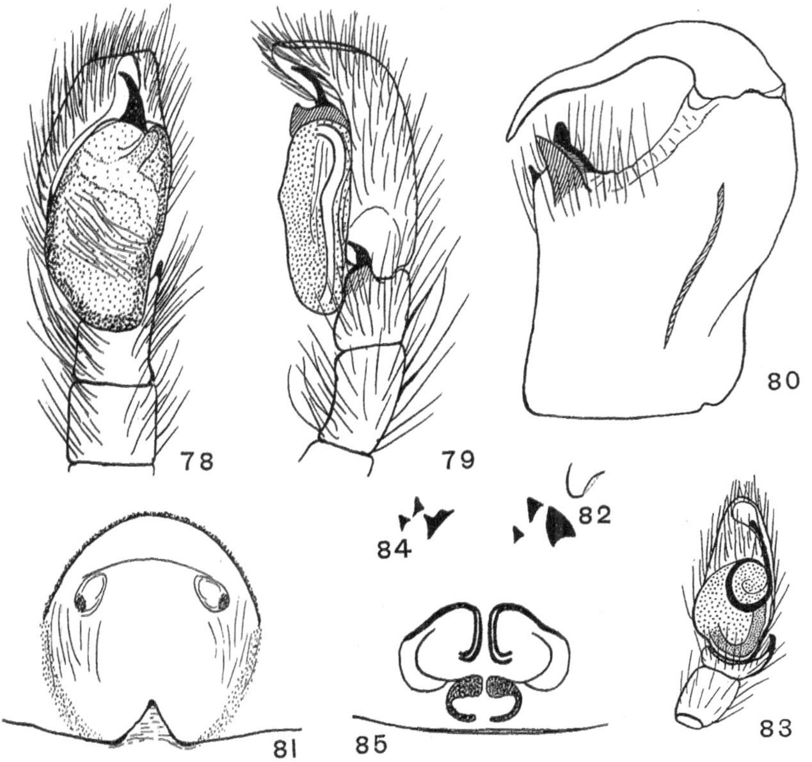

EXTERNAL ANATOMY OF SPIDERS

(Figures 78–85)

78. *Paraphidippus marginatus*, male palpus, ventral view
79. *P. marginatus*, male palpus, retrolateral view
80. *P. marginatus*, left male chelicera
81. *P. marginatus*, epigynum
82. *P. marginatus*, left female cheliceral teeth
83. *Peckhamia picata*, male palpus, ventral view
84. *P. picata*, left male cheliceral teeth
85. *P. picata*, epigynum

and reddish brown distally; other segments of legs reddish brown but progressively lighter toward distal segments. Palpal femora with white scales dorsally, also at distal end of patella on dorsal surface; otherwise segments are reddish brown dorsally, lighter beneath.

Carapace a rich reddish brown, with a white stripe from ALE somewhat below PME and PLE back to posterior declivity but not ordinarily farther, many white scales over dorsum, the most persistent of which are in region of longitudinal thoracic groove and midway between PME; clypeus is white, and the white color continues along ventral margin in a very narrow line; all eyes except AME on black spots. Abdomen: at base a light reddish-brown band, which continues on each side as a lateral stripe; just above this is a basal whitish band, which continues dorsolaterally not quite to posterior end; this delimits a bright somewhat bronzed dorsal area, within which is a series of small obscure light-colored paired spots; ordinarily the venter has a broad central dull reddish-brown stripe, with a narrower lighter stripe on each side.

Female. — Length about 6.7 mm. Legs, 1423; first pair the stoutest. Spines on ventral surface of first two pairs of tibiae and metatarsi essentially same as in male. Cheliceral teeth as shown in Figure 82. Color: legs essentially as in male except somewhat lighter; carapace a rich reddish brown all over except for white scales generally distributed and a very narrow ventral white marginal stripe continuous with the clypeus, which is well clothed in white hair. Color of abdomen much like that of male except that areas are less well defined. Four pairs of light spots in dorsal area usually well defined. On the venter the central reddish stripe usually narrower and the lighter bounding stripes usually broader. Epigynum: when mature there is a fairly deep broad posterior notch at the genital furrow; anteriorly there is a narrow recurved margin and two small openings about four short diameters of one of them apart (Fig. 81).

Distribution. — Probably present throughout the United States, through much of southern Canada and northern Mexico. One of our most common jumping spiders throughout Michigan.

GENUS PECKHAMIA SIMON, 1901

Carapace either fairly low and flat (*scorpionia*) or fairly high and convex (*picata*). Ocular region longer than wide, occupies from one half to two thirds of length of carapace. Anterior row of eyes recurved; AME almost touching one another, at least twice as large as ALE; PME much closer to ALE than to PLE, which are large and and situated near margin of carapace. Chelicerae small, vertical; with a weakly compound retromarginal tooth. Definitely antlike.

KEY TO THE SPECIES OF *PECKHAMIA*

1. Anterior tibiae with three pairs of ventral spines; first legs much enlarged; two white spots in depression behind PLE; carapace fairly high *P. picata*, p. 183
1. Anterior tibiae with only one pair of matched ventral spines; first legs only moderately enlarged; carapace without definite white spots, lower and flatter than in *picata* . . . . *P. scorpionia*, p. 185

Peckhamia picata (Hentz)

(Figures 83–85)

Synemosyna picata Hentz, 1845.
Synageles picata Peckham, 1888, 1892.
P. picata Simon, 1901.
P. picata Peckham, 1909.

Male. — Length, 3.5 mm. Legs, 4123; first pair much enlarged, especially femora, patellae, and tibiae. Spines: three pairs of ventral spines on first tibiae; second tibiae with a pair of terminal ventral spines and two more unmatched ones along retromargin; first pair of metatarsi with two pairs of stout spines; second pair of metatarsi with two pairs of weak spines. Chelicerae small, vertical, with a small retromarginal compound tooth and two small promarginal teeth (Fig. 84). Posterior row of eyes only slightly wider than anterior row; eye space longer than wide in ratio of 9 : 8; occupies nearly nine seventeenths of length of carapace, which is longer than wide in ratio of 17 : 9; PME nearer to ALE than to PLE in ratio of 1 : 2; posterior eyes occupy ten elevenths of width of carapace at that level. Carapace rises gently to PME, remains nearly level to opposite posterior eyes, where there is a shallow groove, then there is a sharp rise to a low hump, from which the posterior declivity descends steeply at first and then very gently to posterior border. Anterior end of abdomen bulbous and separated from remainder of this part of body by a marked groove; whole dorsum covered by a shield. Palp: delicate; patella about twice as long as tibia, which has a stout spur as its retrolateral apophysis; embolus makes a complete revolution and then extends as a long slender spine nearly to distal end of tarsus (Fig. 83). Color: first pair of legs a deep amber except last two segments, which are yellowish; second pair of legs with dark-brown tarsi and metatarsi, other segments light yellow with black prolateral stripe; other legs amber with prolateral stripes less conspicuous; ocular region a glistening granular black; lateral sides of carapace a

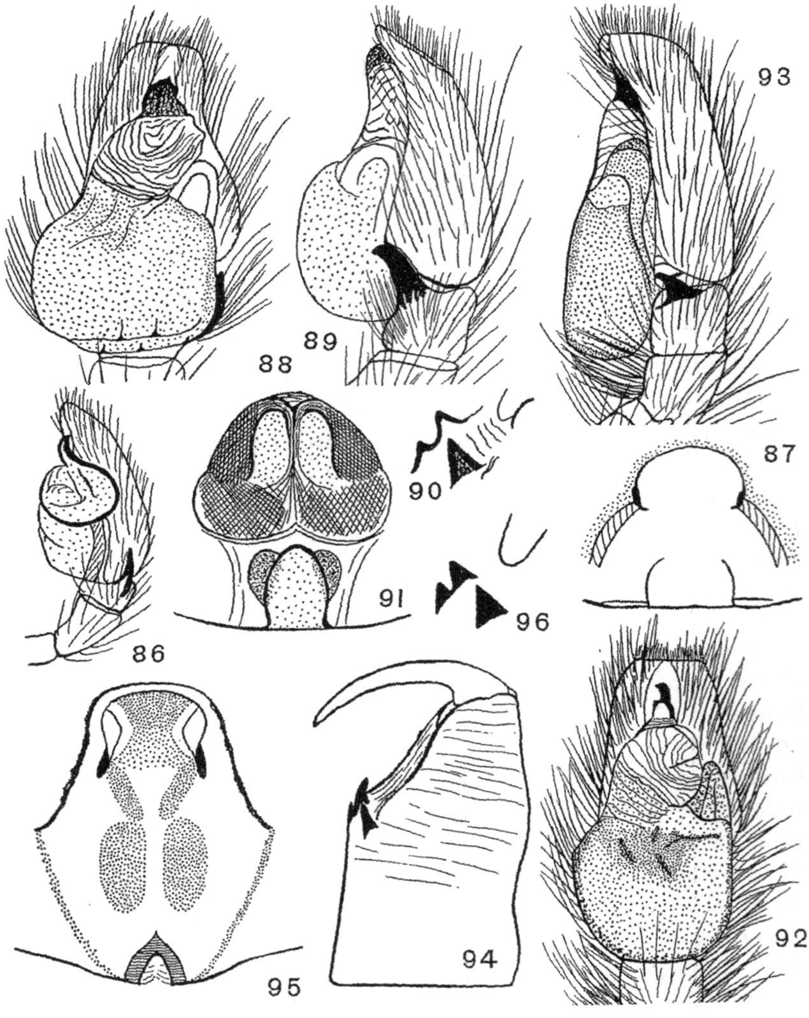

EXTERNAL ANATOMY OF SPIDERS (Figs. 86–96)

86. *Peckhamia scorpionia,* male palpus, retrolateral view
87. *P. scorpionia,* epigynum
88. *Phidippus altanus,* male palpus, ventral view
89. *P. altanus,* male palpus, retrolateral view
90. *P. altanus,* left male cheliceral teeth
91. *P. altanus,* epigynum
92. *Phidippus audax,* male palpus, ventral view
93. *P. audax,* male palpus, retrolateral view
94. *P. audax,* left male chelicera
95. *P. audax,* epigynum
96. *P. audax,* left female cheliceral teeth

rich reddish brown, with three or four large obscure black irregular patches on each side; two white spots close together in the depression just behind the posterior eyes; bulbous anterior part of abdomen a light brown, remainder a rich dark brown; venter almost black, except a yellow patch just posterior to genital furrow.

Female. — Females have not yet been collected in Michigan. The Peckhams (1909) give the following data: length about 4.8 mm.; legs, 4231. Other features essentially like those of male. Epigynum as shown in Figure 85 (taken from specimen in Museum of Comparative Zoölogy at Harvard College).

Distribution. — Believed to be spread throughout the entire United States. Collected only twice in Michigan; Clare County and Emmet County.

Peckhamia scorpionia (Hentz)

(Figures 86–87)

Synemosyna scorpionia Hentz, 1845.
Synageles scorpionia Peckham, 1888.
P. scorpionia Peckham and Peckham, 1909.

Description from specimens collected at Sea Cliff, New York, by Mr. Nathan Banks and loaned by the Museum of Comparative Zoölogy at Harvard College.

Male. — Length, 3.2 mm. Legs, 4123 ?; first pair somewhat enlarged, especially the femora; anterior tibiae only a little longer than patellae. Spines: anterior tibiae with two pairs of matched ventral spines, an odd one on retromargin, and none terminal; second tibiae seem to have no ventral spines; first and second metatarsi with two pairs of ventral spines each. Chelicerae with a single relatively large retromarginal fissidentate tooth, and probably two small promarginal teeth; mildly concave medially. Anterior row of eyes just slightly wider than posterior row; eye space longer than wide in ratio of about 8 : 7; occupies a little less than two thirds of length of carapace, which is longer than wide in ratio of 33 : 18. PME closer to ALE than to PLE in ratio of 7 : 20; posterior eyes occupy about ten elevenths of width of carapace at that level. Carapace rises very gently to opposite PME, then falls gently to a point opposite PLE, and steeply to about one tenth of its length from posterior margin. Anterior third of abdomen bulbous and separated from remainder by a marked constriction; whole dorsum covered by a shield. Palp: moderately robust; patella nearly twice as long as body of tibia, which has a single stout elongated retrolateral apophysis; embolus makes a complete revolution and ends in a hook (Fig. 86). Color: all legs and palp a deep amber; first legs somewhat darker than others (only a slight indication of the black stripes mentioned by the Peckhams, 1909); carapace a mahogany brown with eyes on black spots and a few white hairs around anterior row of eyes; also indications of a small median spot of white scales at beginning of posterior declivity, with two smaller white spots lateral to the median one. Abdomen also mahogany brown dorsally, venter much lighter, with lateral sides striped with alternate light and dark lines.

Female. — Length about 3.0 mm. Much like male except in respect to secondary sexual characters. Carapace lower and flatter. Legs generally lighter-colored. Dorsal abdominal shield restricted to dorsal part of bulbous anterior end of abdomen. A white band crosses dorsum at the groove; on venter is also a broad light procurved band just posterior to genital furrow. Epigynum: obscure; a slightly swollen area bounded in front by a recurved margin enclosing a yellowish undifferentiated area; just anterior to the genital groove is another yellowish area partly surrounded by a delicate margin (Fig. 87). I do not see this organ as it was illustrated by the Peckhams, 1909.

Distribution. — Reported from many isolated localities extending over nearly all the United States and southeastern Canada. Collected only once in Michigan; Calhoun County, May, 1935.

Genus PHIDIPPUS C. L. Koch, 1846

Medium to large spiders; usually hairy. Carapace high, robust, wide, and convex laterally. Ocular quadrangle from one third to two thirds wider than long; occupies two fifths of length of carapace or more; wider behind than in front. Eyes relatively small, AME less than twice as large as ALE; PME about half as far from ALE as from PLE, which do not occupy nearly the width of the carapace at that level.

KEY TO THE SPECIES OF *PHIDIPPUS*

MALES

1. Chelicerae with a tubercle on front surface; a central dorsal abdominal somewhat triangular white spot as well as other light spots and bars; palpal bulb terminates in a short beaklike spine springing from a broad base (Fig. 92) .... *P. audax*, p. 189
1. Not with combination of characters given above 2

2. Chelicerae without a tubercle on front face of basal segment; dorsal region of abdomen with a broad rust-red margin; palpal bulb terminates in a short broad spine (Fig. 88) .... *P. altanus*, p. 188
2. Not with combination of characters given above 3
3. Abdomen uniformly reddish brown dorsally; first pair of legs with many white hairs; palpal bulb terminates in a stout sickle-shaped spine (Fig. 97) .... *P. brunneus*, p. 191
3. Not with combination of characters given above 4
4. First pair of legs notably elongated; clypeus without white hairs; abdomen usually with glistening silvery scales and red spots in central stripe; palpal bulb terminates in a slender curved spine of moderate length *P. clarus*, p. 193
4. Not with combination of characters given above 5
5. Clypeus with long white hair; carapace with two broad white dorsolateral stripes; palpal bulb terminates in a somewhat twisted slender spine with a broad base .... *P. insigniarius*, p. 195
5. Not with combination of characters given above 6
6. Abdomen and carapace reddish dorsally with two or three pairs of small white spots within two dark stripes; palpal bulb terminates in a long uniformly slender spine, which is turned ventrally (Fig. 111) *P. pius*, p. 197
6. Not with combination of characters given above 7
7. Legs and cephalic region unusually hirsute; abdomen with two somewhat indistinct dark stripes, within which are four pairs of whitish spots; palpal bulb with an elongated rugulose area beyond which a short broad spine extends (Fig. 113) .... *P. purpuratus*, p. 199
7. Legs and cephalic region not unusually hirsute; carapace and abdomen at least partly reddish; palpal bulb with distal rugulose area short and broad, terminating in a spine of moderate length and stoutness *P. whitmanni*, p. 200

FEMALES

1. Epigynum more or less completely separated into right and left halves by a longitudinal septum 2
1. Epigynum not separated into right and left halves by a longitudinal septum 5
2. A large species, about 10 mm.; dorsum of abdomen with marginal red or yellow oblique bars, central region with small glistening scales; epigynum with a deep posterior marginal notch with lateral extension and a shallow depression separated into halves by a well-defined longitudinal septum (Fig. 91) .... *P. altanus*, p. 188
2. Not with combination of characters given above 3
3. All legs rather light brown; carapace uniformly reddish brown overlaid with white scales; abdomen fairly uniformly light reddish brown; epigynum much longer than wide, with a broad posterior marginal notch and shallow depressed anterior region separated into halves by a nearly suppressed longitudinal septum (Fig. 101) .... *P. brunneus*, p. 191
3. Not with combination of characters given above 4
4. Carapace brownish, with many whitish hairs; very prominent tufts of black hair between PME and PLE; much whitish hair on palps and clypeus; abdomen with a black median dorsal stripe bifurcated anteriorly to

enclose a white spot, also with yellowish basal band and two bright red marginal stripes .... *P. insigniarius*, p. 195

4. All legs a reddish brown, palps with much long white hair on last three segments; carapace a reddish brown with many white scales; abdomen a dull brownish dorsally with two median darker stripes usually bearing four pairs of small white spots; epigynum with a deep but rather narrow posterior marginal notch, and centrally with two rather deep depressions separated by a somewhat suppressed longitudinal septum (Fig. 116) *P. purpuratus*, p. 199

5. Abdomen nearly black, with a central dorsal white spot usually followed by two pairs of white bars and more obscure whitish spots; epigynum considerably longer than wide, with a fairly deep-bordered posterior marginal notch and a shallow depression without any median longitudinal septum (Fig. 95) .... *P. audax*, p. 189

5. Not with combination of characters given above 6

6. All legs more or less darkly annulated; abdomen with two dorsal dark stripes bearing several pairs of small whitish spots; venter usually with a narrow dark central stripe and a broader darker stripe on each side reaching to base of spinnerets; epigynum nearly as wide as long *P. clarus*, p. 193

6. Not with combination of characters given above 7

7. All legs fairly uniformly yellowish; carapace also generally yellowish except black ocular area; abdomen yellowish, with two central dorsal narrow black stripes usually with two pairs of small white spots; epigynum with a broad shallow posterior marginal notch and a scalloped anterior border just behind which are two small shallow depressions (Fig. 112) .... *P. pius*, p. 197

7. All femora darkened, other segments barred; a light stripe from AME nearly across ocular area; dorsum red, sometimes with white spots; venter usually black, with two light stripes converging to spinnerets; epigynum nearly as wide as long, broadly oval (Fig. 118) *P. whitmanni*, p. 200

Phidippus altanus Gertsch

(Figures 88–91)

Male. — Length about 9.0 mm. Legs, 1423; first pair considerably the stoutest; all legs fringed but none heavily. Spines: first pair of tibiae with three pairs of ventral spines not evenly matched; those on promargin occupy just a little more than half of length of segment whereas those on retromargin occupy about two thirds of length; the second pair of tibiae have one pair of terminal spines and two others on the retromargin not matched by corresponding prolateral spines. First two pairs of metatarsi with the usual two pairs of ventral spines. Distal outer corners of maxillae drawn out into very definite hooklike extensions. Posterior row of eyes wider than anterior row in ratio of 20 : 17; eye space occupies six thirteenths of length of carapace, which is longer than wide in ratio of 13 : 11; posterior eyes occupy almost four fifths of width of carapace at that level. Fang groove with a single tooth of moderate size on retromargin and two teeth on promargin, the inner of which is very small (Fig. 90). Sternum very convex near posterior end. Palp: patella considerably longer than body of tibia; tibia with a retrolateral spur nearly as long as body and curved only a little toward the bulb, which is swollen at the base, where it is almost quadrilateral; embolus short, stout, irregularly blunt at distal end (Figs. 88–89). Color: legs generally dark brown, with many darker spots and annulations together with white scales; fringe on first patella and a portion of first tibia are partly composed of whitish hairs; palp brownish with a sparse black fringe dorsally and ventrally at proximal end of femur; chelicerae iridescent; carapace dark brown; abdomen with a broad nearly black central stripe, within which are two pairs of short oblique rust-red bars; completely surrounding this central dark stripe is a broad bright rust-red margin (short recumbent hairs); long whitish hairs grow out of this red margin; lateral and ventral surfaces dark brown, with the usual four narrow-beaded light stripes.

Female. — Known only from Dr. Gertsch's original description. Length, 10.5 mm. Color: like that of male except as follows: palps and clypeus with long white hair; legs with more white hair and white scales than male; abdomen with red or yellow dorsal margin around a larger dark central stripe overlaid with shiny black scales. Epigynum: not figured by Dr. Gertsch but said to be like that of *P. ardens;* my Figure 91 was drawn from specimen given by the author of the species.

Distribution. — Described from Yellowstone National Park; also known from Alberta, Saskatchewan, and now from Michigan, but taken in this state only once, in Midland County, by R. R. Dreisbach, May, 1939.

Phidippus audax (Hentz)

(Figures 92–96)

Attus audax Hentz, 1844.
P. tripunctatus Emerton, 1891.
P. audax Peckham and Peckham, 1909.

Male. — Length about 9.6 mm.; great variation in size among different individuals. Legs, 1423; first pair of legs heavily fringed and somewhat enlarged; others progressively less fringed and less hairy toward last pair. Spines difficult to see because of heavy fringes. Distal outer corner of maxillae drawn out into distinct tubercles. Carapace very broad and blunt anteriorly. Posterior row of eyes wider than anterior row in ratio of 47 : 40; eye space occupies sixteen thirty-ninths of length of carapace, which is longer than wide in ratio of 39 : 31. Posterior row of eyes occupies about three fourths of width of carapace at that level. Chelicerae robust with rugate surface, with prominent tubercle on front face of basal segment near distal end, with a slender retromarginal tooth and two promarginal teeth, the inner of which is very small (Fig. 94). Sternum very convex near posterior end. Palp: very hairy, patella somewhat longer than tibia, which has a small pointed apophysis turned distally toward the bulb, which is swollen basally; the embolus is a stout short hooklike spine (Figs. 92–93). Color: legs generally dark brown or black, more or less streaked with light brown, and with much long hair; first pair of legs with white fringes on prolateral surface of distal end of femora, prolateral side of patellae, proximal end of metatarsi and tarsi, elsewhere fringes are black and gray. Carapace a dark brown; abdomen generally a dark brown with a narrow white basal band, which continues for a short distance dorsolaterally; near middle of dorsum is a white spot, which is frequently somewhat triangular and is continued laterally into white pencil lines; farther back are usually two pairs of narrow bands in a somewhat dorsolateral position, and there may be additional small white spots or bands. Venter usually dark brown, with four obscure narrow-beaded light stripes. Palpal femur has white scales and hair dorsally and the patella only white scales. Chelicerae very iridescent-greenish.

Female. — Length about 11.0 mm., with great variation among different individuals. Legs, 4123; first pair the stoutest; fringed as in male. First tibiae with three pairs of ventral spines, occupying only a little more than one half of length of segment; second tibiae with one pair of terminal ventral spines and two others along retromargin, occupying two thirds of length of segment; first and second metatarsi with two pairs of ventral spines. Fang groove with teeth as shown in Figure 96. No hooks or tubercles on outer distal corners of maxillae. Otherwise essentially like the male except for color. Tufts of long stiff black hair in eye region. Color: basal abdominal band often lacking, as well as one or more pairs of the narrow white bands behind the central dorsal white spot. Epigynum: longer than wide; a bordered, fairly deep notch in the posterior margin; somewhat depressed in anterior half, where there are small secondary depressions or possibly openings, as shown in Figure 95.

Distribution. — Probably spread over the entire United States and most of southern Canada. Apparently our most common *Phidippus* in Michigan; taken in many localities in both peninsulas.

Phidippus brunneus Emerton

(Figures 97–101)

P. brunneus Peckham and Peckham, 1909.

Male. — Length about 7.0 mm. Legs, 1423; first pair the stoutest; all legs with weakly developed fringes, the first most heavily fringed. Spines: as usual except that those of first leg seem to be weaker than usual in other species (mostly broken from my only Michigan male). Maxillae (Fig. 100) with mildly developed hooks on outer distal corners (hardly more than tubercles). Posterior row of eyes wider than anterior row in ratio of 20 : 17; posterior row occupies four fifths of width of carapace at that level; eye space occupies twenty-three fifty-sixths of length of carapace, which is longer than wide in ratio of 28 : 25. Fang groove with a single simple stout retromarginal tooth and two promarginal teeth, the inner of which is very small (Fig. 99). Palp: patella considerably longer than tibia; tibial retrolateral apophysis short, stout at base, distally tapered and rather strongly curved toward bulb; bulb basally much swollen, terminates in a stout sickle-shaped hook, blunt at distal end (Figs. 97–98). Sternum strongly convex near posterior end. Color: legs all dark reddish brown except dorsally lighter with a narrow median dark stripe, especially on femora; first legs with many white prolateral hairs; chelicerae iridescent-greenish; palps with white scales dorsally on femora and patellae; carapace generally dark brown, with eyes on black areas; abdomen with a narrow whitish basal band, which extends dorsolaterally to about the middle; dorsum entirely covered by a reddish-brown area, which is said to look uniformly red in life. Venter is uniformly brownish except for the usual four narrow-beaded light stripes.

Female. — Length about 10.5 mm. Legs, 4132; first pair considerably the stoutest. Spines essentially as in male. Distal outer

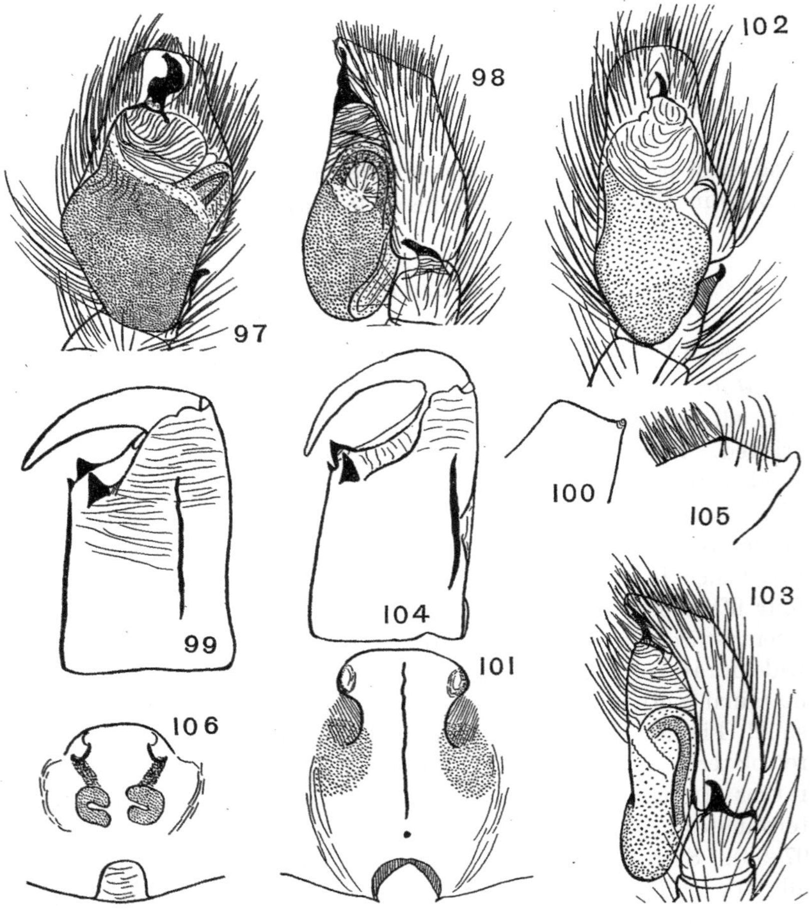

EXTERNAL ANATOMY OF SPIDERS

(Figures 97–106)

97. *Phidippus brunneus,* male palpus, ventral view
98. *P. brunneus,* male palpus, retrolateral view
99. *P. brunneus,* left male chelicera
100. *P. brunneus,* distal end of maxilla
101. *P. brunneus,* epigynum
102. *Phidippus clarus,* male palpus, ventral view
103. *P. clarus,* male palpus, retrolateral view
104. *P. clarus,* left male chelicera
105. *P. clarus,* distal end of maxilla
106. *P. clarus,* epigynum

corners of maxillae not drawn out into hooks. Other features essentially like those in male except the color. Color: all legs light reddish brown, with considerable hair of two kinds, short white scales and mixed long black and white hair; palps with much long white hair and sparse long black hair; chelicerae covered basally with white scales, elsewhere in front rugose and green-iridescent; carapace generally reddish brown overlaid with white scales; long white hairs and a few black ones are generally distributed, with long black hairs crowded somewhat in region of eyes, where whole ocular area is much darker; clypeus with many white scales and some long white hairs; abdomen generally and fairly uniformly light reddish brown dorsally, lighter on ventral surface. Epigynum: with a fairly deep, broad bordered notch in posterior margin; just anterior to notch organ is swollen; still farther anterior is a depression of moderate depth, through which a suppressed longitudinal septum partly divides the depressed area into halves (Fig. 101).

Distribution. — Well known from New England and New York; also known from Michigan and Illinois. Collected in Michigan only a few times and only in the central and southern part of the Southern Peninsula.

Phidippus clarus Keyserling

(Figures 102–106)

P. clarus Keyserling, 1884.
P. insolens Peckham, 1888.
P. multiformis Emerton.
P. clarus Peckham, 1909.
Dendryphantes clarus Petrunkevitch, 1911.

Male. — Length about 7.3 mm. Legs, 1423; first pair much the longest and stoutest, also moderately fringed; other legs lightly fringed. Spines: first pair of tibiae with three pairs of ventral spines, rather weak, those along the promargin occupy considerably less than one half of length of segment, whereas those along the retromargin occupy somewhat more than one half of length of segment; second pair of tibiae with two pairs of ventral spines fairly well matched and another farther back on retromargin; first two metatarsi with two pairs of ventral spines. Outer distal corners of maxillae drawn out into blunt robust extensions (Fig. 105). Posterior row of eyes wider than anterior row in ratio of 7 : 6; eye space occupies twenty-three fifty-eighths of length of carapace, which is longer than wide in ratio of 58 : 49; posterior eyes occupy five sevenths of width of carapace at that level. Fang groove with a fairly stout retromarginal tooth and two promarginal teeth, the inner and smaller of which is almost a part of the larger one (Fig. 104). Palp: patella considerably longer than body of tibia, which has a simple retrolateral apophysis, which is turned retrolaterally and moderately curved toward the bulb; bulb not greatly swollen in the individuals examined; embolus ends in a short slender hook (Figs. 102–103). Color: all legs a dark reddish brown, with light-colored regions above, within which are narrow darker stripes, especially true of femora; palp with dorsal white scales from femur to base of tarsus; also a fringe, a part of which is composed of long white hairs; carapace very dark brown or black; no white on clypeus; abdomen with a white basal band, which extends laterally only a short distance; from basal band a broad central dark stripe extends to spinnerets; central stripe usually with silvery scales and three pairs of red spots in posterior two thirds; on each side of central stripe is a dorsolateral stripe covered with short light hair, which looks reddish in older specimens and grayish in younger ones; abdomen ventrally dark-colored, with four faint narrow light stripes.

Female. — Length about 10.0 mm. Legs, 4132; first pair considerably the stoutest; all with sparse fringes, somewhat more ample on first. Ventral spines on first tibiae and metatarsi like those of male; second metatarsi with ventral spines like those of male, but ventral spines on tibia very irregular even on the right and left sides of the same individuals. No hooks on outer distal corners of maxillae. Color: legs like those of male, except that they are lighter; first legs and palps with a good deal of white hair; carapace a rather bright reddish brown (appears quite reddish in life), much darker on ocular region; a sparse covering of short light hair and long light hair, especially at the lateral sides; the abdomen in my Michigan specimens is usually a fairly uniform reddish gold except for two faint darker central stripes, within which appear a few pairs of small white spots; the venter almost always shows a narrow central dark stripe reaching back about two thirds of length and two other broader and darker stripes reaching to base of spinnerets and meeting there. Epigynum with features shown in Figure 106.

Distribution. — Probably distributed throughout most of the United States; fairly common throughout southern Michigan, but not yet collected in the Northern Peninsula.

Phidippus insigniarius C. L. Koch

(Figures 107–109, 126)

P. insigniarius Banks, 1907.
P. insigniarius Peckham and Peckham, 1909.

Male. — Length about 6.7 mm. Legs, 1423; first pair well fringed with white hairs; others similarly fringed but less amply. Spines: first tibiae with three pairs of ventral spines of moderate size; on prolateral side spines occupy less than one half of length of segment, and on retromarginal side they occupy a little more than one half of segment; second tibiae with one pair of terminal spines and one other on retromargin near middle of segment; first two pairs of metatarsi with two pairs of ventral spines as usual. Outer distal corners of maxillae extended into small blunt apophyses or tubercles. Posterior row of eyes wider than anterior row in ratio of 32 : 27; eye space occupies two fifths of length of carapace, which is longer than wide in ratio of 5 : 4; posterior eyes occupy three fourths of width of carapace at that level. Fang groove with a single simple retromarginal tooth and two promarginal teeth, the inner of which is very small (Fig. 109). Palp: patella considerably longer than tibia; tibia with a retrolateral apophysis, stout at base, pointed and spurlike distally, acutely turned toward bulb, with a greater angle than usual between it and body of tibia; embolus fairly stout at base but somewhat spirally twisted, terminating in a slender process of moderate length (Figs. 107–108). Color: legs all generally dark brown, with lighter dorsal surfaces and narrow dark dorsal stripes irregularly distributed; fringes on legs white, and white scales appear on several segments, most persistently on prolateral and dorsal surfaces of first two pairs of tibiae and metatarsi; carapace dark brown, almost black, with prominent tufts of black hair behind and below PME; chelicerae brightly iridescent-green; clypeus covered with long white hair; two broad lateral white stripes extend from clypeus to posterior declivity, but do not unite there; abdomen has a basal white band, which extends dorsolaterally not quite halfway toward posterior end; generally the whole dorsum is bright reddish brown (red in life); two nearly black stripes extending from the light basal band posteriorly to about the middle, where they unite and then continue as a single stripe with lateral extensions to the anal tubercle. There is thus a dorsal abdominal pattern different from that of any other spider in this genus in Michigan. The ven-

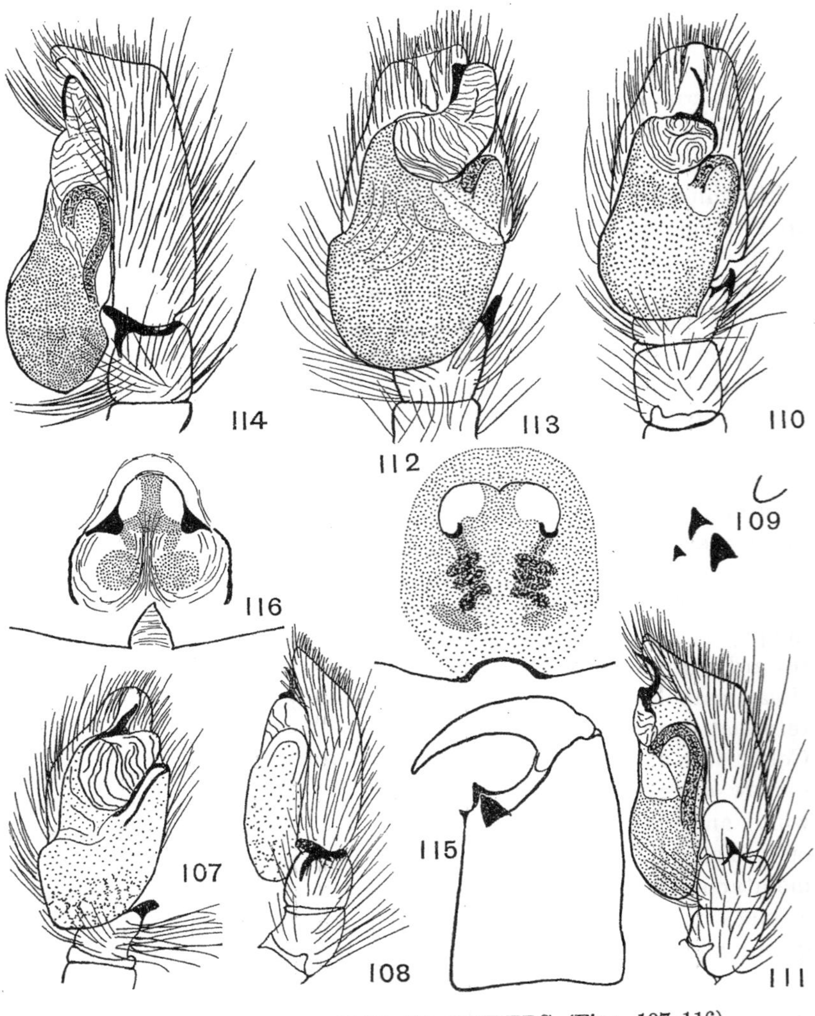

EXTERNAL ANATOMY OF SPIDERS (Figs. 107–116)

107. *Phidippus insigniarius*, male palpus, ventral view
108. *P. insigniarius*, male palpus, retrolateral view
109. *P. insigniarius*, left male cheliceral teeth
110. *Phidippus pius*, male palpus, ventral view
111. *P. pius*, male palpus, retrolateral view
112. *P. pius*, epigynum
113. *Phidippus purpuratus*, male palpus, ventral view
114. *P. purpuratus*, male palpus, retrolateral view
115. *P. purpuratus*, left male chelicera
116. *P. purpuratus*, epigynum

ter has a broad central dark stripe from genital furrow to near base of spinnerets; on each side of the dark stripe is a narrower light stripe; the lateral sides are dark; thus, when the specimen is viewed from the ventral aspect, five rather conspicuous stripes are seen, three dark and two light ones.

Female. — No females yet taken in Michigan. According to Peckham, length varies from 8 to 10 mm. Legs, 4132; first pair fringed as in male. Palp: whole organ covered with long white hairs tinged with red at extremities of patellae and tibiae. Colors of other parts of body seem to be very similar to those of male. Epigynum: not figured by the Peckhams (1909), but said to be very similar to that of *P. johnsoni* (Peckham). During the completion of this paper Miss Elizabeth B. Bryant kindly loaned me for study a specimen of each sex from the collections of the Museum of Comparative Zoölogy at Harvard College. From the female I have drawn the epigynum to show its essential features (Fig. 126).

Distribution. — Known from the Atlantic coastal region from New England to South Carolina; also reported from several Middle Western states, and westward to British Columbia. Some of these records appear to be doubtful. Taken only once in Michigan, by R. R. Dreisbach; Midland County, May, 1941.

Phidippus pius Scheffer

(Figures 110–112)

P. pius Scheffer, 1905.
P. pius Peckham, 1909.

Male. — Length about 6.1 mm. Legs, 1423; first pair considerably the longest but not greatly thickened. Spines: first pair of tibiae with three pairs of moderately robust ventral spines not so crowded toward distal end as usual, promarginal series occupy almost exactly one half and retromarginal series only slightly more than one half of length of segment, also more nearly matched than usual; second pair of tibiae with irregular number of ventral spines, on left side only one promarginal terminal and on right side only one, the promarginal middle one; first two metatarsi with two pairs of ventral spines as usual. Distal outer corners of maxillae with small pointed extensions. Posterior row of eyes wider than anterior row in ratio of 7 : 6; eye space occupies almost four ninths of length of carapace, which is longer than wide in ratio of 89 : 71; posterior row of eyes occupies fourteen seventeenths of width of carapace at that level; PME nearer to ALE than to PLE in ratio of 2 : 3. Fang groove with a single simple retromarginal tooth of moderate size, promargin with two teeth, the inner of which is small. Palp: patella somewhat longer than tibia, which has a short pointed apophysis only gently turned toward bulb; embolus relatively long, slender, and curved ventrally (Figs. 110–111). Color: all legs generally an amber color; all femora and patellae with black prolateral stripes; posterior two pairs of femora and patellae also with black retrolateral stripes; first pair of tibiae dark brown on distal half except on dorsal surface; second pair dark on both prolateral and retrolateral surfaces; last two pairs mostly dark everywhere but on dorsal surfaces; remaining segments mostly light at ends but dark in the middle; carapace a deep orange, with sparse short white hairs generally distributed, clustered somewhat around eyes, which are on black spots. Abdomen reddish dorsally, lighter at sides and ventrally; beginning about one third from base are two dark stripes, within which is a series of two or three pairs of small whitish spots.

Female. — Length about 9.0 mm. Legs, 4123; nearly equally stout with first slightly enlarged. Spines essentially like those of male, but with some variation. Outer distal corners of maxillae smooth. Eyes essentially like those of male. Color: all legs yellowish, with no dark markings; palps yellowish and hairy; all legs somewhat hairy; carapace yellow, with ocular area orange and eyes on black spots; a few long black hairs over whole dorsal surface, with tufts of a few black hairs each in regions of ALE, PME, and PLE. Abdomen with an almost uniformly pale yellowish ground color somewhat lighter laterally and beneath; two central dorsal narrow black stripes begin near the base and extend almost to anal tubercle; two small white spots usually show in posterior halves of the dorsal stripes. Epigynum: posterior marginal notch broad and shallow; near anterior border is a narrow scalloped margin, and behind the margin are two small depressions (Fig. 112).

Distribution. — Described from Kansas; also known from Oklahoma and Michigan. Collected twice in this state, both times in the southern part; Jackson and Hillsdale counties.

Phidippus purpuratus Keyserling

(Figures 113–116)

P. purpuratus Keyserling, 1884.
P. mystaceus Emerton, 1891.
P. borealis Banks, 1895.
P. galathea Peckham, 1901.
P. purpuratus Peckham and Peckham, 1909.

Male. — Length about 9.6 mm. Legs, 1423; first pair robust, much the stoutest but not much the longest; first pair strongly fringed with moderately long hair, others less heavily fringed. Spines: rather weak for legs so robust; first pair of tibiae with three pairs of ventral spines, those along promargin occupy distal half of segment, those along retromargin occupy about two thirds of segment; second pair of tibiae with one pair of terminal ventral spines and one other at about middle on retromargin. First two pairs of metatarsi with the usual two pairs of ventral spines. Distal outer corners of maxillae with very definite small hooks strongly cornified. Posterior end of sternum not so strongly convex as in many species. Posterior row of eyes wider than anterior row in ratio of 48 : 39; eye space occupies about three sevenths of length of carapace, which is longer than wide in ratio of about 7 : 6; posterior row of eyes occupies five sixths of width of carapace at that level; PME nearer to ALE than to PLE in ratio of 1 : 2. Fang groove with a single simple retromarginal tooth of moderate size and two promarginal teeth, the inner of which is very small (Fig. 115). Palp: patella somewhat longer than body of tibia, which has a single simple apophysis turned quite definitely toward the bulb; bulb terminates distally in a broad rugose area, on prolateral side of which the embolus emerges as a short stout spine (Figs. 113–114). Color: all legs and palps a rich dark reddish brown; fringes mostly black except on prolateral side of first patellae, where they are whitish; chelicerae iridescent-green in front and laterally; fang moderately robust and not strongly curved; some white scales on legs, especially the first pair; color of carapace similar to that of legs, with long black hair in tufts chiefly around ALE and PME, shorter black hair fairly uniformly distributed; ocular area darker with eyes on black spots. Abdomen generally lighter-colored; usually an indistinct light basal band, which extends laterally for about one third of distance to spinnerets, a little beyond its termination is a small isolated light oblique mark; centrally there are two faint dark stripes, within which are usually four pairs of whitish spots; a fairly complete covering of whitish scales all over abdomen, but these are easily brushed off; laterally and ventrally color is darker; venter with a central dark stripe bounded by two narrow-beaded light stripes and containing near the middle two more incomplete narrow-beaded stripes.

Female. — Length about 11.0 mm. Legs, 4132. In general, like the male in all essential features of spines and fang groove. Outer distal corners of maxillae without hooks or extensions. Color: legs very similar to those of male except somewhat lighter; palps with much long white hair on last three segments, a tuft of black hair on prolateral side of tarsus and many white scales, especially on dorsal surface. Carapace colored as in male, but with more white scales. Abdomen dull brownish dorsally, with basal light band very narrow and often broken; centrally there are usually two darker stripes, with the four pairs of white spots clearer than in male; many white scales over dorsum and lateral sides; venter and lateral sides very similar to those of male. Epigynum: a fairly deep posterior marginal notch; at some distance anterior to the notch two rather deep and fairly large depressions, only partly separated by a suppressed and incomplete septum; the depressions connect anteriorly with a central depression, which has a very low narrow anterior margin (Fig. 116).

Distribution. — Well known from New England and New York; also known from several localities in the Midwest, and from Missouri, Utah, and Texas. Collected in Michigan at several localities in both peninsulas; fairly common.

Phidippus whitmanni Peckham

(Figures 117–121)

P. rufus Peckham, 1888.
P. whitmanni Peckham and Peckham, 1909.

Male. — Length about 6.9 mm. Legs, 1423; first pair only moderately longer and stouter than others; first pair well fringed (said to vary considerably among different individuals), others weakly fringed. Spines: first pair of tibiae with three pairs of ventral spines, those on promargin occupy slightly less than half of length of segment, whereas those on retromargin occupy almost exactly three fifths of length of segment. Second pair of tibiae with a pair of terminal ventral spines, right and left tibiae irregular with respect to others. First two pairs of metatarsi with two pairs of ventral spines

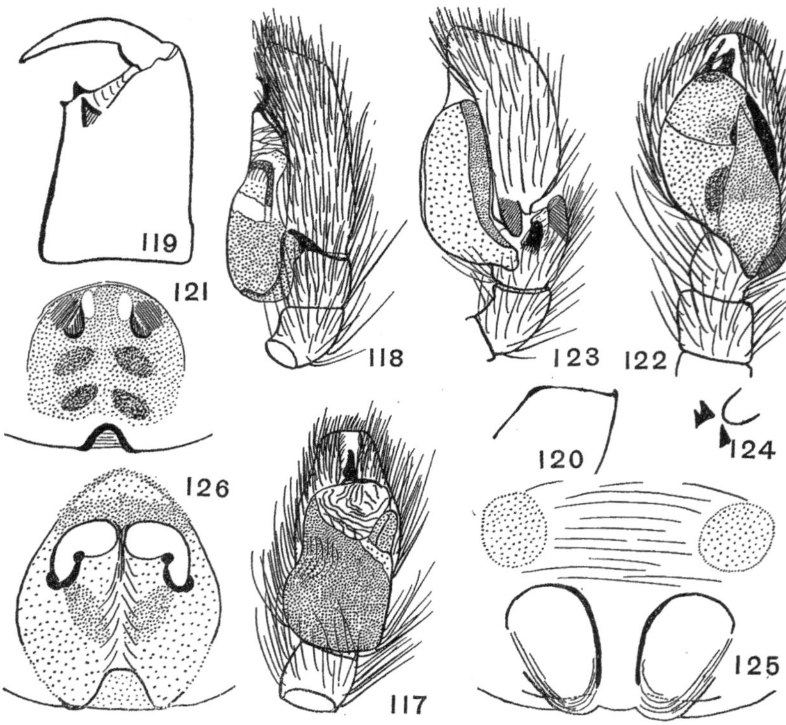

EXTERNAL ANATOMY OF SPIDERS
(Figures 117–126)

117. *Phidippus whitmanni*, male palpus, ventral view
118. *P. whitmanni*, male palpus, retrolateral view
119. *P. whitmanni*, left male chelicera
120. *P. whitmanni*, distal end of maxilla, male
121. *P. whitmanni*, epigynum
122. *Phlegra fasciata*, male palpus, ventral view
123. *P. fasciata*, male palpus, retrolateral view
124. *P. fasciata*, left male cheliceral teeth
125. *P. fasciata*, epigynum
126. *Phidippus insigniarius*, epigynum

as usual. Outer distal corners of maxillae with small hooklike extensions (Fig. 120). Posterior row of eyes longer than anterior row in ratio of about 6 : 5; eye space occupies almost half of length of carapace which is longer than wide in ratio of 49 : 41; posterior eyes occupy about thirteen sixteenths of width of carapace at that level.

Fang groove as usual, with a single retromarginal tooth and two promarginal teeth, the inner of which is small (Fig. 119). Palp: patella as usual, somewhat longer than body of tibia; tibial apophysis the usual spur of moderate size, broad basally, pointed distally, and distinctly turned toward bulb; anterior tarsal rugulose area triangular, with embolus a rather slender spine curved ventrally throughout its whole length (Figs. 117–118). Color: legs a reddish brown, with distal ends of femora and tibiae much darker; first leg with much white hair for fringes and quite well supplied with white scales; other legs much the same but less completely fringed; palps also with much long white hair and white scales, especially on dorsal side; chelicerae without much iridescence; carapace reddish generally, with many white scales, especially along lateral sides, ocular area darker. Abdomen generally reddish over dorsal area, with an obscure light basal band, which extends only a short distance on each side; there are faint dark central broken stripes; about midway there is a faint light oblique bar on each side and farther posterior another pair of smaller transverse bars; just above anal tubercle is a light area; venter generally reddish brown, with three central stripes, a median stripe and two others somewhat more lateral.

Female. — Length about 10.0 mm. Legs, 4123. Females have not been available during the writing of this description, but they are described by the Peckhams (1909) as very similar to the males in most respects. A light stripe extends from between AME to about opposite PLE. The black venter has two light stripes converging toward spinnerets. Epigynum: general shape oval, with a deep-bordered posterior marginal notch, an anterior obscure margin, behind which are small light areas (probably openings) and spermathecae (Fig. 121).

Distribution. — Well known from New England and New York; also reported from several localities in Ohio, Michigan, Illinois, and Wisconsin. Collected only once in Michigan, by R. R. Dreisbach; Lake County, June, 1940.

Genus PHLEGRA Simon, 1876

Carapace long, widest behind the middle, rather low and flat. Ocular space occupies only about one third of length of carapace. PME about midway between ALE and PLE. Metatarsi shorter than tarsi.

Phlegra fasciata (Hahn)

(Figures 122–125)

Salticus fasciatus Hahn, 1831.
P. fasciata Simon, 1901.
P. leoparda Peckham and Peckham, 1909.

Male. — Length about 6.0 mm. Legs, 4312; all pairs nearly equally stout. First pair of tibiae has three pairs of ventral spines, with all but the terminal spines on promargin moved far lateral in position and elongated; second pair of tibiae with considerable irregularity in my only mature male, on right side as described by the Peckhams (1909), with one pair of terminals and two others on retromargin, whereas on left side there are three pairs, as on first pair of tibiae; first two pairs of metatarsi with two pairs of long robust spines. Maxillae dilated distally but smooth over outer distal corners. Posterior row of eyes only slightly wider than anterior row; eye space occupies one third of length of carapace; posterior row of eyes occupies slightly more than four fifths of width of carapace at that level; PME just slightly closer to PLE than to ALE. Carapace longer than wide in ratio of 49 : 32; widest opposite interval between second and third coxae. Fang groove with a single simple rather slender retromarginal tooth and two slender closely crowded promarginal teeth (Fig. 124). Palp: patella and tibia about equal in length and both short; tibia with two retrolateral apophyses of moderate length, the smaller is contiguous to a strongly chitinized basal tarsal apophysis; tarsal bulb considerably swollen basally, at distal end of bulb is what is considered a slender embolus, and beside it a shorter stouter apophysis (Figs. 122–123). Color: all legs and palp generally dark brown, more or less streaked except tarsi, all of which are yellowish; carapace generally a rich dark-mahogany brown, two conspicuous white stripes extend from ALE nearly straight back to posterior margin; clypeus well covered with stiff bluish hairs; a narrow white border extends from clypeus on each side along margin for some distance; ocular area and lateral sides, often described as red, are simply lighter and a different shade of brown in my alcoholic specimens; sternum light brown. Abdomen with three white stripes, less conspicuous than those on carapace; elsewhere dark brown; a weakly developed dorsal shield. Venter dull brown, with four light obscure narrow-beaded stripes.

Female. — Length about 8.0 mm. Legs, 4312. Like the male in all essentials except as noted below. The reddish ocular area, reddish lateral areas of head, and bluish hair on clypeus all lacking; white stripes on carapace somewhat more extensive; usually a little spot of white between PME. Legs lighter brown and with dark annulations at distal and proximal ends of most segments. Epigynum posterior border scalloped; a short distance anterior two shallow depressions are partly encircled by well-defined margins; near anterior boundary two spermathecae lie more than twice the diameter of one of them apart (Fig. 125).

Distribution. — Quite well known from New England, New York, Ohio, and a few localities westward as far as the Rocky Mountains. Collected in Michigan only twice; both specimens are males; Calhoun and Mason counties.

Genus SALTICUS Latreille, 1884

Carapace long, narrow in front, a little wider behind, moderately high and convex above; about two thirds as wide as long. Eye region occupies a little less than half of length of carapace; wider than long in ratio of about 11 : 9; PME either equidistant from ALE and PLE or a little nearer to PLE; width of posterior row considerably less than width of carapace at that level. Legs nearly equally stout. Sternum considerably narrowed in front. Lip a little longer than wide. Chelicerae considerably produced, porrect, and divergent in males; vertical and normal in females; only one simple retromarginal tooth near base of fang.

Salticus scenicus (Clerck)

(Figures 127–130)

Araneus scenicus Clerck, 1757.
Epiblemum faustum Hentz, 1832.
E. scenicum Emerton, 1891.
S. scenicus Simon, 1901.
S. scenicus Peckham and Peckham, 1909.

Male. — Length about 4.0 mm. Legs, 1423; first pair only slightly stouter than others. Ventral spines of first and second pairs of tibiae entirely lacking. Outer distal corners of maxillae not extended into hooks or tubercles. Eyes with usual generic features. Chelicerae strongly produced, porrect, with a long and somewhat sinuous fang; fang groove with a single large retromarginal tooth near base of fang

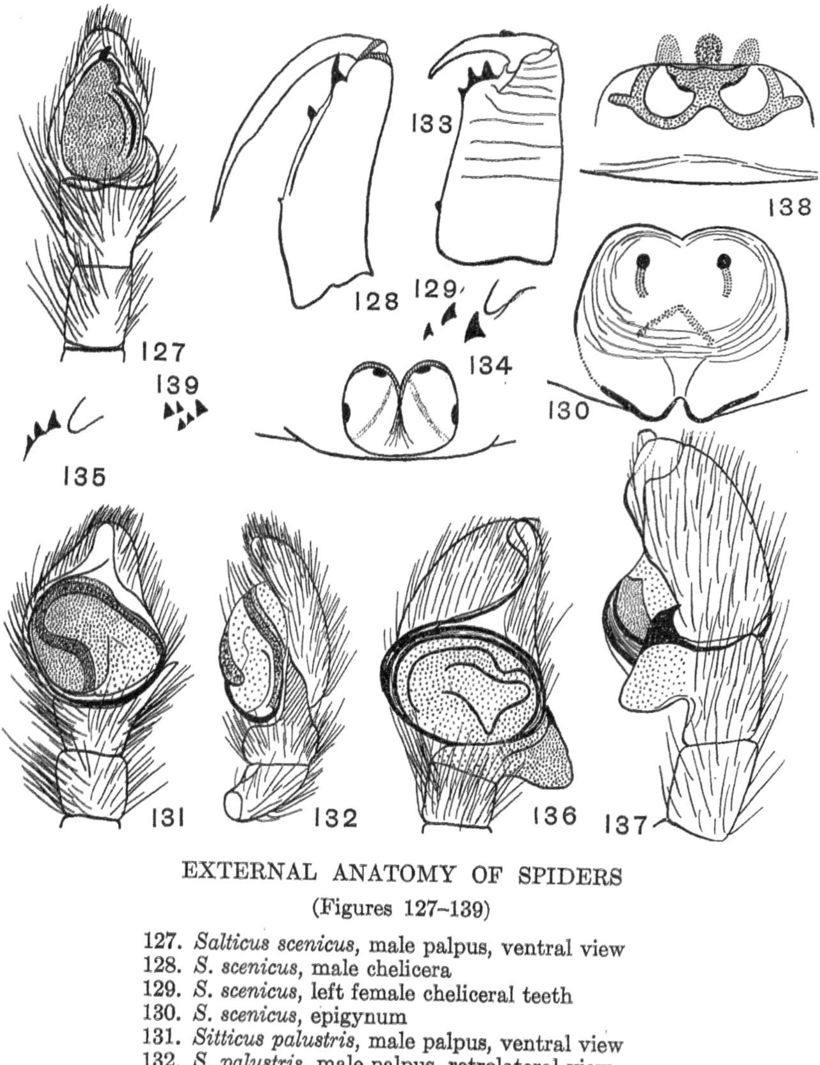

EXTERNAL ANATOMY OF SPIDERS

(Figures 127–139)

127. *Salticus scenicus*, male palpus, ventral view
128. *S. scenicus*, male chelicera
129. *S. scenicus*, left female cheliceral teeth
130. *S. scenicus*, epigynum
131. *Sitticus palustris*, male palpus, ventral view
132. *S. palustris*, male palpus, retrolateral view
133. *S. palustris*, left male chelicera
134. *S. palustris*, epigynum
135. *S. palustris*, left female cheliceral teeth
136. *Synemosyna formica*, male palpus, ventral view
137. *S. formica*, male palpus, retrolateral view
138. *S. formica*, epigynum
139. *S. formica*, left female cheliceral teeth

and two promarginal teeth, one of moderate size about one third from base of fang and a smaller one much farther removed (Fig. 128). Palp: patella somewhat longer than body of tibia; tibial apophysis a robust hook strongly excavated contiguous to base of tarsus; tarsal bulb moderately swollen, terminates in a short spinous embolus at distal end (Fig. 127). Color: all legs generally an amber color, with many dark annulations and streaks; palps amber, with many white scales, especially along dorsal surface and also with a moderate supply of long black hair; legs also with many white scales. Carapace generally dark brown, with ocular area black, sometimes with a lighter center showing two crescent-shaped black spots, but these are seen only when hair and scales have been rubbed off; usually a white central spot just above AME, a pair of dorsal white somewhat triangular spots just behind PLE; also a marginal white stripe all the way from white clypeus to posterior border; often a covering of iridescent light-colored scales over remainder of ocular area. Abdomen generally brownish, with a good deal of iridescence (said to be red in some individuals); a white basal band and two pairs of white oblique bars with more white near base of spinnerets.

Female. — Length 5.4 mm. Legs, 4132. Spines as described for the male. Chelicerae vertical and of normal size. Fang groove with a single stout tooth on retromargin and two promarginal teeth, the inner of which is smaller (Fig. 129). Color: lighter and less brilliant than in male, with less iridescence. Epigynum: relatively large; with a fairly deep posterior notch; a deep depression with scalloped anterior border (Fig. 130).

Distribution. — Reported from the greater part of North America, Europe, and North Africa. Fairly common all over the State of Michigan.

Genus SITTICUS Simon, 1901

Fang groove with no retromarginal teeth (in the only Michigan species), or with a single tooth in west coast species. Carapace fairly high and wide. Eye region considerably wider than long, anterior row about equal in width to posterior row; occupies slightly more than two fifths of length of carapace; PME slightly closer to ALE than to PLE; third row not quite so wide as carapace at that level. Anterior coxae widely separated; sternum oval and somewhat truncated anteriorly; labium as wide as long.

Sitticus palustris (Peckham)

(Figures 131–135)

Attus palustris Peckham, 1883.
S. palustris Simon, 1901.
A. sylvestris Emerton, 1891.
S. palustris Peckham and Peckham, 1909.

Male. — Length about 3.9 mm. Legs, 1423; first pair only a little stouter than others and with no appreciable fringes. Spines: first pair of tibiae with three pairs of ventral spines (the Peckhams record only two pairs, but I have failed to find any of my specimens which have less than one ventral terminal and two other pairs of matched spines, some have three pairs, and I consider this condition to be normal); second pair of tibiae with much irregularity, the most frequent condition being two ventral retromarginals, either with or without a terminal one on promargin; first two pairs of metatarsi with two pairs of ventral spines. Outer distal corners of maxillae without hooks or tubercles. Posterior row of eyes only slightly wider than anterior row; eye region occupies nearly four ninths of length of carapace, which is longer than wide in ratio of 17 : 13; posterior row of eyes occupies nearly fourteen fifteenths of width of carapace at that level; PME nearer to ALE than to PLE in ratio of 5 : 6. Chelicerae normal; fang groove with no retromarginal teeth and with three promarginal teeth closely crowded together (Fig. 133). Palp: patella a little longer than body of tibia; tibial retrolateral apophysis a stout spur, pointed distally and longer than body of tibia; bulb wider than long, containing a long looped tubule; embolus arises near base on retrolateral side and curves around more than two thirds of circumference of bulb (Figs. 131–132). Color: legs and palps generally a light brown, with many dark spots and streaks, also with many white scales dorsally and prolaterally, especially near the joints; palpal femur snow white, also white fringes on patella and tibia; carapace dark brown with scaly iridescence in ocular area, a white margin on clypeus continues on each side to posterior border; a broken white central narrow stripe from posterior declivity to opposite PME; another broken white stripe on each side from near posterior margin to PME and this extends beyond to surround anterior eyes, where there are some reddish hairs. Abdomen: a dull brownish, with a light basal band which extends laterally as a narrow stripe sometimes almost to the anal tubercle; also white spots at end of first quarter, two larger transverse elongated spots making almost a bar across the middle; venter with a covering of dirty white hairs.

Female. — Length, 5.44 mm. Legs, 4123. Spines: show considerable irregularity; apparently one terminal ventral on promargin on first tibia and two other pairs, but those on promargin are moved over so far laterally as to appear almost as much lateral as ventral; second tibiae appear usually to have a single terminal ventral spine on promargin and two others on retromargin. First two pairs of metatarsi have the usual two pairs of ventral spines on each segment. Fang groove with teeth as shown in Figure 135. Color: essentially like that of male but with less striking contrasts. Other general features essentially like those of male. Epigynum: no notch in posterior margin; a single large depression with, perhaps, a suppressed septum in the center (Fig. 134).

Distribution. — Well known from New England, New York, New Jersey, Ohio, Illinois, and Michigan. Collected quite frequently in numerous localities in both peninsulas.

Genus SYNEMOSYNA Hentz, 1845

Most antlike of our Michigan Salticidae. Carapace long and slender; with a deep transverse groove nearly in middle; with posterior end narrowed to a tubular form, thus apparently adding to length of abdomen. Eye space occupies about one third of length of carapace. PME a little closer to ALE than to PLE. Anterior row of eyes strongly recurved. Abdomen also with a dorsal groove (deep in males) just in front of middle.

Synemosyna formica Hentz

(Figures 136–139)

Janus gibberosus C. L. Koch, 1846.
S. formica Emerton, 1891.
S. formica Simon, 1901.
S. formica Peckham and Peckham, 1909.

During the writing of this description I had for study both males and females from Massachusetts and Indiana; they belong to the Museum of Comparative Zoölogy at Harvard College.

Male. — Length, 4.48 mm. Legs, 4312; nearly equally stout. Spines weak and somewhat difficult to see. Chelicerae (see fig. 743,

p. 673, Gertsch-Comstock, 1940) somewhat developed, divergent, and medially somewhat excavated; fang groove with several small teeth on promargin, and a small tooth together with a larger compound tooth on retromargin. Posterior row of eyes wider than anterior row in ratio of 38 : 35; eye space wider than long in ratio of 19 : 17; occupies just a little more than one third of length of carapace, which is longer than wide in ratio of 47 : 19; PME nearer to ALE than to PLE in ratio of 6 : 9; posterior row occupies more than nine tenths of width of carapace at that level; anterior row of eyes strongly recurved; AME close together; diameter of AME greater than that of ALE in ratio of 7 : 3; diameter of PLE greater than that of PME in ratio of 6.5 : 2. Cephalic part of carapace rather high, gently arched to just behind PLE, then carapace dips steeply into deep median transverse groove; beyond groove carapace rises to a height about three quarters as great as elevation of head; then carapace descends again to cylindrical posterior end. Pedicel with a large strongly chitinized dorsal plate and a smaller ventral plate. Abdomen longer than wide in ratio of almost 3 : 1; just beyond first third there is a deep transverse groove, and behind this the posterior half is somewhat bulbous; a well-developed dorsal shield and a strongly chitinized ventral plate in front of the genital furrow extend dorsally to meet the dorsal shield. Palp: patella about as long as tibia, which has a very stout retrolateral apophysis, which is greatly extended ventrally and strongly chitinized anteriorly, where it terminates in a sharp point; the bulb is strongly swollen; the embolus is a long slender filament coiled about the bulb and then extended distally in a long loop (Figs. 136–137). Color: all legs an amber color, progressively darker toward the last pair; first legs with a conspicuous prolateral black stripe on patellae, tibiae, and metatarsi; a less conspicuous brown retrolateral stripe on tibiae and metatarsi; body generally a rich mahogany brown (black on several of my immature specimens); all eyes except AME on black spots, ocular area elsewhere lighter; often a light spot in middle of the transverse groove and another at each end of the groove near the margin. Abdomen usually has a light band across the bottom of the transverse groove continued onto the broad light area on venter.

Female. — Length about 5.7 mm. Essentially like the male except for minor features. Fang groove with teeth as shown in Figure 139. Somewhat lighter-colored than male, and the light spots are much more conspicuous. Dorsal shield reduced to a small cap. Epigynum: a simple area much wider than long, with a series of tubules making an intricate pattern (Fig. 138).

Distribution. — Believed to occur throughout the entire United States and southeastern Canada. Collected in Michigan several times, but only in Calhoun and Midland counties.

GENUS TUTELINA SIMON, 1901

The generic status of the species included here is uncertain, but there is no space in this paper for the discussion of the problem. There are objections to every suggestion thus far made regarding the taxonomic treatment of these species. Tentatively the following features may be considered common and sufficient to establish the genus: carapace of moderate height or lower and flatter (*hartii*); abdomen of moderate length and cylindrical. Legs 1423 in male; 4123 in females. Posterior row of eyes somewhat wider than anterior row; eye space occupies less than one half of carapace; PME closer to ALE than to PLE. Fang groove with an elongated keel-like ridge on promargin and a simple small tooth on retromargin. More or less iridescence on both abdomen and carapace. Male palps built in accord with pattern exhibited by the male of *T. elegans* (Figs. 140–141); epigyna constructed in accord with pattern exhibited by the females of *T. elegans* (Fig. 143).

KEY TO THE SPECIES OF *TUTELINA*

MALES

1. First tibia with a tuft of hair on ventral surface near distal end; embolus a moderately stout spine enlarged near distal end and springing from a bulbous base; tibial apophysis a small pointed spur (Fig. 140) *T. elegans*, p. 211

1. Not with combination of features given above 2

2. Most antlike of the Michigan species; carapace more rounded than usual; tibial retrolateral apophysis a stout spur narrowed and then turned ventrally near distal end; embolus a fairly stout pointed spur (Fig. 144) *T. formicarius*, p. 213

2. Not with combination of features given above 3

3. A short stiff ventral fringe on anterior tibiae extending onto distal end of patella; carapace unusually low and flat; abdomen with a lighter (often reddish) margin nearly encircling dorsum; embolus a stout spine not at all spatulate distally . . . . *T. hartii*, p. 214

3. Resembles *T. elegans;* tibial apophysis short, broad at base, distally pointed; embolus a blunt stout spine, constricted a little in middle, somewhat spatulate distally .... *T. similis,* p. 216

FEMALES

1. Epigynum with a very deep narrow posterior marginal notch and two nearly parallel short black spots on each side near the middle (Fig. 146); no basal abdominal band; legs conspicuously striped *T. formicarius,* p. 213
1. Not with combination of features given above 2
2. Epigynum with a broad, moderately deep posterior marginal notch and a single elongated procurved black spot on each side somewhat in front of middle (Fig. 150); abdomen dorsally dark but marginally lighter (often reddish), with a broad encircling band; legs not conspicuously striped *T. hartii,* p. 214
2. Not with combination of features given above 3
3. Epigynum with a broad and shallow posterior marginal notch; openings to spermathecae more or less oblique (Fig. 154); anterior tibiae with not more than two pairs of ventral spines .... *T. similis,* p. 216
3. Epigynum with a broad and shallow posterior marginal notch slightly scalloped; openings to spermathecae with recurved anterior boundaries (Fig. 143); anterior tibiae ordinarily with three pairs of ventral spines *T. elegans,* p. 211

Tutelina elegans (Hentz)

(Figures 140–143)

Attus elegans Hentz, 1845.
A. superciliosus Hentz, 1845.
Icius elegans Keyserling, 1884.
T. elegans Peckham and Peckham, 1909.

Male. — Length, 4.0 mm. Legs, 1423; first pair much the longest and somewhat the stoutest, but all slender; first pair weakly fringed, but with a conspicuous ventral distal tuft of black hair on first tibiae. Spines: first pair of tibiae with three pairs of ventral spines; second pair of tibiae irregular, but usually with a pair of terminal ventral spines and two others along retromargin; first two pairs of metatarsi with two pairs of ventral spines. Outer distal corners of maxillae extended into small hooks (Fig. 142). Posterior row of eyes a little wider than anterior row; eye space occupies about four ninths of length of carapace, which is longer than wide in ratio of about 3 : 2; posterior row of eyes occupies five sixths of width of carapace at that level. Fang groove with a single rather small retromarginal tooth; promargin with no true teeth, but with an elongated ridge strongly chitinized and with a row of short black stiff bristles.

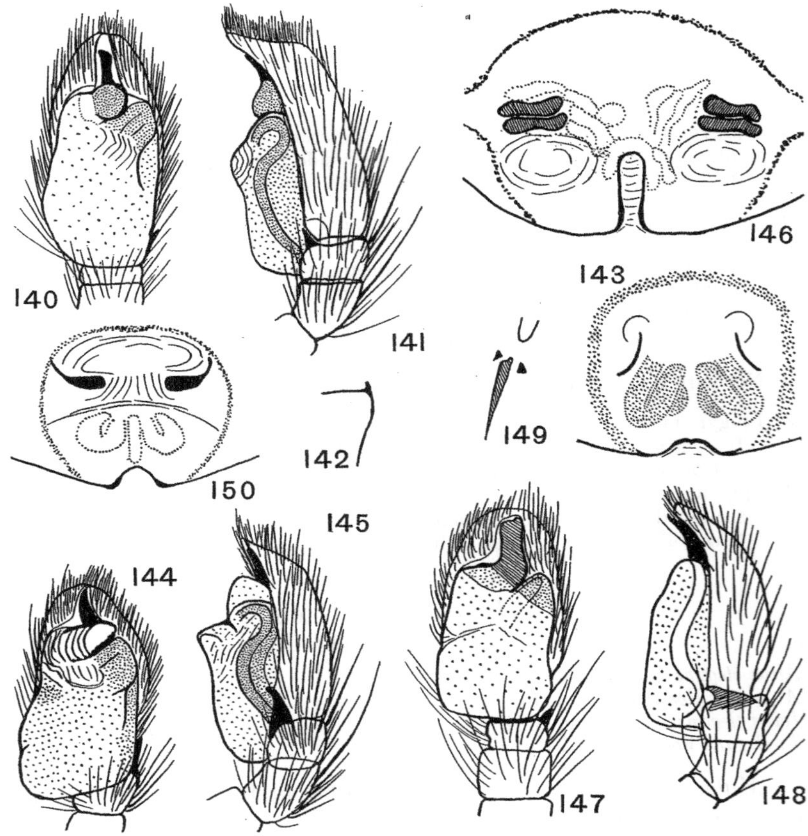

EXTERNAL ANATOMY OF SPIDERS

(Figures 140–150)

140. *Tutelina elegans*, male palpus, ventral view
141. *T. elegans*, male palpus, retrolateral view
142. *T. elegans*, distal end of left male maxilla
143. *T. elegans*, epigynum
144. *Tutelina formicarius*, male palpus, ventral view
145. *T. formicarius*, male palpus, retrolateral view
146. *T. formicarius*, epigynum
147. *Tutelina hartii*, male palpus, ventral view
148. *T. hartii*, male palpus, retrolateral view
149. *T. hartii*, left male cheliceral teeth
150. *T. hartii*, epigynum

Palp: both tibia and patella short, the latter somewhat the longer; tibial apophysis a short, sharply pointed spur; a short but fairly stout embolus arises from bulbous base (Figs. 140–141). Color: legs generally yellowish; dorsal black stripes along most segments. several segments also with prolateral and retrolateral black stripes. Greenish-yellow iridescent scales generally over abdomen and carapace. Most of ocular area black; ocular tufts black, with few white scales; posterior declivity dark brownish, with remainder of carapace a golden yellow overlaid with white scales sparsely distributed; a narrow white marginal stripe. Abdomen generally a dark brown, only a little lighter ventrally.

Female. — Length about 4.5 mm. Legs, 4123. Spines: first and second tibiae may have no spines, but they appear usually to have one or two, none terminal. Outer distal corners of maxillae smooth, without hooks. Other features essentially like those in male. Color: legs and carapace like those in male; abdomen sometimes with a light basal band; otherwise dark like the male, but more iridescence is present. Epigynum: rather inconspicuous and easily confused with others, especially when immature; a fairly broad-bordered notch in middle or posterior margin; a fairly well-defined recurved anterior border; two small openings about twice the long diameter of one of them apart (Fig. 143).

Distribution. — Well known from New England, New York, the eastern coastal states, several southern states, the Midwest as far westward as Nebraska and Oklahoma. Collected in Michigan from several localities in Emmet, Midland, Oceana, and Calhoun counties.

Tutelina formicarius (Emerton)

(Figures 144–146)

Icius formicarius Emerton, 1891.
I. formicarius Banks, 1895.
I. formicarius Peckham and Peckham, 1909.

Male. — Length, 5.0 mm. Legs, 1423; first pair only a little more robust than others; without definite fringes. Spines: first pair of tibiae with three pairs of weak ventral spines; second pair of tibiae with a pair of terminal ventral spines, two more along retromargin and only one more along promargin; first two pairs of metatarsi with two pairs of weak ventral spines. Outer distal corners of maxillae with robust hooks. Carapace fairly robust and rounded above. Posterior row of eyes wider than anterior row in ratio of 12 : 11; ocular space occupies about three sevenths of length of carapace, which is longer than wide in ratio of 35 : 23; posterior row of eyes occupies about six sevenths of width of carapace at that level; PME closer to ALE than to PLE in ratio of 13 : 18. Fang groove with a long chitinized keel-like ridge on promargin and a single simple tooth on retromargin. Palp: patella somewhat longer than short body of tibia; tibial retrolateral apophysis a stout spur, which extends parallel to tibia and near tip turns ventrally; embolus a short curved pointed spur (Figs. 144–145). Color: generally a dark brown; all legs broadly striped but less conspicuously than in *T. elegans;* many white scales on palps and legs, especially the first pair; a white marginal stripe on carapace; a poorly developed dorsal abdominal shield. Some iridescence, especially on abdomen; venter dark except the area in front of the genital furrow, which is much lighter.

Female. — Length about 5.7 mm. Legs, 4123. Essentially like the male except in minor features. Color is more brilliant, at least in those specimens which I have examined. Epigynum: most distinctive in all Michigan species; the posterior marginal notch is very long and narrow; near the middle on each side are two short dark parallel bands (Fig. 146).

Distribution. — Well known from southern New England and New York; also known from Ohio and Michigan. Collected only twice in Michigan, once in Leelanau County and once in Cheboygan County.

Tutelina hartii (Emerton)

(Figures 147–150)

Icius hartii Emerton, 1891.
I. hartii Peckham and Peckham, 1909.

Male. — Length, 4.2 mm. Legs, 1423; first pair considerably the stoutest, especially the femora, which are also flattened laterally, and patellae and tibiae; first pair with fringes, weak on femora but heavy on patellae and tibiae. Spines: first tibiae with three pairs of ventral spines, those on promargin crowded onto distal half, those on retromargin occupying more than distal half; ventral spines on second pair of tibiae same as first, except that the most proximal on promargin are lacking; first two pairs of metatarsi with two pairs of ventral spines each. Whole outer edge of maxillae strongly chitinized and outer distal corners drawn out into short tubercles. Posterior row of eyes only slightly wider than anterior row; ocular space occupies about four ninths of length of carapace; posterior row occupies about five sixths of width of carapace at that level; PME closer to ALE than to PLE in ratio of 5 : 6. Fang groove with a single simple small retromarginal tooth; promargin with a small tooth and a chitinized elongated ridge (Fig. 149). Palp: patella somewhat longer than tibia when viewed dorsally; tibial apophysis a short pointed spur directed ventrally toward bulb; tarsus somewhat swollen basally; bulb almost equally quadrilateral; embolus a stout spine, blunt and slightly notched distally (Figs. 147–148); scantily fringed with long white hair. Color: all legs light brownish, darkened prolaterally and retrolaterally but not definitely striped; a scant covering of white scales; fringe on first pair dark brown. Carapace a medium brown, with dorsal surface lighter except ocular area, which is darker; a narrow white marginal stripe; white scales dorsal to first row of eyes. Dorsum generally a reddish color, with a darker figure within the reddish area; lateral sides and venter almost black; some iridescence shows on both carapace and abdomen. Whole carapace lower and flatter than usual in the genus.

Female. — Length about 4.2 mm. Legs, 4123; first leg essentially like that of male, but the fringe less well developed. Ventral spines of first two pairs of tibiae and metatarsi like those of male, except that on first tibiae they are not crowded toward distal half and the terminal spines on ventral side of second metatarsi are lacking. Outer edges of maxillae strongly chitinized, but not provided with tubercles at outer distal corners like those of male. Color: very similar to that of male; there is more iridescence, especially on abdomen, where there is a pinkish white basal band that extends dorsolaterally somewhat brokenly to posterior end. Epigynum: a broad, moderately deep posterior marginal notch; a little distance in front of middle on each side is a single dark procurved spot (Fig. 150).

Distribution. — Well known in New England and New York; also reported from Ohio, Michigan, Illinois, Wisconsin, and Nebraska. Probably collected in Michigan only four times, always in the vicinity of Albion, Calhoun County, but two females are doubtfully assigned to the species.

Tutelina similis (Banks)

(Figures 151–154)

Icius similis Banks, 1895.
I. similis Peckham and Peckham, 1909.

Male. — Length about 4.0 mm. Legs, 1423; first legs only slightly enlarged and scantily fringed except patella and tibia, where long white hair is well developed (the Peckhams state that males usually have black fringes). Spines: first pair of tibiae usually with two pairs of ventral spines, most proximal pair lacking; second pair of tibiae with only two ventral spines, both retromarginal; first pair of metatarsi with usual two pairs of ventral spines, but second pair of metatarsi apparently has only one pair (terminal); all spines slender, fragile, difficult to see. Outer distal corners of maxillae drawn out into blunt hooks. Posterior row of eyes only slightly wider than anterior row; eye space occupies nearly eight seventeenths of length of carapace, which is longer than wide in ratio of 84 : 55; posterior row of eyes occupies twelve thirteenths of width of carapace at that level; AME closer to ALE than to PLE in ratio of 5 : 6. Fang groove with a small retromarginal tooth and a small promarginal tooth followed by a long low strongly chitinized ridge or keel (Fig. 153). Palp: patella about twice as long as body of tibia, which has a short basally stout distally pointed apophysis, which is directed diagonally toward the distal end of the tarsus; tarsal bulb swollen basally (Figs. 151–152); embolus a stout spine slightly narrowed in the middle, curved ventrally, somewhat spatulate apically (I do not see the distal end as it was illustrated by J. H. Emerton in the Peckhams' [1909] figures). Color: all legs yellowish, with narrow black stripes less distinct than in *T. elegans;* carapace generally reddish brown, with numerous white scales and a few white hairs; ocular area almost black; just above AME are two incurved tufts of brown hairs mixed with a few white scales and hairs; dorsum of abdomen a darker brown with a few white scales, a few light hairs and many glistening iridescent scales; venter is lighter, with many of the glistening scales.

Female. — Length about 4.8 mm. Legs, 4123. Spines: apparently only a distal terminal pair of ventral spines on first pair of tibiae; second pair of tibiae with none on ventral surface; first pair of metatarsi with two pairs of ventral spines; second pair with one pair of ventral spines (terminal). No hooks or tubercles on outer distal

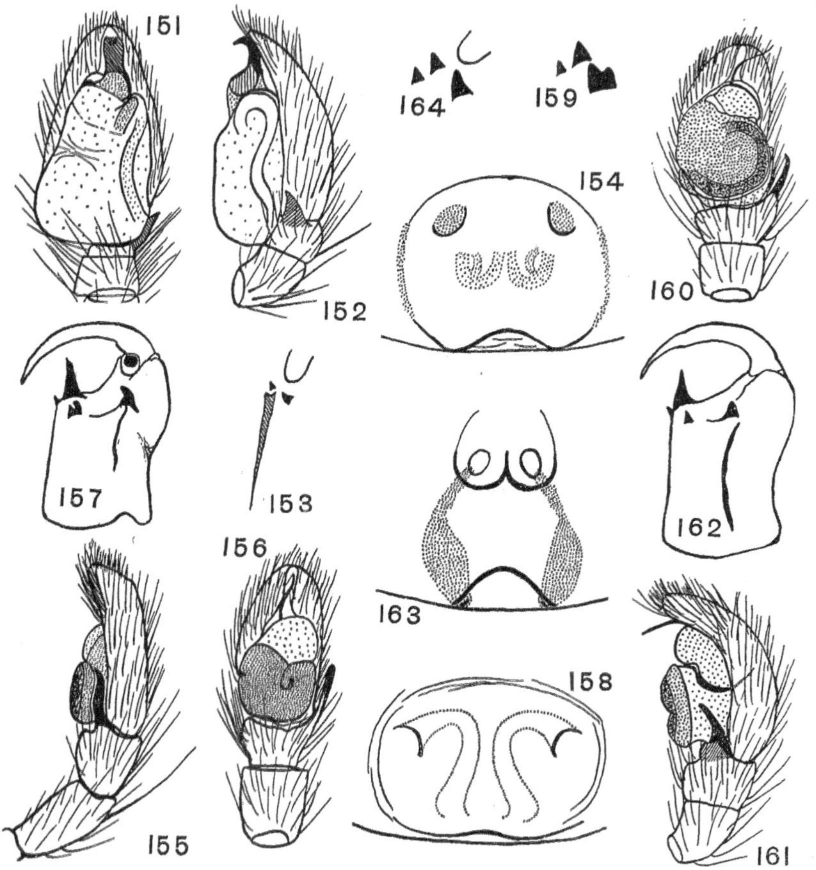

EXTERNAL ANATOMY OF SPIDERS (Figs. 151–164)

151. *Tutelina similis*, male palpus, ventral view
152. *T. similis*, male palpus, retrolateral view
153. *T. similis*, left male cheliceral teeth
154. *T. similis*, epigynum
155. *Zygoballus bettini*, male palpus, retrolateral view
156. *Z. bettini*, male palpus, ventral view
157. *Z. bettini*, left male chelicera
158. *Z. bettini*, epigynum
159. *Z. bettini*, left female cheliceral teeth
160. *Zygoballus nervosus*, male palpus, ventral view
161. *Z. nervosus*, male palpus, retrolateral view
162. *Z. nervosus*, left male chelicera
163. *Z. nervosus*, epigynum
164. *Z. nervosus*, left female cheliceral teeth

corners of maxillae. Color as described for the male, except that legs are more distinctly striped; carapace darker; no hair tufts above AME; abdomen with a faint light basal band, at least in some individuals, darker with fewer iridescent scales. Epigynum: a distinct posterior marginal notch; small openings somewhat oblique in position (Fig. 154).

Distribution. — Known from New England, New York, Michigan, Wisconsin, Illinois, Nebraska, Washington State, New Mexico, Arizona, and British Columbia. Collected in Michigan from several localities in the Lower Peninsula. Females are easily confused with those of *T. elegans*, and they have sometimes been referred to the wrong species.

Genus ZYGOBALLUS Peckham, 1885

Carapace high, a little longer than wide, widest at level of PLE, steeply declivitous behind these eyes. Ocular region about one fifth wider than long, posterior row of eyes somewhat wider than anterior row, ocular region occupies about three fifths of length of carapace. PME nearer to ALE than to PLE; latter row as wide as carapace at that level.

KEY TO THE SPECIES OF *ZYGOBALLUS*

MALES

1. First coxae at least five times as long as wide; anterior face of chelicerae with a stout tooth near base of fang; fang groove with a stout retromarginal, usually hammerheaded, tooth .... *Z. bettini*, p. 218
1. First coxae much less than five times as long as wide; anterior face of chelicerae without stout tooth near base of fang (sometimes a small tooth); fang groove with a stout retromarginal tooth, which is usually not at all hammerheaded .... *Z. nervosus*, p. 220

FEMALES

1. Epigynum with a broad, deep posterior marginal notch; with a single large median opening far from posterior margin and within the large opening two small openings to spermathecae .... *Z. nervosus*, p. 220
1. Epigynum with a very shallow posterior marginal notch; with no large median opening but with the two small openings far apart and directed obliquely .... *Z. bettini*, p. 218

Zygoballus bettini Peckham

(Figures 155–159)

Male. — Length about 4.0 mm. Legs, 1423; first pair much elongated, including coxae and trochanters, with femora considerably thickened. Spines: first pair of tibiae with three pairs of ventral spines, none terminal and not well matched; those on promargin more robust and closer together than those on retromargin; second pair of tibiae with three single unmatched ventral spines along retromargin; first two pairs of metatarsi with two pairs of ventral spines each. Outer distal corners of maxillae without hooks or tubercles, but with whole outer surface somewhat excavated. Positions of eyes generic. Chelicerae strongly developed, somewhat porrect, strongly divergent; with a prominent tooth on anterior face near base of fang, which is quite sinuous; fang groove with a small bifid tooth and a large somewhat hammerheaded excavated tooth on retromargin; also a large elongated promarginal tooth, at base of which is a small one (Fig. 157). Palp: patella somewhat longer than body of tibia; tibial retrolateral apophysis a slender spur reaching to transverse division of tarsus; tarsus divided into a lighter anterior smaller part and a darker reddish-brown posterior part; embolus a slender elongated spine (Figs. 155–156). Color: first pair of legs, chelicerae, and palps all a rich reddish brown, all other legs yellowish, with a few reddish-brown spots and regions, especially on fourth legs. Carapace a rich reddish brown overlaid with white scales; eyes located on black spots; clypeus with many white scales, which extend over anterior face of chelicerae. Abdomen a lighter reddish brown dorsally, with a light basal band which extends laterally for a short distance on each side; also two pairs of somewhat oblique light bars farther posterior and fine light lines over posterior half.

Female. — Length about 4.5 mm. Legs, 4123; first pair somewhat enlarged as in male, but coxae and trochanters not much elongated. Ventral spines of first two pairs of tibiae and metatarsi as recorded for the male, except that second tibiae have only two on retromargin. Chelicerae normal; fang groove with two simple promarginal teeth and a stout bifid retromarginal tooth (Fig. 159). Color: first pair of legs with only femora a rich reddish brown, other segments of first legs and all other legs and palps also yellowish, with a few reddish-brown spots, mostly at distal ends of segments of legs. Carapace essentially like that of males. Abdomen lighter-colored, with more light spots dorsally; in addition to light basal band there is a series of paired central light spots and dark spots alternating from base to near anal tubercle, but these are variable in extent and number in different individuals. Epigynum: somewhat wider than long; very simple, with a shallow posterior marginal notch and two small openings far apart and oblique in position (Fig. 158).

Distribution. — Well known from New England southward to Florida and westward to Nebraska and Texas. Collected in several localities in the southern part of the Southern Peninsula of Michigan; sometimes found in large numbers in a small area, as they were near Homer, Calhoun County, June, 1935.

Zygoballus nervosus (Peckham)

(Figures 160–164)

Eris nervosus Peckham, 1891.
Z. terrestris Emerton, 1891.
Z. nervosus Peckham and Peckham, 1909.

Male. — Length about 3.5 mm. Legs, 1423; first pair somewhat enlarged, especially the femora. Spines: first pair of tibiae with three pairs of ventral spines; second pair of tibiae with only two ventral spines, both retromarginal; first two pairs of metatarsi with two pairs of ventral spines each. Distal outer corners of maxillae without hooks or tubercles; outer surfaces only mildly excavated. Positions of eyes generic. Chelicerae strongly developed, porrect, divergent, deeply excavated along fang groove; either with or without a small tooth on front face near base of fang; fang groove with a large retromarginal tooth not hammerheaded, as in the male of *Z. bettini*, and a small simple retromarginal tooth near inner end of groove; also a single stout promarginal tooth with a small one at its base (Fig. 162). Palp: segments generally more robust than in *Z. bettini;* patella somewhat longer than body of tibia, but both segments about as wide as long; tibial apophysis a slender spur, but does not extend more than halfway to transverse division of bulb; embolus a slender considerably elongated filament (Figs. 160–161). Color: all legs generally a light amber, first pair of femora a reddish brown and other femora darkened a little prolaterally and retrolaterally. Palps and chelicerae a reddish brown. Carapace reddish brown, with eyes on black spots, with numerous white scales; clypeus covered with white scales; considerable iridescence in ocular area. Abdomen a rich reddish brown dorsally, with several faint light chevrons in posterior half; a light basal band, which continues laterally for more than half of distance to spinnerets.

Female. — Length about 4.2 mm. Legs, 4123; first pair enlarged as in male. Ventral spines of first two pairs of tibiae and metatarsi essentially like those in male. Chelicerae normal; fang groove with two simple promarginal teeth and a single simple retromarginal tooth (Fig. 164). Color: legs much as in male, but with many reddish-brown spots and darkened areas. Carapace very similar to that of male. Abdomen lighter than in male; in some individuals whole dorsum a mottled reddish brown and light yellow; in others a light basal band followed by a series of oblique light bars with two large light spots behind them; still farther back there are often several indistinct chevrons. Epigynum: much longer than wide; a broad and fairly deep posterior marginal notch; a single obscure central opening with two smaller openings within (in well-developed and cleaned individuals [Fig. 163]).

Distribution. — Well known from New England and New York, southward as far as Virginia and westward as far as Illinois. Fairly common in the Southern Peninsula of Michigan, where it has been taken in several localities.

Albion College
Albion, Michigan

GENERAL REFERENCES

Bryant, Elizabeth B. 1942. Descriptions of Certain North American Phidippus (Araneae). Am. Mid. Nat., 28 (3): 693–707.

Cambridge, F. 1897–1905. Arachnida. Araneida and Opiliones. *In* Biologia Centrali-Americana, Vol. II.

Chickering, Arthur M. 1939. Anyphaenidae and Clubionidae of Michigan. Pap. Mich. Acad. Sci., Arts, and Letters, 24 (1938), Part II: 49–84.

—— 1940. The Thomisidae (Crab Spiders) of Michigan. *Ibid.*, 25 (1939): 189–237.

Comstock, John H. 1940. The Spider Book. Revised and Edited by W. J. Gertsch, Ph.D. Doubleday, Doran and Company, Inc., New York.

Peckham, George W., and Peckham, Elizabeth G. 1909. Revision of the Attidae of North America. Trans. Wis. Acad. Sci., Arts, and Letters, 16, Part I, No. 5: 355–646.

Petrunkevitch, A. 1911. A Synonymic Index-Catalogue of Spiders of North, Central and South America with All Adjacent Islands, Greenland, Bermuda, West Indies, Tierra del Fuego, Galapagos, etc. Bull. Am. Mus. Nat. Hist., Vol. 29.

PETRUNKEVITCH, A. 1928. Systema Aranearum. Trans. Conn. Acad. Arts and Sciences, 29: 1–270.

—— 1939. Catalogue of American Spiders. Part I. *Ibid.*, 33: 133–338.

SIMON, EUGÈNE. 1892–1903. Histoire naturelle des araignées. Deuxième édition. Vols. I–II. Librairie encyclopédique de Roret, Paris.

NOTES ON NAIADES FROM THE GREEN, SALT, AND TRADEWATER RIVERS IN KENTUCKY

WILLIAM J. CLENCH

Museum of Comparative Zoölogy
Harvard University

and

HENRY VAN DER SCHALIE

University of Michigan

SHORTLY before his death in 1927 Dr. A. E. Ortmann published a paper on the mussels of the Green River drainage in Kentucky. On the basis of the factual information set forth in his report (1926 : 186) he was able to stress particularly that "there is a sharp line between Cumberland and Green Rivers in southern Kentucky, separating two apparently old faunas, the Ohioan and the Cumberlandian." Additional information presented here indicates that this observation is correct.

In 1927, the year following Ortmann's publication, Mr. William J. Clench and Dr. Peter Okkelberg visited the Green River, collecting for the Museum of Zoology, University of Michigan. A special effort was made to obtain specimens from the lower portion of that drainage, a region hitherto unexplored and one obviously in need of examination (see map, Plate VIII in Ortmann's 1926 paper). Also, sufficient collections were made in the Tradewater and Salt rivers to establish the fact that an Ohioan fauna was present in these important tributaries to the Ohio. Ortmann was well aware of the urgency of further investigation in the Salt River when he wrote: "There is yet in Kentucky the Salt River system, between the Green and Kentucky Rivers. . . . However, it is to be expected, that this drainage also has a fauna similar to those of the Kentucky and Green Rivers, that is to say, an Ohioan fauna, without typical Cumberlandian elements."

TABLE I

MUSSEL DISTRIBUTION OF THE GREEN RIVER DRAINAGE OF KENTUCKY, TOGETHER WITH THOSE OF THE TRADEWATER AND SALT RIVERS

Species	Nolin River		Barren River drainage										Mud River		Rough River		Lower Green River				Tradewater River, 4 mi. w. of Providence, Webster Co.	Salt River	
	near Glendale, Hardin Co.	7 mi. n. of Upton, Hardin Co.	Bay's Fork, Claypool, Warren Co.	Trammel Cr., near Allen Springs, Allen Co.	Beaver Cr., 2 mi. s. of Game, Barren Co.	West Fork Drake's Cr., Massey Mill, Warren Co.	Drake's Cr., 7 mi. s. of Bowling Green, Warren Co.	Drake's Cr., 1 mi. e. of Mt. Victor, Warren Co.	Barren River, 2 mi. se. of Austin, Barren Co.	Barren River, 2 mi. n. of Meador, Allen Co.	Barren River, 1 mi. n. of Claypool, Warren Co.	Barren River, 4 mi. e. of Bowling Green, Warren Co.	Wolf Lick Cr., 4 mi. sse. of Dunmore, Logan Co.	Mud River, 3 mi. e. of Dunmore	8 mi. e. of Dundee, Ohio Co.	at Dundee, Ohio Co.	nw. of Rio, Hart Co.	Rochester, Butler Co.	6 mi. e. of Calhoun, McLean Co.	at Calhoun, McLean Co.		Beech Fork, Bardstown, Nelson Co.	Rolling Fork, New Haven, Nelson Co.
2. Fusconaia subrotunda	..	..	..	..	..	..	..	..	..	..	..	+	..	..	..	..	+	..	..	..	..	..	..
3. flava	..	+	+	+	..	+	+	..	+	..	..	..	..	..	..	..	+	..	..	..	+	..	+
4. undata	..	..	..	..	..	..	..	..	..	..	..	..	..	..	+	+	..	+	..	+	..	..	..
5. Megalonaias gigantea	..	..	..	..	..	..	..	..	..	..	+	..	+	+	+	+	+	+	..	+	+	+	+
6. Amblema costata	+	+	+	..	+	+	+	+	+	+	+	+	+	+	+	+	+	..	..	+	+	+	+
7. peruviana	..	..	..	..	..	..	..	..	..	..	..	..	..	..	..	..	..	+	..	+	..	..	..

8. Quadrula pustulosa	..	..	..	..	..	+	..	+	+	..	..	+	..	..	+	+	+	+	..	+	+	+	+
9. quadrula	..	..	..	..	..	..	..	..	..	..	..	..	..	..	..	+	..	+	..	+	+	+	+
10. nodulata	..	..	..	..	..	..	..	..	..	..	..	..	..	..	..	+	..	+	..	..	+	..	..
12. cylindrica	..	..	..	..	..	+	..	..	..	..	..	+	..	..	..	..	..	..	..	..	..	..	..
13. Cyclonaias tuberculata	..	..	+	..	..	+	+	..	+	+	..	+	..	..	..	..	+	..	..	..	..	..	..
14. Plethobasus cooperianus	..	..	..	..	..	..	..	..	..	..	..	..	..	..	+	..	..	..	..	..	..	..	..
15. Tritigonia verrucosa	..	..	+	+	+	+	+	..	+	+	+	..	+	+	+	+	..	+	..	+	+	+	+
16. Pleurobema cordatum	..	..	..	..	..	+	..	..	..	..	..	+	..	..	..	..	..	..	+	+	..	..	..
17. cordatum catillus	..	..	..	..	..	..	..	..	..	..	..	+	..	..	..	..	+	..	..	..	..	..	..
18. cordatum coccineum	..	..	..	..	..	..	..	..	+	..	..	..	..	..	+	..	..	..	..	..	..	..	..
19. cordatum pyramidatum	..	..	..	..	..	..	..	..	..	..	..	..	..	..	..	..	..	+	..	..	..	..	..
20. clava	..	..	..	..	..	..	..	..	..	..	..	+	..	..	..	..	+	..	..	..	..	..	..
21. Elliptio crassidens	..	..	..	..	..	..	..	..	..	..	..	..	..	..	..	..	..	+	..	+	+	..	..
22. dilatatus	+	+	+	..	..	+	+	+	+	..	..	+	..	..	+	+	+	+	..	+	..	..	+
24. Arcidens confragosus	..	..	..	..	..	..	..	..	..	..	..	..	..	+	..	+	..	..	..	..	+	..	..
25. Lasmigona costata	+	+	+	..	..	+	+	+	+	+	+	+	..	+	+	+	..	+	..	..	..	+	+
26. complanata	..	..	..	..	..	..	..	..	..	..	..	..	..	..	..	+	..	+	..	..	..	+	..
27. Anodonta imbecillis	..	..	..	..	..	..	..	..	..	..	..	..	..	..	..	..	..	..	..	..	..	..	+
28. grandis	..	..	+	..	..	..	..	..	..	..	..	..	..	..	..	..	..	..	..	..	..	..	..
30. Alasmidonta calceolus	..	..	..	..	..	+	+	..	+	..	..	..	..	..	..	..	..	..	..	..	..	..	..
31. marginata	..	..	..	..	..	+	+	..	..	..	..	..	..	..	..	..	..	..	..	..	..	..	..
32. Strophitus rugosus	+	+	..	+	..	+	+	..	+	..	..	+	..	..	..	..	..	..	..	..	..	..	..
33. Ptychobranchus fasciolaris	..	..	..	..	..	+	+	+	+	+	..	+	..	..	+	+	+	+	..	+	..	..	+
34. Obliquaria reflexa	..	..	..	..	..	..	..	..	..	..	..	..	..	..	..	+	..	+	..	+	+	..	..
35. Cyprogenia irrorata	..	..	..	..	..	..	..	+	..	..	..	+	..	..	..	..	+	+	..	..	..	..	+
37. Obovaria subrotunda	..	..	..	..	..	+	..	..	..	..	..	..	..	..	..	..	..	+	..	+	..	..	+
38. Actinonaias carinata	..	..	+	..	+	+	+	+	+	+	+	+	..	..	+	+	+	+	..	+	+	..	+
39. Plagiola lineolata	..	..	..	..	..	..	..	..	..	..	..	..	..	..	..	..	..	+	..	..	..	..	..
40. Leptodea fragilis	..	..	..	..	+	..	..	+	..	..	..	+	..	+	..	+	..	..	..	+	+	+	..
41. Proptera alata	..	..	+	..	+	+	+	+	+	+	+	+	+	+	..	+	..	+	+	+	..	+	+
43. Carunculina glans	..	..	+	..	..	+	+	..	..	..	..	..	..	..	..	..	..	+	..	..	..	..	..
45. Micromya nebulosa	..	..	..	..	..	+	..	..	..	..	..	..	..	..	..	..	..	..	..	..	..	..	..
46. lienosa	..	+	+	..	..	+	+	+	..	..	..	+	..	+	..	..	..	+	..	..	..	..	..
47. ortmanni	..	..	+	..	..	+	+	+	..	..	..	+	..	..	..	..	+	..	..	..	..	..	..
49. Ligumia recta latissima	..	..	..	..	..	..	..	+	..	..	..	..	..	..	..	..	..	+	..	..	..	..	..
50. Lampsilis anodontoides	..	..	..	..	..	..	..	..	..	..	..	..	..	..	+	+	..	..	..	..	..	..	+
51. anodontoides fallaciosa	..	..	+	..	..	..	..	..	..	..	..	..	+	..	..	..	..	+	..	+	+	+	..
52. siliquoidea	+	+	+	+	..	+	+	+	..	..	..	..	+	+	+	+	+	+	..	+	+	..	+
53. ovata	..	+	+	..	..	..	..	+	..	..	..	..	..	..	..	..	+	+	..	+	..	+	..
54. ovata ventricosa	..	..	..	..	+	+	+	..	..	..	..	+	..	..	+	+	..	..	..	..	+	..	+
55. fasciola	..	..	+	..	..	+	+	..	..	..	..	+	..	..	..	..	..	..	..	..	..	..	..
56. Truncilla donaciformis	..	..	..	..	..	..	..	..	..	..	..	..	..	..	..	..	..	..	..	..	+	..	..
57. truncata	..	..	..	..	..	+	..	..	..	..	..	..	..	..	+	+	..	+	..	+	+	..	+
58. Dysnomia triquetra	..	..	..	..	..	..	..	..	..	..	..	..	..	..	..	..	+	..	..	..	..	..	..
61. sulcata	..	..	..	..	..	..	..	..	..	..	..	..	..	..	..	..	..	+	..	..	..	..	..

With the data now available we need no longer merely suppose that the Salt River contains a mussel assemblage similar to that of the Green River. Furthermore, the collection made in the Tradewater clearly shows the absence of Cumberlandian species in that drainage system.

In order to simplify and clarify the presentation of the pertinent factual data contributed by the collections of Clench and Okkelberg, a distribution chart has been prepared (Table I). The species list is similar to that given by Ortmann for the Green River. The column headings are so arranged that collections of species (indicated by +) in the tributaries and the upper reaches of the river are noted first. All the records are new and serve to supplement those given by Ortmann in his Green River report.[1]

An examination of the chart shows some interesting conditions when the species are considered in the light of information recorded by Ortmann. Several listed by him were not found by Clench and Okkelberg. Chief among these are: *Fusconaia ebenus, Plethobasus cyphyus, Anodontoides ferussacianus, Carunculina parva, Micromya fabalis, Ligumia subrostrata, Truncilla donaciformis*, and *Dysnomia flexuosa*. It is important in this connection to note that all but the last of these were given by Ortmann on the strength of the list presented by S. F. Price (1900). Although all eight species are Ohioan and likely to occur in the Green River, we must state that they have not been discovered. Several others not taken by Clench and Okkelberg are known to be in the Green River, but are collected only rarely. They are: *Quadrula metanevra, Anodonta imbecillis* (taken in Salt River), *Lastena lata, Anodonta suborbiculata, Obovaria retusa*, and *Dysnomia torulosa*. For some curious reason Ortmann failed to list *Lasmigona complanata* as a possibility in the Green River. Perhaps the fact that so few collections were available to him from the lower portion of that drainage accounts for this omission. The chart clearly shows that *L. complanata* is confined to the lower Green. Another addition to the list presented by Ortmann is *Dysnomia sulcata*. He states (1926 : 185): "All these representatives have been found in Green River, with the exception of *Dysnomia sulcata*." Al-

[1] The following species of Ortmann's list are not represented in the table: (1) *Fusconaia ebenus*, (11) *Quadrula metanevra*, (23) *Lastena lata*, (29) *Anodontoides ferussacianus*, (36) *Obovaria retusa*, (42) *Carunculina parva*, (44) *Micromya fabalis*, (48) *Ligumia subrostrata*, (59) *Dysnomia torulosa*, (60) *D. flexuosa* = *foliata*.

though he was forced to assume the absence of *D. sulcata* in this river, later collections now reveal its presence.

Unless one is familiar with the animals under discussion, it will be difficult to appreciate just what species are regarded as typifying a "Cumberlandian" naiad fauna. Even among conchologists this matter is not altogether clear because the exact ranges of certain mussels must be determined before it can be ascertained to which major faunal assemblage a doubtful species belongs. However, Ortmann (1924: 42–44) has suggested a list of species to be considered Cumberlandian. The names are given here so that one may see by comparison the striking contrast between this compilation of Cumberlandian elements and those in the distribution chart that contains a purely Ohioan or Interior Basin fauna. The following list comprises the species that are unquestionably restricted to the upper Cumberland River:

Fusconaia barnesiana
Quadrula intermedia
Pleurobema oviforme
Alasmidonta minor
Ptychobranchus subtentum
Actinonaias pectorosa
Carunculina moesta
Medionidus conradicus
Pegias fabula
Dromus dromas
Micromya nebulosa
Micromya vanuxemensis
Micromya taeniata
Micromya trabalis
Dysnomia brevidens
Dysnomia lenior
Dysnomia turgidula
Dysnomia florentina
Dysnomia capsaeformis
Dysnomia arcaeformis
Dysnomia haysiana
Dysnomia lewisi

Three additional Cumberlandian species are in the upper Tennessee River drainage, but are absent from the Cumberland River:

Lexingtonia dolabelloides
Lasmigona holstonia
Conradilla caelata

The naiad fauna of the Salt and Tradewater rivers is clearly Ohioan, as is shown by the chart. The addition of the Tradewater River to the list of rivers lacking a trace of Cumberlandian elements is of particular interest since it reveals that the line separating these two distinct faunal assemblages, the Ohioan and the Cumberlandian, is evidently just as sharp as Ortmann anticipated. It would be interesting to have an explanation of the presence of such clearly defined distribution patterns from those who maintain that passive agents other than fishes are responsible for the distribution of the mussels now found in those rivers.

LITERATURE CITED

Ortmann, A. E. 1926. The Naiades of the Green River Drainage in Kentucky. Ann. Carnegie Mus., 17: 167–188.

—— 1924. The Naiad-Fauna of Duck River in Tennessee. Am. Mid. Nat., 9: 18–62.

Price, S. F. 1900. Mollusca of Southern Kentucky. Nautilus, 14: 75–79.

THE AMPHIBIANS AND REPTILES OF THE UNIVERSITY OF MICHIGAN BIOLOGICAL STATION AREA IN NORTHERN MICHIGAN

CHARLES W. CREASER

INTRODUCTION

IN 1908 the University of Michigan established the Biological Station at Douglas Lake, Cheboygan County, Michigan. Since then a great deal of work has been done on the vertebrate fauna of the region, and it is now possible to evaluate and summarize results. This paper, which is the first of a series intended to cover the general subject, is devoted to herpetology. Incidentally it will serve the beginning student as a guide in investigating the herpetology of any given area. A wide range of literature on taxonomy, distribution, and life history of amphibians and reptiles has been included, so as to make possible a broad understanding of the relationship of local species to those of North America as a whole.

Many zoologists use specimens of the amphibian and reptilian fauna of the station for studies in related fields, such as parasitology and ecology. For their benefit notes are introduced on the life histories of animals observed in the area and on the location of habitats where they occur in greatest abundance. A number of papers in the literature cited will be of assistance in both field and laboratory studies.

No attempt has been made to compile a complete bibliography; in general, citations are made only of important standard works of recent date.

THE AREA OF THE BIOLOGICAL STATION

The region selected for this report is limited to Cheboygan and Emmet counties. For the most part it is a natural geographic area;

within it is all the shoreline of the northern tip of the Southern Peninsula from Little Traverse Bay to Hammond Bay. The southern line of these counties crosses the tip from Little Traverse Bay, Lake Michigan, almost to Rogers City, on Lake Huron. Bois Blanc Island (Mackinac Co.), off the northeast shore of Cheboygan County, is within the natural area. A small district near Rogers City is also outside this county.

Because of unsatisfactory winter-storage facilities only a very small collection of specimens is kept at the station, but the local fauna is fully represented at the Museum of Zoology of the University of Michigan. The taxonomy of the species here listed follows that in use at the museum.

This study has been aided by Dr. G. R. La Rue, former director of the station, and Dr. A. H. Stockard, the present director. Mrs. Helen T. Gaige, of the University of Michigan, identified specimens and read the manuscript. Many other persons, including staff members, students, and assistants, have supplied valuable material.

GENERAL LITERATURE

As a basis for the taxonomy of the amphibians and reptiles, and also as a source for English names, the recent check list of Stejneger and Barbour (1939) for these groups in North America has been used. Material on all the amphibians and reptiles known to occur in the northeastern part of North America may be found in the volumes by Pratt (1923) and Jordan (1929). The distribution of certain species in North America is treated by Schmidt in an interesting article, "Herpetological Evidence for the Post-glacial Eastward Extension of the Steppe in North America" (1938a). The journal *Copeia* is the only one covering the field exclusively. It contains many papers on the life histories of the amphibians and reptiles, together with reviews of the literature.

REGIONAL LITERATURE

In the regional literature of vertebrate fauna one may mention first *The Herpetology of Michigan* (1928), by Ruthven, Thompson, and Gaige. The authors have kindly made available to me material they have gathered for a revised edition of this booklet. Another valuable work is that by F. N. Blanchard, whose *Natural History of Vertebrates* (1935) gives the forms and life histories of amphibians and reptiles and also contains a study of the literature.

The herpetology of the region of the Biological Station has received considerable attention. The first publication, by Ruthven (1911a), recorded the presence of fourteen species. Four more were added by Ruthven, Thompson, and Thompson (1912), and the annotated check list by Ellis (1917), who studied the ecology of the herpetological fauna, contains six not previously noted. Creaser (1928) found one new species. An article by Blanchard (1928a) lengthened the list by four, but the printer inadvertently omitted *Natrix s. sipedon.* Ruthven, Thompson, and Gaige (1928) made another addition. The present paper gives two more species known to live in the region, making a total of thirty-two.

Class AMPHIBIA

The only general taxonomic work on all the amphibia of North America is the *Batrachia of North America,* published by Cope in 1889. A more recent account of this class is Noble's *Biology of the Amphibia,* which appeared in 1931. For a classification of amphibians giving characteristics of the orders see Blanchard (1935), page 31. Moore's studies (1939, 1940 a–b, 1942) on the temperature tolerance and rates of development of the eggs of amphibia are important contributions to our knowledge of ecology and distribution.

A handbook by Logier, *The Amphibians of Ontario* (1937), and an article by Smith, "The Amphibians of Kansas" (1934), present material that enables one to make interesting comparisons with the fauna of this region.

Order Caudata — Tailed Amphibians

Four families are represented in the salamander fauna of the Douglas Lake region, three by a single genus and the fourth by but two. Only six species have been found. They are rather uniformly distributed throughout the state, and their occurrence here is of no zoogeographic importance, except that all are near the northward extension of the range of the order in eastern America. It may also be recorded that *Ambystoma tigrinum,* the only other salamander known to occur in Michigan but not collected in this region, ranges northward to Manistee County in the north-central part of the state.

The reason for its failure to extend its range is not now evident, but it is a clearly established fact that the central part of lower Michigan is the northern limit for several species of amphibians and reptiles. The range of *Ambystoma tigrinum*, however, extends into Canada in the area just to the west of Michigan.

There is no general publication on the salamanders of North America, and the special literature on the subject is widely scattered. A list of the species occurring in North America has been published by Stejneger and Barbour (1939), and keys for northeastern forms, together with descriptions of them, may be found in the works of Cope (1889) and Bishop (1943) and in the popular books by Pratt (1923) and Jordan (1929). Dunn's volume (1926) on the family Plethodontidae, which is here represented by two species, is an excellent treatment. The distribution and life history of Michigan fauna have been ably presented by Ruthven, Thompson, and Gaige (1928), who give keys for the identification of species in this region. Keys to the Michigan species, as well as life histories and a discussion of methods of research, appear in a helpful volume by Blanchard (1935). Schmidt's leaflet, *The Salamanders of the Chicago Area* (1930), covers the salamanders of this region. A fine work on the salamanders of New York, with detailed descriptions of their life histories and habits, is that by Bishop (1941), which contains a lengthy bibliography.

Several important papers on life histories of species occurring in the Douglas Lake district have been published by Dr. Blanchard and his students. They are frequently cited here in connection with names of species.

Family I. Necturidae

1. *Necturus* Rafinesque

1. *Necturus maculosus maculosus* (Rafinesque), mudpuppy

Occurrence. — Abundant. The station beach, to which many dead ones drift; boathouse; Hook Point, Sedge Point, Douglas Lake; Crooked Lake; Big Stone Bay.

Life history. — Young found in beach pools at Douglas Lake, and eggs near Hook Point, Douglas Lake; East Maple River Bridge; Indian River on Indian River; Crooked Lake in early July; and Cheboygan River near the paper mill, July 2, 1937.

Literature. — Several subspecies for North America are listed by Viosca (1937) and Stejneger and Barbour (1939). Eycleshymer (1906), Ruthven (1911b), Smith (1911a), Ellis (1917), Pearse (1921), Bishop (1926, 1932), Blanchard (1928a), Hamilton (1932), Senning (1940).

Family II. Salamandridae

2. *Triturus* Rafinesque
2. *Triturus viridescens viridescens* (Rafinesque), common newt

Occurrence. — Sedge Point Pool, Hook Point, and Douglas Lake; Fontinalis Run; Munro Lake; Mud Lake.

Life history. — Breeding aquatic forms and larvae in Sedge Point Pool in later June and early July, also Hook Point (1939). Terrestrial subadults, Sedge Point, Hook Point (1939), Smith's Bog (1939), and Munro Lake; larvae, Mud Lake. Eggs, June and July on vegetation in Sedge Point Pool. Dr. Thomas (1937) has studied this newt in connection with the life histories of certain of its parasites.

Literature. — Gage (1891), Jordan (1893), Ruthven (1911b), Pope (1924), Noble (1926, 1929), Blanchard (1928a), Morgan and Grierson (1932), Hamilton (1932), Lipsett and Piatt (1936), Thomas (1937 a–b), Reinke and Chadwick (1940), Adams (1940).

Family III. Ambystomidae

3. *Ambystoma* Tschudi
3. *Ambystoma jeffersonianum* (Green), Jefferson's salamander

Occurrence. — Rather rare. Shore of North Fish-Tail Bay, Sedge Point Pool; Smith's Bog (1939). Lake Michigan beaches in Emmet County; Duncan Bay, near Cheboygan (July 18, 1940).

Life history. — Aquatic larvae in Sedge Point Pool in July and in Bryant's Bog.

Literature. — Piersol (1910), Ruthven (1911b), Smith (1911b), Ellis (1917), Blanchard (1928a), Mohr (1931), Clanton (1934), Moore (1939).

4. *Ambystoma maculatum* (Shaw), spotted salamander

Occurrence. — Scarce. Shore of North Fish-Tail Bay, mouth of Carp Creek on Burt Lake, Maple River Cove, Nichol's Bog.

Literature. — Eycleshymer (1895), Smith (1907, 1910, 1911b), Breder (1927), Blanchard (1928a, 1930b, 1935), Moore (1939).

Family IV. Plethodontidae

4. *Plethodon* Tschudi
5. *Plethodon cinereus* (Green), red-backed salamander, dusky salamander

Occurrence. — Very abundant; Carp Creek Valley; Hook Point; Burt Lake hardwoods, and most moist woods of the region.

Life history. — Eggs in rotten logs in late June and early July. Both color phases are found in the same log. Eggs taken on June 24 hatched in the laboratory on July 15. Adults are eaten by the shrew (*Sorex cinereus*), which is abundant in this region.

Literature. — Piersol (1909, 1914), Cochran (1910), Ruthven (1911b), Ellis (1917), Dunn (1926), Blanchard (1928b, 1935), Hamilton (1932).

5. *Hemidactylium* Tschudi
6. *Hemidactylium scutatum* (Schlegel), four-toed salamander

Occurrence. — Rare. Adults in Bryant's Bog and Reese's Bog, and near Vincent Lake, June, 1935.

Life history. — Eggs, Bryant's Bog, June 20, 1925. Aquatic larvae, Bryant's Bog in summer.

Literature. — Bishop (1919), Blanchard (1922, 1923, 1928a, 1931, 1933b, 1934, 1933c), Branin (1935).

ORDER SALIENTIA — FROGS AND TOADS

Three families of tailless amphibia exist in the region. There are five genera and ten species, two of which are very rare. Near here *Acris crepitans*, *Rana catesbeiana*, *Rana palustris*, and *Pseudacris nigrita* reach the northern limits of their range. *Rana septentrionalis*, which occurs throughout the Northern Peninsula, has not been taken in the northern tip of the Southern Peninsula. *Bufo fowleri* is the only other Michigan species missing from the fauna of this area, although the sand dunes in which it is found on the west side of the state run to this region in an almost continuous line and provide an apparently suitable habitat.

There are several general works on the frogs and toads of North America besides that of Cope (1889). The taxonomy here used is that of the check list of Stejneger and Barbour (1939). Dickerson, *The Frog Book* (1907), discusses in detail the life histories and characteristics of frogs and toads. The more recent *Handbook of Frogs and Toads* (1933), by Wright and Wright, also covers the field in detail. Both of these works contain keys for the identification of Amphibia. Keys for the forms found in the State of Michigan are given by Ruthven, Thompson, and Gaige (1928), Blanchard (1935), and Schmidt (1929). Wright and Wright (1924) have published a key to the eggs of Salientia, and Wright (1929) has described the North American tadpoles.

These books contain excellent résumés of the literature on the fauna of special localities; citations from them will not be made here except when the material is especially pertinent for this region.

FAMILY V. BUFONIDAE

6. *Bufo* Laurenti
7. *Bufo americanus americanus*, American toad

Occurrence. — Abundant throughout the region; found in numbers on the lake beaches and under driftwood.

Life history. — Tadpoles and transforming young in early July at Sedge Point Pool and in the pools near East Point.

Literature. — Ruthven (1911b), Ellis (1917), Blanchard (1928a), Moore (1939). On the evidence of Gaige (1932) Stejneger and Barbour (1939) list two subspecies of the common American toad. The other subspecies is found near James Bay in northern Ontario.

Mrs. Gaige reports that the record of *Bufo fowleri* for Emmet County in the Douglas Lake area as given by Ruthven, Thompson, and Gaige (1928) is an error. The most northern record is now Mason County.

Family VI. Hylidae

7. *Acris* Duméril and Bibron
8. *Acris crepitans* Baird, cricket frog

Occurrence. — Rare. Specimens on record at the University of Michigan Museum (Ruthven, Thompson, and Thompson (1912)). Recorded as very abundant in the lowland near the west side of Reese's Bog by Ellis (1917). Not taken by Blanchard or by anyone else since then.
Literature. — Not found by Blanchard (1928a). Specimens recorded by Ruthven, Thompson, and Thompson (1912) and by Ruthven, Thompson, and Gaige (1928). Now recorded by them (unpublished) as found in Michigan north to Montcalm County in southern Michigan and from Delta County in the Upper Peninsula. Wright (1931).

8. *Pseudacris* Fitzinger
9. *Pseudacris nigrita triseriata* (Wied), swamp tree frog

Occurrence. — Rare. A few specimens have been taken in the hardwoods near Burt Lake on Colonial Point; North Fish-Tail Bay (Blanchard, 1928a).
Literature. — Blanchard (1928a) records one specimen of this species as *Pseudacris feriarum.* Ruthven, Thompson, and Gaige (1928) list no specimen in the northern half of the Lower Peninsula. Wright and Allen (1908), Wright (1914), Blanchard (1935, p. 46).

9. *Hyla* Laurenti
10. *Hyla crucifer crucifer* Wied, spring peeper

Occurrence. — Fairly common in woods and near bog situations on the north shore of Douglas Lake between Sedge Point and Robert's Point; Reese's Bog; Penney Lake; Burt Lake hardwoods on Colonial Point.
Literature. — Wright (1914), Ellis (1917), Blanchard (1928a), Wright and Wright (1933).

11. *Hyla versicolor versicolor* (Le Conte) tree toad

Occurrence. — Common about camp. Often found on the cabins and under window blinds and canvas automobile covers.
Life history. — Transforming young at Smith's Bog. A number of breeding males in chorus after dark, June 20, 1927, were recorded by Blanchard (1928a). The orange-tailed tadpoles have been found in Sedge Point Pool and Bryant's Bog. Eggs taken in Sedge Point Pool, June 27, 1935, hatched in the laboratory in six days.
Literature. — Ruthven, Thompson, and Thompson (1912), Wright (1914), Ellis (1917), Blanchard (1928a), Wright and Wright (1933).

Family VII. Ranidae

10. *Rana* Linné
12. *Rana sylvatica cantabrigensis* Baird, wood frog

Occurrence. — Rather common in the bog forests on the north shore of Burt Lake; Maple River outlet; Sedge Point Pool; Burt Lake hardwoods.

Literature. — Ruthven (1911b), Wright (1914), Ellis (1917), Cummins (1920), Blanchard (1928a), Moore (1939), 1942.

13. *Rana catesbeiana* Shaw, bullfrog

Occurrence. — Now very rare; recently taken (1932–41) near Twin Lakes, Cheboygan County; Indian River; and North Fish-Tail Bay, Douglas Lake (1939). Ellis (1917) records it as very abundant along Bessey Creek in 1913 and as found near Lancaster Lake, Munro Lake, on both branches of Maple River, and at Cecil Bay, Emmet County. Not taken by Blanchard (1928a) near the station.

Life history. — Thousands of dead transforming tadpoles lay on the shore of the South Twin Lake, Cheboygan County, in late July, 1935.

Literature. — Ruthven, Thompson, and Gaige (1938) record a single specimen from the Northern Peninsula. Evidently this region is near the northward limit of its range. Wright (1914), Ellis (1917), Blanchard (1928a), Frost (1935), Pollister and Moore (1937), Thomas (1939), Moore (1939, 1942), Raney (1940).

14. *Rana clamitans* Latreille, green frog

Occurrence. — Abundant near Sedge Point Pool; Fontinalis Run; Mud Lake; and in small numbers in all suitable habitats.

Life history. — Eggs were found at Mud Lake and Sedge Point Pool, July 4, 1940, and specimens brought into the laboratory at that time of year frequently lay eggs in captivity. Green frog eggs in the gastrula stage were taken on July 22, 1940, in a quarry pool near Afton, Michigan. They were placed in a round battery jar, which at noon on July 23 reached a temperature of 29.5° C. All hatched on July 26. Tadpoles in all stages of transformation may be found in July at Sedge Point Pool, Fontinalis Run, and Hook Point.

Literature. — Ruthven (1911b), Ellis (1917), Blanchard (1928a), Thomas (1939), Moore (1939).

15. *Rana palustris* Le Conte

Occurrence. — Rare. Known only from a single location in Cheboygan County, collected by Dr. Metzelaar, which was recorded by county only and listed by Ruthven, Thompson, and Gaige (1928). One specimen only has been taken from the Northern Peninsula.

Literature. — Moore (1939, 1942).

16. *Rana pipiens* Schreber, leopard frog

The status of the name *Rana pipiens* for our local frog is in question. Stejneger and Barbour (1939) give *Rana brachycephala* for frogs in this region. Gaige, University of Michigan Museum, prefers to list it as *Rana pipiens* until a full study of the situation is made.

Occurrence. — Abundant in wet grassy places near breeding sites such as Sedge Point Pool, Maple River outlet, Hook Point; Fontinalis Run, and the north shore of Burt Lake.

Life history. — The tadpoles are found in Sedge Point Pool, Fontinalis Run, and like places. They transform in the middle of July. Force (1933) has studied the growth of the species in this region.

Literature. — Ruthven (1911b), Ellis (1917), Blanchard (1928a), Creaser and Gorbman (1939), Moore (1939), Thomas (1939), Shumway (1940).

Class REPTILIA

The only general descriptive review of the Reptilia of North America is that of Ditmars (1907, 1937), but other useful publications are the catalogue of Baird and Girard (1853) and, of recent date, the check list of Stejneger and Barbour (1939), which included the type locality and the range of each form. There is also a valuable classification of living reptiles by Blanchard (1935). The work of Stejneger and Barbour has been used as the basis for the taxonomy of the Reptilia here reported.

Large sections of several general publications are devoted to Reptilia. One may consult Stejneger (1895) for an account of the poisonous snakes of North America. Klauber (1936) and Gloyd (1940) have made special studies of the rattlesnakes. Cope (1900) describes the crocodilians, lizards, and snakes of North America. A comprehensive work is Blanchard's *Key to the Snakes of the United States, Canada, and Lower California* (1924, 1939). Another key may be found in Pope's *Snakes Alive and How They Live* (1937). Ruthven, Thompson, and Gaige (1928) give a detailed account of the Reptilia of the State of Michigan, as does Blanchard (1935). Logier's *Reptiles of Ontario* (1939) is of much comparative interest. The local lists published by Ellis (1917) and Blanchard (1928a) have notes on all species they list.

Conant's well-prepared book, *The Reptiles of Ohio* (1938), may be used to advantage in studying the many species of the Douglas Lake area. Other valuable publications on the snakes of the United States and Canada are those of Ditmars (1939) and Schmidt and Davis (1941).

Order Squamata

SUBORDER SAURIA — LIZARDS

The only general review of the lizards of North America published since that of Cope (1900) is the *Key to the Lizards of the United States and Canada* (1935), by Burt. Camp (1923) made a classification of the lizards, and Taylor (1935) studied the genus *Eumeces*, the sole genus represented in the station area. It is of interest to note that the only species recorded for Michigan occurs in some numbers throughout even the northern part of the Lower Peninsula, which is near the northern limit of its range.

Family VIII. Scincidae

11. *Eumeces* Wiegmann
17. *Eumeces fasciatus* (Linné), blue-tailed skink

Occurrence. — Not uncommon in certain small areas. Duncan Bay; Burt Lake; north end of Black Lake; Grace Harbor at Hammond Bay; and near the headwaters of Laperell Creek. For a long time in the station's history this species was known only from a single specimen collected by Ellis (1917) near the north end of Burt Lake.

Life history. — Six eggs were found on July 27, 1939, at Grace Harbor in the moist sawdust of the interior of a decaying dock log, the outer shell of which was fairly solid. Dr. Blanchard collected eggs from Duncan Bay and Grace Harbor. These he allowed to hatch, and the young were released on the station area and have been re-collected there (1935–41). They seem to be established in and about the old log laboratory, since unmarked young in their second summer were collected there on August 10, 1937.

Literature. — Ruthven (1911a), Ellis (1917), Blanchard (1928a), Noble and Mason (1933), Taylor (1935).

SUBORDER SERPENTES — SNAKES

The snake fauna of the region is composed of ten species, seven of which are not uncommon about the station and throughout the region. The failure of *Coluber constrictor flaviventris, Elaphe obsoleta obsoleta, Elaphe vulpina,* and *Thamnophis butleri,* all present in the southern part of the Lower Peninsula, to extend to this region is of zoogeographic interest.

Several of the species reach the northern limits of their range within this region, as do some of the Amphibia. These species are *Heterodon contortrix, Lampropeltis t. triangulum, Storeria dekayi, Thamnophis s. sauritus,* and *Sistrurus c. catenatus.*

Publications with keys and descriptions useful in the study of snakes of this region are Blanchard (1925, 1935); Ruthven, Thompson, and Gaige (1928); Pope (1937); Conant (1938); Ditmars (1939); Schmidt and Davis (1941). Walls made an interesting contribution in his paper (1940) on the eyes of snakes. Numerous life-history and taxonomic studies are cited in the present paper under the names of the species to which they refer.

Family IX. Colubridae

12. *Diadophis* Baird and Girard
18. *Diadophis punctatus edwardsii* (Merrem), eastern ringneck snake

Occurrence. — Not uncommon inside decaying logs and under driftwood along the sandy Great Lakes beaches; Big Stone Bay, Grace Harbor; Black Lake;

north end of Burt Lake; Pine Point and Grapevine Point, Douglas Lake; mouth of Carp Creek.

Life history. — The eggs are laid from early July until August in the moist interior of decaying logs on the exposed beaches and also under boards. Blanchard (1927, 1930a) has given a detailed account of the life history of this snake in this region.

Literature. — Ruthven, Thompson, and Thompson (1912), Ellis (1917), Langlois (1925), Blanchard (1928a, 1930a, 1942).

13. *Heterodon* Latreille

19. *Heterodon contortrix* (Linné), puffing adder

Occurrence. — Rare, the northern limit of its range. Though reported by Ellis in 1917 as fairly common throughout the region, in the past fifteen years it has become so rare that it is collected about the station only at intervals of several years. A very small one was found near the boathouse on June 23, 1935, and an adult at the north end of Burt Lake on July 12, 1939. Two were taken during the summer of 1942.

Literature. — Ruthven, Thompson, and Thompson (1912), Ellis (1917), Blanchard (1928a).

14. *Opheodrys* Fitzinger

20. *Opheodrys vernalis vernalis* (Harlan), smooth green snake

Occurrence. — Fairly common in grassland, in the open aspen cover, and along roads about the station.

Life history. — The eggs have been found under stones and logs along Lake Michigan beaches in late July and August. Ellis (1917) collected a set of six eggs on August 14, 1914. Blanchard (1933a) has described the eggs and young in detail.

Literature. — Ruthven (1911b), Ellis (1917), Langlois (1925), Blanchard (1928a), Dymond and Fry (1932).

15. *Lampropeltis* Fitzinger

21. *Lampropeltis triangulum triangulum* (Lacépède), milk snake

Occurrence. — Generally distributed throughout the region of the station, which is the northern limit of its range.

Life history. — Several clutches of eggs were collected by Blanchard from the interior of logs on the shore of Black Lake, and elsewhere. A female laid nine eggs in a snake box in the laboratory on July 14, 1940.

Literature. — Ruthven (1911b), Ellis (1917), Blanchard (1921, 1928a).

16. *Natrix* Laurenti

22. *Natrix sipedon sipedon* (Linné), water snake

Occurrence. — One of the most abundant snakes in this region; found along lake shores and streams and near ponds throughout the region.

Life history. — On three occasions snakes of this species have been observed along Grapevine Point catching and attempting to eat rather large individuals of *Necturus.* It is frequently seen out in the lake catching fish, and an analysis of stomach content shows that the small fish of the shore, including the dead ones of the beach drift, constitute a large part of its food. Blanchard and Finster (1933) have studied its life history, using marking methods in their experiments. Omitted from Blanchard's 1928 list.

Literature. — Ruthven (1911), Ellis (1917), Langlois (1925), Clay (1938).

17. *Storeria* Baird and Girard
23. *Storeria dekayi* (Holbrook), De Kay's snake

Occurrence. — Very rare. Taken first by Blanchard since 1928 and therefore not included in his list of that date. This is the northern limit of its range. A large specimen was collected by Miss Frances Waite at Nigger Creek on July 6, 1940, and a large one under a log on the shore of Duncan Bay on July 18, 1940. A pregnant female was taken near Duncan Bay late in June, 1941, and was held in the laboratory until August 5, when it gave birth to seven young.

Literature. — Shields (1929), Noble and Clausen (1936), Clausen (1936), Noble (1937), Conant (1938), Trapido (1939, 1940).

24. *Storeria occipito-maculata* (Storer), red-bellied snake

Occurrence. — A few are found each year. Station area; north end of Black Lake; Lake Michigan and Huron beaches. They are secretive and usually lurk under some stick, stone, board, or other cover.

Life history. — Langlois (1925) records nine young born in captivity at the station on August 20, 1923, and describes the event in detail. Blanchard (1928a, 1937), Trapido (1940).

18. *Thamnophis* Fitzinger
25. *Thamnophis sauritus sauritus* (Linné), ribbon snake

Occurrence. — Rather common in restricted habitats where grass borders a lake, pool, or stream. North Fish-Tail Bay, Sedge Point, Smith's Bog. This region is the northern limit of its range.

Life history. — The numerous young are born in late July and August.

Literature. — Ruthven (1908), Ellis (1917), Blanchard (1928a).

26. *Thamnophis sirtalis sirtalis* (Linné), garter snake

Occurrence. — Very common throughout the region.

Life history. — The very numerous young are born in late July and August. Blanchard and Finster (1933) have recorded some interesting facts on the wanderings of marked specimens near the station.

Literature. — Ruthven (1908, 1911b), Ellis (1917), Burt (1928), Blanchard (1928a), Blanchard and Finster (1933), Harrison (1933), Rahn (1940).

Family X. Crotalidae

19. *Sistrurus* Garman
27. *Sistrurus catenatus catenatus* (Rafinesque), massasauga, rattlesnake

Occurrence. — Not uncommon on Bois Blanc Island (Mackinac County) near Cheboygan; Hammond Bay in Presque Isle County; reported from Lake Huron shore east of Cheboygan and from a farm on the headwaters of Laperell Creek, Cheboygan County. Not listed by Blanchard (1928a).

Literature. — Stejneger (1895), Atkinson and Netting (1927), Creaser (1928), Klauber (1936), Schmidt (1938a), Gloyd (1940).

Order Testudinata — Turtles

No general account of all the turtles of North America has yet appeared except that found in Ditmars' *Reptile Book* (1907) and in the revised edition, *The Reptiles of North America* (1937), which has good descriptions and some keys. An excellent work dealing with a smaller area is Pope's *Turtles of the United States and Canada* (1939). Keys for certain regions of North America have been published by Pratt (1923) and Jordan (1929). The nomenclature used here is in accordance with that of the check list of Stejneger and Barbour (1939).

Among other works of value for a study of this area are Cahn's extensive survey (1937) of the turtles of Illinois and Schmidt's illustrated leaflet (1938b) on those in the Chicago area. Schmidt's work contains a key. Species occurring in the Douglas Lake region are interestingly described in Conant's "The Reptiles of Ohio" (1938) and Babcock's *The Turtles of New England* (1919) and *Field Guide to New England Turtles* (1938) and by Lagler (1943).

At the present time the turtle fauna of the Douglas Lake region is not plentiful. Only five of the nine Michigan species are here represented, and so few specimens of two of these have been collected as to suggest that the northern limits of their abundance is somewhat to the south. The two species that are common are found throughout the Upper Peninsula.

Clemmys insculpta occurs in the Lower Peninsula northward from Muskegon and Iosco counties and in the Upper Peninsula, but intensive collecting in this region for twenty-five years failed to reveal its presence until a specimen was taken on July 1, 1936, at the north end of Burt Lake.

Sternotherus odoratus, Clemmys guttata, Graptemys geographica, and *Amyda spinifera* are confined to the southern half of the Lower Peninsula. *Terrapene carolina* may well be added to this list, and *Emys blandingii* might be included, since a few individuals are found in the Biological Station area. Their failure to occur here is an interesting problem in the zoogeography of Michigan.

Family XI. Chelydridae

20. *Chelydra* Schweiger
28. *Chelydra serpentina* (Linné), snapping turtle

Occurrence. — A few large specimens are collected each year. Carp Creek, north end of Burt Lake; Douglas Lake; Nigger Creek.

Life History. — A nest with eggs was discovered at the mouth of Fontinalis Run on the north end of Burt Lake on June 28, 1934. It had just been uncovered, presumably by a red squirrel, which left the site upon our approach. The eggs were in two and three layers. The nest was less than six inches deep, in the moist clean sand of the beach just above the ordinary wave level and partly under a half-buried log. Most of the eggs were untouched by the discoverer, and the remains of about six more were about the place. Two months later several were opened, and two very small living embryos were found.

Literature. — Ellis (1917), Blanchard (1928a).

Family XII. Testudinidae

21. *Clemmys* Ritgen
29. *Clemmys insculpta* (Le Conte), wood turtle

Occurrence. — Very rare, one specimen taken at the north end of Black Lake on July 1, 1936, by collectors from the parasitology laboratory. It is now in the University of Michigan collection at Ann Arbor.

22. *Emys* Duméril
30. *Emys blandingii* (Holbrook), Blanding's turtle

Occurrence. — Very rare. Marl Bay (August 7, 1924), Riggsville (July 3, 1929), Maple River (late June, 1935). This region is at the northern limit of the turtle's range, although there is a record from Marquette County in the Upper Peninsula.

Literature. — Blanchard (1928a), Schmidt (1938b).

23. *Terrapene* Merrem
31. *Terrapene c. carolina* (Linné), box turtle

Occurrence. — Our only record is that of Ellis (1917), who reported collecting a specimen in the sand near the station in August, 1913. Thirty years of collecting have failed to reveal another specimen.

Literature. — Ellis (1917), Blanchard (1928a), Hildebrand (1929).

24. *Chrysemys* Gray
32. *Chrysemys picta marginata* (Agassiz), western painted turtle

Occurrence. — Very common about suitable places in the region. Sedge Point Pool, Bessey Creek; Fontinalis Run, Nigger Creek; French Farm Lake; Maple River.

Life history. — On July 1, 1938, a female was observed in the act of laying eggs in a flask-shaped hole that she had made in the side of a sand road a few feet from French Farm Lake. She was removed to the laboratory, but no eggs were laid during the summer. Destroyed nests were found on the top of the steep-sloped sand dunes near this lake.

Literature. — Ruthven (1911b), Ellis (1917), Blanchard (1928a), Bishop and Schmidt (1931, p. 134).

University of Michigan Biological Station
Douglas Lake, Cheboygan County, Michigan

LITERATURE CITED

ADAMS, A. ELIZABETH. 1940. Sexual Condition in *Triturus viridescens*. III. The Reproductive Cycle of the Adult Aquatic Form of Both Sexes. Am. Journ. Anat., 66: 235–275.

ATKINSON, D. A., AND NETTING, M. GRAHAM. 1927. The Distribution and Habits of the Massasauga. Bull. Antivenin Inst. Am., 1 (2): 40–44.

BABCOCK, H. L. 1919. The Turtles of New England. Mem. Boston Soc. Nat. Hist., 8 (3): 325–431, 16 pls.

—— 1938. Field Guide to New England Turtles. New England Mus. Nat. Hist., Nov., 1938. Guide No. 2.

BAIRD, S. F., AND GIRARD, C. 1853. Catalogue of North American Reptiles in the Museum of the Smithsonian Institution. Part I, Serpents, 1853. *In* Smithsonian Inst., Miscellaneous Collections, 2 (1862): xvi + 172.

BISHOP, S. C. 1919. Notes on the Habits and Development of the Four-toed Salamander, *Hemidactylium scutatum* (Schlegel). New York State Mus. Bull., 219–220, 251–282, 7 pls.

—— 1926. Notes on the Habits and Development of the Mudpuppy, *Necturus maculosus* (Rafinesque). *Ibid.*, 268: 5–60, 11 pls.

—— 1932. The Spermatophores of *Necturus maculosus* Rafinesque. Copeia, 1: 1–3.

—— 1941. The Salamanders of New York. New York State Mus. Bull., 324: 1–365.

—— 1943. Handbook of Salamanders, xii + 500, 400 illus. *In* Handbooks of American Natural History, Vol. III. Comstock Publishing Co., Ithaca, New York.

—— AND SCHMIDT, F. J. W. 1931. The Painted Turtles of the Genus *Chrysemys*. Field Mus. Nat. Hist., Publ. 293, Zool. Ser., 18: 123–139, figs. 1–27.

BLANCHARD, FRANK N. 1921. A Revision of the King Snakes: Genus *Lampropeltis*. Bull. U. S. Nat. Mus., No. 114, 260 pp.

—— 1922. Discovery of the Eggs of the Four-toed Salamander in Michigan. Occ. Pap. Mus. Zool., Univ. Mich., 126: 1–3.

—— 1923. The Life History of the Four-toed Salamander. Am. Nat., 57: 262–268.

—— 1925. A Key to the Snakes of the United States, Canada, and Lower California. Pap. Mich. Acad. Sci., Arts, and Letters, 4 (1924), Part II, 1–65.

—— 1927. Eggs and Young of the Eastern Ring-neck Snake, *Diadophis punctatus edwardsii*. Pap. Mich. Acad. Sci., Arts, and Letters, 7 (1926): 279–292, pls. 13–19.

—— 1928a. Amphibians and Reptiles of the Douglas Lake Region in Northern Michigan. Copeia, 167: 42–51.

—— 1928b. Topics from the Life History and Habits of the Red-backed Salamander in Southern Michigan. Am. Nat., 62: 156–164.

Blanchard, Frank N. 1930a. Further Studies of the Eggs and Young of the Eastern Ring-neck Snake, *Diadophis punctatus edwardsii*. Bull. Antivenin Inst. Am., 4 : 2–10.

—— 1930b. The Stimulus to the Breeding Migration of the Spotted Salamander *Ambystoma maculatum* (Shaw). Am. Nat., 64 : 154–167.

—— 1933a. Eggs and Young of the Smooth Green Snake, *Liopeltis vernalis* (Harlan). Pap. Mich. Acad. Sci., Arts, and Letters, 17 (1932) : 493–508, pls. 51–56.

—— 1933b. Spermatophores and the Mating Season of the Salamander *Hemidactylium scutatum* (Schlegel). Copeia, 1 : 40.

—— 1933c. Late Autumn Collections and Hibernating Situations of the Salamander *Hemidactylium scutatum* (Schlegel) in Southern Michigan. *Ibid.*, 4 : 216.

—— 1934. The Spring Migration of the Four-toed Salamander, *Hemidactylium scutatum* (Schlegel). *Ibid.*, 1 : 50.

—— 1935. Natural History of Vertebrates (except Birds), pp. 1–82. Revised Edition, 1938, Edwards Brothers, Ann Arbor, Michigan.

—— 1937. Data on the Natural History of the Red-bellied Snake, *Storeria occipito-maculata* (Storer) in Northern Michigan. Copeia, 3 : 151–62, figs. 1–4.

—— 1942. The Ring-neck Snakes, Genus *Diadophis*. Bull. Chicago Acad. Sci., 7 (1) : 144.

—— and Blanchard, F. C. 1931. Size Groups and Their Characteristics in the Salamander *Hemidactylium scutatum* (Schlegel). Am. Nat., 65 : 149–164.

—— and Finster, Ethel B. 1933. A Method of Marking Snakes for Future Recognition, with a Discussion of Some Problems and Results. Ecol., 14 : 334–347.

Branin, M. Selyn. 1935. Courtship Activities and Extraseasonal Ovulation in the Four-toed Salamander, *Hemidactylium scutatum* (Schlegel). Copeia, 4 : 172–175.

Breder, Ruth B. 1927. The Courtship of the Spotted Salamander. Zool. Soc. Bull., 30 : 51–56.

Burt, C. E. 1928. Sexual Dimorphism in the Tail Length of the Common Garter Snake, *Thamnophis sirtalis sirtalis* (Linné). Copeia, 166 : 13–14.

—— 1935. A Key to the Lizards of the United States and Canada. Trans. Kansas Acad. Sci., 38 : 255–305.

Cahn, Alvin R. 1937. The Turtles of Illinois. Nos. 1–2. Biol. Monogr., 16 : 1–218, 31 pls., 20 maps.

Camp, C. L. 1923. Classification of the Lizards. Bull. Am. Mus. Nat. Hist., 48 (11) : 289–481.

Clanton, W. 1934. An Unusual Situation in the Salamander *Ambystoma jeffersonianum* (Green). Occ. Pap. Mus. Zool., Univ. Mich., 290 : 1–14, pl. 1.

CLAUSEN, H. J. 1936. Observations on the Brown Snake *Storeria dekayi* (Holbrook), with Special Reference to the Habits and Birth of the Young. Copeia, 2: 98–102.

CLAY, WILLIAM M. 1938. A Synopsis of the North American Water Snakes of the Genus *Natrix*. Copeia, 4: 173–182.

COCHRAN, M. ETHEL. 1910. The Biology of the Red-backed Salamander (*Plethodon cinereus erythronotus* Green). Biol. Bull., 20: 323–349, 10 figs.

CONANT, ROGER. 1938. The Reptiles of Ohio. Am. Mid. Nat., 20: 1–200.

COPE, E. D. 1889. The Batrachia of North America. Smithsonian Institution, U. S. Nat. Mus., Ser. No. 45. Bull. U. S. Nat. Mus., No. 34, 525 pp., 86 pls., 119 figs.

—— 1900. The Crocodilians, Lizards, and Snakes of North America. Ann. Rep. Smithsonian Inst. for 1898, pp. 151–1294.

CREASER, CHARLES W. 1928. The Northward Distribution of Venomous Snakes in Michigan. Pap. Mich. Acad. Sci., Arts, and Letters, 8 (1927): 369–370.

—— AND GORBMAN, AUBREY. 1939. Species Specificity of the Gonadotropic Factors in Vertebrates. Quart. Rev. Biol., 14: 311–331.

CUMMINS, HAROLD. 1920. The Rôle of Voice and Coloration in Spring Migration and Sex Recognition in Frogs. Journ. Exper. Zool., 30: 325–343.

DICKERSON, MARY C. 1906. The Frog Book. North American Toads and Frogs, with a Study of the Habits and Life Histories of Those of the Northeastern States, 253 pp., 300 photographs. Doubleday, Page and Co., New York. New Edition, 1938.

DITMARS, R. L. 1907. The Reptile Book, xxxii + 472, pls. 1–136. Doubleday, Page, and Co., New York. New Revised Edition, 1937, under the title *The Reptiles of North America*. Doubleday, Doran & Co.

—— 1939. A Field Book of North American Snakes, xii + 305, 48 illus. Doubleday, Doran & Co., New York.

DUNN, E. R. 1926. The Salamanders of the Family *Plethodontidae*, 441 pp. Smith College, Northampton, Mass.

DYMOND, J. R., AND FRY, F. E. J. 1932. Notes on the Breeding Habits of the Green Snake (*Liopeltis vernalis*). Copeia, 2: 102.

ELLIS, MAX M. 1917. Amphibians and Reptiles of the Douglas Lake (Michigan) Region. Ann. Rep. Mich. Acad. Sci., 19: 45–63.

EYCLESHYMER, A. C. 1895. The Early Development of Amblystoma, with Observations on Some Other Vertebrates. Journ. Morphol., 10: 343–418, 22 pls.

—— 1906. The Habits of *Necturus maculosus*. Am. Nat., 40: 123–136.

FORCE, EDITH B. 1933. The Age of the Attainment of Sexual Maturity of the Leopard Frog *Rana pipiens* (Schreber) in Northern Michigan. Copeia, 3: 128–131.

FROST, S. W. 1935. The Food of *Rana catesbeiana* Shaw. Copeia, 1: 15–18.

GAGE, SIMON HENRY. 1891. The Life History of the Vermilion-spotted Newt (*Diemyctylus viridescens* Rafinesque). Am. Nat., 25: 1084–1110, 1 pl.

GAIGE, HELEN T. 1932. The Status of *Bufo copei*. Copeia, 3: 134.

GLOYD, HOWARD K. 1940. The Rattlesnakes, Genera *Sistrurus* and *Crotalus*. Chicago Acad. Sci. Spec. Publ. No. 4, pp. 1–266, 10 figs., 31 pls., 22 maps.

HAMILTON, WILLIAM J., JR. 1932. The Food and Feeding Habits of Some Eastern Salamanders. Copeia, 2: 83–86.

HARRISON, MARJORIE B. 1933. The Significance of Knobbed Anal Keels in the Garter Snake, *Thamnophis sirtalis sirtalis* (Linnaeus). Copeia, 1: 1–3.

HILDEBRAND, S. F. 1929. Review of Experiments on Artificial Culture of Diamond-back Terrapin. Bull. U. S. Bur. Fish., Doc. No. 1060, 45: 25–70.

JORDAN, DAVID STARR. 1929. Manual of the Vertebrate Animals of the Northeastern United States, xxxi + 446. Thirteenth Edition, Completely Revised. World Book Co., Yonkers-on-the-Hudson, New York.

JORDAN, EDWIN O. 1893. The Habits and Development of the Newt (*Diemyctylus viridescens*). Journ. Morphol., 8: 269–366.

KLAUBER, LAWRENCE. 1936. A Key to the Rattlesnakes, with Summary of Characteristics. Trans. San Diego Soc. Nat. Hist., 8 (20): 185–276, figs. 1–112.

LAGLER, KARL F. 1943. Food Habits and Economic Relations of the Turtles of Michigan with Special Reference to Fish Management. Am. Midland Nat., 29 (2): 257–312.

LANGLOIS, T. H. 1925. Notes on Some Michigan Snakes. Pap. Mich. Acad. Sci., Arts, and Letters, 4 (1924): 605–610.

LIPSETT, JAMES C., and PIATT, JEAN. 1936. Experimental Hybridization of *Triturus v. viridescens* and *Triturus v. symmetrica*. Copeia, 1: 1–3.

LOGIER, E. B. S. 1937. The Amphibians of Ontario. Royal Ont. Mus. of Zool., Toronto. Handbook No. 3, pp. 1–16.

—— 1939. The Reptiles of Ontario. Royal Ont. Mus. of Zool., Toronto. Handbook No. 4, pp. 1–63.

MOHR, CHARLES E. 1931. Observations on the Early Breeding Habits of *Ambystoma jeffersonianum* in Central Pennsylvania. Copeia, 3: 102–104.

MOORE, JOHN A. 1939. Temperature Tolerance and Rates of Development in the Eggs of Amphibia. Ecol., 20 (4): 459–478.

—— 1940a. Adaptive Differences in the Egg Membranes of Frogs. Am. Nat., 74: 89–93.

—— 1940b. Stenothermy and Eurythermy of Animals in Relation to Habitat. *Ibid.*, 74: 188–191.

—— 1942. The Role of Temperature in Speciation of Frogs. Biological Symposia, 6: 189–213. Jaques Cattell Press, Lancaster.

MORGAN, ANN H., AND GRIERSON, MARGARET C. 1932. Winter Habits and Yearly Food Consumption of Adult Spotted Newts, *Triturus viridescens*. Ecol., 13: 54–62, 4 figs.

NOBLE, G. K. 1926. The Long Island Newt: A Contribution to the Life History of *Triturus viridescens.* Am. Mus. Novitates, 228: 1–11.

—— 1929. Further Observations on the Life History of the Newt *Triturus viridescens. Ibid.*, 348: 1–22.

—— 1931. The Biology of the Amphibia, xiii + 577. McGraw-Hill Book Co., Inc., New York.

—— 1937. The Sense Organs Involved in the Courtship of *Storeria, Thamnophis* and Other Snakes. Bull. Am. Mus. Nat. Hist., 73: 673–725.

—— AND CLAUSEN, H. J. 1936. The Aggregation Behavior of *Storeria dekayi* and Other Snakes. Ecol. Monogr., 6: 269–316.

—— AND MASON, E. R. 1933. Experiments on the Brooding Habits of the Lizards Eumeces and Ophisaurus. Am. Mus. Novitates, 619: 1–29, 6 figs.

PEARSE, A. S. 1921. Habits of the Mud-Puppy *Necturus*, an Enemy of Food Fishes. Dept. Comm., U. S. Bur. Fish., Econ. Circ. 49, 8 pp.

PIERSOL, W. H. 1909. The Habits and Larval State of *Plethodon cinereus erythronotus.* Trans. Can. Inst., 8: 469–493, 14 figs.

—— 1910. Spawn and Larva of *Ambystoma jeffersonianum.* Am. Nat., 44: 732–738, 4 figs.

—— 1914. The Egg-laying Habits of *Plethodon cinereus.* Trans. Roy. Can. Inst., Toronto, 10: 121–126.

POLLISTER, ARTHUR W., AND MOORE, JOHN S. 1937. Tables for the Normal Development of *Rana sylvatica.* Anat. Record, 68: 489–496.

POPE, CLIFFORD H. 1937. Snakes Alive and How They Live, xii + 238. Viking Press, New York.

—— 1939. Turtles of the United States and Canada, xviii + 343, 99 illus. Alfred A. Knopf, New York.

POPE, PHILIP H. 1924. The Life-History of the Common Water Newt (*Notophthalmus viridescens*), together with Observations on the Sense of Smell. Ann. Carnegie Mus., Pittsburgh, 15: 305–362.

PRATT, HENRY SHERRING. 1923. A Manual of the Land and Fresh Water Vertebrate Animals of the United States (Exclusive of Birds), xv + 422, figs. 1–184, 1 map. P. Blakiston's Sons and Co., Philadelphia. Second Edition, 1935, xvii + 416.

RAHN, HERMANN. 1940. Sperm Viability in the Uterus of the Garter Snake, *Thamnophis.* Copeia, 2: 109–115.

RANEY, EDWARD C. 1940. Summer Movements of the Bullfrog, *Rana catesbeiana* (Shaw), as Determined by the Jaw-Tag Method. Am. Mid. Nat., 23: 733–745.

REINKE, E. E., AND CHADWICK, C. S. 1939. Inducing Land Stage of *Triturus viridescens* to Assume Water Habitat by Pituitary Implantations. Proc. Soc. Exper. Biol. and Med., 40 : 691–693.

—— —— 1940. The Origin of the Water Drive in *Triturus viridescens*. I. Induction of the Water Drive in Thyroidectomized and Gonadectomized Land Phases by Pituitary Implantations. Journ. Exper. Zool., 83 : 223–235.

RUTHVEN, A. G. 1908. Variations and Genetic Relationships of the Garter Snakes. Bull. U. S. Nat. Mus., 61 : 1–201.

—— 1911a. Amphibians and Reptiles in a Biological Survey of the Sand Dune Region on the South Shore of Saginaw Bay, Michigan. Geol. and Biol. Surv. Mich., Publ. No. 4, Biol. Ser. No. 2, pp. 257–272.

—— 1911b. Notes on Michigan Reptiles and Amphibians, III. Ann. Rep. Mich. Acad. Sci., 13 : 114–115.

—— THOMPSON, CRYSTAL, AND THOMPSON, HELEN. 1912. The Herpetology of Michigan. Mich. Geol. and Biol. Surv., Publ. No. 10, Biol. Ser. No. 3, pp. 10–166, pls. I–XX, figs. 1–55.

—— THOMPSON, CRYSTAL, AND GAIGE, H. T. 1928. The Herpetology of Michigan, Michigan Handbook Series, 3 : 1–228, 19 pls. University Museums, University of Michigan.

SCHMIDT, KARL P. 1929. The Frogs and Toads of the Chicago Area. Field Mus. Nat. Hist, Zool. Leaflet No. 11, Chicago, 1929, pp. 1–15.

—— 1930. The Salamanders of the Chicago Area. Field Mus. Nat. Hist., Zool. Leaflet No. 12, Chicago, 1930, pp. 1–16.

—— 1938a. Herpetological Evidence for the Post-glacial Eastward Extension of the Steppe in North America. Ecol., 19 : 396–407.

—— 1938b. The Turtles of the Chicago Area. Field Mus. Nat. Hist., Zool. Leaflet No. 14, Chicago, 1938, pp. 1–24.

—— AND DAVIS, D. DWIGHT. 1941. Field Book of Snakes of the United States and Canada, xiii + 365. G. P. Putnam's Sons, New York.

SENNING, WM. C. 1940. A Study of Age Determination and Growth of *Necturus maculosus* Based on the Parasphenoid Bone. Am. Journ. Anat., 66 : 483–495.

SHIELDS, F. B. 1929. Notes on the Young of *Storeria dekayi*. Copeia, 171 : 52–54.

SHUMWAY, WALDO. 1940. Stages in the Normal Development of *Rana pipiens*, I. The External Form. Anat. Rec., 78 : 139–147.

SMITH, BERTRAM G. 1907. The Breeding Habits of *Ambystoma punctatum* (Linn.), Am. Nat., 41 : 381–390, 1 fig.

—— 1910. The Structure of the Spermatophores of *Ambystoma punctatum*. Biol. Bull., 18 : 204–211, 2 figs.

—— 1911a. The Nests and Larvae of *Necturus*. Biol. Bull., 20 : 191–200, 7 figs.

—— 1911b. Notes on the Natural History of *Ambystoma jeffersonianum*, *A. punctatum*, and *A. tigrinum*. Bull. Wis. Nat. Hist. Soc., 9 : 14–27.

SMITH, HOBART M. 1934. The Amphibians of Kansas. Am. Mid. Nat., 15: 377–528.

STEJNEGER, LEONARD. 1895. The Poisonous Snakes of North America. U. S. Nat. Mus. Rep., 1893, pp. 337–487.

—— AND BARBOUR, THOMAS. 1939. A Check List of North American Amphibians and Reptiles, xvi + 207. Fourth Edition. Harvard University Press, Cambridge.

TAYLOR, EDWARD H. 1935. A Taxonomic Study of the Cosmopolitan Scincoid Lizards of the Genus *Eumeces*, with an Account of the Distributional Relationships of Its Species. Kansas Univ. Sci. Bull., 23: 1–643, 84 figs., 43 pls.

THOMAS, LYELL H. 1937a. *Bothriocephalus rarus* n. sp., a Cestode from the Newt *Triturus viridescens* Raf. Journ. Parasitol., 23: 119–132.

—— 1937b. Environmental Relations and Life History of the Tapeworm *Bothriocephalus rarus* Thomas. *Ibid.*, 23: 133–152.

—— 1939. Life Cycle of a Fluke, *Halipegus eccentricus* n. sp., Found in the Ears of Frogs. *Ibid.*, 25: 207–221, 2 pls.

TONER, G. C. 1940. Delayed Hatching in the Snapping Turtle. Copeia, 4: 265.

TRAPIDO, HAROLD (MS). 1939. Studies on the Taxonomy and Natural History of the Snakes of the Genus *Storeria* in the United States and Canada, M. A. thesis, Cornell University.

—— 1940. Mating Time and Sperm Viability in *Storeria*. Copeia, 2: 107–109.

VIOSCA, PERCY, JR. 1937. A Tentative Revision of the Genus *Necturus*, with Descriptions of Three New Species from the Southern Gulf Drainage Area. Copeia, 2: 120–138.

WALLS, GORDON L. 1940. Ophthalmological Implications for the Early History of the Snakes. Copeia, 1: 1–8.

WRIGHT, A. A., AND WRIGHT, A. H. 1933. Handbook of Frogs and Toads, pp. 1–231. Comstock Publ. Co., Ithaca, New York. Second Edition, 1942.

WRIGHT, A. H. 1914. North American Anura. Life Histories of the Anura of Ithaca, New York. Carnegie Inst. of Washington, Publ. No. 197: 1–98, 21 pls.

—— 1929. Synopsis and Description of North American Tadpoles. Proc. U. S. Nat. Mus., Vol. 74, vii + 70, 9 pls.

—— 1931. Life Histories of the Frogs of Okefinokee Swamp, Georgia, 497 pp. The Macmillan Co., New York.

—— AND ALLEN, A. A. 1908. Notes on the Breeding Habits of the Swamp Cricket Frog, *Chorophilus triseriatus* Wied. Am. Nat., 42: 39–42.

—— AND WRIGHT, A. A. 1924. A Key to the Eggs of the Salientia East of the Mississippi River. *Ibid.*, 58: 375–381.

PROPERTIES OF PARAFFIN RELATING TO MICROTECHNIQUE

WILFRID TAYLOR DEMPSTER

IN THE literature of petroleum technology, which appears to have escaped the attention of biologists, various properties of paraffin have been investigated. In spite of this work an adequate exposition of the behavior of paraffin in its microtechnical applications must involve further study. The following account relates primarily to new observations, although brief consideration will be given to pertinent aspects of the petroleum literature. Attention is directed especially to certain properties of the solid state of paraffin that are involved in microtome sectioning. Related studies on the mechanics of sectioning and on the distortions due to sectioning have appeared separately (Dempster, 1942 a–b; 1943).

Commercial paraffins are mixtures of long straight chain hydrocarbons having in common the type formula C_nH_{2n+2}. For practical purposes the term "paraffin" is applied (Gurwitsch and Moore, 1927; Berne-Allen, 1936) to mixtures containing various numbers and amounts of homologous paraffin hydrocarbons that vary in chain length from 19 to 34 carbon atoms. Mixtures of still longer hydrocarbon chains are commonly referred to as petroleum ceresins or rod waxes. The type of compounds and the amount of each that enters into a particular paraffin sample are determined by the source of the petroleum from which wax is extracted and by the processes of refining and blending. Purified hydrocarbon samples of essentially uniform molecular weight as well as synthetic paraffin hydrocarbons have been prepared, but these are laboratory rather than commercial products.

A low melting-point (m.p.) commercial paraffin consists predominantly of hydrocarbons having low melting points, whereas a high melting-point paraffin is composed of or contains a significant percentage of heavy molecule hydrocarbons of high melting point. Mixtures of long and short chain compounds show intermediate melting points. In view of the similar molecular structure, paraffin constituents mix in all proportions, and such mixtures are characterized by a range of solidifying temperature rather than by a discrete fusion point. On fusing, paraffins tend to form solid solutions, though in certain respects, as will be discussed later (p. 255), there is some separation of heavier and lighter components.

PARAFFIN COOLING CURVES

By use of a technique similar to that outlined in the standards of the American Society for Testing Materials (A. S. T. M., 1939) data representing the cooling of paraffin in relation to time were obtained

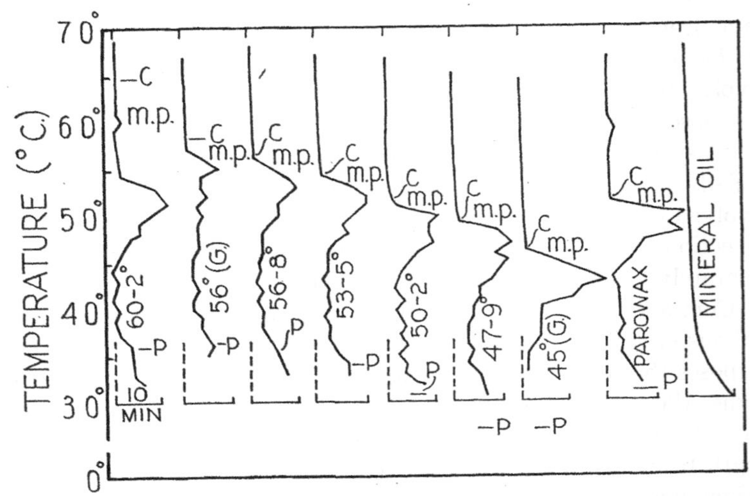

FIG. 1. Cooling curves of various types of embedding paraffins. The numbers on the curves refer to melting-point ranges indicated by the distributers. Parowax, from Standard Oil Company; 45° and 56° (G.), from Grübler; others are Bioloid of Will Corporation. C = cloud point; m.p. = melting point; P = plastic point as determined by a shearing test. The abscissa for each curve indicates the number of minutes required for a temperature drop of 1° C.

for a group of common embedding paraffins over temperatures that extended both above and below the range of solidification. These data, when plotted as the number of minutes required for paraffin to cool through each successive degree of temperature, are shown in the curves of Figure 1.

At the cloud point C solid paraffin begins to precipitate as a fine or feathery clouding. Shortly below this, at the melting point, the continuing process of solidification results in the liberation of heat of fusion, and the rate of cooling becomes significantly prolonged. Technically (A. S. T. M. standards) the melting point is defined as the "temperature at which melted paraffin wax, when allowed to cool, first shows a minimum rate of temperature change." Below the melting point considerable heat is liberated, and meanwhile the ratio of solid to liquid paraffin increases until the mass finally becomes solid throughout. The humps on the curves correspond with a range of temperature of fifteen or more degrees over which large amounts of heat are dissipated. The size and the shape of the humps represent a visual measure of the quantity of heat lost and its rate of dissipation. When the paraffin mass is entirely or nearly solid, stirring of the mass, as required by the method of testing, is unsatisfactory, and the lower parts of the curves below the point of solidification are unreliable. Scott-Harley (1939) tested a number of paraffin waxes with a special method that did not involve stirring. At some point, varying ordinarily from five to ten degrees below the melting point and characteristic for each paraffin, a rapid but transitory loss of heat occurred; this heat loss presumably corresponds with the secondary peaks shown in some of the curves recorded here. This point was considered the temperature of final solidification. Ideally, deviations in the thermal curve below the solidification point are associated with peculiarities of the thermal properties of the solid state of paraffin (cf. Lord, 1939, and p. 257 of the following discussion).

THE STRUCTURE OF SOLID PARAFFIN

Several authors (Groth, 1910; Buchler and Graves, 1927; Gruse, 1928; Katz, 1930, 1932) have described paraffin crystals precipitated out of solutions of common solvents as either flat platelets or needles. These and a variety of modifications may be readily observed with the microscope; the differences in crystal structure vary with such factors as solution concentration, temperature of precipitation, and rate of solvent evaporation. Thin layers of melted paraffin, if allowed to cool slowly upon a glass surface, likewise may show needles and stratification of platelets.

Under conditions of mass crystallization, however, as in ordinary embedding procedures, certain peculiarities in crystal structure that apparently have not previously been recorded may be observed. A mass of about twenty grams of Bioloid embedding paraffin (Will

Corporation, Rochester, N. Y.) with a melting point of 50–2° C., chilled in a paper box by immersion in water of 20° C., will serve for a description of the typical paraffin structure. It will be apparent that the description applies to crystalline aggregates and not to true crystals in the physical sense of perfectly homogeneously formed bodies. A formation of complex crystalline aggregates rather than true crystals must be expected when molecules of different sorts, longer and shorter hydrocarbon chains, for instance, do not segregate to form definitive crystal types. For the sake of simplicity, however, the term "crystal" will be used in a loose sense to refer to these aggregates.

Relatively uncompressed sections of paraffin from the microtome show a mosaic of similar crystals cut in various planes. The section illustrated in Figure 1, Plate I, indicates that the crystals are rodlike or "needle-shaped," with apparently bluntly pointed or rounded ends; they may be relatively straight or may be curved; and some appear to be distorted or twisted by contiguous crystals. Two or more crystals may be fused, and sometimes branched patterns may be seen.

The crystals vary a little in size, and the more typical ones may have a cross-section diameter of a tenth millimeter and a length at least two to four times this diameter. Branching crystals approaching a millimeter in length may be found. Paraffin cooled over several hours in a Dewar flask containing warmed water produces crystals several millimeters long, with proportional increase in diameter. Rapid cooling, however, does not induce uniform formation of small crystals. Cooling of a mass of paraffin in ice water results in the formation of a peripheral zone of about two millimeters consisting of tiny crystals with a predominant orientation perpendicular to the surfaces exposed to cooling. Crystals of the peripheral one- to three-tenths millimeter are diminutive and apparently consist of both needles and plate formations. The central mass is composed of crystals variously oriented and of essentially the same size as those in masses cooled in water at 20° C. or in air at room temperature. The peripheral zone of hardened paraffin, because of its low thermal conductivity, acts as an effective insulation and permits typical crystal growth in the interior of a mass.

Characteristically, as viewed in cross section (Fig. 2, Pl. I), paraffin crystals appear to have a core area around which four to six or more concentric lamellae of varying translucency may at times be seen. Sometimes the axial region is not clearly differentiated, but often it appears as an area ten, twenty, or more micra in diameter. In longitudinal section the central core not only traverses the length of the crystal, but in united forms appears to continue from one crystal to another. Striations running parallel to the central core and indicating the lamellar structure of the crystals are likewise evident in longitudinal section. Katz (1930) showed photographs of needle crystals precipitated from solutions of wax in which axial cores appeared; lamellation is not evident in these figures, but in some cases photographs of the ends of crystals indicate hooklike and recurrent spines in profile that are suggestive of incomplete lamellation. As seen in cross section, crystal outlines are at times deformed into five-, six-, or seven-sided figures by contiguous crystals.

If one of the poorer paraffin solvents,[1] such as amyl acetate, is flooded over a paraffin section, crystals appear as in Figure 3, Plate I. The softer paraffin is dissolved from the crystals and leaves a skeleton of harder paraffin. If the hard and soft paraffins of a mixture are indicated by *A* and *B* respectively, the concentric lamellation shown may be explained as follows: As the temperature of warm liquid paraffin is slowly lowered through the fusion range, tiny needle-like crystal nuclei of *A* form a basis for crystal growth. Heat of fusion flows from the solidified needle surface to elevate the temperature of the adjacent liquid paraffin above the melting point of fraction *B*. Only *A* precipitates on the crystal nucleus, and a crystal core is produced. More heat of fusion is liberated, and the precipitation of *B* is further delayed. When all of *A* within a convenient diffusion range has precipitated, the liquid adjacent to the core is richer in *B* than it was formerly. With *A* being added to the solid in smaller amounts, less heat of fusion leaves the needle surface, and the softer material may now lay down a lamella of wax. Heat of fusion again raises the temperature locally, so that only *A* can congeal at the crystal surface. Thus a crystal grows with alternate concentric lamellae produced by

[1] Relative solubility of 50–2° C. m.p. paraffin in a group of common solvents at 26–7° C., expressed as the number of grams of paraffin dissolved in 100 c.c. of solvent, was found to be as follows: carbon bisulphide, 78.0 g. (most active solvent); toluene, 25.0 g.; benzene, 21.5 g.; xylene, 20.8 g.; trichlorethylene, 19.0 g.; chloroform, 17.0 g.; ether, 8.8 g.; cedar-wood oil, 3.8 g.; amyl acetate, 0.9 g.; terpineol, 0.45 g.; dioxan, 0.2 g.; n-butyl alcohol, 0.1 g.; ter-butyl alcohol, 0.1 g. Paraffin of 60–2° C. m.p. was but slightly less soluble, and relative solubility followed the same order.

the alternate precipitation of hard and soft material around a needle-like core of paraffin. In a cooling melt of paraffin under routine conditions growth of individual crystals from crystal nuclei to fully formed lamellated crystals of ordinary size requires only a minute or so. Adjacent crystals increase in size till they abut on one another separated only by a thin layer at the mutual interface. Sometimes the lamellar structure at the periphery of one crystal may be definitely incorporated into the structure of the contiguous crystals.

Crystallographic examination of the different embedding paraffins studied, those indicated in Figure 1, shows them to have the same basic structure. Clear-cut lamellation was not observed in Parowax crystals. A distinctly different organization was seen only in crystals of 60–2° C. m.p. Bioloid.

The admixture of paraffin and adulterant waxes ordinarily produced crystals that could not be interpreted in terms of the crystal pattern described above. This was true of mixtures containing beeswax, bayberry wax, or high melting-point microcrystalline petroleum wax (Barnsdall Special, 160–5° F. m.p. and 180–5° F. m.p.). Mixtures containing diglycol stearate or glyceryl monostearate showed, however, a meshwork of typical though somewhat distorted paraffin crystals in the interstices of which were microcrystalline or granular stearate crystals. This is most easily seen when paraffin constitutes about half or less of the mass. Apparently — this statement is suggested by microscopic observations of wax mixtures during crystallization in a watch glass — paraffin crystals exhibit their characteristic structure in a mixture only when paraffin needle nuclei are first to appear at the cloud point. If other types of wax crystals appear first in a cooling melt, as occurs with adulterants of higher melting point, or in the 60–2° C. m.p. Bioloid, the paraffin amalgamates in crystalline masses of variable pattern; these patterns have not been studied.

Although a picture of the structure of solid paraffin may be of interest in itself, an understanding of crystal patterns is especially important since it allows one to study and analyze such plastic changes in paraffin as accompany the sectioning process (Dempster, 1942a). Likewise, the mottling or "crystallization" defect common to solid paraffin finds an explanation (p. 261).

Within the interstitial spaces of an embedded tissue — blood-vessel lumina, epithelial vesicles, interfascial clefts — paraffin crystals do not show characteristic structure, even when the space is large enough to accommodate several normal-sized crystals (Fig. 1, Pl. I). Usually microcrystals or irregular lamellation, oriented in relation to tissue elements and without a definite pattern, fill the interstitial spaces. Lamellated paraffin cannot be detected within cell outlines. If a section containing tissue is observed with a polarizing microscope, the interstitial paraffin appears varicolored and brilliant, whereas tissue elements are dark (Richards, 1942). A section through paraffin-infiltrated celloidin (4 per cent) also shows darkly in polarized light; yet the celloidin has a waxy texture, and, with heat, paraffin is melted from it. It may thus be reasonably assumed that paraffin infiltrated and adsorbed on and in the cells is present in an amorphous rather than a crystalline form and hence does not show. If the pattern of paraffin congelation is modified by contiguous tissue elements, crystal growth per se would seem to be of little significance as a tissue-distorting influence. Crystals existing at the fusion range are very soft and cannot be expected to exert forces of great magnitude. During the change from the liquid to the solid state, however — that is, within the fusion range — paraffin decreases in volume by from 5 to 10 per cent (Carpenter, 1926). Under this influence infiltrated tissue would undoubtedly be subjected to collapsing and deforming strains.

TRANSITION POINTS

In an important paper Carpenter (1926) showed that certain properties of solid paraffin were different above and below a transition temperature lying 10° ± C. below the melting point. Observations of mass crystallization in a clock glass, as the temperature dropped through the solidification range, showed that, after needle formation began, the paraffin acquired an initial opacity and a dull surface. Then, as the temperature dropped through a range of ten to fifteen degrees below the melting point, the mass became translucent and the surface shiny. At the transition point the coefficient of expansion suddenly increased, and the contraction in volume of cooling paraffin was momentarily arrested. Solubility of wax in acetic acid was relatively less at temperatures below the transition point than above. Air solubility in solid paraffin ordinarily was minimal at a point about ten degrees below the melting point.

Confirmatory studies on the transition point have been made, and usually both purified fractions and commercial samples entered into the studies. Katz (1932) proved that at a temperature 5–10° C. below the melting point crystals precipitated from solutions as needles, and that below this temperature platelets were formed. Scott-Harley (1939) and Lord (1939) have found clear evidence of transition points by analyses of paraffin cooling curves. Measurements of electrical conductivity by Jackson (1935) showed a transition point ten to fifteen degrees below the melting point. Müller (1932), using X-ray analysis, measured lattice dimensions of paraffins over the thermal range between room temperature and the melting points of the paraffins tested. Transition points characterized by a change from one type of molecular symmetry to another were seen.

Each variety of paraffin has a discrete and characteristic transition point that may be somewhat closer to or farther from the melting point than the transition points of other paraffins. The differing properties above and below the transition temperature are considered to be due to changes in the molecular or microcrystalline structure of the paraffin constituents. These changes are probably effected by alterations in the positions of hydrogen atoms relative to the carbon atoms of the hydrocarbon chains — changes akin to those recognized in stereoisomers. In some cases transition points are not evident; sometimes a mixture of two samples, each showing transition points, is not characterized by a transition point.

Carpenter described a change in crystal form from needles to platelets as wax in a clock glass changed from the opaque to the translucent form. This change is readily seen if the surface of a melt is studied microscopically during congelation. Microtome sections through the surface of such a mass disclose a mosaic of lamellated plates rather than the typical needles described above; one or two millimeters deeper, however, sections reveal typical lamellated needles. The transition from needles to plates thus appears to be a surface phenomenon limited to the mesh of superficial needles that are oriented in a plane at the surface. Below the surface crystals are indifferently oriented and do not coalesce to form a new crystal pattern.

PARAFFIN PLASTICITY

Plastic deformation may be defined as a change in the shape of a solid without the appearance of rupture or discontinuity, occurring as a response to a deforming force (i.e. greater than the gravity effects due to the weight of the material itself) and persisting after the force has been discontinued. A substance that is easily molded by a small force or is more extensively deformed by a larger one is relatively more plastic than a more resistant material. To say that paraffin exhibits a high degree of plasticity means that, at some temperature under consideration, paraffin as contrasted with some other material is readily and permanently molded without the occurrence of fractures by the application of a stated tensile, compressive, or shearing force.

Temperature is an important condition in considering plastic changes. A stick of warm paraffin stretches like taffy; a similar stick at room temperature may break without apparent necking. Similarly, below some critical temperature, paraffin compresses but little with loading; at a higher temperature the same load causes a noticeable distortion (Dempster, 1942a). The difference between the plastic properties of paraffin is brought out most strikingly, however, by a shearing test.

A large block of clear paraffin about fifteen millimeters thick and with flat sides is clamped between vise jaws with three millimeters of the upper block edge projecting. A metal rod with a square cross section is placed so that it straddles both vise jaws at 90° and held so that a plane face is forward and perpendicular to the vise jaws (Fig. 2). When the rod is shoved forward along the vise jaws all the projecting paraffin is scraped away. If the paraffin, vise, and rod have been maintained at a testing temperature until an even temperature has been established and if a constant speed of rod movement is obtained, either of two types of chip is produced (Fig. 2). Above a critical temperature a thick, compressed, and relatively smooth-surfaced chip is produced; below, a segmented chip, made of transverse prisms with a serrated upper surface, is formed. The temperature of transition between one and the other of the types of chip, which on repeated testing varies not more than a half degree, may be called the plastic point.

On the curves of Figure 1, *P* indicates the plastic points of the different paraffins as tested by this method. These points do not bear a constant relation to the melting points of the respective paraffins, though the low melting-point paraffins in general have lower plastic points than the high melting-point paraffins do. Paraffins of 45° C. and 47–9° C. m.p. have the same plastic point; the 60–2° C. m.p., 56° C. m.p., and 56–8° C. m.p. paraffins have a common high plastic point; the 53–5° C. m.p. paraffin is but slightly lower, and the 50–2° C. m.p. paraffin is intermediate between the higher and the lower values. What is particularly significant is that the plastic points correlate better than melting points with curves representing the degree of compression of microtome sections (Dempster, 1943). Paraffins with similar plastic points are similarly compressed by the microtome knife, and those with high plastic points are compressed less than paraffins with lower plastic transition points. This correla-

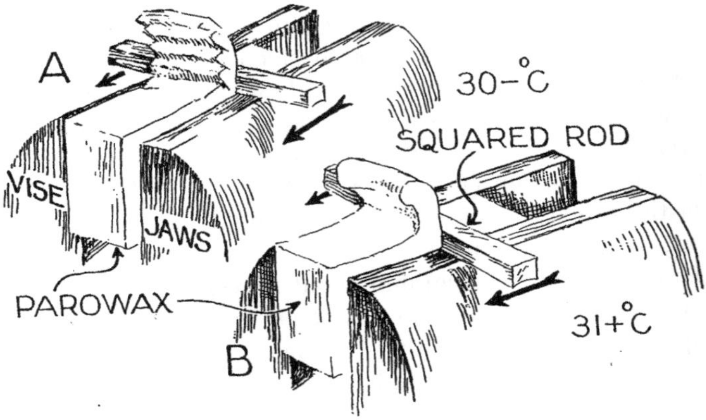

FIG. 2. Types of paraffin chip removed in a shearing test. Sketch A shows the type of chip produced at temperatures below the plastic transition point; sketch B, the suddenly changed character of a chip at a slightly higher temperature

tion will explain an observation of Aumonier (1938) that a paraffin of higher melting point produced more compressed microtome sections than a paraffin of lower melting point.

It is assumed tentatively, in the absence of data on transition temperatures for the paraffins dealt with here, that plastic points are dependent on the changes in paraffin that occur at characteristic transition temperatures, but that energy supplied by the testing forces results in plastic manifestations at a temperature (plastic point) actually lower than the true transition point.

Plastic point, determined by some such method as that mentioned, should, if the method is suitably standardized, be of much more value to microscopists as a practical correlate with sectioning behavior of paraffins than the melting-point data now used.

THE CONGELATION OF PARAFFIN

It is a standard embedding procedure to cool a solidifying mass rapidly, and it is assumed that such cooling prevents "crystallization," that is, produces an amorphous mass or reduces the size of crystals significantly. This assumption is based (Kappers, 1907) on analogies with the crystallization of metals and electrolytes in which thermal conductivity is high and in which solidification occurs at a discrete temperature rather than over a fusion range. In metals the intercrystalline junctions are harder than the crystalline material proper (Rosenhain and Haughton, 1935) and exert greater stress to distorting forces. A metal composed of small crystals and having a proportionately large number of intercrystalline junctions is different in its properties from the same metal composed of large crystals; in paraffin, however, intercrystalline junctions are soft, and readjustments in a mass under stress occur in large part at such regions (Dempster, 1942a). There is no evidence of work hardening in paraffin, as contrasted with metals, and a thoroughly kneaded mass of paraffin showing little evidence of structural organization sections like a freshly cast mass. It is thus questionable whether small crystals of paraffin, or amorphous paraffin, have any advantage in microtechnique over larger ones. If each crystal is contiguous with all adjacent crystals so that no interstices remain, the paraffin will be clear and homogeneous; such a mass will react in a uniform manner to deforming forces such as those of the sectioning process.

As Carpenter (1926) has shown, clear solid paraffin may contain from seven to fifteen volumes of air dissolved in one hundred volumes of paraffin. If the air is completely dissolved and is evenly distributed through a mass, the paraffin should be clear. The so-called "crystallization" of paraffin, as the term is used by microscopists, appears to be the naked-eye visualization of crystals — a mottling effect due to intercrystalline discontinuities or air spaces. This defect appears if air or volatile solvents are thrown out of solution during the congelation of paraffin or if paraffin is stressed at temperatures well below the melting point, i.e. stress opacity (Nádai, 1931; Dempster, 1942a). Petroleum oils may remain in some paraffins owing to inadequate refining and are an unavoidable source of mottling.

Congelation during routine paraffin embedding may be visualized as follows: Air dissolved in liquid paraffin is trapped within the mass when a congealed film appears at the surface. At the solid periphery of such a mass some air is released (because of minimal solubility in the paraffin just congealed) to be redissolved primarily in the still liquid region. Contraction (i.e. reduction in volume) of the solid periphery initially increases the pressure on the enclosed fluid mass, and this should enhance air solubility. The solid peripheral zone grows in thickness with further congelation. Contraction, occurring primarily at the interface between the solid and the liquid phases, results in a decreased total volume of the mass, and a compensatory depression of the surface layer of congealed paraffin begins. As this layer becomes thicker and at the same time more resistant to contracting forces, the liquid interior of the mass is then subjected to decreasing pressures. Such pressure will tend to volatilize dissolved air from the liquid phase at the core of the paraffin mass, and the central region may show air-filled intercrystalline spaces and visible mottling after complete solidification. The peripheral zone, if not stressed too much by contraction forces, is clear. Cooling of a mass of paraffin by immersion in cold water, after a congealed film has formed over the surface of the melted paraffin, has often been recommended, but, it may be inferred from the foregoing discussion, rapid cooling provokes increased mottling.

Routine embedding procedures too commonly allow the introduction of the mottling defect, and when mottling occurs, especially when discontinuities of the matrix are close to embedded tissue, crumpled and defective sections result. Procedures that eliminate mottling or reduce it to an insignificant amount are the following: (1) Small bits of tissue (2–3 mm. diam.) may be embedded in small buttons, a gram or less, of paraffin slowly pipetted to a glass plate. Mottling defects are negligible for very small masses of paraffin, since the solid surface layer is never too thick or tough to compensate fully for contraction. (2) Larger pieces of tissue may be embedded in a somewhat deep container of paraffin if the mass is cooled from the bottom while the surface is kept fluid by playing a hot spatula over it. This allows the escape of air that is released during the solidification process. (3) A considerable excess of paraffin may be used during embedding in a tall vessel, and the mass may be cooled, chiefly from the bottom. This method provides an excess of liquid in which released air might dissolve. On solidification all but the lower part of the mass containing the tissue may be cut away to be reused later;

the matrix surrounding the tissue is clear of mottling. This procedure like that in (2) is somewhat slow, and the paraffin crystals may be definitely larger than those in smaller masses cooled more rapidly. (4) Precooling of larger bits of infiltrated tissue in air may be followed by immersion in a container of molten paraffin. When the mass is cooled from below, mottling of the matrix is so far away from the tissue that the defect is of no importance.

SUMMARY

Unadulterated paraffin when congealed in an embedding mass is composed of contiguous needle-like crystalline aggregates, each of which consists of a central core and many concentric lamellae made of alternate depositions of hard and soft paraffin. This crystalline mass forms the matrix for embedded tissue. Except for the most peripheral layer rapid cooling of a mass does not materially reduce the size of crystals. If each crystal is in intimate contact with other crystals so that no interstices occur, the paraffin is clear and homogeneous, and good microtome sections may be made. Mottling or "crystallization" is a defect due to air spaces between crystals, which makes crystals visible grossly. Suggested procedures that eliminate or reduce mottling are outlined. Within tissue interstices typical needle structure of paraffin is supplanted by irregular lamellation, but within cells the paraffin is amorphous. Cooling curves are presented to show characteristics of a group of typical embedding paraffins. The literature on paraffin transition points is cited in relation to a discussion of the plastic behavior of paraffin. Above and below a characteristic temperature for each paraffin the mode of response to a distorting force becomes altered. This temperature, for standardized conditions of testing, may be called the plastic point, and it may be used as a measure of the relative hardness of paraffin. The sectioning behavior of paraffin is more closely related to the plastic point of a sample than to the melting point.

UNIVERSITY OF MICHIGAN

LITERATURE CITED

A. S. T. M. 1939. Standard Method of Test for Melting Point of Paraffin Wax, pp. 180–182. Book of A. S. T. M. Standards, Part 3, 1175 pp. Am. Soc. Testing Materials, Philadelphia.

Aumonier, F. J. 1938. Notes on the Distortion of Paraffin Sections. Journ. Roy. Microsc. Soc., 58: 253–257.

Berne-Allen, A. 1936. The Solubility of Refined Paraffin Waxes in Petroleum Fractions, 83 pp. Columbia University Dissertation, New York.

Buchler, C. C., and Graves, G. D. 1927. The Petroleum Waxes. Indust. and Eng. Chem., 19: 718–724.

Carpenter, J. A. 1926. The Physical and Chemical Properties of Paraffin Wax, Particularly in the Solid State. Journ. Inst. Petrol. Technol., 12: 288–315.

Dempster, W. T. 1942a. The Mechanics of Paraffin Sectioning by the Microtome. Anat. Rec., 84: 241–267.

—— 1942b. Distortions Due to the Sliding Microtome. *Ibid.*, 84: 269–274.

—— 1943. Paraffin Compression Due to the Rotary Microtome. Stain Technol., 18: 13–24.

Groth, P. 1910. Chemische Krystallographie, 3. Teil, 804 pp. Wm. Engelmann, Leipzig.

Gruse, W. A. 1928. Petroleum and Its Products, 377 pp. McGraw-Hill Book Co., New York.

Gurwitsch, L., and Moore, H. 1927. The Scientific Principles of Petroleum Technology, 470 pp. D. Van Nostrand Co., New York.

Jackson, W. 1935. Conductivity-Temperature Studies on Paraffin Waxes. Trans. Faraday Soc., 31: 827–835.

Kappers, C. U. Ariëns. 1907. Auf welchem Grund beruht es, dass die schnelle Abkühlung des Paraffins für histologische Einbettungen günstig ist? Zeitschr. f. wiss. Mikrosk., 24: 254–257.

Katz, E. 1930. The Crystallisation of Paraffin Wax. Journ. Inst. Petrol. Technol., 16: 870–888.

—— 1932. The Crystallisation of Paraffin Wax, Pt. II. *Ibid.*, 18: 37–52.

Lord, H. D. 1939. Thermometric Determination of Transition Points in Paraffin Wax. Journ. Inst. Petrol. Technol., 25: 263–276.

Müller, A. 1932. An X-Ray Investigation of Normal Paraffins near Their Melting Points. Proc. Roy. Soc. Lond., Ser. A, 138: 514–530.

Nádai, A. 1931. Plasticity, a Mechanics of the Plastic State of Matter, 349 pp. McGraw-Hill Book Co., New York.

Richards, O. W. 1942. The Effective Use and Proper Care of the Microtome, 88 pp. Spencer Lens Co., Buffalo, New York.

Rosenhain, W., and Haughton, J. L. 1935. An Introduction to the Study of Physical Metallurgy, 368 pp. Constable & Son, London.

Scott-Harley, C. R. 1939. Transition Effects in Paraffin Wax. Journ. Inst. Petrol. Technol., 25 · ?[illegible]8-251.

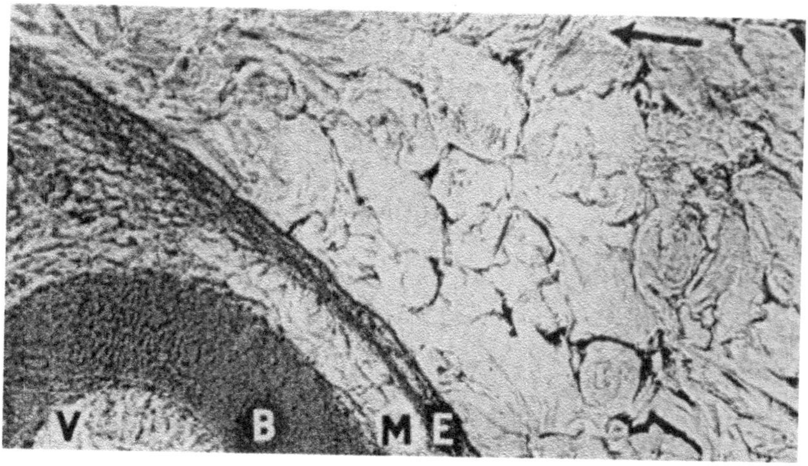

Fig. 1. Photomicrograph of a paraffin section from the microtome (12-micron section; 56–58° C. m.p. Bioloid paraffin; × 90). The mosaic to the right shows crystals cut crosswise, obliquely, and longitudinally. Within the embryo tissue at the left (as at *V* or *M*, for instance) paraffin crystals are small and are altered. Abbreviations: B, brain wall; E, epidermis; M, mesenchyme; V, ventricle of brain. The arrow indicates the direction of knife movement in the making of the section

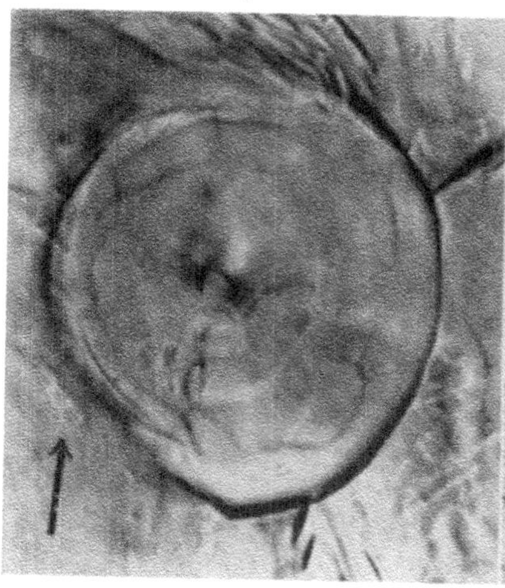

Fig. 2

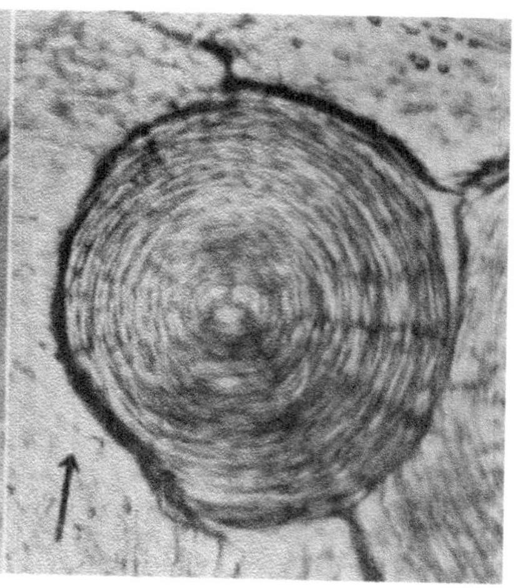

Fig. 3

Fig. 2. A cross section of a relatively undistorted paraffin crystal (50–52° C. m.p. Bioloid paraffin; 25 μ) showing a central core and a suggestion of concentric lamellae. × 500

Fig. 3. The same crystal after flooding with amyl acetate. Lamellation of hard paraffin is shown

THE THREAD-WAISTED WASPS (HYMENOPTERA-SPHECINAE) OF MICHIGAN, WITH KEYS AND DISTRIBUTION

ROBERT R. DREISBACH

THE subfamily Sphecinae, commonly called thread-waisted wasps, belongs to the family Sphecidae and the superfamily Sphecoidea. The Sphecoidea are separated from the other superfamilies, except the Apoidea, by two characters, namely: (1) The pronotum is developed on the dorsal lateral margin into lobes called tubercles; (2) The posterior margin of the pronotum never touches the tegulae. This superfamily may be distinguished from the Apoidea, or bees, by the hind tarsi, which are slender and in which the first joint is not broadened or thickened, and by the simple unbranched hairs. The Sphecinae in particular can be segregated from the bees by virtue of their petiolate abdomen.

The Sphecidae differ from the other families of the Sphecoidea by the presence of a prepectus, which distinguishes them from the Bembecidae and the Cerceridae; by the absence of an elongate process on the mesosternum; and by the slender femora, which separate them from the Ampulicidae. The petiole of the abdomen and the large body size at once make it easy to separate the subfamily Sphecinae from the other subfamilies of the Sphecidae.

Keys, descriptions, and synonymy (with two exceptions) for this subfamily as found in North America are given in the literature cited at the end of this paper.

The two exceptions in the synonymy noted above apply to the species *Chlorion rubidorsum* (Costa) and the subgenus *Ammobia.* Dr. H. T. Fernald, in his *The Digger Wasps Belonging to the Subfamily Chlorioninae,* 1906, refers to *C. rubidorsum* (Costa) as *C. bifoveolatum* (Taschenberg). In a communication to the writer he cites the reference to the original description as follows: "*Enodia rubidorsum* Costa, *Ann. Mus. Zool. Napoli,* I, 1862, p. 69." The name "Am mobia," used in this paper for a subgenus of Chlorion, is given as "Proterosphex" in Dr. Fernald's paper. Further study of this genus convinced him that "Ammobia" is the proper name.

In comparing the number of species and varieties of Sphecinae reported in Michigan with records for other states we find 28 species and varieties listed for Michigan, 19 for New York, 19 for Connecticut, 23 for New Jersey, and 26 for North Carolina. In addition to the 28 species and varieties actually collected there are 7 more species which, to judge from their present known distribution, most certainly will be captured in Michigan at some future date, and for this reason they are inserted in the keys.

The thread-waisted wasps of Michigan are divided into three tribes and five genera, as follows:

Tribe I. Chlorionini — genus *Chlorion* (and subgenera).
Tribe II. Sphecini — genera *Podalonia* and *Sphex*.
Tribe III. Sceliphronini — genera *Chalybion* and *Sceliphron*.

The sexes are easily distinguished in this group, as they are in most of the Hymenoptera. The males have 13-jointed antennae and 7 dorsal abdominal segments, and the inner margins of the eyes are strongly converging. The females have 12-jointed antennae and 6 dorsal abdominal segments, and the inner margins of the eyes are for the most part parallel. The sting and the male genitalia, if exserted, also separate the sexes at a glance.

Some of the principal technical terms used in the keys are:

Prepectus, a narrow area along the anterior margin of the episternum of the mesothorax, just behind the tubercles and just below the tegulae, bounded posteriorly by a suture.

Stigmatal groove, a groove extending from the insertion of the petiole to the ventral end of the stigma, or spiracle. This groove passes just dorsal to the swelling above posterior coxa and curves upward to the stigma.

Mesepimeron, a narrow plate extending from just below the insertion of the hind wings to the posterior coxa.

Mesopleuron, the broad plate just below the insertion of the front wings.

Metapleuron, the lateral plate of the metathorax, lying just above and parallel to the mesepimeron.

Metanotal flange, a thin, almost membranous process, with a texture similar to that of the tegula. It starts below and extends back of the hind wing on the metanotum. It is best seen from the side with the wings in normal resting position.

Species marked with an asterisk (*) have not yet been collected in this state.

KEY TO THE MICHIGAN GENERA OF SPHECINAE

1. Second cubital cell receiving only the first recurrent vein; third cell receiving second recurrent vein; claws with one to six teeth near base; tibiae spinous; tarsal comb present in female except in subgenus Isodontia .... Tribe Chlorionini, *Chlorion*
1. Second cubital cell receiving both recurrent veins, or second recurrent vein interstitial with second transverse cubital 2
2. Claws simple without a tooth beneath; tarsal comb present in female; abdomen generally very elongate, petiole of one or two segments; cubital vein of hind wings usually originates beyond transverse median vein Tribe Sphecini 3
2. Claws with single minute tooth beneath, rarely without tooth; tarsal comb absent in female; abdomen with one-segmented petiole; cubital vein in hind wings interstitial (or nearly so) with transverse median vein; metathorax with large U-shaped area above Tribe Sceliphronini 4
3. Petiole consisting of two linear segments; comb teeth on hind tibial spine closely packed together, slender; spiracle on second petiole segment (first abdominal segment) at or behind middle .... *Sphex*
3. Petiole consisting of one segment; dorsal plate of first abdominal segment more or less bell-shaped; comb teeth on hind tibial spine not crowded, stout; spiracle on first abdominal segment in front of middle, at least not behind middle, of segment .... *Podalonia*
4. Species black and yellow, not metallic; clypeus flat, usually bidentate at apex; petiole of abdomen about twice as long as metathorax; transverse median vein in front wings not interstitial with basal vein, but uniting with median vein a little before origin of basal nervure *Sceliphron*
4. Species metallic blue or violaceous; clypeus usually tridentate anteriorly; petiole of abdomen not longer than metathorax; transverse median vein in front wings interstitial with basal vein .... *Chalybion*

KEY TO THE MICHIGAN SPECIES OF THE GENUS *CHLORION*

1. Second cubital cell of fore wing higher than broad 2
1. Second cubital of fore wing as broad as or broader than high 8
2. Claws with one tooth .... Subgenus *Chlorion* 3
2. Claws with two or more teeth 4
3. Body bright blue or green .... *C. cyaneum* Dahlb.
3. Body bronze blue .... *C. cyaneum aerarium* Patton
4. Claws with two teeth, clypeus with median truncated lobe and sinus on each side .... Subgenus *Palmodes* 5
4. Claws with three to six teeth; clypeus with median emargination Subgenus *Priononyx* 7
5. Abdomen black or at most only faintly brownish .. *C. laeviventris* (Cress) *
5. Abdomen more or less ferruginous or yellow 6

6. Tip of abdomen black .... *C. abdominalis* (Cress)
6. Abdomen entirely ferruginous or yellow; wings fuliginous; female with six comb teeth .... *C. rufiventris* (Cress) *

7. Abdomen more or less ferruginous or yellow; ventral plate of sixth abdominal segment of male excavated on posterior margin *C. rubidorsum* (Costa) *
7. Abdomen black or dark brown; ventral plate of sixth abdominal segment of male not excavated as in preceding species .. *C. atratum* (Lepel)

8. Stigmatal groove absent .... Subgenus *Isodontia* 9
8. Stigmatal groove present .... Subgenus *Ammobia* 11

9. Mandible with two teeth .... *C. aztecum* (Sauss) *
9. Mandible with three teeth 10

10. Legs black .... *C. harrisi* Fernald
10. Legs more or less yellow .... *C. auripes* Fernald

11. Abdomen black; wings dark with violet reflections *C. pennsylvanicum* (L.)
11. Abdomen mostly red, with apical segments black .. *C. ichneumoneum* (L.)

KEY TO THE MICHIGAN SPECIES OF THE GENUS *SPHEX*

1. Females 2
1. Males 19

2. Collar long or short, but transversely ridged 3
2. Collar short, smooth 4

3. Medium to large species; meso- and metapleural pubescent bands strong, not sericeous .... *S. procerus* (Dahlb.)
3. Medium to small species; meso- and metapleural pubescent bands weakly pubescent, pleura whitish sericeous .... *S. politus* (Cress) *

4. Propodeal side ridges running strongly forward 5
4. Propodeal side ridges running slightly or not at all forward 6

5. Thorax and pleura dull black; pleura rough between punctures; veins of wings, especially anterior part, deep yellow, apical part clouded; propodeal side ridges running strongly forward *S. placidus placidus* (Sm.)
5. Thorax and pleura glossy, shining; pleura smooth between punctures; veins of wings yellowish; propodeal side ridges running forward on anterior part only .... *S. fernaldi* Murray

6. Wing veins light golden yellow, except that costa is dark-colored from just in front of stigma to its terminus .... *S. harti* Fernald
6. Wing veins not yellow 7

7. Episternal suture with backward notch, forming a horizontal V *S. aureonotatus* (Cam.)
7. Episternal suture not forming such a notch 8

8. Episternal suture short, ending about opposite prothoracic lobe 9
8. Episternal suture long, extending well down toward under side of body 13

9. Legs partly red; scutellum practically without ridges .. *S. aberti* (Hald.) *
9. Legs entirely black; scutellum longitudinally ridged 10

10. Metanotal flange large, reddish; prothoracic lobe, lower sides of collar, streaks on the meso- and metapleura, and large spot in front of middle coxa, reddish; eyes convergent below .... *S. leopardus* Fernald
10. Metanotal flange small or absent; prothoracic lobe and rest of thorax black; eyes converging only slightly below 11

11. Tegula reddish yellow; base of wing yellowish except subcosta; thorax evenly and generally pubescent on sides; strong meso- and metapleural bands present .... *S. junceus* (Cress)
11. Tegula black; base of wing not yellowish; thorax only slightly pubescent; mesopleural pubescent band narrow; metapleural pubescent band absent 12

12. Very slight metanotal flange; base of costa of fore wing amber in reflected light .... *S. kennedyi* Murray
12. No metanotal flange; base of costa in fore wing black in reflected light *S. urnarius* (Dahlb.)

13. Legs partly ferruginous; thorax densely pilose *S. breviceps breviceps* (Sm.) *
13. Legs black 14

14. Collar rising almost vertically from neck, then turning backward to form a rather long, almost flat, dorsal surface; no metanotal flange; scutellum longitudinally ridged .... *S. arvensis* (Dahlb.)
14. Collar rising slopingly from neck to highest point; dorsal surface much shorter 15

15. Pilosity of head and thorax black 16
15. Pilosity of thorax white, although head pilosity may be black 17

16. Wings light yellow on basal half, somewhat fuliginous on apical half; first abdominal segment and about one half of second red, remainder of abdomen black .... *S. fernaldi* Murray
16. Entire wings deep fuliginous; second abdominal segment mainly red, with posterior border black .... *S. nigricans* (Dahlb.)

17. Head pilosity black; inner margin of eyes parallel; no metanotal flange *S. mediatus* (Cress)
17. Head pilosity white; inner margin of eyes slightly converging below; a moderate metanotal flange 18

18. Pleura with sericeous hairs short and inconspicuous *S. pilosus* subsp. *nudus* Murray
18. Pleura heavily sericeous .... *S. pilosus* subsp. *pilosus* Fernald

19. Episternal suture with a backward notch forming a wide horizontal V; clypeus with a sharp tooth .... *S. aureonotatus* (Cam.)
19. Episternal suture running straight downward, no notch 20

20. Episternal suture extending only to or just below the prothoracic lobe . 21
20. Episternal suture extending down to under side of body 27

21. Collar long, broad, flat above, transversely ridged, rising rather sharply from neck .... *S. procerus* (Dahlb.)
21. Collar shorter, not flat above, not turning back at a sharp angle 22

22. Pilosity of thorax black .... *S. fernaldi* Murray
22. Pilosity of thorax white 23

23. Clypeal margin broadly truncate, sometimes with weak central emargination; legs partly ferruginous .... *S. aberti* (Hald.) *
23. Clypeal margin not broadly truncate, always with central emargination 24

24. Metanotal flange very large, reddish; prothoracic lobe, lower sides of collar, streaks on meso- and metapleura, and a large spot in front of middle coxa, reddish; eyes convergent below ..... *S. leopardus* Fernald
24. Metanotal flange small or absent; prothoracic lobe and pleura entirely black 25

25. Mesonotum entirely smooth, no transverse ridges; costa of fore wing amber in reflected light; first (second petiole segment) and second abdominal segment red, remainder of abdomen entirely black without bluish tint; no metanotal flange .... *S. kennedyi* Murray
25. Mesonotum with a few transverse ridges, although it may be almost smooth; costa of fore wing black in reflected light; first two or three abdominal segments may be red, remainder of abdomen black with bluish tint; no metanotal flange 26

26. First and second abdominal segments red with black dorsal band, remainder of abdomen black; mesopleural pubescent band grading into sericeous in front; metapleural band weakly developed *S. urnarius* (Dahlb.)
26. First, second, and third abdominal segments red, generally with a black dorsal band, remainder of abdomen black; mesopleural pubescent band heavy, sharply defined; metapleural pubescent band well developed *S. junceus* (Cress)

27. Collar finely transversely ridged .... *S. politus* (Cress) *
27. Collar smooth or with only faint traces of ridges 28

28. Entire body clothed with black hairs throughout 29
28. Entire body not clothed with black hairs throughout 31

29. Wings strongly tinged with yellow, both veins and surface between the veins; hyaline to beyond the cells, the outer part partly fuliginous; dull black .... *S. placidus placidus* (Sm.)
29. Wings more or less fuliginous; more or less shining 30

30. Wings somewhat fuliginous; body shining, glossy between the punctures; clypeus shining, with a few punctures .... *S. fernaldi* Murray
30. Wings strongly fuliginous; body between punctures not glossy, hardly shining, dull; clypeus dull, more heavily punctured *S. nigricans* (Dahlb.)

31. Collar rising almost vertically from the neck, then turning backward to form an almost flat long dorsal surface; head pilosity black *S. arvensis* (Dahlb.)
31. Not as in preceding, collar rising slopingly from neck to highest point; head pilosity white 32

32. Legs partly ferruginous .... *S. breviceps breviceps* (Sm.) *
32. Legs black 33

33. Wing veins yellow; metanotal flange moderately large .. *S. harti* Fernald
33. Wing veins dark brown to black; metanotal flange very small 34

34. Pleura with sericeous hairs short and inconspicuous
S. pilosus subsp. *nudus* Murray
34. Pleura more or less heavily sericeous .... *S. pilosus* subsp. *pilosus* Fernald

KEY TO THE MICHIGAN SPECIES OF THE GENUS *PODALONIA*

1. Females 2
1. Males 5

2. Abdomen entirely black; without a metanotal flange ... *P. luctuosa* (Sm.)
2. Abdomen red and black or entirely red 3

3. Metanotal flange very large, with strong emargination; clypeus bulging only slightly in center, with many punctures, but these separated by several times their diameter; mesepimeron without transverse striae except traces just before posterior edge. .... *P. violaceipennis* Lepel
3. Metanotal flange small or, if large, without a strong emargination; clypeus strongly bulging in center 4

4. Mesopleuron with a great many tiny punctures, surface glossy; more or less conspicuous short brownish hairs on surface of mesopleuron; abdomen red and black or, rarely, entirely red .... *P. sericea* Murray
4. Mesopleuron with very few tiny punctures, surface not glossy; brownish sericeous hairs if present inconspicuous .... *P. robusta* (Cress)

5. Metanotal flange very large, with strong emargination; pilosity of head and thorax black; abdomen red and black ..... *P. violaceipennis* Lepel
5. Metanotal flange large, small, or none, but if large, without strong emargination; pilosity of head black but pilosity of thorax may be white 6

6. No metanotal flange; pilosity of head and thorax black; clypeus broadly transverse .... *P. luctuosa* (Sm.)
6. Metanotal flange present; clypeus narrowly transverse; at least part of pilosity of thorax generally white 7

7. Mesopleuron glossy, with abundant white sericeous hairs; metapleuron and mesepimeron without prominent transverse ridges *P. sericea* Murray
7. Mesopleuron not glossy, and with few or no sericeous hairs; metapleuron and mesepimeron with prominent transverse ridges .. *P. robusta* (Cress)

The genera *Sceliphron* and *Chalybion* consist of but a single species each, namely, *S. caementarium* (Drury) and *C. cyaneum* Klug., respectively.

The distribution records are taken from insects in the museums of the University of Michigan and Michigan State College, the collections of Mr. George Steyskal, Mr. Curtis Sabrosky, and the author.

As is usual in the distribution records of insects, there are no specimens from a good part of the territory covered. Nor are there records of the collection of Sphecinae from the thirteen following counties: Iron, Menominee, Delta, Arenac, Sanilac, Shiawassee, St. Clair, Barry, Eaton, Cass, Branch, Hillsdale, and Lenawee. Eleven more counties are represented by only one species. On the other hand, Midland County is represented by 18 of the 28 species and varieties; Oakland, by 15; Washtenaw and Huron, by 13; and Cheboygan, Manistee, Ingham, and Berrien, by 12 each. There have been collected 15 species in the Upper Peninsula, 24 in the Lower Peninsula, and one on Isle Royale. Two species, *Sphex arvensis* (Dahlb.) and *S. pilosus* subsp. *pilosus* Fernald, have been found only in the Upper Peninsula. *Chlorion auripes* Fernald, *C. cyaneum* Dahlb., *C. cyaneum aerarium* Patton, *C. harrisi* Fernald, *C. ichneumoneum* L., *C. pennsylvanicum* L., *Sphex aureonotatus* Cam., *S. leopardus* Fernald, *S. nigricans* (Dahlb.),* and *S. placidus placidus* (Sm.) have been collected only in the Lower Peninsula. The remainder have been taken in both the Upper and the Lower peninsulas.

ANNOTATED LIST OF SPECIES AND VARIETIES

CHLORION ABDOMINALIS (Cress). — Dates of capture: July 6 to August 9. L. P.:[1] Cheboygan, Crawford, Roscommon, Iosco, Osceola, Muskegon, Wayne.

CHLORION ATRATUM Lepel. — Dates of capture: July 16 to September 5. U. P.:[1] Schoolcraft and Mackinac. L. P.: Cheboygan, Charlevoix, Alpena, Benzie, Kalkaska, Crawford, Alcona, Manistee, Missaukee, Iosco, Mason, Osceola, Gladwin, Oceana, Midland, Huron, Gratiot, Lapeer, Ingham, Livingston, Oakland, Van Buren, Kalamazoo, Jackson, Washtenaw.

CHLORION AURIPES Fernald. — Dates of capture: July 20 to August 22. L. P.: Midland, Livingston, Oakland, Washtenaw, Wayne.

CHLORION CYANEUM Dahlb. — Dates of capture: June 13 to August 25. L. P.: Gratiot, Allegan, Berrien.

CHLORION CYANEUM AERARIUM Patton. — Dates of capture: July 26 to September 5. L. P.: Manistee, Wexford, Lake, Midland, Kalamazoo, Washtenaw, Wayne.

CHLORION HARRISI Fernald. — Dates of capture: June 12 to Septem-

[1] L. P., Lower Peninsula; U. P., Upper Peninsula.

ber 3. L. P.: Cheboygan, Crawford, Manistee, Wexford, Roscommon, Iosco, Lake, Osceola, Clare, Gladwin, Oceana, Midland, Huron, Saginaw, Clinton, Genesee, Ingham, Livingston, Oakland, Washtenaw, Wayne, Berrien.

CHLORION ICHNEUMONEUM L. — Dates of capture: July 6 to September 21. L. P.: Manistee, Lake, Clare, Gladwin, Oceana, Isabella, Midland, Huron, Ottawa, Genesee, Ingham, Livingston, Oakland, Kalamazoo, Calhoun, Jackson, Washtenaw, Wayne, St. Joseph.

CHLORION PENNSYLVANICUM L. — Dates of capture: July 24 to September 12. L. P.: Iosco, Midland, Bay, Huron, Genesee, Ingham, Livingston, Oakland, Kalamazoo, Jackson, Washtenaw, Wayne, St. Joseph.

SPHEX ARVENSIS (Dahlb.). — Dates of capture: July 15 to August 29. U. P.: Dickinson, Chippewa.

SPHEX AUREONOTATUS Cam. — Dates of capture: July 9 to September 5. L. P.: Midland, Ingham, Oakland, Wayne, Berrien.

SPHEX HARTI Fernald. — Dates of capture: July 3 to September 23. U. P.: Chippewa. L. P.: Leelanau, Manistee, Midland, Huron, Saginaw, Ingham, Oakland, Wayne, Berrien.

SPHEX JUNCEUS (Cress). — Date of capture: August 23. H. T. Fernald, in his work on North American and West Indian digger wasps of the genus *Sphex*, says this species has been collected in Michigan.

SPHEX KENNEDYI Murray. — Dates of capture: June 13 to September 24. U. P.: Gogebic, Mackinac. L. P.: Cheboygan, Presque Isle, Alcona, Manistee, Iosco, Mason, Osceola, Midland, Bay, Huron, Montcalm, Gratiot, Saginaw, Oakland, Van Buren, Washtenaw, Berrien.

SPHEX LEOPARDUS Fernald. — Dates of capture: July 21 to September 14. L. P.: Midland, Oakland.

SPHEX MEDIATUS (Cress). — Dates of capture: July 2 to September 2. U. P.: Keweenaw, Baraga, Marquette, Schoolcraft, Luce, Chippewa. L. P.: Cheboygan, Roscommon.

SPHEX NIGRICANS (Dahlb.). — Dates of capture: June 19 to August 9. L. P.: Allegan, Oakland, Kalamazoo, Washtenaw, Wayne, Berrien.

SPHEX PILOSUS subsp. NUDUS Murray. — Dates of capture: July 3 to September 24. Isle Royale. U. P.: Dickinson, Marquette, Alger, Chippewa, Mackinac. L. P.: Charlevoix, Lapeer.

SPHEX PILOSUS subsp. PILOSUS Fernald. — Date of capture: August 2. U. P.: Ontonagon.

SPHEX PLACIDUS PLACIDUS (F. Smith). — Dates of capture: September 3 to 21. L. P.: Oakland, Washtenaw, Wayne.

SPHEX PROCERUS (Dahlb.). — Dates of capture: June 2 to September 27. U. P.: Schoolcraft, Mackinac. L. P.: Cheboygan, Charlevoix, Otsego, Montmorency, Benzie, Grand Traverse, Crawford, Oscoda, Manistee, Wexford, Roscommon, Lake, Osceola, Gladwin, Oceana, Midland, Huron, Muskegon, Ionia, Ingham, Oakland, Jackson, Washtenaw, Berrien.

SPHEX URNARIUS (Dahlb.). — Dates of capture: May 10 to September 25. U. P.: Baraga, Dickinson, Marquette, Alger, Schoolcraft, Mackinac. L. P.: Cheboygan, Presque Isle, Charlevoix, Leelanau, Antrim, Otsego, Montmorency, Benzie, Kalkaska, Crawford, Oscoda, Alcona, Manistee, Roscommon, Ogemaw, Iosco, Lake, Osceola, Clare, Gladwin, Oceana, Mecosta, Isabella, Midland, Bay, Huron, Muskegon, Gratiot, Saginaw, Ionia, Genesee, Lapeer, Ingham, Oakland, Van Buren, Kalamazoo, Washtenaw, Wayne, Berrien, St. Joseph.

PODALONIA LUCTUOSA Sm. — Dates of capture: May 13 to September 20. U. P.: Dickinson, Marquette, Schoolcraft, Chippewa, Mackinac. L. P.: Cheboygan, Charlevoix, Wexford, Mason, Lake, Osceola, Isabella, Midland, Huron, Gratiot, Saginaw, Lapeer, Ingham, Oakland, Wayne.

PODALONIA ROBUSTA (Cress). — Dates of capture: June 26 to September 25. U. P.: Keweenaw, Gogebic, Alger, Chippewa, Mackinac. L. P.: Cheboygan, Otsego, Benzie, Grand Traverse, Kalkaska, Crawford, Manistee, Missaukee, Ogemaw, Midland, Huron, Saginaw.

PODALONIA SERICEA Murray. — Dates of capture: July 4 to August 21. U. P.: Chippewa. L. P.: Van Buren, Berrien. The specimen from Van Buren County is a paratype and is in the University of Minnesota collection.

PODALONIA VIOLACEIPENNIS (Lepel). — Dates of capture: May 29 to September 7. U. P.: Chippewa, Mackinac. L. P.: Cheboygan, Charlevoix, Leelanau, Kalkaska, Alcona, Manistee, Missaukee, Osceola, Gladwin, Midland, Huron, Kent, Ingham, Van Buren, Calhoun, Wayne.

SCELIPHRON CAEMENTARIUM (Drury). — Dates of capture: June 30 to September 1. U.P.: Baraga, Mackinac. L. P.: Cheboygan, Charlevoix, Grand Traverse, Crawford, Manistee, Roscommon, Iosco, Oceana, Midland, Bay, Huron, Montcalm, Tuscola, Genesee, Ingham, Livingston, Oakland, Macomb, Van Buren, Kalamazoo, Jackson, Washtenaw, St. Joseph, Monroe.

CHALYBION CYANEUM Fab. — Dates of capture: June 1 to August 17. U. P.: Mackinac. L. P.: Cheboygan, Grand Traverse, Manistee, Newaygo, Midland, Huron, Montcalm, Gratiot, Saginaw, Genesee, Lapeer, Ingham, Macomb, Van Buren, Kalamazoo, Calhoun, Washtenaw, Wayne, Berrien.

MIDLAND, MICHIGAN

REFERENCES

FERNALD, H. T. 1906. The Digger Wasps of North America and the West Indies Belonging to the Subfamily Chlorioninae. Proc. U. S. Nat. Mus., No. 1487, 31 : 291–423.

—— 1927. The Digger Wasps of North America of the Genus Podalonia (Psammophila). Proc. U. S. Nat. Mus., No. 2681, 71, Art. 9 : 1–42.

—— 1934. The North American and West Indian Digger Wasps of the Genus Sphex (Ammophila Auct.). E. O. Painter Printing Co., Deland, Florida.

HUTSON, J. C. 1919. The North American Species of the Genus Sceliphron. Trans. Am. Entomol. Soc., 45 : 203–227.

MICKEL, C. E. 1917. Synopsis of Sphecoidea of Nebraska. Nebraska Univ. Studies, 17 (4) : 25–89.

MURRAY, W. D. 1938. Some Revisions in the Genus Sphex, with One New Species, a New Subspecies, and a New Name (Hymenoptera, Sphecidae). Ann. Entomol. Soc. Am., 31 (1) : 17–42.

—— 1940. Podalonia (Hymenoptera: Sphecidae) of North and Central America. Entomologica Americana, 20 (1) : 1–77.

PORTER, B. A. 1926. American Wasps of the Genus Sceliphron Klug. Proc. U. S. Nat. Mus., No. 2650, 70, Art. 1 : 1–22.

ROWHER, S. A. 1916. Sphecoidea, pp. 645–697 in Guide to the Insects of Connecticut. Part III: The Hymenoptera, Wasp-like Insects of Connecticut, by H. L. Viereck and Others. State of Connecticut, Public Document No. 47. State Geological and Natural History Survey. Bulletin No. 22.

BANDING AS AN AID IN STUDYING THE ACTIVITIES OF THE LITTLE BROWN BAT, *MYOTIS LUCIFUGUS LUCIFUGUS* *

HAROLD B. HITCHCOCK

ALTHOUGH bats are by no means uncommon in Michigan, much remains to be discovered about their activities. Where, for example, do the hundreds of little brown bats that form colonies in buildings during the summer pass the winter? Is it the same bat that we always find behind the cottage shutter? How far do bats wander during their nocturnal feeding flights? These are just a few of the things about which little is yet known. Bats of the commonest species, *Myotis lucifugus lucifugus,* are easily caught in almost any neighborhood where there are buildings, and they can be permanently marked by means of numbered bands. Using this technique Griffin (1940) learned about the homing abilities and migrations of bats in the New England area; Mohr (1934) conducted similar work in Pennsylvania, employing fingerling tags. Hitchcock and Reynolds (1942) studied homing of banded bats in Ontario, across the border from Michigan.

During the summer of 1942 an opportunity was afforded to the writer to study bats at the Biological Station of the University of Michigan, at Douglas Lake, Cheboygan County. With the exception of one specimen of *Myotis keenii septentrionalis,* all those captured were of the species named above. In all ninety were tagged with bird bands issued by the Fish and Wildlife Service of the Department of the Interior, Washington, D.C. The bands were clamped around the hind leg, where apparently they cause no inconvenience or injury.

The easiest place in which to collect bats at the Station is behind the shutters of the faculty cabins, especially after the middle of August. Here they can be reached from the ground and gathered

* This study was made through the coöperation of the University of Michigan Biological Station at Douglas Lake, Cheboygan County, Michigan.

by hand. Another haunt is in the roof of one of the automobile shelters, where some could usually be extracted by means of forceps from cracks between the beams, and squeakings indicated many more out of sight between the ridgepole and the roofing. A small colony was located in the door of one of the cabins, where a sprung piece of metal sheathing permitted access to a narrow space behind the boards. At the very close of the season a colony was discovered in the attic of the caretaker's residence, where forty, representing perhaps half of the population at the time, were captured.

The first banding was done on July 24, when two adult males and one adult female were caught behind the shutters of faculty cabins. Both males were subsequently recaptured twice in the same area; the female was retaken once.

The next banding was done on August 20, when 19, including 2 of the first lot, were caught behind the shutters. The increase was due to the addition of young individuals, the distribution being as follows: adult male, 1; adult female, 3; young male, 7; young female, 8. On this occasion the bats were released not at the Station, but near Petoskey, about sixteen miles in a straight line from Douglas Lake. They were carried there by car and freed on the afternoon of August 21. On August 23, when the next collecting was done, one of these bats, a young male, was recaptured behind the shutter of a faculty cabin. The only other bat from this group that is known to have returned was found on August 31 behind the same shutter where it had been first captured on July 24.

Homing experiments previously reported have dealt with bats captured either in colonies or in caves, and have demonstrated that both immature and adult bats are able to home for distances considerably greater than those recorded here. The present experiment is of interest, however, in showing that they will return to an area even after scattering from the colony. Little is known of the speed of homing, so that the fact that one came back before the third night is of some interest. It is unfortunate that the last opportunity of making a complete roundup of bats was on August 24, soon after the closing of the Station. During the last week of August Professor Charles W. Creaser made several inspections of shutters still in place, and on the 31st recovered one of the bats that had been released at Petoskey, as noted above.

On August 24 all the known hiding places of bats at the Station were visited, and fifty were collected. These were taken 325 miles by car and released on August 25 near Port Sanilac, Michigan, some 185 miles from Douglas Lake in a straight line. It will be interesting to check over the bats at the Station this summer (1943) to see whether any of the lot return to their old haunts.[1]

Because all the bats handled were banded, information on their whereabouts during the winter may be gained. The little brown bat is known to hibernate in subterranean retreats. Caves and passages in abandoned mines frequently shelter them by the hundreds. Since there are no caves near Douglas Lake, bats of this species summering there probably migrate. Do they go to caves in Indiana, or in Missouri, where large numbers hibernate? Bats from the Biological Station may help to answer this question if someone collects a banded specimen and reports the place and date of capture to the Fish and Wildlife Service.

MIDDLEBURY COLLEGE
MIDDLEBURY, VERMONT

LITERATURE CITED

GRIFFIN, D. R. 1940. Migrations of New England Bats. Bull. Mus. Comp. Zoöl., 86: 217–246, pls. 1–5.

HITCHCOCK, H. B., AND REYNOLDS, K. 1942. Homing Experiments with the Little Brown Bat, *Myotis lucifugus lucifugus* (Le Conte). Journ. Mammal., 23: 258–267.

MOHR, C. E. 1934. Marking Bats for Later Recognition. Proc. Penna. Acad. Sci., 8: 26–30.

[1] These studies were continued during the summer of 1943 by Miss E. Mohrmann, working under Dr. C. W. Creaser. None of the bats banded in 1942 were recaptured, but some seventy-five new specimens were tagged.

A CHECK ON THE FIN-CLIPPING METHOD FOR ESTIMATING FISH POPULATIONS *

LOUIS A. KRUMHOLZ

THE first instance of marking fish for experimental purposes is recorded by Izaak Walton in *The Compleat Angler* (1653). According to him, Sir Francis Bacon studied the age and homing instincts of salmon "by tying a riband, or some known tape or thread, in the tail of some young Salmons which have been taken in weirs as they have swimmed towards the salt water; and then by taking part of them again, with the known mark, at the same place, at their return from the sea, which is usually about six months after." Since his time there has evolved the use of button, strap, and belly tags as means of identification.

Calderwood (1902) indicated that Fraser marked salmon in 1829 by removing the adipose fin. Fin clipping as a means of marking fish has been employed extensively in studies of salmon, trout, and other fresh-water fish, mostly, however, in determining migration, growth, and survival of plantings.

Recently David H. Thompson, of the Illinois Natural History Survey, used fin clipping as one of the procedures in making an estimate of the population of black crappies, *Pomoxis nigro-maculatus* (LeSueur), in Lake Senachwine, a backwater of the Illinois River, near Henry, Illinois. It is unfortunate that the results of these experiments and a description of the technique employed have not been published. Underhill (1941) used the same formula as Thompson in determining the size of a breeding population of chub suckers, but marked the fish by removing two or three adjoining scales from the side. Thompson's procedure (slightly modified) has been employed by the Michigan Institute for Fisheries Research in making population estimates of four Michigan lakes, but only on Twin Lake in Oscoda County was it feasible to check the method by poisoning.

* Contribution from the Institute for Fisheries Research of the Michigan Department of Conservation.

METHODS

During the first part of the check on Thompson's method the anal fin was clipped. Later the left pectoral, instead of the anal fin, was removed. The fins were cut about one-eighth inch from the flesh to lessen risk of infection and because the mark was intended to last only for the duration of the experiment. All fish taken in the nets each day were weighed, measured, marked, and returned to the water at a central station, indicated by a buoy (Thompson did not

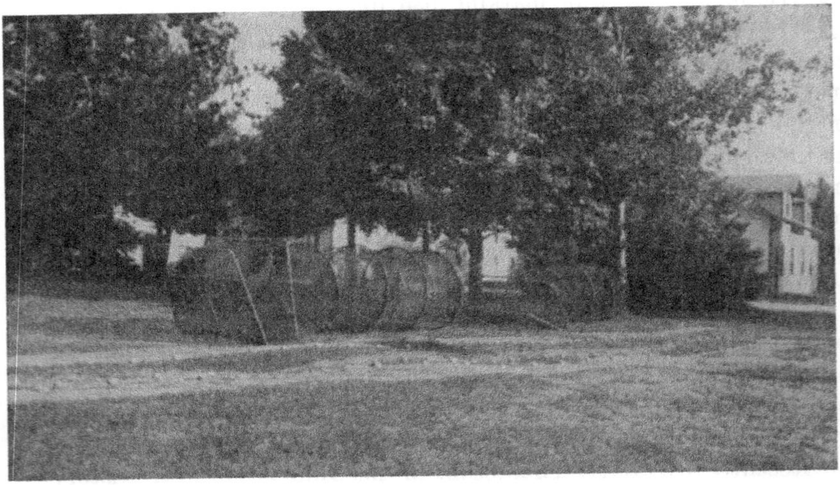

FIG. 1. Net A, special fyke net for trapping small fish

release the fish at a central station), to afford equal opportunity for the dispersal of the fish in any direction. Marked fish that were subsequently taken in nets were recorded as "returns" and again released at the central station. Those marked fish that died in the nets were also regarded as returns, and were subtracted from the total number of marked fish in the lake. Any of these fish that were found dead outside the nets were not counted as returns, but were subtracted from the number of marked fish. A complete circuit of the shoreline in quest of dead fish was made twice each day. No fish were marked during the last ten days of net operation.

Four different sizes of fyke nets, designated by the letters A, B, C, and D in the text, with the same letters indicating their locations in the lake (Fig. 2), were used in the population study of the north basin of Twin Lake.

Net A. — A special fyke net, developed in Wisconsin for trapping small fish, equipped with two galvanized pipe frames five feet wide and four feet deep at the front (Fig. 1). The first frame has a vertical center bar for attachment of the lead. Four other hoops are wooden, and each has a diameter of four feet. The lead was 150 feet long and four feet deep. The entire net and lead was hung with three-eighths-inch square mesh and treated with copper oleate.

Net B. — A large tarred fyke with a front hoop diameter of four feet hung with one-inch square-mesh netting. It was equipped with wings twenty feet long and four feet deep, with one and one-half-inch square mesh. A lead one hundred feet long and four feet deep with two-inch square mesh was used with this net.

Net C. — A small tarred fyke of one-inch square mesh with a three-foot front opening, with twenty-foot wings of one and one-half-inch square mesh. A fifty-foot lead was used with this net.

Net D. — A small fyke with a three-foot front opening equipped with twenty-foot wings. The entire net was hung with three-eighths-inch square netting and treated with copper oleate. No lead was used with this net.

DESCRIPTION OF TWIN LAKE

Twin Lake, which consists of two basins, is located in T. 25 N., R. 2 E., Section 25, about eight miles south and two miles west of Mio, Michigan. The north basin, which was used for the experiment, is about seven and three-quarters acres in extent and has a maximum depth of thirty-five feet (Fig. 2). It has a rather definite sandy shoreline, and the deeper parts of the bottom are covered with pulpy peat. The slope of the basin is rather steep, except at the south end. Many trees that have fallen into the water or were dropped there in lumbering have become water-soaked and are found nearly everywhere along the shoreline. These deadheads restricted the setting of nets to some extent and made seining impracticable. Aquatic plants were fairly abundant in the shallow areas.

The south basin (10.1 acres), somewhat larger than the north basin, has a maximum depth of forty feet. The shoreline is more irregular than that of the north basin, but the bottom types of the two depressions are quite similar. There are many deadheads around

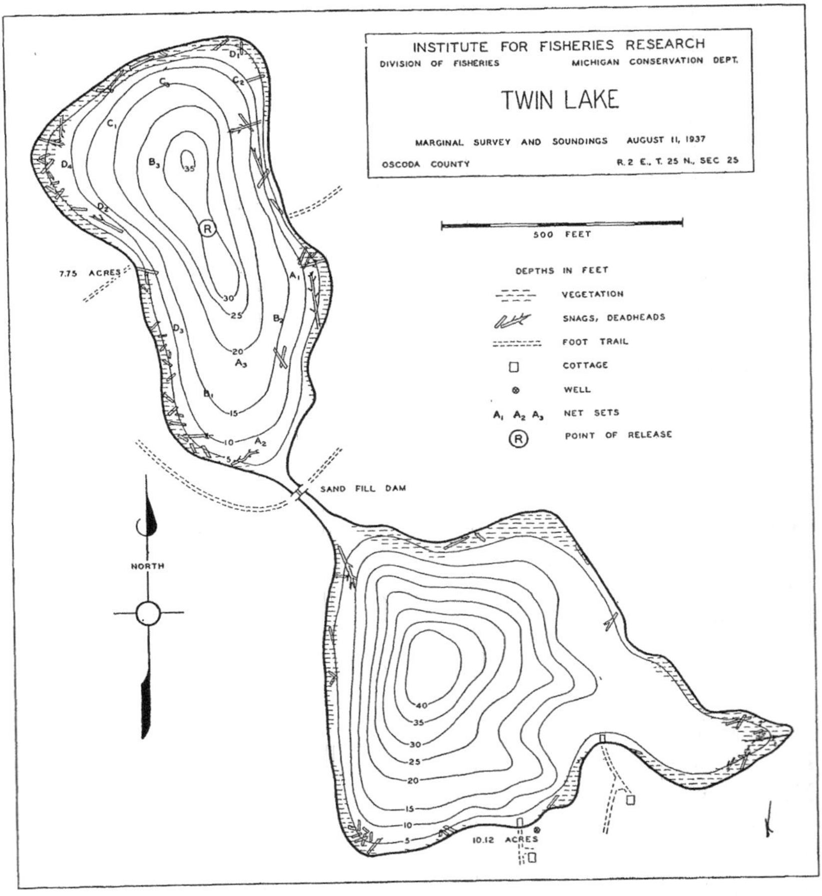

FIG. 2. Twin Lake, Oscoda County, Michigan (outlines and depth contours taken from a map by the United States Forest Service)

the shore of the south basin also, but the aquatic vegetation was not so abundant as that in the north basin.

The Huron National Forest kindly provided Civilian Conservation Corps labor, under the direction of Mr. Gifford Adams, to construct a barrier dam between the two basins during March and April, 1941. An eight-foot opening was left in the middle of the dam to allow free passage of the fish from one basin to the other prior to the poisoning. The central opening of the dam was filled in on August 28.

PROCEDURE

The nets were set in the locations indicated in Figure 2. All sets were made with the lead running diagonally from the shore. The nets were moved and reset as follows: original set, on July 29 (A_1, B_1, C_1, and D_1 on the map); second set, on August 6 (A_2, etc.), after the nets had had time to dry; second change, on August 13 (A_3, etc.), with nets again allowed time to dry before resetting; an additional change with net D, on August 20.

The number of fish caught, the number marked, and the number of returns are shown in Table I. These data furnished the material for an estimate of the population of the fish over forty-five millimeters in total length. Those of a smaller size passed through the meshes of the nets and therefore could not be included in the estimate.

The mathematics of the method of making an estimate of the total fish population that Thompson used has been described by Schnabel (1938, formula 2). It can be proved algebraically that this formula gives yields that are high, but correction for this error is too laborious to be considered practicable. The formula we used, as modified from Schnabel, is $P = \frac{\Sigma \cdot A \cdot B}{\Sigma \cdot C}$, in which P is the population, A is the number of fish examined on any one day, B is the number of marked fish in the lake, and C is the number of returns.

Investigations that involve the use of marked fish in netting operations are subject to certain uncontrollable errors. Some of the more frequent causes of error are: (1) Fish marked during the experiment may die or may be caught by anglers; (2) The gear may be so selective as to limit the size of the fish taken; (3) The duration of the experiment may be so short as to give misinformation; and (4), If the investigation is extended over a considerable time, small fish will grow to a size large enough to be taken by the nets.

The opening in the middle of the dam between the two basins allowed fish to pass freely from one to the other. Observations at various times during the investigation showed that they actually passed both out of and into the north basin, and it is assumed that the movement was approximately equal and did not have any marked effect on the population estimate.

An intensive creel census, during which the entire catch of each fisherman was examined at the end of the day's fishing, was in

TABLE I

DATA AND ESTIMATE OF THE TOTAL POPULATION OF FISH OVER FORTY-FIVE MILLIMETERS IN LENGTH OF THE NORTH BASIN OF TWIN LAKE, OSCODA COUNTY, MICHIGAN, JULY 29–AUGUST 23, 1941

Date	A Number of fish examined	Number of fish marked	Number of marked dead fish	B Number of marked fish in lake	Product A · B	Sum of products	Number of returns	C Sum of returns	Estimated population
July									
30	53	53	..	..	...	....	..	..	...
31	55	53	..	53	2,915	2,915	2	2	1,458
August									
1	67	64	..	106	7,102	10,017	3	5	2,003
2	59	55	..	170	10,030	20,047	2	7	2,864
4	85	75	3	225	19,125	39,172	6	13	3,013
5	94	79	..	297	27,918	67,090	3	16	4,193
6	53	50	..	376	19,928	87,018	1	17	5,119
7	115	109	15	426	48,990	136,008	5	22	6,182
8	59	56	3	520	30,680	166,688	4	26	6,411
9	53	49	13	573	30,371	197,059	4	30	6,567
11	53	46	51	609	32,277	229,336	5	35	6,552
12	68	63	1	604	41,074	270,410	2	37	7,308
13	45	41	2	666	29,970	300,380	4	41	7,326
14	38	37	..	705	26,790	327,170	..	41	7,980
15	45	..	..	742	33,390	360,560	3	44	8,195
16	28	..	..	742	20,776	381,336	..	44	8,667
18	40	..	1	741	29,640	410,976	2	46	8,934
19	20	..	..	741	14,820	425,796	..	46	9,256
20	30	..	..	741	22,230	448,026	5	51	8,785
21	27	..	..	741	20,007	468,033	1	52	9,001
22	42	..	..	741	31,122	499,155	1	53	9,418
23	20	..	..	741	14,820	513,975	..	53	9,698

operation on both basins of the lake during the whole summer. The census clerk was instructed to take added precaution in checking in order that accurate records of any fish marked during the experiment caught by anglers might be accounted for in the compilation of the population data. Inasmuch as no marked fish were recorded in the anglers' catches, there was no loss by this means. No allowance was made in the estimate for unmarked fish either caught by anglers or found dead in the lake.

The nets used were very selective. Nets A and D caught many small fish, but only a few large ones. On the other hand, nets B and C could not retain fish as small as those netted by A and D, but caught about 80 per cent of all fish taken during the study. Net B, although the mesh size was considerably larger than that of net A, caught more fish than the combined catches of the three other nets. It has been suggested by Van Oosten (1935) that fishes may react negatively to darkness and therefore avoid an impounding net darkened by small meshes. The data from the fish taken in all nets were lumped together on the assumption that the differences in the catches of the individual nets would counteract one another and would give a fairly reliable estimate of the population.

On August 29, 1941, the waters of the north basin of Twin Lake were treated with powdered derris root (5 per cent rotenone). The powder was mixed into a thin suspension and poured over the surface waters in the wake of an outboard motor in accordance with the technique developed by the Michigan Institute for Fisheries Research (Eschmeyer, 1937; Greenbank, 1941). In addition to the application of derris to the surface waters, a new method of introducing the chemical into the deeper waters was tried with satisfactory results. The apparatus consisted of a ten-inch funnel attached by the small end to a five-eighths-inch garden hose. Two ten-pound sash weights were used to keep this end of the hose well under water. A fifteen-gallon drum fitted with a five-eighths-inch spigot was set in the middle of the boat and the hose attached. The drum was filled with a suspension of derris, and the boat was moved over the deeper parts of the lake. The suction caused by the inverted funnel and the forward motion of the boat pulled the suspension out of the drum into the deep water. The process of dispersion in deep water was rather slow owing to the small diameter of the hose and the relatively great amount of friction. A hose three-fourths to one inch in diameter would be more satisfactory than the one used in the experiment.

The water temperature at the surface was 70° F. at 8:45 A.M., when the application of derris was begun, and distressed fish were seen rising at various points within fifteen minutes. Two thirds of the derris was applied to the surface and one third was put in the deeper water. The application was completed by 12:45 P.M.

Individual lengths, weights, and scale samples were taken from

TABLE II

NUMBER AND WEIGHT IN POUNDS BY SPECIES OF FISHES HAVING A TOTAL LENGTH GREATER THAN FORTY-FIVE MILLIMETERS RECOVERED FROM THE NORTH BASIN OF TWIN LAKE AFTER TREATMENT WITH DERRIS*

Date	Largemouth Bass		Bluegill		Pumpkinseed		Bluegill × pumpkinseed hybrids		All fish	
	Number	Weight	Number	Weight	Number	Weight	Number	Weight	Number	Weight
August										
29	107	31.56	3,470	247.25	155	11.63	83	13.38	3,815	303.82
30	20	3.88	1,320	54.31	50	3.25	17	2.00	1,407	63.44
31	24	5.63	3,002	99.56	94	6.56	34	5.19	3,154	116.94
September										
1	3	0.88	638	38.88	10	0.88	5	1.25	656	41.89
2	1	0.31	374	21.75	6	0.50	5	1.31	386	23.87
3	3	0.56	78	6.63	2	0.25	...	.....	83	7.44
4	1	0.12	46	4.00	1	0.12	...	.....	48	4.24
5	...	.....	33	2.12	...	.....	...	.....	33	2.12
6	...	.....	11	0.75	...	.....	...	.....	11	0.75
7	...	.....	11	0.94	1	0.12	...	.....	12	1.06
Totals	159	42.94	8,983	476.19	319	23.31	144	23.13	9,605	565.57

* In addition there were 1,773 miscellaneous fry less than 45 mm. long having a total weight of 11.88 pounds.

all the largemouth black bass, *Huro salmoides* (Lacépède), recovered and from representative samples of bluegills, *Lepomis m. macrochirus* Rafinesque, pumpkinseeds, *Lepomis gibbosus* (Linnaeus), and bluegill × pumpkinseed hybrids. These were the only kinds of fish in the lake. The fish for which individual data were not taken were counted and weighed. A complete pickup of all dead fish was continued for ten days after the poisoning. Table II shows the numbers and weights of those recovered.

DISCUSSION

Table III gives the numbers and percentages of the fish of different size groups taken during the netting and poisoning operations. The figures for netting are for the aggregate catches of all four nets. It is evident that the nets were selective, since most of the fish were more than one hundred millimeters in total length.

Estimates of the populations of the individual species and the numbers of each taken during the poisoning are given in Table IV. The numbers collected by means of poisoning do not corroborate

TABLE III

NUMBER AND PERCENTAGES OF FISHES OF DIFFERENT SIZES IN SAMPLES TAKEN BY NETTING AND POISONING FROM THE NORTH BASIN OF TWIN LAKE, OSCODA COUNTY, MICHIGAN

Total length millimeters	Mean length inches	Netting		Poisoning	
		Number of fish	Percentage	Number of fish	Percentage
Bluegills					
0–50	1.6	15	2.3	87	6.7
51–100	3.1	97	14.7	274	21.1
101–150	5.0	235	35.8	614	47.3
151–200	6.7	287	43.6	286	22.1
201–250	8.2	22	3.3	36	2.8
251–300	10.4	2	0.3	..	..
Pumpkinseeds					
0–50	1.3	1	0.4	8	2.5
51–100	3.1	19	7.5	121	38.5
101–150	5.0	177	70.2	141	44.9
151–200	6.5	54	21.4	38	12.1
201–250	8.4	1	0.4	6	1.9
Largemouth black bass					
51–100	2.9	3	12.0	21	11.5
101–150	4.5	5	20.0	26	14.4
151–200	6.6	13	52.0	38	20.9
201–250	9.0	4	16.0	82	45.0
251–300	10.5	..	..	11	6.0
301–350	13.2	..	..	4	2.2
Bluegill × pumpkinseed hybrids					
51–100	3.1	3	3.2	12	9.0
101–150	5.4	23	24.8	19	14.2
151–200	6.7	67	72.0	99	73.8
201–250	8.1	..	..	4	3.0
All fish					
0–50	1.6	16	1.5	95	4.9
51–100	3.1	122	11.9	428	22.2
101–150	5.0	440	42.8	800	41.5
151–200	6.7	421	41.0	461	23.9
201–250	8.9	27	2.6	128	6.7
251–300	10.5	2	0.2	11	0.6
301–350	13.2	..	..	4	0.2
Totals	....	1,028	....	1,927	....

TABLE IV

ESTIMATES OF POPULATIONS OF INDIVIDUAL KINDS AND THE NUMBER OF EACH RECOVERED DURING THE POISONING IN THE NORTH BASIN OF TWIN LAKE, OSCODA COUNTY, MICHIGAN

Data are given only for fish over forty-five millimeters long.

Species	*Estimated population*	*Fish recovered during poisoning*
Largemouth black bass	51	159
Bluegills	16,832	8,983
Pumpkinseeds	1,360	319
Bluegill × pumpkinseed hybrids	837	144
Totals	19,080	9,605

the individual population estimate for any one species. Had the nets been fished for a longer time and had special effort been made to catch only one species, the individual estimates would probably have been much closer. The investigation was intended to give a total-population estimate only of fish over forty-five millimeters in total length.

The total population as estimated from the netting operations was 9,698 fish as compared with 9,605 taken in the poisoning. There is no way of knowing how nearly complete the recovery of poisoned fish was inasmuch as many fish, particularly small ones, may sink at once to the bottom and disintegrate there. Fortunately the preferred habitat of small fish is along shore, so that recovery of a large percentage is possible if one is careful. Some dead fish are eaten by reptiles, birds, and mammals. In this investigation 86 per cent of all marked fish assumed to be alive in the lake were recovered in the poisoning. Prior to the poisoning some of the marked ones might have escaped through the open channel into the south basin, and some might have died without being recovered. The total population of fish more than forty-five millimeters in total length as estimated in Table I is only one per cent higher than that found by means of the poisoning.

The estimate from netting operations was very close to that obtained by poisoning in this first check on the fin-clipping method for estimating fish populations. Further studies of this type are needed to prove definitely the accuracy of the method. At the present time the supply of rotenone-containing roots has been drastically curtailed because of the war, and no suitable substitute has been found. Other checks of this method will be made when conditions permit.

Institute for Fisheries Research
Ann Arbor, Michigan

LITERATURE CITED

Calderwood, W. L. 1902. A Contribution to the Life History of the Salmon, as Observed by Means of Marking Adult Fish. Ann. Rep. Fish. Board Scotland, 1901, 20, Part II: 55–100.

Eschmeyer, R. William. 1937. Some Characteristics of a Population of Stunted Perch. Pap. Mich. Acad. Sci., Arts, and Letters, 22 (1936): 613–628.

Greenbank, John. 1941. Selective Poisoning of Fish. Trans. Am. Fish. Soc., 70 (1940): 80–86.

Schnabel, Zoe Emily. 1938. Estimation of Total Fish Population of a Lake. Am. Math. Journ., 45 (1938): 348–352.

Underhill, A. Heaton. 1941. Estimation of a Breeding Population of Chub Suckers. Trans. Fifth North Am. Wildlife Conf. (1940): 251–256.

Van Oosten, John. 1935. Logically Justified Deductions concerning the Great Lakes Fisheries Exploded by Scientific Research. Trans. Am. Fish. Soc., 65 (1935): 71–75.

Walton, Izaak. 1653. The Compleat Angler or the Contemplative Man's Recreation. London.

NATURAL ENEMIES OF CRAYFISHES IN MICHIGAN *

KARL F. LAGLER AND MARY JANE LAGLER

IT IS the purpose of this paper to bring together records of specific predator and prey relationships for the crayfishes of Michigan and to point out, so far as possible, the interactions of vulnerability and distribution, habitat, and habits. It is generally known that crayfish are of considerable importance in the food chains of a large number of animals representing the principal vertebrate classes. Moreover, there are noteworthy differences in the degree of this importance among the various predators and in life-history stages of certain of them. Some of the natural enemies of crayfish are economically significant; others have little or no such moment. It is clearly demonstrable, however, that crayfish not only contribute directly to our resources but by their incidence in the food of fish-eating vertebrates also buffer valuable food and game fishes that might otherwise be "destroyed."

The taxonomy of crayfish and notes on their distribution and ecology in the present paper are from Creaser (1931). Many of the identifications were made by Dr. Karl E. Goellner, who also kindly read the manuscript and made suggestions for its improvement. Names of fish are from Hubbs and Lagler (1941); of amphibians and reptiles, from Stejneger and Barbour (1936); of birds, from Van Tyne (1938); and of mammals, from Burt (MS). In accordance with ornithological practice the vernacular names of birds are regarded as equivalent to technical names when begun with capital letters.

Information given here on specific interrelations between vertebrates and crayfishes is largely from food studies based on analyses of the contents of digestive organs or of droppings and is mostly original. Data for the snakes and the fish-eating birds are from the publications or manuscript reports on fish predation by the senior author, in collaboration with J. Clark Salyer, II.

* Contribution from the Department of Zoölogy, University of Michigan.

SPECIFIC RELATIONSHIPS

Of special interest are the exact relationships of predators to particular prey species. Parasitologists and students of the natural history of crayfishes will find here answers to many questions and clues to others that they have directed to us. Workers in fish management may obtain indications helpful in choosing crayfishes best suited for introduction into waters where they are lacking. The choice will obviously be those species that are most abundant, exhibit a wide habitat range, and show desirability as fish food and ability to protect preferred fish species from common enemies by buffering. The decapod species will be taken up individually and relationships based on actual records disclosed.

Of the eight species of crayfish that are recorded for Michigan only five figure in the food of vertebrates. The other three species (*Cambarus rusticus*, *C. fodiens*, and *C. blandingii acutus*) are uncommon in those habitats from which our predator material has been obtained and, so far as we know, are not eaten by any higher animal in the area. One of these, *C. fodiens*, is a burrowing species, and its vulnerability to vertebrate predation is reduced by its fossorial habit.

Cambarus virilis Hagen

Cambarus virilis has a wide distribution in Michigan, being known from all the major drainages. Although occurring abundantly in both lentic and lotic habitats, it is reported to have a preference for streams in which there are stones that provide hiding places. It inhabits both trout and nontrout waters. In all its places of abode it is subject to predation by many animals; those for which we have records are listed and discussed below.

Fishes. — *C. virilis* has been found to compose only a small amount of the food of 11 spotted gars (*Lepisosteus productus*) from lakes in the southern part of the state (Lagler, Obrecht, and Harry, 1943). It is perhaps of even less importance in the food of longnose gars (*Lepisosteus osseus oxyurus*), occurring only once in 136 specimens from southern Michigan (Lagler and Hubbs, 1940). The lower incidence in the latter gar species is probably due to the fact that the longnose gar is less well adapted to the bottom-feeding habit requisite for capturing crayfish than is the spotted gar, which has a shorter, broader snout; neither of these gars is by any means a regular bottom feeder. By contrast, *C. virilis* is very prominent in the food of the bowfin (*Amia calva*) in the same region. A total of 67 was present in 204 bowfins (Lagler and Hubbs, 1940; Lagler and Applegate, 1942); they indicate a particular vulnerability of crayfish to a bottom-feeding predacious fish.

Although we have records of crayfishes in the food of three species of trout, none are for *C. virilis*. It may be assumed, however, that because of its general availability it is eaten by the common brook trout (*Salvelinus f. fontinalis*), the brown trout (*Salmo trutta fario*), and the coast rainbow trout (*Salmo gairdnerii irideus*).

C. virilis is one of the crayfishes often eaten by basses; both large-mouth bass (*Huro salmoides*) and smallmouth bass (*Micropterus d. dolomieu*) prey on it.

Amphibians. — No specimens of *C. virilis* were found in a large series of mudpuppies (*Necturus maculosus*) from a lake in southern Michigan (Lagler and Goellner, 1941). It seems probable, however, that bottom-feeding *Necturus* are predators of this crustacean, particularly in such habitats as the Huron River near Dexter, Michigan, where both occur.

Reptiles. — Examination of a large series of water snakes (*Natrix s. sipedon*) from all over the Lower Peninsula has failed to show this crayfish in the food. Numerous garter snakes (*Thamnophis s. sirtalis*) collected about lakes, ponds, and streams also contained none, although they very often held fish.

Larger aquatic turtles as a group are feeders upon crayfishes (Lagler, 1943). Two specimens of *C. virilis* were identified among the remains of several crayfishes eaten by 113 musk turtles, mostly from lakes in southern Michigan. The snapping turtle (*Chelydra serpentina*) contained 127 crayfish of this species in 323 specimens from over the entire state, and 66 Blanding's turtles (*Emys blandingii*) from the Lower Peninsula held 36. It is attacked sparingly by the Midwestern painted turtle (*Chrysemys picta marginata*); only two were found in 413 individuals. Perhaps a more common predator is the spiny soft-shelled turtle (*Amyda s. spinifera*), since two specimens of *C. virilis* were present in only 15 of them. The species was not identified among the crayfishes eaten by the geographic turtle (*Graptemys geographica*), and it is not known that they are preyed upon by the spotted turtle (*Clemmys guttata*) or the wood turtle (*Clemmys insculpta*).

Birds. — Several birds eat crayfish extensively. These crustaceans are of very great import to mergansers (Salyer and Lagler, 1940). Both the American Merganser (*Mergus merganser americanus*) and the Hooded Merganser (*Lophodytes cucullatus*) feed upon *C. virilis*, particularly in winter. It was especially numerous in the food of American Mergansers from the shallow waters of Saginaw Bay, where it abounds. Other records that we have are for the American Bittern (*Botaurus l. lentiginosus*), Eastern Green Heron (*Butorides v. virescens*), Great Blue Heron (*Ardea h. herodias*), American Golden-eye (*Bucephala clangula americana*), and the Eastern Belted Kingfisher (*Megaceryle a. alcyon*).

Mammals. — Although several fur bearers in the state are natural enemies of *C. virilis*, there are actual predation records for only one, the otter (*Lutra c. canadensis*), reported by Lagler and Ostenson (1942). Indications are that this crayfish is readily available to the otter in trout-stream habitats, even in winter. These streams ordinarily do not freeze over in winter, and otter easily obtain crayfish from their hiding places under stones. Other probable mammalian predators of *virilis* are the opossum (*Didelphis v. virginiana*), raccoon (*Procyon l. lotor*), and mink (*Mustela vison*).

Cambarus immunis Hagen

Cambarus immunis is known from scattered locations in the Lower Peninsula. It is most abundant in stagnant shallow waters, but also occurs in lakes and slow-moving streams. Habitat preference is a water area with a soft bottom in which it may partially bury itself. This choice of environment strikingly reduces incidence as prey of several animals which eat crayfish from other habitats, but which do not frequent small ponds of this region. Examples are gars (*Lepisosteus spp.*), American Merganser (*Mergus merganser americanus*), and Hooded Merganser (*Lophodytes cucullatus*). The habitat choice serves, however, to increase vulnerability to some turtles which are common associates of *C. immunis* in these surroundings. General scarcity in situations other than that most preferred is probably responsible for the small number of individuals occurring in the food of vertebrates as well as for the fewness of vertebrates which are known to be their natural enemies.

Fishes. — In the food of the bowfin (*Amia calva*), *C. immunis* ranks third after *C. virilis* and *C. propinquus* among the crayfish species preyed upon in southern Michigan (Lagler and Hubbs, 1940; Lagler and Applegate, 1942). The only other fish recorded as a predator of *C. immunis* is the smallmouth bass (*Micropterus d. dolomieu*), although it may safely be assumed that the largemouth bass (*Huro salmoides*) and other predacious fish species are numbered among its natural enemies.

Amphibians. — We have no records of predation on *C. immunis* by an amphibian, although frogs, toads, and mudpuppies are known to eat crayfish (Tack, 1941; Lagler and Goellner, 1941).

Reptiles. — Examinations of an extensive series of water snakes (*Natrix s. sipedon*) and garter snakes (*Thamnophis s. sirtalis*) from aquatic habitats have failed to disclose a relationship between ophidians and this species.

Owing to the somewhat less general abundance of *C. immunis* and to its more restricted habitat preference, it has fewer turtles among its natural enemies than other crayfishes have (Lagler, 1943). Only 21 specimens of *C. immunis* were found in 323 snappers (*Chelydra serpentina*), but not all of these turtles were from habitats frequented by this crustacean. It was the only one identified in the food of 27 geographic turtles (*Graptemys geographica*), mostly from the slow-moving portions of larger rivers in the southern part of the Lower Peninsula; seven were eaten. Nine of 22 crayfish in 15 spiny soft-shelled turtles (*Amyda s. spinifera*) were *C. immunis.*

Birds. — Marsh and wading birds, in addition to the Eastern Belted Kingfisher (*Megaceryle a. alcyon*), are this crayfish's principal avian enemies. Among the birds with definite records are the Great Blue Heron (*Ardea h. herodias*), Eastern Green Heron (*Butorides v. virescens*), and the American Bittern (*Botaurus l. lentiginosus*).

Mammals. — The otter (*Lutra c. canadensis*) feeds sparingly on *C. immunis* (Lagler and Ostenson, 1942). That the distribution of these two animals overlaps so slightly in the state and that the habitats of their principal choice differ so markedly probably accounts for the low incidence in the stomachs of otter.

Cambarus propinquus Girard

The most common crayfish in Michigan, found practically throughout the state, is *Cambarus propinquus.* It lives in a wide variety of lake, pond, and stream habitats, often in association with

C. virilis, and it is an outstanding component of the food of most animals that eat crayfish.

Fishes. — The longnose gar (*Lepisosteus osseus oxyurus*) is an unimportant predator of *C. propinquus* even as it was of *C. virilis*. Only one of each of these crayfish species was in the food of 136 of such gars from southern Michigan. Two hundred and four bowfins, however, had eaten a few more individuals of *C. propinquus* (a total of 75) than of *C. virilis* (a total of 67) (Lagler and Hubbs, 1940; Lagler and Applegate, 1942).

Rainbow trout (*Salmo gairdnerii irideus*) are prominent among the predators of *C. propinquus* in stream habitats. Creaser (1931) reports finding many specimens during stomach examinations of this trout from various localities in Michigan. We have additional records for 10- to 11-inch brook trout (*Salvelinus f. fontinalis*).

Amphibians. — The mudpuppy (*Necturus maculosus*) is the only amphibian predator of *C. propinquus* that we know (Lagler and Goellner, 1941). This crayfish occurred in 17 of 24 lake-dwelling mudpuppies that contained crayfish, and is probably also taken in larger streams that it shares with juvenile and adult mudpuppies.

Reptiles. — Crayfish-eating chelonians are among the most significant predators of this species (Lagler, 1943). Several members of *C. propinquus* were present in snapping turtles (*Chelydra serpentina*) from the Lower Peninsula, and 67 were found in 66 Blanding's turtles (*Emys blandingii*) from the same region. For Blanding's turtle there are almost twice as many records for *propinquus* as for *virilis*, which numbered 36. An intergrade between Bell's turtle and the Midwestern painted turtle (*Chrysemys picta: bellii* × *marginata*) from the Upper Peninsula was found to have eaten crayfish of the species *C. propinquus*, and one of 413 specimens of Midwestern painted turtles (*Chrysemys picta marginata*) from the Lower Peninsula had done likewise. Nine were in the food of 15 spiny soft-shelled turtles (*Amyda s. spinifera*) from southern Michigan. The species may also have been represented among the unidentifiable remains of many crayfish in 113 musk turtles (*Sternotherus odoratus*) and 27 geographic turtles (*Graptemys geographica*).

Birds. — American Mergansers (*Mergus merganser americanus*) eat crayfish of this species in both trout and nontrout waters, but fewer members of *C. propinquus* than of *C. virilis* are eaten by the Hooded Merganser (*Lophodytes cucullatus*). *C. propinquus* is considerably more important, however, than *C. virilis* in the food of the American Golden-eye (*Bucephala clangula americana*). *C. propinquus* is second only to *C. immunis* in the food of the marsh-inhabiting American Bittern (*Botaurus l. lentiginosus*). Numerically it is approximately equal with *C. virilis* as a constituent in the diet of the Eastern Green Heron (*Butorides v. virescens*) and the Great Blue Heron (*Ardea h. herodias*). Many are also eaten by the Eastern Belted Kingfisher (*Megaceryle a. alcyon*).

Mammals. — Although probably of basic importance in the food of several mammals of this region, this crayfish is definitely known to be utilized only by the otter (*Lutra c. canadensis*) (Lagler and Ostenson, 1942).

Cambarus robustus Girard

Cambarus robustus occurs in scattered regions in the principal river systems of Michigan that drain into Lake Huron and Lake Erie and in the Grand River drainage. It characterizes the fauna of the swifter and shallower parts of watercourses and, like *C. virilis*, it seems to prefer streams with stones under which it may hide. The species is not very numerous, and hence its natural enemies are poorly known. There are no records among the Amphibia, Aves, or Mammalia.

Fishes. — One member of this species was found in a bowfin (*Amia calva*) of a sizable series studied by Lagler and Hubbs (1940), but none was present in specimens examined by Lagler and Applegate (1942).

Reptiles. — Among many crayfish eaten by snapping turtles (*Chelydra serpentina*) there was only one specimen of *C. robustus* (Lagler, 1943), but five of 22 crayfish in the food of 15 soft-shelled turtles (*Amyda s. spinifera*) were of this species.

Cambarus diogenes Girard

A preëminently burrowing species, *Cambarus diogenes* is known from widely separated areas throughout the state. During certain seasons it leaves its burrows and lives in lakes, ponds, and streams. Because of its secretive habits it does not appear to be a staple item in vertebrate fare, although it is eaten by several animals.

Fishes. — To date the northern pike (*Esox lucius*) is the only fish that has been shown to prey extensively (though locally) on *C. dio-*

genes. We have studied a large series of pike of all sizes from newly impounded waters over the old Seney marshes of the Upper Peninsula and have repeatedly found this crayfish in stomachs. It is eaten only sparingly by the bowfin (*Amia calva*) (Lagler and Applegate, 1942).

Reptiles. — Reptilian predators are confined to the turtles; we find the common snapping turtle (*Chelydra serpentina*) and intergrades between Bell's turtle and the Midwestern painted turtle (*Chrysemys picta: bellii* × *marginata*) preying to a small extent on this species (Lagler, 1943).

Birds. — *C. diogenes* was represented only once in several American Bittern (*Botaurus l. lentiginosus*), Eastern Green Heron (*Butorides v. virescens*), and Eastern Belted Kingfisher (*Megaceryle a. alcyon*) that have been found to prey on crayfish.

Mammals. — Again the only mammal for which we have a definite record of having eaten crayfish is the otter (*Lutra c. canadensis*), and in the series studied (Lagler and Ostenson, 1942) *C. diogenes* was identified only once.

GENERAL RELATIONSHIPS

Besides the specific records of predator–prey relationships of crayfishes that have just been given, there are many others that are at present grouped as general relationships. In part these are based on literature reports of "crayfishes" in the food of Michigan vertebrates, either in Michigan or elsewhere in their range. The rest are from analyses of stomach contents in which were found fragments of crayfish in such condition that identification of species was impossible. These general relations are yet to be explored for their species components in Michigan.

Fishes. — Among the fishes, most predacious forms feed to some extent on crayfish. The kinds included, besides those already mentioned in the species accounts, are: bullheads (*Ameiurus spp.*), mud pickerel (*Esox vermiculatus*), white bass (*Lepibema chrysops*), yellow perch (*Perca flavescens*), pikeperch and saugers (*Stizostedion spp.*), rainbow darter (*Poecilichthys c. caeruleus*), warmouth bass (*Chaenobryttus gulosus*), larger sunfishes (*Lepomis spp.*), crappies (*Pomoxis spp.*), northern rock bass (*Ambloplites r. rupestris*), and burbot (*Lota lota maculosa*).

Amphibians. — Crayfish in Michigan may be preyed upon by

amphibians other than the mudpuppy (*Necturus maculosus*), for which we have already given records. Frogs (*Rana catesbeiana*, *R. clamitans*, *R. pipiens*, and *R. palustris*) and the American toad (*Bufo americanus*) are known to eat crayfish elsewhere than in this state (Tack, 1941).

Reptiles. — In addition to the several species of turtles that are predators of these crustaceans we have one Michigan record each for the common water snake (*Natrix s. sipedon*) and for the queen snake (*Natrix septemvittata*).

Birds. — Many birds other than those already listed prey on crayfish. Prominent among these by their occurrence in Michigan are: Pied-billed Grebe (*Podilymbus p. podiceps*), American Egret (*Casmerodius albus egretta*), Snowy Egret (*Leucophoyx t. thula*), Little Blue Heron (*Florida c. caerulea*), Black-crowned Night Heron (*Nycticorax nycticorax hoactli*), Black Duck (*Anas rubripes*), Pintail (*Anas acuta tzitzihoa*), Blue-winged Teal (*Anas discors*), Shoveller (*Spatula clypeata*), Lesser Scaup Duck (*Nyroca affinis*), Buffle-head (*Bucephala albeola*), Old Squaw (*Clangula hyemalis*), White-winged Scoter (*Melanitta fusca deglandi*), Ruddy Duck (*Oxyura jamaicensis rubida*), Red-breasted Merganser (*Mergus serrator*), Eastern Red-tailed Hawk (*Buteo jamaicensis borealis*), Northern Red-shouldered Hawk (*Buteo l. lineatus*), American Rough-legged Hawk (*Buteo lagopus s.-johannis*), Broad-winged Hawk (*Buteo p. platypterus*), Marsh Hawk (*Circus cyaneus hudsonius*), Eastern Pigeon Hawk (*Falco c. columbarius*), Sandhill Crane (*Grus canadensis tabida*), King Rail (*Rallus e. elegans*), Virginia Rail (*Rallus l. limicola*), Killdeer (*Charadius v. vociferus*), Ring-billed Gull (*Larus delawarensis*), Barn Owl (*Tyto alba pratincola*), Eastern Screech Owl (*Otus asio naevius*), Great Horned Owl (*Bubo v. virginianus*), and Northern Barred Owl (*Strix v. varia*).

Mammals. — Besides the otter (*Lutra c. canadensis*), already mentioned, other valuable Michigan fur bearers are natural enemies of crayfish. Dearborn (1932) has shown that these crustaceans compose a substantial amount of the food of the opossum (*Didelphis v. virginiana*), the raccoon (*Procyon l. lotor*), and the mink (*Mustela vison*). An interesting seasonal variation was noted for the mink; crayfish comprised 68 per cent of the bulk of its food in the summer and only 8 per cent in the winter. Errington (1941) has found that the muskrat (*Ondatra zibethica*) eats crayfish.

Although crayfish are eaten by man in some parts of the country, the practice is not common. Some crayfish are used as bait, however, by fishermen and by others as food for fishes being artificially propagated.

University of Michigan

REFERENCES

Barrows, Walter Bradford. 1912. Michigan Bird Life, xiv + 822. Lansing: Michigan Agricultural College.

Creaser, Edwin P. 1931. The Michigan Decapod Crustaceans. Pap. Mich. Acad. Sci., Arts, and Letters, 13 (1930): 257–276.

Dearborn, Ned. 1932. Food of Some Predatory Fur-Bearing Animals in Michigan. Univ. Mich. School of Forestry and Cons., Bull. No. 1: 1–52.

Errington, Paul L. 1941. Versatility in Feeding and Population Maintenance of the Muskrat. Journ. Wildlife Management, 5 (1): 68–89.

Evermann, Barton Warren, and Clark, Howard Walton. 1920. Lake Maxinkuckee: A Physical and Biological Survey, 1: 1–660. Indianapolis: The Department of Conservation, State of Indiana.

Fisher, A. K. 1893. The Hawks and Owls of the United States in Their Relation to Agriculture. U. S. Dept. Agric., Div. Ornithol. and Mammal., Bull. No. 3: 1–210.

Forbes, Stephen Alfred, and Richardson, Robert Earl. 1920. The Fishes of Illinois. Second Ed., cxxxvi + 357. Springfield: State of Illinois Department of Registration and Education.

Howell, Arthur H. 1932. Florida Bird Life, xxiv + 579. Florida Department of Game and Fresh Water Fish, in Cooperation with U. S. Bur. Biol. Surv.

Hubbs, Carl L., and Lagler, Karl F. 1941. Guide to the Fishes of the Great Lakes and Tributary Waters. Bull. Cranbrook Inst. Sci., 18: 1–100, figs. 1–118.

Lagler, Karl F. 1943. Food Habits and Economic Relations of the Turtles of Michigan with Special Reference to Fish Management. Am. Midland Nat., 29 (2): 257–312.

—— and Applegate, Vernon C. 1942. Further Studies of the Food of the Bowfin (*Amia calva*) in Southern Michigan, with Notes on the Inadvisability of Using Trapped Fish in Food Analyses. Copeia, 1942 (3): 190–191.

—— and Goellner, Karl E. 1941. Notes on *Necturus maculosus* (Rafinesque), from Evans Lake, Michigan. Copeia, 1941 (2): 96–98.

—— and Hubbs, Frances V. 1940. Food of the Long-nosed Gar (*Lepisosteus osseus oxyurus*) and the Bowfin (*Amia calva*) in Southern Michigan. Copeia, 1940 (4): 239–241.

Lagler, Karl F., and Ostenson, Burton T. 1942. Early Spring Food of the Otter in Michigan. Journ. Wildlife Management, 6 (3) : 244–254.

—— Obrecht, Carl B., and Harry, George V. 1943. The Food and Habits of Gars (*Lepisosteus spp.*) Considered in Relation to Fish Management. Investigations of Indiana Lakes and Streams, 2 (1942) : 117–135.

Ortmann, A. E. 1906. The Crawfishes of the State of Pennsylvania. Mem. Carnegie Mus., 2 (10) : 343–523.

Pirnie, Miles David. 1935. Michigan Waterfowl Management, xxi + 328. Lansing: Department of Conservation.

Roberts, Thomas S. 1932. The Birds of Minnesota, I: xxii + 691. Minneapolis: The University of Minnesota Press.

Salyer, J. Clark, II, and Lagler, Karl F. 1940. The Food and Habits of the American Merganser during Winter in Michigan, Considered in Relation to Fish Management. Journ. Wildlife Management, 4 (2): 186–219.

Stejneger, Leonhard, and Barbour, Thomas. 1936. A Check List of North American Amphibians and Reptiles, xvi + 207. Cambridge: Harvard University Press.

Tack, Peter Isaac. 1941. The Life History and Ecology of the Crayfish *Cambarus immunis* Hagen. Am. Midland Nat., 25 (2) : 420–446.

Trautman, Milton B. 1940. The Birds of Buckeye Lake, Ohio. Misc. Publ. Mus. Zool., Univ. Mich., 44: 1–466.

Turner, Clarence L. 1926. The Crayfishes of Ohio. Ohio State Univ. Bull., 30 (11) : 145–195.

Van Tyne, Josselyn. 1938. Check List of the Birds of Michigan. Occ. Pap. Mus. Zool., Univ. Mich., 379: 1–44, 1 map.

Weed, Clarence M., and Dearborn, Ned. 1903. Birds in Their Relations to Man, viii + 380. Philadelphia and London: J. B. Lippincott Company, 1903.

ANGLERS' CATCHES FROM PORTIONS OF CERTAIN MICHIGAN TROUT STREAMS IN 1939 AND 1940, WITH A DISCUSSION OF INDICES TO ANGLING QUALITY *

DAVID S. SHETTER

IN MOST fish-yield studies so far reported it has been impossible to calculate the catch in pounds per unit of effort or per unit of water area because the fish removed by angling were not weighed. There were two good reasons for omitting the weighing: (1) Many of the early creel censuses were made by relief labor, which could not always be depended upon to take accurate measurements; hence the simpler the data requested the more reliable would be the information obtained; (2) At lakes particularly a large number of anglers leave the water at noon and at dusk. Creel-census clerks were required to record only the number of fish taken, the species, and an estimate of the average size of the catch. Unless the data to be written down were kept at a minimum, the census taker could not interview all the fishermen.

In 1939 and 1940 crews of the Civilian Conservation Corps were provided by state or United States Forest Service camps. These men were carefully selected and trained in their duties in advance of the fishing seasons. Where this help was not available, creel-census clerks were hired by the Conservation Department. All the work was carefully supervised. Intensive censuses of this type were limited to portions of seven trout streams in 1939 and to five in 1940, from one to twelve miles in length. Enough men were used to enable them to obtain all the information required without inconveniencing the anglers.

Delay in arrival of balances prevented weighing the catches until June 15, 1939, after which weights were secured with few exceptions.

* Contribution from the Institute for Fisheries Research of the Michigan Department of Conservation.

Prior to that date nearly all the fish were measured individually. On all streams there were times when weights could not be recorded because fishermen disregarded the signs asking them not to clean their fish before they had been weighed. Rarely were anglers seen without being interviewed by the clerks. The probable catch of these anglers was assumed to be average for the two-week period.

In order to make the final figures on yield as complete and as accurate as possible the following procedure was employed. Weight–length curves for each species and for each stream were prepared from all fish that were weighed and measured. From these curves the weights of fish of known lengths could be closely approximated. If only the number of fish caught by an angler was known, the average length for each species for the period was used, and the corresponding weights on the curves were assigned to this catch. The catch of anglers seen but not questioned was considered average in all respects for the period and was also added to the known total.

Errors in the calculated yields may be present because of inaccurate measurements of the fish, variation in the weight–length relationship of trout during the fishing season, and dependence on the catch of anglers not interviewed. However, they are to some extent compensating and were present on each stream, so that results should be reasonably accurate and comparable.

Areal measurements of the portions of streams under census were determined either from plane-table maps or by chaining the length and width.

YIELD TO THE ANGLERS

Tarzwell (1938) reported the yield of rainbow trout in Tonto and Horton creeks, Arizona, as from 29.5 to 54.2 pounds per acre in 1936 and 1937, with an angling pressure ranging from 125 to 241 man-hours per acre per season. What part of the catch consisted of hatchery trout is not stated, but some planting was done in both streams, possibly with legal-sized fish. In determining the natural production of a stream the weight of planted hatchery trout caught should be deducted from the total poundage recorded. Tarzwell's figures may therefore be too high.

From data published by Surber (1940) it may be calculated that the yields of brook and rainbow trout per acre in 1938 and 1939 in the St. Mary River, Virginia, were 7.55 and 9.64 pounds, respectively.

TABLE I

DATA FOR THE 1939 AND 1940 SEASONS ON PORTIONS OF MICHIGAN TROUT STREAMS COVERED BY INTENSIVE CREEL CENSUSES

Stream and county	Water area in acres	Total hours of angling	Angling hours per acre per season	Number of legal trout caught	Total pounds of legal trout removed	Number of legal trout caught per hour	Pounds of legal trout caught per hour	Pounds of legal trout caught per acre	Net yield * per acre	Percentage of unsuccessful anglers
1939										
Pine River (Lake)	57.5	15,226.50	265	7,459	1,517.71	0.49	0.100	26.4	23.1 †	48
Pigeon River (Otsego)	65.4	6,754.75	103	3,213	725.38	0.48	0.107	11.1	? ‡	56
North Branch of Au Sable River (Crawford)	54.5	7,050.00	129	2,914	673.06	0.41	0.095	12.3	11.6 †	63
Little Manistee River (Lake)	17.7	2,333.25	132	862	249.51	0.37	0.107	14.1	11.4 †	59
Canada Creek (Presque Isle)	21.2	873.00	41	278	67.64	0.32	0.078	3.2	?	57
White River (Newaygo)	10.4	3,080.00	296	1,853	318.11	0.60	0.103	30.6	?	41
Hunt Creek (Montmorency)	4.3	780.50	175	492	66.98	0.63	0.086	15.6	?	55
1940										
North Branch of Au Sable River (Crawford)	54.5	6,675.50	123	2,128	574.30	0.33	0.086	10.6	9.6 §	63
White River (Newaygo)	17.5	2,131.75	122	448	148.50	0.22	0.070	8.5	7.5 §	79
East Branch of Tahquamenon River (Chippewa)	6.3	725.00	115	689	172.70	1.02	0.238	27.4	22.7 †	41
Fishdam River (Delta)	3.0	589.50	197	471	204.90	0.80	0.348	68.3	64.7 §	42
Hunt Creek (Montmorency)	4.6	901.25	189	406	60.35	0.45	0.067	12.8	12.8 †	66

* Total poundage minus weight of catch of legal-sized hatchery plantings.
† All legal-sized hatchery plantings were marked.
‡ Estimates impossible because not all legal-sized hatchery plantings were marked.
§ Based on percentage of legal-sized hatchery plantings taken in preceding years.

Angling pressures ranged from 103 to 130 man-hours per acre per season. Hatchery fish were planted as advanced fingerlings from 3.75 to 6.00 inches in length during late summer and fall in the years 1935, 1936, and 1938. Since they were in the stream at least one year before entering the catch their contribution to the poundage produced may be regarded as production by the stream.

In 1939 the total yield from sections of seven Michigan trout streams was 15.7 pounds per acre (Table I). Considerable variation is evident. The highest yield in pounds per acre of trout was from the White River (30.6), followed in order by the Pine River (26.4), Hunt Creek (15.6), the Little Manistee River (14.1), the North Branch of the Au Sable River (12.3), the Pigeon River (11.1), and Canada Creek (3.2).

In 1940 sections of only five streams were covered by intensive creel censuses because of a reduction of Civilian Conservation Corps manpower. The total yield from them was 15.3 pounds per acre, but yields of legal trout per acre varied greatly in different waters. The highest yield was recorded for the Fishdam River (68.3), followed by the East Branch of the Tahquamenon River (27.4), Hunt Creek (12.8), the North Branch of the Au Sable (10.6), and the White River (8.5). The marked difference in the yield recorded for the White River in 1939 and 1940 demonstrates the great variation between a stream area that is a good trout habitat during the entire year and one which is capable of supporting trout only during spring and early summer.

Calculations of the yield of "wild" trout (net yield) were possible because all hatchery-reared trout present in the areas under census were either jaw-tagged or fin-clipped or the percentage of hatchery fish in the catch could be estimated on the basis of returns from previous plantings of marked trout on many of these streams (Shetter and Hazzard, 1942). From the available data concerning the pounds per acre of wild trout removed by angling it may be computed that from 80.5 to 94.8 per cent of the total poundage so removed was composed of trout of natural origin, even though generous plantings of legal-sized fish were made.

INDICES TO ANGLING QUALITY

Two indices for comparing angling quality are presented below and in Table I. One index, which has been widely used in the past, is the number of legal trout caught per hour; the other is the number of pounds of legal trout caught per hour. After trials in computing several other indices to angling quality, such as the percentage of successful anglers, size of trout, and so on, the simplest and most accurate measure appears to be the number of pounds of legal trout caught per hour of angling. This figure includes the factors of number and size of fish, which are generally considered equally important in determining the quality of angling.

Depending on which criterion is used to judge the creel census data, the streams will change their position in the listings of angling quality. This is illustrated by Table II.

TABLE II

RATINGS OF STREAMS BASED ON THE NUMBER AND ON THE POUNDS OF TROUT CAUGHT PER HOUR IN 1939 AND 1940

	Judged on the basis of number per hour		*Judged on the basis of pounds per hour*	
1939	Hunt Creek	(0.63)	Little Manistee River	(0.107)
	White River	(0.60)	Pigeon River	(0.107)
	Pine River	(0.49)	White River	(0.103)
	Pigeon River	(0.48)	Pine River	(0.100)
	North Branch of Au Sable River	(0.41)	North Branch of Au Sable River	(0.095)
	Little Manistee River	(0.37)	Hunt Creek	(0.086)
	Canada Creek	(0.32)	Canada Creek	(0.078)
1940	East Branch of Tahquamenon River	(1.02)	Fishdam River	(0.348)
	Fishdam River	(0.80)	East Branch of Tahquamenon River	(0.238)
	Hunt Creek	(0.45)	North Branch of Au Sable River	(0.086)
	North Branch of Au Sable River	(0.33)	White River	(0.070)
	White River	(0.22)	Hunt Creek	(0.067)

It should be pointed out that the average sizes of the trout caught in the various streams contributed noticeably to their order in the lists. In Table III will be found the average lengths and the average weights of the various species of trout captured in the creel-census areas in 1939 and 1940. Where there are no entries in the table, the species were not present in the streams specified or were not captured by anglers. If the ratings of the streams on the basis of pounds of trout captured per hour of angling are compared with the other available creel-census data and the average sizes given in the table,

TABLE III

Average Lengths and Average Weights of Trout Taken by Angling from Portions of Michigan Trout Streams Covered by Intensive Creel Censuses during the 1939 and 1940 Trout Seasons (Lengths in Inches, Weights in Ounces)

Stream and year	Brook trout			Brown trout			Rainbow trout			All trout	
	Number	Average length	Average weight	Number	Average length	Average weight	Number	Average length	Average weight	Number	Average weight
1939											
Pine River	1,446	7.9	2.97	12	8.9	5.33	6,001	8.3	3.32	7,459	3.26
Pigeon River	2,337	8.1	3.29	228	9.1	5.13	648	9.0	4.24	3,213	3.61
North Branch of Au Sable River	2,430	7.8	2.83	464	10.5	8.04	20	9.5	7.80	2,914	3.69
Little Manistee River	273	8.7	4.66	254	9.5	5.02	335	8.2	4.32	862	4.63
Canada Creek	265	8.4	3.86	2	7.2	2.15	11	7.3	5.02	278	3.89
White River	1,853	7.8	2.75	...	...	...	...	...	...	1,853	2.75
Hunt Creek	492	7.5	2.18	...	...	...	...	...	...	492	2.18
1940											
North Branch of Au Sable River	1,759	8.9	3.16	355	11.1	10.00	14	9.1	6.17	2,128	4.32
White River	143	8.3	4.21	141	9.5	6.01	164	9.5	5.65	448	5.30
Fishdam River	446	9.1	6.76	25	11.0	10.56	...	...	...	471	6.96
East Branch of Tahquamenon River	689	8.5	4.01	...	...	...	...	...	...	689	4.01
Hunt Creek	406	7.6	2.38	...	...	...	...	...	...	406	2.38

it will be seen that, in order to arrive at a high pounds-per-hour index, either a large number of trout must be taken per unit of effort or the size of the trout caught must be above average. Examination of the tabular material reveals a marked variation between Hunt Creek and the Little Manistee River in 1939. On Hunt Creek the highest catch per hour (0.63 fish) was recorded, yet the fish were of the least average weight (2.18 ounces), and the calculation of the pounds-per-hour index placed it second from the bottom (0.086) when quality of angling was judged on the latter basis. On the Little Manistee River the catch per hour was only 0.37 fish, yet the average size of all trout caught was over twice that of fish caught in Hunt Creek, and the calculation of the pounds-per-hour index was increased accordingly (0.107) and more nearly to the true position as regards general angling quality.

It is the author's belief that the pounds of fish caught per unit of angling effort is the more valid index in comparing the quality of the angling of different bodies of water or in comparing the results of the angling in different seasons or parts of the season on the same body of water, and this index will be employed in the future whenever it is possible to obtain the weight of anglers' catches.

ACKNOWLEDGMENTS

The author wishes to acknowledge the coöperation of the Civilian Conservation Corps in taking creel-census records of most of the streams. The United States Forest Service provided enrollees for censuses on the following rivers: White, Little Manistee, the East Branch of the Tahquamenon, and Fishdam. State-operated camps supplied census clerks for work on the Pigeon River, Canada Creek, and the North Branch of the Au Sable River. Dr. Albert S. Hazzard assisted in planning and directing the study and in the preparation of the report.

Institute for Fisheries Research
Ann Arbor, Michigan

LITERATURE CITED

SHETTER, DAVID S., AND HAZZARD, ALBERT S. 1942. When Shall We Plant "Keeper" Trout? Michigan Conservation, 11 (4) : 3–5. Published by Michigan Department of Conservation, Lansing, Michigan.

SURBER, EUGENE W. 1940. Lost: 10, 839 Fingerling Trout! An Appraisal of Planting Fingerling Trout in St. Mary River, Virginia. Progressive Fish-Culturist, U. S. Bur. Fish. Memo., I–131, No. 49 (March–April, 1940) : 1–13.

TARZWELL, CLARENCE M. 1938. Factors Influencing Fish Food and Fish Production in Southwestern Streams. Trans. Am. Fish. Soc., 67 : 246–255. 1937.

GEOGRAPHY

SOME APPLICATIONS OF AERIAL PHOTOGRAPHS TO GEOGRAPHIC INVENTORY

JOSEPH A. RUSSELL *
Syracuse University

FRED W. FOSTER †
Michigan State College

KENNETH C. McMURRY
University of Michigan

AN AERIAL photograph is a reasonably permanent record of a part of the earth's surface at a certain time. Its use as a surveying method and as an aid to a more complete understanding of the nature of the land is justified by the detail which it provides, the accuracy of location it makes possible, the speed it permits in the accomplishment of various tasks, and its economy as compared with other techniques. Aerial photographs contain a vast amount of geographical, sociological, and geological data. This encyclopedia of information does not become available, however, unless methods are developed to translate the shadows, tones, geometric forms, and photographic textures into a vocabulary which speaks in terms of previous experience.

One such method is the establishment of regional keys or legends for the natural and cultural phenomena shown on the pictures. Common knowledge may make it easy to identify certain features, such as lakes and streams, forest as distinguished from cleared land, orchards, buildings, and communication networks. The more subtle differences in texture and tone need explanation. The development of such a system offers a new inventory method to the geographer.

* Mr. Russell's field and office work in connection with this paper was made possible by his appointment as Predoctoral Field Fellow of the Social Science Research Council, 1935–36.

† Part of the material for this paper was collected when Mr. Foster was an Earhart Fellow at the University of Michigan, 1935–36.

There are direct and associative keys. Direct keys are valuable for reading the nature of an object from the tone, texture, shape, or shadow of its image. Thus the billowy texture and dappled tone of northern Michigan upland hardwoods (maple, beech, birch, poplar, and so on) are easily distinguished from the smooth, uniformly dark appearance of swamp conifers (cedar, spruce, balsam, tamarack). Associative keys make possible the deduction of the nature of an object from an interpretation of its surroundings, and must be used when identification of an object alone is difficult.

Most of the phenomena in which the geographer is interested are best recognized by combining direct and associative ways. The forest type and the density of the trees, the drainage system, certain information about cultivated land, the presence and the nature of buildings, and the communication pattern are among the quantitative facts which can be read by applying direct keys, but the quality of most of them must be implied by associative facts. The location of surface features, the boundaries of broad soil types, and the functions of various structures can be deciphered in this manner. The procedure followed in determining the use of a structure may be shown by pointing out the photographic differences which appear between a rural school, a country church, and a farmhouse. Direct keys identify all three as buildings, but other than that there is little to differentiate them unless the church steeple shows prominently. Positive identification of each depends, then, upon associative keys, such as those presented on pages 333–334.

Both direct and associative keys must be firmly based on field experience and should be applied with an understanding of the land-use technique in the area concerned. The first step in setting up the keys is a careful analysis of several areas to determine the characteristic vegetation associations, the correlation between vegetation and site, crop combinations, house types, and other regional landscape phenomena. All these aspects should be studied on the photographs until identifiable differentiating tones from black to white, variations in texture, and geometric forms become apparent. Identifications should then be made on the basis of the photographs alone, and the results carefully checked in the field. The regionality of the keys will be recognized if attempts are made to utilize them in areas having different ecological conditions.

The field worker who is interested in interpretation will probably have neither the time nor the ability to use the methods employed by the photogrammetric engineer. Where the research is of such a nature as to demand exactness it will usually be found that the preliminary work of correcting for distortion and other discrepancies of a technical nature has already been accomplished. The establishment of ground control and the application of the radial-line or slotted-templet method of plotting are tasks that will normally be left to the specialist. With the photographs available through the many governmental agencies and private concerns numerous ways can be devised for mapping physical and cultural data of value to the geographer. The system will vary with the nature of the problem and the character of the terrain. The methods discussed below have all been used in the field with satisfactory results.

It is often desirable to map directly on the print. In rural areas with a great many roads or in places where varied cover is interspersed with open country, traverses can be made with little regard to pace or compass. Enough landmarks can usually be identified to enable one to maintain a constant check on position. Photographs are seldom available for use during the season in which they are taken. It is accordingly necessary to correct them for the numerous changes that have occurred since they were made, such as new field lines, road relocation, and recent timber operations. Ink or brightly colored pencils are best for making identification symbols.

As a variation of this method the photographs may be used on the plane table. With the table compass set to the north magnetic pole and the print oriented to true north, mapping proceeds in the usual manner. This system is of particular value in heavily wooded country or in regions which lack roads, railroads, or other cultural markings.

When it is undesirable to write on the prints transparent material is placed over the photograph. Sheets of cellulose acetate offer a fairly good surface for writing and drawing. Clear acetate is perfectly transparent, but does not take pencil well, and ink gives poor results unless the sheet is kept free of dirt and oily substances. The acetate with a matte surface is much better suited to marking, but lacks complete transparency even when in direct contact with the photograph. When the elements being mapped present a great deal of detail or when there is little contrast on the print it is often necessary to lift the sheet away from the photograph so that the objects beneath may be recognized. By fastening two pieces of acetate over one another special items can be mapped separately, as, for instance, soil and vegetation.

For one reason or another it may be impossible to take the photographs into the field. Under these circumstances the major pattern is easily traced on vellum. The resulting outline when properly oriented on the plane table offers a very satisfactory base for mapping.

When inventory is concerned with some selected item, such as vegetation boundaries in heavily wooded sections, it is often desirable to run an ordinary plane-table or compass traverse. Then photographs carried along provide a ready reference and aid greatly in plotting, especially if the distances are being paced.

Temporary or permanent mosaics are often available, but they are too cumbersome for direct mapping. Their chief value is in the composite picture they present. Work is more easily planned and the relationships, problems, and complexities of any region are better understood when they are employed. The following method has proved satisfactory: The pictures are assembled before the worker goes into the field; once the area is visualized as a whole, the photographs are separated and used for mapping; when the task is completed they are reassembled and employed as a basis for correlation.

The stereoscope is an invaluable aid to field and office interpretation. With a small pocket-type instrument the vertical aspects of a location are examined as it is mapped in the horizontal plane. Surface features and other areal variations along the traverse may be viewed under the stereoscope and their extensions plotted in detail. In this manner slope, landform boundaries, and different sorts of vegetation can be mapped with considerable accuracy. The lens type of stereoscope gives a great deal of vertical exaggeration and covers only a small part of the stereoscopic field at any one time. The instrument using mirrors has a much wider field, but relief is less pronounced and hence less decipherable.

The professional geographer is interested in four phases of the landscape: How much? What kind? Where? Why are conditions as they are? The first question is quantitative, the second qualitative, the third locational, and the fourth logical. Their solution involves: (1) determination and location of the boundaries of each type of phenomenon; (2) determination of its nature; and (3) reasons for its being as it is in regard to both position and nature. In general, it may be said that controlled aerial photographs and mosaics are superior to other field methods in defining location and boundaries; that the usual field methods are superior to photographic techniques in the characterization of the nature of an object, although the regional keys allow certain definite identifications; and that each technique has advantages (and disadvantages) in explaining the nature and position of the object under consideration. Photographs allow for a greater part of the landscape to be studied at a single view, and therefore introduce broad relationships which might be obscured because the student on the ground "sees the trees and not the woods."

Geographic field work may take the form of a reconnaissance survey to establish broad types of distributions or of a microstudy intended to set up the minute geographic relationships which constitute the broader patterns. Aerial photographs are of value chiefly in the first, broad type of field investigation, although many of the problems of detail can be recognized and solved through an interpretation of the prints.

A prime requisite for the translation of aerial photographs is a knowledge of the processes involved in the production of the particular natural or cultural feature that is to be identified. The trained observer may note a geometrical pattern associated with a field, but it has no significance unless he recognizes it as the intersection of the "dead" furrows, left at the time of plowing. (Such dead furrows ordinarily signify grain fields.) With this and other aids it may be possible to draw some conclusions about the kind of crop in the field when the picture was taken.

More generally, familiarity with the area under investigation is essential. Though the railroad, with its characteristic curves, is identifiable in any place, many other cultural forms are not. Farmsteads, road patterns, field shapes, and other features have different appearances in various parts of the world. Hence any problem of interpretation is partly solved by a knowledge of the physical and cultural conditions in the area investigated.

It is further essential that the observer have the proper perspective. Since most observations are made from the ground it is difficult to orient the viewpoint with a vertical position. The reading of aerial photographs requires that the appearance of an object from above shall be as easy a means of identification as the aspect of the same object in the horizontal plane. Appreciation of this fact from the very beginning will aid greatly.

Several examples of direct and associative keys are presented below. These by no means exhaust the possibilities of photographic interpretation, but are intended to be suggestive of the type of information required for identification of objects. In applying such keys the interpretation of an aerial photograph is accomplished through an evaluation of four factors: (1) the flat two-dimensional shape; (2) the tone, from black to white; (3) the texture, from smooth to rough; and (4) the shadows cast by the objects to be identified.

1. *Shape.* — Since it is customary to observe phenomena on a horizontal or, at best, an oblique plane, it is difficult suddenly to transfer the study of an object to the vertical perspective. For this reason it is necessary to develop a familiarity with the ground plan of the various types of objects, and to be able to translate this plan into a horizontal concept.

2. *Tone.* — The tone range from black to white in which an object appears is the result of the amount of light transmitted from the surface of the object to the camera. If the amount reflected is large, the object will appear light; if small, a darker shade prevails. This variation is due to differences in the angle of the rays with respect to the camera and in the nature and surface of the object.

The elevation of the sun above the horizon affects the tone, as does the position of the lens with respect to the sun. If the sun is behind the airplane, the features pictured ahead of the camera appear white or light-colored, whereas the same objects will be photographed in a darker shade if they are to the rear of the plane at the time of exposure.

There is considerable variation in the reflective quality of the different types of material and surface. Smooth surfaces reflect more light than rough ones, and dark objects less than light ones. For instance, roads are nearly always lighter than surrounding fields, and plowed or trampled fields are not so dark as those in which crops are standing. In addition, the varying intensity of development of individual prints is reflected to some extent in the shading of the photograph.

For these reasons it is difficult to use tonal variations to identify objects far apart on the same print or on neighboring prints. Tone is valuable in showing differences within a small area, in indicating changes in texture, and in the location of shadows.

3. *Texture.* — "Texture" means the number of tonal changes in any given part of the picture. Highly variegated or dappled sections indicate a rough surface or one which lacks uniformity in some other respect, whereas a smooth, even appearance shows homogeneity of surface and reflective quality. Striking differences appear between upland hardwoods and swamp conifers, for example. The broad leaves and the roughly globular shape of the tree crowns give the deciduous areas a spotty look, while the uniformly dark, spired evergreens are smooth in texture.

4. *Shadows.* — Shadows are dead-black images which intrude upon otherwise lighter shades. The form which casts the shadow can usually be located in the ground plan, but small objects such as telephone poles have only the shadow as a means of identification.

Shadows are important in photographic interpretation because the outline formed on the ground may be more readily identifiable than the plan, tone, or texture of the object itself. The shadow covers the side of the elevation which is away from the light. For this reason it is best to orient the photograph with the shaded part toward the observer. It is also desirable to place the light used in study in a position comparable to that of the sun at the time the picture was taken. This will usually put the top of the picture toward the south. The slight disadvantage which results is compensated for by the increased perception of relief. If the picture is improperly oriented, hills appear as depressions.

Shadows have value in determining the location, shape, height, or depth of any object, and the nature of the object through any of these facts. They are particularly important in areas of accentuated vertical differentiation or in winter. The more marked the break in surface, the sharper, and therefore the more distinguishable, will be the resulting shadow. In winter the dark tone stands out conspicuously against the white surface.

Shadows which are too long cover the details of the picture over considerable areas; therefore photography is limited to a few hours around noon. Objects cast shadows far out of proportion to their size when exposures are made early in the morning or late in the day, and, in addition, black out much significant detail. This is an obstacle especially during the low-sun period, so that desirable photographic flying time in winter is more strictly limited than in summer.

KEYS TO THE INTERPRETATION OF AERIAL PHOTOGRAPHS IN THE NORTHEASTERN PART OF THE UNITED STATES[1]

Topography

Relative relief, location of relief features, and direction of slope are often discernible by interpreting shadows, vegetation associations, drainage patterns, location and direction of roads, and cultivation characteristics:

1. Identification of relief through shadows: All shadows indicate unevenness of surface, either natural or cultural; shape of shadow must be evaluated to determine type of projection which casts it. See last paragraph on this page for manner of deducing cuts and fills.
2. Identification of relief through vegetation
   a. Type of vegetation indicates relative relief if growth habits of the associations are known. Keys to associations have been established which distinguish between upland and lowland species (see below).
   b. Long, more or less continuous strips of trees projecting into otherwise nearly completely cultivated areas suggest land too steep to farm, particularly where tentacles extend from a continuous strip of vegetation into farmed area. Scattered fields also often indicate relief.
3. Identification of relief through drainage
   a. Streams meander on flat land and flow in straight narrow valleys where gradient is steep.
   b. Streams flow from high to low elevations, showing direction of slope.
   c. Streams usually indicate areas lower than surrounding country.
   d. Soil-drainage patterns, shown as dark markings in cultivated land, express direction of slope.
   e. Erosion scars, light dendritic or washboard tracings, indicate slope. Associated fan-shaped accumulations mark base of slope.
4. Identification of relief through roads and railroads
   a. Secondary roads are laid out with numerous curves, hairpin turns, and the like, in order to avoid hills and to descend steep slopes along easy angles.
   b. First-class roads and railroads are engineered without reference to minor topographic forms, but such features are distinguishable by a study of the shadows indicating cuts and fills. If road is cut through a hill, shadow will fall across road, whereas it will fall away from road if depression has been filled.

[1] Inspiration for several of the keys presented here was obtained from many sources, published and unpublished. The authors have tested and elaborated such keys. A complete bibliography of articles which contain examples of interpretations of aerial photographs is impracticable here.

5. Identification of relief through cultivation
   a. Irregular boundaries of cultivated land in an area otherwise regular in pattern may be attributed to a sharp break in surface.
   b. Arrangement of plow and drill lines, as in contour farming, may suggest relief, but since farm techniques vary widely this is not conclusive.
   c. Absence of cultural patterns in area of moderately dense population may point to stretches of rough, broken land, associated with a patchy distribution of forest, bush, swamp, marsh, and lakes. A continuous strip of vegetation in area of advanced cultivation may mark a break in surface.

Certain physiographic forms may be identified as follows:

1. Fault lines
   a. Sharp continuous soil change is evidenced by differences in tonal quality of soil, in natural vegetation, or in stand of crops.
   b. Drainage displacement is apparent.
   c. Scraps may be identified.
2. Anticlines
   a. Drainage is radial.
   b. Arching of streams follows bedding planes.
3. Sedimentary rocks
   a. Erosion of these forms may look like hill contours from the air.
   b. Contacts between overlying strata are often marked by linear distributions of denser vegetation.
4. Sinkholes
   a. Shadows, if photograph is oriented properly, reflect characteristic funnel shape of depression.
   b. Holes may be vegetation-choked.
5. Kettle holes
   a. Round lakes or marshes in morainic country indicate kettle holes.
   b. Shadows accentuate depression.
   c. Concentric circles of vegetation may mark successive stages in lake filling.
6. Stream capture
   a. Presence of dry gaps is indicative.
   b. Decapitated stream system may be observed.
7. Alluvial fans
   a. Characteristic shape may be identified by soil tone different from that of surrounding territory.
   b. Fans are found in areas of sharp relief, where there will be evidence of abrupt flattening of grade.
   c. Stream which formed fan spreads out at base of fan; distributary channels may be seen.
   d. Spring line may appear as dark saturated zone along a line of equal elevation.

8. Ravines
   a. Broken bank lines may be apparent, usually accompanied by change in vegetation or by abrupt shift from cultivation to woods.
   b. Streams are not always visible because of trees, but when visible they fill entire floor of valley.
   c. Dendritic pattern of trees outlines streams and tributaries.
9. Moraines
   a. Unevenness of surface is revealed by road routes, gullied slopes, shadows, cuts, and fills.
   b. Tonal variation is great because of heterogeneity of soil types.
   c. Morainic hollows, kettle holes, and so on are darker than surrounding surface area, owing either to damp soil or to marsh vegetation. If still water-filled, lake is, of course, visible.
   d. Cultivated areas are interspersed with woodlots, and both are irregular in form.
   e. Stream courses are haphazard and winding; if outwash plain is associated with morainic belt stream will flow more regularly through this type of surface.
10. Outwash plains
   a. Road pattern is more uniform than in morainic area.
   b. Tonal similarity is due to greater homogeneity of material, but tonal variation is still that characteristic of glacial regions.
   c. Streams may be incised and banks wooded.
   d. Fields are more regular and agricultural land use more complete than in morainic area.
   e. Pitted outwash marked with small lakes and poorly drained depressions is strikingly revealed if land has been cleared.
11. Drumlins
   a. Stoss and lee slopes often give a "tadpole" appearance.
   b. Valleys between drumlins are poorly drained and hence darker; they are often uncleared or in hay or pasture.
   c. Fields are oriented with trend of drumlins; long, narrow areas of cultivation are characteristic.
   d. Numerous shadows appear at road cuts when highway is transverse to drumlins.
12. Eskers
   a. Sinuous nature of eskers is readily discernible, especially if they are forested and in an otherwise cleared zone.
   b. Eskers often break or dislocate pattern of cultivation.
   c. Gravel pits are common; shadows and white or light tone aid in their identification.
   d. Roads are usually cut through eskers and are hence often in shadow at point of crossing.
13. Youthful and mature surfaces: though youthful and mature surfaces are best identified through shadow and tone, distribution of cleared areas often aids in establishing stage of erosion. Location of fields and farms and their size and relative density reveal approximate ex-

tent of level land in valleys or on interfluves. Furthermore, irregularities of occurrence are a key to amount and degree of slope.

Drainage

Drainage features may be identified as follows:

1. Water bodies (except streams)
   a. Flat uniformity of tone occurs, its darkness being dependent on depth of water. Light tones indicate shallow water; darker tones, deeper water. This fact aids in identification of subaqueous forms such as bars and reefs.
   b. In some instances water is white if sun's rays are reflected directly into camera.
   c. Shore line is distinguished by abrupt change in tone and texture from land to water.
   d. Small lakes enclosed with concentric rings of varying tone and texture reveal vegetation sequence in course of extinction of lake.
   e. Waves and breakers are often visible.
2. Rivers, streams, and creeks
   a. Tone shades from black to white, depending on reflection of light.
   b. Most streams wind, and vary in width. This variation distinguishes them from roads or other continuous phenomena. In areas of particularly steep gradient where stream course is relatively straight there will be erosion scars, lack of vegetation on such scars, bank lines, and other signs of V-shaped valley.
   c. Streams are usually paralleled by a dense ribbon of vegetation, especially noticeable in cultivated country.
   d. Bridges occur; type can be identified by close inspection of shadow.
   e. When water itself is nearly invisible, through overhanging of vegetation, increased density of trees along stream may be a clue, especially if occasional fleck of light is reflected from water.
3. Rapids and falls
   a. Light tone of water appears between two darker areas, caused by turmoil of rapid motion.
   b. Banks may narrow.
   c. If streams are used for navigation there may be evidence of portage trails. If streams are canalized, locks, together with lock house and other evidences of use, may be discerned.
4. Braided streams
   a. Wide bank lines appear.
   b. Small winding meandering stream in valley bottom is characteristic; the many channels are darker than rest of valley bottom.
   c. Light-toned sand and gravel islands occur between channels.
   d. No cultivation is observable in valley bottom.
5. Flood plains
   a. Bank line borders plain on one side and stream on the other.

b. As streams meandered over flood plains deposition may have been unequal and not of uniform composition. This unevenness is reflected in orientation of fields, which are aligned with reference to position of old meanders.

c. Sometimes plains are densely cultivated, but cultivation lines (boundaries) are limited by their extent.

d. Vegetation is absent except along bank lines or streams.

6. Bank lines

a. In cultivated areas bank lines appear as discontinuous winding zones of vegetation between two areas of cultivated land, between flood plain and upland, or between cultivated areas and marsh or swamp zones. Lines roughly follow course of streams.

b. Drainage cuts and gullies may be present.

c. Definite cultivation boundaries due to steep banks are observable.

d. Bank lines usually separate two soil zones of different tone.

7. Oxbow lakes and meander scars

a. Dark concentric markings are observable, tone due to soil dampness and marsh vegetation.

b. Oxbow lakes have tone and texture typical of water, whereas coarser texture appears if vegetation chokes scar.

c. Oxbow lakes and meander scars are usually associated with flood plain of meandering stream, but in areas of uplift they may be well above stream.

d. Inside of meander may be a slip-off slope, which in cultivated areas will often be wooded. Undercut bank on outside of meander may show light tone because of absence of vegetation.

8. Canals

a. Tone and texture are those characteristic of water.

b. Regular width and constant direction distinguish canals from streams.

c. Banks of dirt, built-up roads, and so on may indicate artificial construction.

d. Signs of use — locks, dams, and boats — may be apparent.

10. Dams and similar constructions

a. Backwater resembles lake.

b. Sharp break occurs in looks of water at dams, for white flow appears below dams.

c. Dams show a light line, which will be straight in contrast to sinuous form of rest of shore line of backwater.

d. Some indication of relief of area may be derived from shore line of backwater, since it is at a contour line.

11. Soil drainage: Dry soils are usually lighter in tone; damp soils are darker.

Soils

Broad generalizations about soils are possible from aerial photographs. No detailed identifications should be attempted on this basis alone unless the observer is a specialist, although in the field the problem of soil boundaries may be expedited by their use. Vegetation associations revealed by photographs also give some clue to the nature of the soil, since it is one of the important factors in plant ecology.

1. Broad generalizations
   a. Damp soils show darker than dry soils in same general area.
   b. Sands are light gray; clays, dark gray; loams, intermediate in color, but generally more mottled in texture.
   c. Soils in glaciated areas are almost invariably highly variegated in tone because of heterogeneous nature of till.
2. Correlation of vegetation and soil
   a. Marsh and swamp vegetation indicates peat and muck.
   b. Cultivation may mark loamy soils.
   c. Upland hardwoods are often expressive of loam.
   d. Certain species of conifers occur on excessively wet soils, others, on excessively dry soils; consequently, if species are distinguishable, they may give clues to soil type.

Mineral Resources

Most minerals are subsurface phenomena, and therefore difficult to locate with certainty on a photograph. An expert, however, is able to recognize signs which indicate areas that might prove profitable fields for more detailed surveys.

Aerial photographs have been used by petroleum geologists to determine geological structure. Both exploratory and detailed surveys have been aided by the application of direct and associative keys. It is not within the scope of this paper to present exact descriptions of the appearance of most geological phenomena, because others more experienced in the use of photographs for this purpose are already working in the field. One example is given to indicate the type of information required in their identification:

1. Salt domes
   a. Circular area which is faintly lighter or darker than surrounding area is characteristic.
   b. Drainage, soil erosion, and road pattern help determine domelike relief.

Weather and Climate

No details of value about the weather and the climate are obtainable from aerial photographs, although it is possible to develop a very general picture of the climate of an area on the basis of its vegetation, land forms, house types, and drainage.

Coasts and Shores

Shore-line phenomena, both subaqueous and emergent, are readily studied on aerial photographs. Unless the water surface is light-struck (dead white) it photographs dark, the tone of the image varying from black to gray with the changes in depth. Deep water shows dark; shallows appear light. On this basis submerged forms can be identified by their shape and position.

1. Submarine deltas
   a. Submarine deltas occur where streams enter other bodies of water.
   b. Characteristic shape may be defined by light tone of water body, due to its shallowness on delta.
   c. It is often possible to tell direction of current of water bodies which streams enter by distorted shape of deltas.
2. Spits, bars, offshore bars, coastal reefs, submerged channels, terraces, tidal depths, and so on: All these forms betray themselves by tone changes from shallow to deep water.
3. Mud flats
   a. Flats occur on landward side of barrier beach and appear in photographs as uniform drab-gray slightly textured expanses; no signs of relief.
   b. Complicated system of drainage channels appears as dark tracings.
   c. Mud flats show less coarse texture than marshes because periodic inundation prevents growth of mesophytic vegetation.
4. Beaches
   a. Beaches appear as strips of uniform texture bordering water bodies.
   b. Tone is light away from water, darker near water, as moisture content of sand or shingle increases.
   c. Sand beaches show white; shingle beaches are darker and more mottled.
   d. Fringe of vegetation or sand dunes may occur on landward side of beach.
5. Abandoned beaches
   a. Light tracings which show characteristic sinuous shape of beach lines may be followed, sometimes discontinuously.
   b. Abandoned beaches are often associated with more or less "washboard" appearance; many are arcuate and separated by marsh areas.
   c. All sand areas photograph nearly pure white, lack definite texture, and look very similar to light-struck water.
6. Sand dunes
   a. Appearance is that typical of sand, but modified by light and shadow characteristic of rough terrain.
   b. Drainage lines are absent where streams have cut through dune area.

There is usually a sharp break in direction of stream where it has changed its course to find a place to cut through dunes.

c. Distribution of vegetation is spotty, with irregular breaks where blowouts or slides have cut into sands.

d. Type of dunes may be determined by study of shape. Ridge lines are often horseshoe-shaped or bowed.

7. Sea cliffs and bluffs

a. Absence of well-defined beach lines may indicate formation dipping into sea or wave-cutting at base of vertical cliff.

b. Water shows dark immediately adjacent to shore, and no terrace or other shelf features appear.

c. Choppy white narrow surf lines may occur at contact between land and water.

Plant Life

One of the most important contributions to geographic inventory which can be made through the use of aerial photographs is the delimitation of vegetation zones. Forestry agencies, both private and public, have made forest surveys by interpreting vertical or oblique prints. Results indicate that this method of estimating volume and species is satisfactory for all except the most detailed surveys.

The keys presented here are regional and are intended to be applicable only to northern Michigan, northern New York, and areas which have similar vegetation associations.

1. Woodland in general: Identification of woodland is obvious, but determination of species and stand requires keys.

a. Uniformity of tone and texture indicates uniformity of species, age, and stand.

b. Changes in appearance of stand indicate a change in density, size, or species.

2. Hardwoods

a. Billowy light and dark tone is characteristic. Broadleaf associations usually photograph lighter than conifers because broad leaves reflect more light to camera than do needles of conifers.

b. Individual crowns show light on side toward sun, with a semicircular shadow on side away from sun. These shadows may be interrupted by neighboring trees, but texture of hardwood stand is characteristically mottled, "salt and pepper."

c. Where individual hardwoods stand in fields, some hint as to species may be obtained through shadow; otherwise it is difficult to identify association with accuracy unless observer is intimately acquainted with regional associations. Once common upland hardwood association of an area is established, it is possible to guess at particular association under consideration. Changes in species will usually change photographic appearance of stand in some small but often recognizable way.

3. Conifers
   a. A smooth gray to nearly black photographic tone is characteristic. Degree of smoothness depends partly on size of individual trees and spacing of trees, but coniferous stand never displays mottled texture of hardwoods. Under comparable lighting conditions conifers will always photograph darker than hardwoods.
   b. Conic nature of most conifers is responsible for texture. Various sizes of crown allow some identification of species, as does fact that jack pine, cedar, and balsam are usually darker than spruce.
   c. In winter photographs identification of conifers as compared with hardwoods is unmistakable. In general it is desirable for observer to be familiar with plant ecology of area under study, although species identification may be made of individual solitary trees from inspection of shadow. Once familiarity with site characteristics of coniferous associations in any region is attained, it is possible to make rough identifications on basis of proximity to water bodies or other evidences of drainage or lack of it.
4. Mixed hardwood and conifers: As texture of photograph changes from smooth to billowy it shows that increasing numbers of hardwoods are entering association. This shift from one association to another may be determined by a count of individual crowns, under magnifying glass, if scale of photograph is large enough. Here, too, it is desirable to know local growth characteristics of various associations.
5. Coniferous trees on well-drained land
   a. Tone and texture remain roughly the same for this association as for lowland conifers.
   b. Individual trees are often discernible through study of crowns, pines showing a widespread branching which hydrophytic conifers lack.
   c. Site factors will help determine this association.
6. Estimates of stand
   a. Rough stand estimates may be made from a quick inspection of photograph, according to crowded or open condition of trees. Sometimes in an open stand ground will show as light or dark smooth patches in midst of coarser-textured area.
   b. More accurate, but still rough, estimates may be made by actual count of crowns with aid of high-powered magnifying glass. Height of trees is discernible from shadows if time of day, time of year, and latitude are known.
   c. Winter photographs are best for determining stand because deciduous species are instantly distinguishable from conifers, and shadows are well defined and measurable.
7. Swamps
   a. Coniferous swamps are usually dark gray and are distinguished by extremely smooth grain and by uniformity of tone and texture.

Type of vegetation must be determined in the field, however, and appropriate adjustments in key must be made for region.

b. Lack of evidence of human occupance such as roads, houses, fields, and the like is characteristic. If trails do appear it is probable, in north-central region, that they are lumber trails, and further evidences of this mode of land use should be sought. Logging roads display many branching lines at right angles, or at regular oblique angles, to main trail, giving washboard appearance to photograph.

c. Stream courses which meander because of flatness of land are characteristic of swamps.

d. Drainage ditches may be observed; these will run straight and show white strips along their sides, where dirt has been thrown up and has dried.

e. There is a definite sequence of grain or textural characteristics on photograph, from grassy marsh types through swamp conifers and mixed hardwood sand conifers to upland hardwood associations. The closer the association approaches the upland hardwoods, the more billowy becomes appearance of stand. Rounded crowns of hardwoods are readily distinguishable from small spirelike crowns of conifers.

8. Marshes

a. Uniform tone and texture are characteristic, tone being dark gray and texture only faintly perceptible.

b. Patches of open water may occur, and if they do, there will be evidences of lake-filling, in that circular rings of apparently different vegetation associations will be distinguishable through varying tones and textures. Brush alder and willow will show lighter and more coarsely grained than marsh grass.

c. There will be few if any signs of human occupance.

d. Drainage patterns are often indistinct, but traceable; they may be augmented by straight canals.

9. Grasslands

a. Light-gray tone and smooth texture are typical, but grasslands are haphazardly splotched with darker tones and more definite texture where brush and trees enter the cover.

b. Fence lines are lacking except in areas which have been cultivated and abandoned, where, moreover, mottled tone becomes more apparent.

c. Houses, if any, are spaced more widely than in cultivated areas.

d. Evidences of trails may appear if areas have been utilized in past. Abandoned houses show as dark boxlike forms and are distinguished from occupied dwellings by lack of tone change in area around them.

10. Brush

a. Wide range of tone from light to dark is characteristic, mottled appearance being due to nonuniformity of brush types and heights. Texture smoother than that of trees, but coarser than that of grassland.

b. Brush cover is not complete, and open spaces accentuate mottling.
c. Evidences of lumbering may appear in signs of lumber trails, logging camps, lumber railroad beds, and the like. Old burns are distinguishable from cutover areas by irregularity of outline in contrast to regular outline of cutover after logging operations.
d. Location in relation to surrounding vegetation associations may indicate type of brush: uplands will have young poplar; lowlands along streams and in swamps will have alder–willow association.

11. Burns: Appearance depends on length of time between fire and aerial photography.
   a. New burns
      (1) Dead-white tone is characteristic; no texture is apparent, except that dead trees show as dark spots.
      (2) Outline is irregular.
      (3) Burns spread across vegetation boundaries.
      (4) Region may be confused with some cultivated areas, but does not have fence lines, roads, or other regular elements.
   b. Old burns
      (1) Definite change in tone and texture appears between burns and surrounding vegetation, but area does not have dead-white, textureless quality of new burns.
      (2) Vegetation types begin to conform to prefire vegetation boundaries imposed by edaphic conditions which were not changed by fire.
12. Cutover areas
   a. Numerous trails are evidence of recent logging operations.
   b. Isolated trees stand well above surrounding vegetation cover.
   c. In locations which were predominantly conifer-bearing, especially in lowlands, logging trails will have a pattern reminiscent of backbone of a fish, with its lateral bony appendages.
   d. Logged deciduous forest looks like ground fatty meat that has been poorly mixed.

Characteristics of Occupance

Among the chief characteristics of occupance are buildings, which are immediately recognizable on photographs. The problem in interpretation is the determination of the function of the structures, generally by means of associative keys. Indications as to use may be gathered from a study of the surroundings of the building under consideration. The factors to be considered are: (1) the shadow of the structure, which will give information as to the form of the building, its height, and its structural characteristics; (2) the tone of the ground surrounding the structure, light ground being evidence of use, since the vegetation has been destroyed; (3) the surrounding buildings, piles of materials, communication lines, and the like; (4) the site, position, and situation, including spacing of the structures with reference to the lot and to neighboring structures; and (5) any other helpful associative characteristics.

It is possible because of space limitations to give but a few examples of keys to the interpretation of specific types of buildings:

1. Rural schools
   a. Structures are often square and boxlike, as shown by roof shapes and shadows.
   b. Larger outbuildings are absent.
   c. Well-defined driveways or paths to adjacent cultivated fields are lacking.
   d. Ground surrounding schools usually light because of absence of grass.
   e. Baseball diamonds or other evidence of recreational activity may be observed.
   f. Shadow of buildings may show small bell tower at front.
   g. Single photographs probably will not show many such structures, since it is not likely that two or three schools will occur within a short distance.
   h. Schools are usually found near outskirts of small rural towns.
2. Rural churches
   a. Ground plans are often in shape of cross.
   b. Structure is centrally located on plot.
   c. Outbuildings are absent except in certain rural sections, where there may be parking sheds.
   d. Ground around buildings is dark because of presence of grass.
   e. Steeples or towers are distinguishable by their shadows.
   f. Cemeteries may be discernible in adjacent lots.
   g. Paths and trails to surrounding fields are wanting.
3. Farmhouses
   a. Design, revealed by roof and shadow, shows more variation than is found in barns, schools, or churches.
   b. Houses are smaller than adjacent barns, but larger than other buildings of farm group.
   c. Ground around houses is dark because of presence of grass.
   d. Houses are often set among trees, whereas other farm buildings may be in the open.
   e. Farm driveways do not end at houses, but continue on to barns.
4. Rural barns
   a. Barns are focal points for lanes, driveways, and paths of farmsteads.
   b. Shapes are distinctly boxlike, with peaked or rounded roofs, marked by their shadows.
   c. Silos or haystacks may be indicated by shadows.

d. Ground around barns is usually light-toned because of absence of grass due to trampling by animals and wear of farm machinery.

5. Lumber mills
   a. If mills are dependent on railroad there will be a spur or siding.
   b. If mills are dependent on a stream there will be a millpond near-by for storage of logs. Dam will be distinguishable.
   c. Most lumber mills have a slash burner and tall stacks.
   d. Piles of seasoning lumber show as small rectangular light patches distributed in regular alignment over a large area.
6. Cottages
   a. Structures are located along shore of lake or river.
   b. Sand beaches and shallow water offshore are characteristic.
   c. Adjacent cultivated land is lacking.
   d. Cottages may be closely spaced.
   e. Haphazard placing indicates groups of cottages.
   f. A road will lead in to the group of buildings.
   g. Docks and floats lie offshore, and V-shaped wash left by speed boats may sometimes be perceived in summer.
7. Hunting shacks
   a. Solitary location in wilderness is characteristic.
   b. Cultivated land is lacking.
   c. Trails lead into woods, not to fields or other houses.
8. Abandoned buildings
   a. Where dwellings have disappeared, cellar outlines can usually be noted if cellars were made of stone or cement.
   b. No white ground is visible around houses or barns.
   c. When farms are in partial use there may be evidence of activity around barns but not around dwellings.
   d. Driveways are not marked by use.
   e. No signs of tracks appear where mailman would turn out to reach letter boxes.
   f. Roadways passing houses often have few signs of travel.
   g. Orchards are unkempt, with many trees missing from normal pattern.
   h. Fields are dark-toned and uncultivated.
   i. When abandonment has reached advanced stage, brush will be scattered through fields and small trees will be particularly in evidence along forest–cleared-land boundary.
   j. Roads will show a tendency to cut across field corners at road intersections.

Rural Production

Identification of cultivated land is immediate because of the regularity of pattern and texture it exhibits, in contrast to the coarser textures, irregular outlines, and more homogeneous tones of forest, grassland, or brushland. In photographs the tone of such areas ranges from white to black, but the lighter shades predominate. Tonal changes are usually sharp and may be emphasized by fence lines. Specific elements may be identified as follows:

1. Field lines
   a. Lines appear on photograph at more or less regular intervals, making a sharp break between areas of relatively homogeneous tone and texture.
   b. Stone and rail fences are distinguishable, being large enough to create a photographic image.
   c. Untilled land along field lines shows darker than cropland, and there may be trees, bushes, and brush along fence zones.
   d. Cattle paths and lanes may show along field lines.
2. Cultivated land: This land is distinguishable from uncultivated pasture by presence of drill or plow marks, and characteristic patterns of dead furrows.
3. Crop types: Hints as to nature of crops may be derived from texture of fields. Corn and potatoes produce a coarser texture than small grains. Regional crop keys may be established, and a knowledge of harvesting methods will aid in crop determination if pictures are taken during harvesting season.
4. Circular plowing and harvesting: Two definite tones sometimes show within same field. If inside zone is dark and outside zone is light, field is being harvested; if outside zone is dark and inside one is light, field is being plowed.
5. Orchards
   a. Round full crowns of trees are indicative.
   b. Spacing and alignment of trees are regular.
   c. Usual proximity of orchards to houses and roads differentiates them from planted forests.
   d. Orchards are ordinarily located away from natural forest areas.

Transportation

1. Trails: Chief significance is in what may be found at ends, and in hints they furnish as to movement of people within specific areas.
   a. Thin white lines running haphazardly through fields, forests, and grassland are characteristic.
   b. Extremely winding nature distinguishes trails from second- and third-class roads.
   c. Trails nearly always begin and end at some point associated with human occupance. At least one end will be connected with some more advanced form of communication.
   d. Trees may obscure trails in forested areas.
2. Roads: Roads range from black to white, and appear as lines. In general, concrete photographs lighter than asphalt, and dirt darker than gravel.

a. First-class roads
  (1) Wide, cleared right of way is characteristic.
  (2) Curves are smooth and broad.
  (3) Routes cut across established patterns of field and town.
  (4) Construction is made without reference to minor obstacles such as hills and swamps, which may be distinguished by cuts and fills. If roads are on a fill, shadow will fall away from them; if they run through a cut, shadow will fall across them.

b. Other roads
  (1) Right of way is narrow.
  (2) Sharp corners and turns are common.
  (3) Established patterns of settlement are followed.
  (4) Routes are controlled in part by minor topographic features, and so show numerous curves and bends.

3. Railroads
  a. Width of line is determined by number of tracks, a single-track system being narrower than a first-class automobile road.
  b. Railroads are usually dead white because of presence of stone fill for roadbed.
  c. Direction of line is changed by long smooth curves.
  d. Branch lines are few, and served by long curves.
  e. Roads that cross railroads can be seen going *over* tracks.
  f. Numerous cuts and fills are apparent where railroad engineers have attempted to smooth out trajectory of line.
  g. Right of way of railroads becomes narrower over streams, whereas roads show wider paths across bodies of water.
  h. Stations, sidetracks, trains, factories, and other forms characteristic of railroads may help in identification.

4. Fords across bodies of water
  a. Fords occur at places where river bank breaks down, as shown by a light marking on picture. Break in bank gives access to body of water.
  b. Roads, trails, tracks, and the like converge on stream at this point, disappear into the water, and emerge on opposite side. Sometimes tracks may be seen on bed of stream.

5. Transmission lines
  a. Cleared band through woods, without apparent roads, is indicative. Trails may be discernible.
  b. Change in direction takes place by means of sharp corners rather than engineered curves.
  c. In cultivated country transmission lines may be identified by structural shadows, which are lined up without reference to established patterns.

6. Airports
  a. Large cleared space without fields, fences, or other obstructions is characteristic.
  b. Large building or buildings are grouped near periphery of field.

- *c.* Long light-toned graded runways are visible, usually at odd angles.
- *d.* Planes on field may show, being either dark or light.

7. Docks
   - *a.* Elongated rectangular projections from a shore line into a water body are indicative.
   - *b.* If docks are of a commercial size storage sheds, railroad tracks, and so on will extend along them.
   - *c.* Steamers may be identified by their form, and type of steamer may sometimes be revealed by the superstructure.

Urban Forms

The city, with its preponderance of cultural forms, lends itself more readily to interpretation that does the rural landscape. Custom, tradition, and utilitarian and functional demands have so standardized urban patterns that many features are easily recognized even in the most cursory examination. Downtown parking lots offer an illustration. Open areas, often with small buildings near the street, with two or more entrance ways, and with lines of parked cars, can seldom be confused with any other feature. It is more difficult, however, to distinguish between commercial parking lots and establishments where secondhand cars are sold. If the space is only partly filled, the car sales lot can probably be recognized by the fact that the cars are parked next to the sidewalks.

Urban interpretation can be accomplished with somewhat more certainty than can rural, if a comparable time is spent in the field. The use of shadows and associative keys is of particular importance. Water tanks, smokestacks, overpasses and similar superstructures can best be identified by their shadows. In fact, identification otherwise is often impossible. It can seldom be established with certainty that a building by itself is an industrial enterprise. Docks, railroad sidings, coal piles, water tanks, large parking lots, bare ground, and low-class residences, however, help give character to the factory structure. Recognition of most urban phenomena is aided by a consideration of the surrounding associated features, which indicate the function of the form observed.

1. Trees
   - *a.* Absence of trees aids in separating commercial and industrial areas from residential parts of city.
   - *b.* Size, or lack, of trees is of considerable value in recognition of areas of different periods of development.
   - *c.* A paucity of trees is indicative of a low-class residential district.

d. A great number of trees often reflects middle-class or upper-class residence zones.

2. Railroads
   a. Routes cut across normal street pattern.
   b. Curves are characteristic.
   c. Cars may be seen on tracks even on small-scale photographs.
   d. Sidings are numerous and fan out along right of way. Since their function is to serve establishments rather than allow for two-way traffic, they are often dead-end lines.
   e. Belt-line position about city is common.
   f. Roundhouses appear in almost all cities of any size. Their images usually resemble a section of a doughnut. If railroads cannot be identified immediately, markings can be traced to see if they lead to a roundhouse, with its car shops, coal piles, and networks of tracks.
   g. Large stations are conspicuous because of tracks that enter into or pass through sheds.
3. Oil, gas, and coal storage
   a. Oil storage
      (1) Tanks appear as round white or light-colored spots.
      (2) Shadow gives some idea of their elevation.
      (3) Tanks are grouped along railroad or waterfront.
      (4) Generally more than one size of tank occurs.
      (5) Rectangular embankments surround tanks.
   b. Illuminating-gas storage
      (1) Generally tanks photograph black or some shade of gray.
      (2) Image is ordinarily a much larger spot than that made by oil tanks.
      (3) Superstructures surrounding tanks cast distinctive shadows.
      (4) Name of city or directional marker may be painted on top.
   c. Coal storage
      (1) Railroad or waterfront storage is common.
      (2) Coal is piled in long ridges, sometimes in many parallel ridges, or in circular form, and is deposited with a regularity not found in natural phenomena.
      (3) Superstructure of overhead traveling crane often throws a characteristic shadow.
4. Industrial buildings: Industrial plants are most easily recognized if observer has some familiarity with functional character of many different factories, for use often determines form and pattern of plant layout.
   a. Smokestacks are apparent, and smoke in the vicinity often imparts a hazy tone to print.
   b. Coal piles and traveling cranes are evident.
   c. Enclosed or open conveying systems lead from one building to another.
   d. Water towers, usually built above one of the buildings, can easily be identified.

e. Saw-tooth roofs, frequently with many skylights, are characteristic.
f. Spur tracks and siding may be seen, with many cars on the tracks.
g. A surrounding area with no buildings is common; light spots in this area indicate that much of ground is bare.
h. Some industrial buildings have blotched look because dust-spotted roofs reflect light with varying degrees of intensity.
i. Factory buildings may be laid out in courtyard effect, with many enclosed units.
j. Office buildings are associated with many factories, usually near street, with landscaped lawn.
k. Most factories occupy large area, with considerable room reserved for parking.
l. Factory names or other information may be painted on roof.
m. Adjacent blocks may be little built up, and such residences as do occur are usually low class, stereotyped, and crowded.

5. Streets
a. Alleys are narrower than streets and usually reflect light as gray or white.
b. In contrast to streets, alleys are seldom tree-lined, but are bordered with garages and other small buildings.
c. Streets are grayer than alleys.
d. Main streets may show two grease streaks, whereas side streets show only one.
e. Blackness of grease streaks often shows direction of flow of traffic, particularly at corners.
f. Car tracks show up as light lines in street.
g. Grease streaks swing in toward curb around safety zones.
h. Boulevards have dark segmented center areas.
i. In business districts sidewalks, shown as narrow white lines, are immediately adjacent to street, whereas in residential districts they are separated from street by narrow band of darker hue.
j. In some sections of city change in character or orientation of street pattern may indicate limits of a subdivision or a certain period of building.

6. The commercial core: Tall buildings of commercial core stand out on most photographs. Elevation of buildings is sufficient so that those not directly below camera are photographed with a large portion of wall exposed. Hence there is an impression of depth not found elsewhere.

7. Commercial zone
a. Numerous parking lots or parked cars are evident along streets.
b. Buildings are more or less continuous along street.
c. Buildings are built out to sidewalk.
d. Many skylights and flat roofs appear.
e. Treeless alleys are characteristic.
f. Alleys are immediately adjacent to backs of buildings.
g. Signs on tops of store buildings often cast shadows.

h. Variation in shade or tone of roofs is noticeable.
i. If buildings are on streets that are diagonal or at right angles to long axis of rectangular blocks, alleys set off commercial establishments in a narrow zone at ends of blocks. Residential alleys usually bisect blocks.

8. Apartments
   a. Shadows show elevations not common to surrounding residential area.
   b. Many wings and courts are characteristic.
   c. Landscaping is often apparent in small area between building and street.
   d. Many cars are parked along curb.
9. Public buildings
   a. Public buildings are generally set back from street.
   b. They may occupy central portion of whole block or a considerable part of block.
   c. They often have many wings and courts.
   d. Peaked roofs are common.
   e. Generally much of area around buildings is landscaped.
   f. Parking areas are provided behind buildings.
   g. Curved walks and drives are characteristic.
10. Schools
    a. Playgrounds are large, with tennis courts, ball diamonds, and other evidences of recreational activity.
    b. Buildings are constructed with wings.
    c. Ventilators are usually numerous.
    d. Heating unit with large smokestack is an adjunct of many schoolhouses.
    e. Bare ground around buildings is indicated by large white area.
    f. Ground in front is landscaped.
11. House types: There is some possibility of identifying house types. Bungalows are easily recognized by low-pitched roofs and the dormer windows. Victorian houses with turrets and projecting bay-window areas and flat-topped modernistic houses can also be recognized. It may be stated as a general rule that house quality improves as ridge-top system becomes more complex.
    a. Superior residences
       (1) Lots are large, and generally outlined by hedge or fence.
       (2) Lawns show considerable landscaping.
       (3) Drives curve, and entrance may be at any point along front or side of lot.
       (4) Houses are large, ridge-top system complex.
       (5) Large driveway areas and large backyards, often with summer-houses and tennis courts, are common.
    b. Ordinary residences
       (1) Houses are fairly close together and lots are small.
       (2) Over a considerable distance garages, drives, and entrance walks are in same position with respect to lot.

(3) Houses seldom have more than two ridge lines; roofs are stereotyped.
(4) Backyards are small and houses occupy most of width of lot.
(5) Houses are regimented with respect to street.

c. Inferior residences
(1) Houses are small, with single roof form.
(2) Lots are small, even when buildings are scattered.
(3) There is no landscaping, few or no trees, little grass.
(4) Garages are not so common as elsewhere.

12. Suburban area
a. Building density varies greatly, though many areas have centers of marked density.
b. Reflection of light from freshly turned earth and numerous tracks in vicinity aid in identification of new buildings; hence suburban area can in part be outlined by building activity.
c. Streets may show use along only part of length.
d. Areas with sidewalks but no streets that show steady use are common.
e. Gardens, truck gardens, and greenhouses are evident. Gardens show up as small light-colored rectangular areas, truck gardens look the same but are on a larger scale, and greenhouses appear as large rectangular areas, often pure white. It is sometimes possible to distinguish larger supports that separate major sections of glass.
f. Large areas may show that they have been prepared for building by fact that streets and sidewalks are present, but there are few if any houses. Such streets and walks are often gradually being reclaimed by vegetation and so may fade out of picture.
g. Occasional farmsteads occur, but farm area is usually small.

13. Parks and cemeteries
a. Parks and cemeteries have curved drives and no through streets, and so interrupt street pattern.
b. A great many trees appear; pools are common.
c. White spots indicate tombstones and vaults in cemeteries. Outline of individual lots, or rather raised or fenced areas, are frequently in evidence on prints.
d. Park areas have more open spaces and, of course, no tombstones.

14. Recreational areas
a. Many of the areas devoted to recreation are large, and interrupt street pattern.
b. Much of areas is treeless.
c. Diamond shape of baseball fields, striped football fields, golf courses with their elongated fairways and numerous white sand traps, and rectangular tennis courts may easily be identified.
d. Areas given over to general play have many conspicuous white spots, where bare earth is exposed.
e. Amusement parks can be identified by shadow of ferris wheel, sinuous sweep of roller coaster, peaked top of merry-go-round, and midway.

GEOLOGY

GROUND-WATER EXPLORATION BY EARTH-RESISTIVITY METHODS

KENNETH K. LANDES AND JAMES T. WILSON

INTRODUCTION

SEARCHING for ground-water supplies by electrical-resistivity methods is by no means new; it has been carried on in various parts of the United States, and the methods are described in standard geophysical textbooks.[1] Exploration by these means has been particularly successful in areas floored by thick deposits of unconsolidated sediment such as valley fills. Because most of the Michigan ground-water supplies are obtained from fluvioglacial sediments in the great drift sheets which cover the state, it was believed by writers that earth-resistivity methods could be successfully applied in Michigan, and the Plymouth district was chosen as a test area.

The assembling and operation of the apparatus used in the project was under the direction of the junior author. The writers gratefully acknowledge financial assistance through a grant from the Horace H. Rackham School of Graduate Studies of the University of Michigan.

PLYMOUTH DISTRICT, MICHIGAN

The state geologist, Dr. R. A. Smith, was consulted for advice regarding a "guinea pig" area. He suggested the Plymouth district because the supply of water then available was inadequate to meet the needs of a growing municipality. The city manager of Plymouth, Mr. C. H. Elliott, was very coöperative. He not only provided logs of wells drilled in that vicinity, but also contributed a truck and a helper.

The Plymouth district lies along the northwestern edge of the lake plains, and the entire area under discussion is floored with lake sediments. The Whittlesey beach passes through the city, and three

[1] Jakosky, J. J., *Exploration Geophysics* (Los Angeles, California, 1940), pp. 366–370.

DESCRIPTION OF STATIONS

Station	*Location*	*Direction of line*
Plymouth 1	Northeast corner of intersection of Blanche and Karmada Sts.	N–S
Plymouth 2	650 feet S. 30° E. of Plymouth No. 1	NE–SW
Plymouth 3	825 feet north of Plymouth No. 1	N–S
Plymouth 4	South side of Junction Ave., 80 feet east of Arthur Ave.	E–W
Plymouth 5	West side of Holbrook Ave., at Caster St.	N–S
Plymouth 6	South side of Pearl St., at Holbrook Ave.	E–W
Plymouth 7	500 feet east of the Plymouth–Northville Road on the north side of Middle River Rouge	E–W
Plymouth 8	East side of Starkweather Ave., 525 feet north of Pearl St.	N–S
Canton 1	South side of Warren Road near Haggerty. Near the middle of the south boundary of the NW. ¼ NW. ¼ of Sec. 12, T. 2 S., R. 8 E.	Along Warren Road
Canton 2	North side of Warren Road at Lotz Road. About 500 feet southwest of station No. 3 and about 11 feet higher	E–W
Canton 3	SW. ¼ SE. ¼ NE. ¼ of Sec. 12, T. 2 S., R. 8 E., 405 feet from south and 990 feet from east line of quarter section. On edge of Tonquish Creek	E–W

beaches of glacial lake Maumee lie between the Whittlesey beach and the upland about two miles to the northwest.

The glacial lake sediments abut against the Inner Defiance moraine, which makes the upland north and west of Plymouth. Presumably the lake sediments rest on ground moraine which, in turn, overlies bedrock (Antrim shale). Three wells in Plymouth have gone through the glacial deposits, logging thicknesses of 135, 148, and 239 feet. These wells were all drilled at approximately the same elevation, so that the relief of the bedrock surface is greater than the relief of the present topographic surface.

According to the geologic folio of the Detroit area,[2] the outcropping lake sediments are mainly clays and loams, with narrow bands

[2] Sherzer, W. H., *Detroit, Michigan, Folio No. 205*, U. S. Geological Survey, 1917.

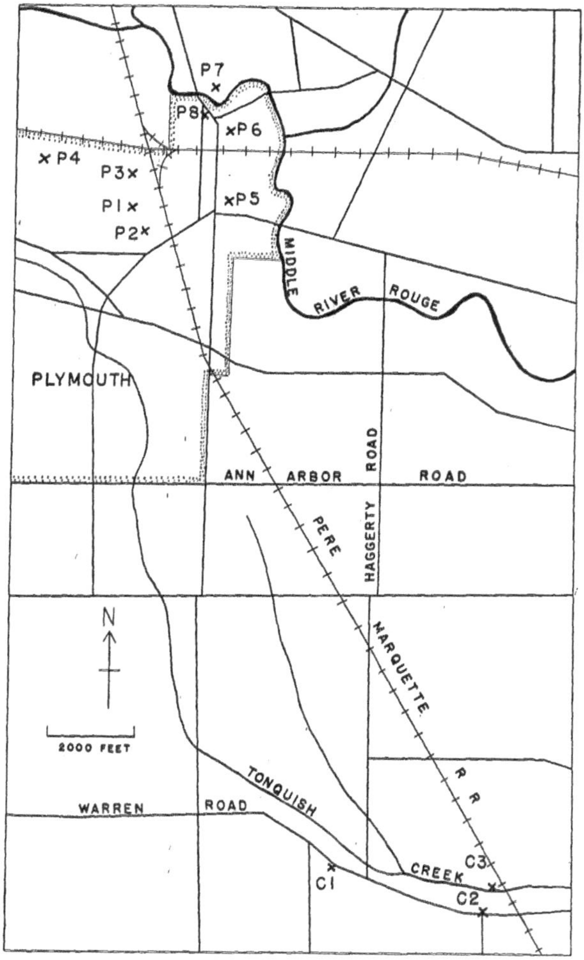

FIG. 1. Plymouth area, showing location of resistivity measurement stations

of fine sand where the old beaches occur. The logs of the drilled wells show similar materials in large part. Fortunately, however, in some places there occur water-bearing sands and even gravels with pebbles up to one-half inch in maximum dimension. The water-bearing sands and gravels are probably deltaic deposits formed by streams draining the adjacent highlands and emptying into the glacial lakes. Fluvio-glacial deposits in the ground moraine immediately overlying the bedrock constitute another possible source of ground-water supply.

Within the city limits of Plymouth eight lines were run with the resistivity apparatus. The location of the stations for them is shown in Figure 1 (P 1–P 8), and the descriptions are given in the table on page 346. One of these (P 1) was at a test well, and two (P 1 and P 7) were about 600 feet from test wells, so that it was possible to calibrate the resistivity measurements with actual well logs. In addition, three stations were occupied in Canton Township to the southeast. These are designated C 1 to C 3 on Figure 1. Stations C 1 and C 3 were also at test wells.

EQUIPMENT

The procedure in the earth-resistivity method is simple. An electric current is passed through the ground between two electrodes,

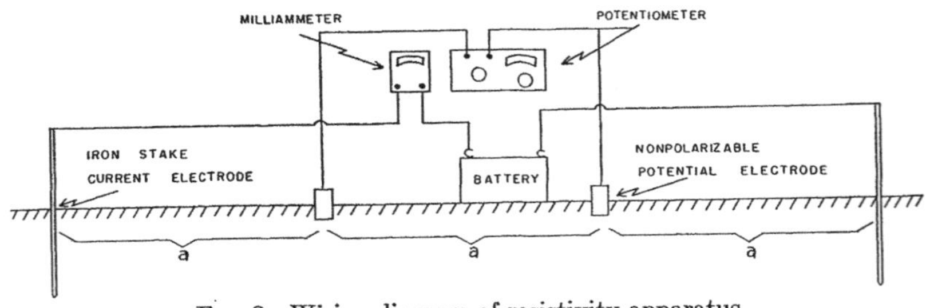

FIG. 2. Wiring diagram of resistivity apparatus

and the resulting potential at the surface is measured between two other electrodes. For all the measurements presented here the electrodes were equally spaced along a straight line, with the outer two being the current electrodes. The arrangement is shown in Figure 2.

If the spacing is a and the ground homogeneous, the resistivity is given by the simple formula:

$$p = 2\pi a \frac{V}{I}$$

where p is the resistivity and V and I are the measured potential and current. This formula is generally used even where the ground is not homogeneous. In case the ground is not homogeneous the computed value of p is usually regarded as a rough measure of the resistivity to a depth *a*. Jakosky[3] notes, however, that in actual practice considerable variation exists. He has found that the vertical penetration ranges from about one fourth to one ninth of the total electrode spacing (3*a*). Many factors are involved in producing the potential measurements at the surface, of which electrode spacing is only one. Therefore this method should be used only where some wells are present for control.

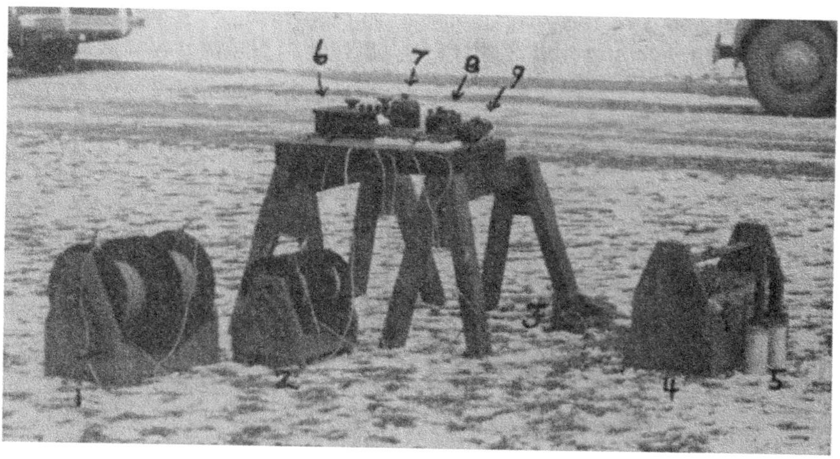

FIG. 3. The measurement apparatus

Explanations of numbers: 1, reels for current leads; 2, reels for potential leads; 3, battery foot switch; 4, battery; 5, spare nonpolarizing electrodes; 6, potentiometer; 7, standard cell; 8, potentiometer galvanometer; 9, milliammeter

A picture of the equipment used is shown in Figure 3. Nos. 1 and 2 on it are the reels for the current and potential leads, respectively. No. 3 is a battery foot switch which enables the operator to have his hands free for adjusting the apparatus on the board. No. 4 is the battery, and spare nonpolarizing electrodes are shown by No. 5. Mounted on a board set on trestles are the potentiometer (6), standard cell (7), potentiometer galvanometer (8), and milliammeter (9).

INTERPRETATION

Figures 4–7 show the resistivity graphs obtained in the Plymouth area plotted beside the logs of wells at or near the station occupied. In general, changes in the lithology of the buried sediments produce

[3] *Op. cit.*, p. 317.

breaks in the resistivity curves at the calculated depth *a*. Inconsistencies between the resistivity curves and the well logs are due to two causes: (1) Although the aquifers (water-bearing sands and gravels) have lower resistivities than the finer non-water-bearing sediments, a saturated clay may have even lower resistivity; and (2) The use of distance *a* as the depth figure is at best only an approximation.

Figure 4 gives the results of readings taken at the intersection of Warren and Haggerty roads in Canton Township. A well drilled at this point encountered 83 feet of clay and below it water-bearing

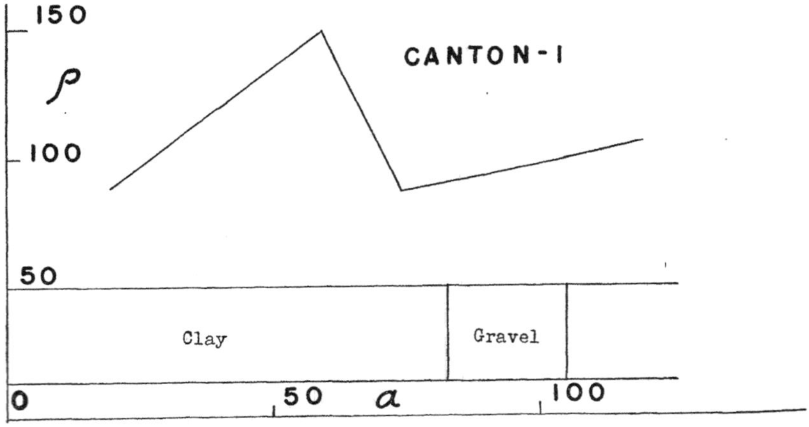

FIG. 4. Resistivity readings at Canton No. 1 plotted against log of test well

gravel down to 105 feet. A marked break in the resistivity curve took place at electrode spacings between 60 and 80 feet. In this well the estimated depths to the top and bottom of the aquifer, based on electrode spacings, are about 20 feet less than actual depths. Obviously, in surveying outward from this well a calibration factor should be applied in calculating depths.

The curve obtained at Canton No. 3 station, shown by the dashed line in Figure 5, indicates increasing resistivity with depth, but with a lessening of the increase at electrode spacings between 60 and 75 feet. The main gravel bed (presumably the principal aquifer) lies between depths of 69 and 88 feet in a well drilled at this point. Again the calculated depth falls short of the real depth.

The resistivity readings obtained at the Canton No. 2 station, 500 feet distant from station No. 3, produce a more complicated curve. The resistivity shows a net increase downward, but levels off at two places, with a zone of lower resistivity between. This station was at an elevation 11 feet higher than Canton No. 3. The difference in elevation practically cancels the calibration of the depth distance suggested by the curve at station No. 3. It would seem that here the upper gravel (as well as the lower part of the overlying clay) is water-bearing, which causes the lower resistivity obtained with an electrode spacing of 60 feet. The lower gravel bed brought about the second leveling off of the curve. The spacing between this

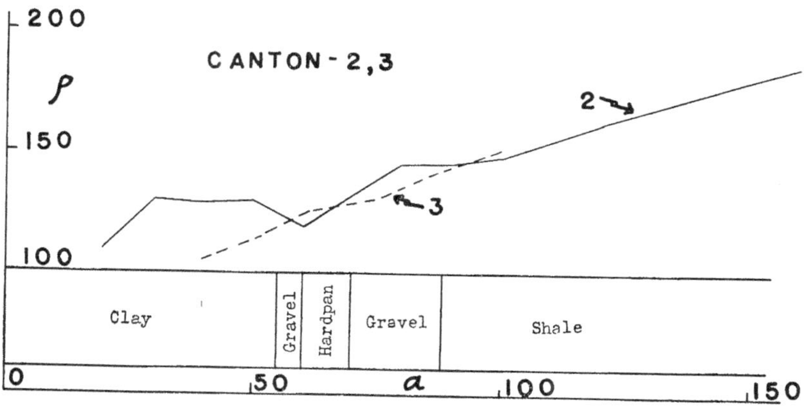

FIG. 5. Resistivity readings at Canton Nos. 2 and 3 plotted against log of well drilled at Canton No. 3

part of the curve and the higher zone of low resistivity suggests that the second gravel lies at a lower level than at Canton No. 3, probably owing to thickening of the "hardpan" layer in the intervening 500 feet.

The curve obtained at Plymouth station No. 1 coincides remarkably well with the log of the well drilled at that point (Fig. 6). At an electrode spacing of 150 feet lower resistivities were encountered. This distance coincides in the log with the top of a fine water sand. The same break was found at station No. 2, 650 feet away, and also at station No. 3, 800 feet distant. Plymouth No. 2 is southeast, and Plymouth No. 3, north, of Plymouth No. 1. However, the downward break of the curve lies at an estimated depth of 120 feet

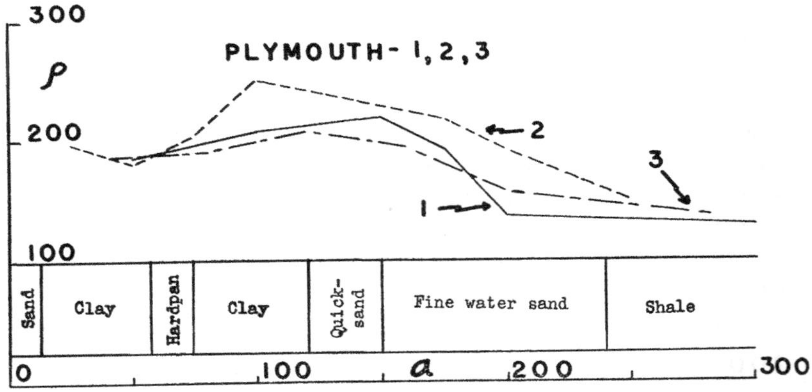

FIG. 6. Resistivity readings at Plymouth Nos. 1–3 plotted against log of test well drilled at Plymouth No. 1

(instead of 150) at Plymouth No. 3, and at 97 feet at Plymouth No. 2. Several interpretations of this are possible. One is that the top of the fine water sand rises to both the north and the south. Another is that the overlying quicksand and clay are saturated with water, in the lower parts at least, at Plymouth Nos. 2 and 3.

The most marked differences in resistivity were obtained at Plymouth No. 7 (Fig. 7). This curve breaks abruptly from exception-

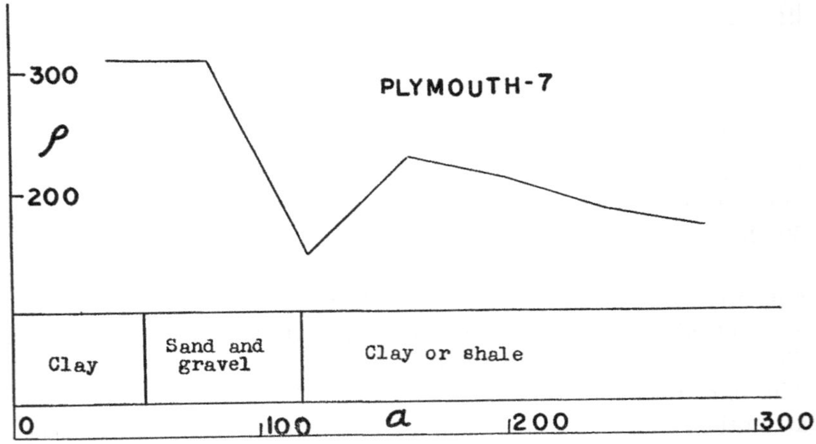

FIG. 7. Resistivity at Plymouth No. 7 plotted against log of well drilled 600 feet to south at Plymouth booster station

ally high resistivity to very low resistivity and then climbs part way back. Evidently a distinct lithologic break occurs underneath the surface in this neighborhood. The very low resistivities encountered suggest the existence of a good aquifer. Subsequent to the taking of these readings a well was drilled about 600 feet distant on the south side of Middle Rouge River on the grounds of the booster station of the Plymouth Water Works. The elevation at the well is approximately the same as that at Plymouth No. 7. The first fifty-one feet of rock was mostly clay. From there to a depth of 116 feet the sediment consists of sand and gravel, much of it coarse, which contains a large supply of water. The aquifer showed such promise that a well has been dug at the site. The capacity of this well is sufficient to make up the present deficiency in the Plymouth water supply.

CONCLUSIONS

The writers believe that electrical-resistivity methods should be used in conjunction with the drilling of test wells in the search for ground-water supplies in the glacial drift of Michigan and elsewhere. It should be remembered at all times that this method is capable only of detecting changes in the resistivity in the buried sediments. It does not find water as such, but it can find low-resistivity layers which will bear checking with the drill. It is also possible to obtain an approximation of the depth of these rock layers of low resistivity.

The cost of a resistivity survey is relatively small. All essential apparatus can be assembled for about $250. The only additional equipment needed is a light pickup truck, but a skilled man is required to read the instruments and interpret the results. Two unskilled assistants complete the party. One such party can run from five to eight lines during the course of a day.

By the use of the resistivity method both time and money can be saved in appreciable amounts. The drill is still necessary, but in an average project the number of holes which it would be necessary to drill could be cut in half. The total cost of an average resistivity survey would not equal the cost of one drilled well.

UNIVERSITY OF MICHIGAN

RESISTIVITY DATA

Plymouth 1

a *feet*	*ohm/feet*
50	186
100	207
150	219
175	193
200	137
250	133
300	128

Plymouth 2

a *feet*	*ohm/feet*
25	199
50	182
75	204
100	250
150	227
175	219
200	193
250	151

Plymouth 3

a *feet*	*ohm/feet*
40	189
80	193
120	206
160	194
200	157
240	147
280	137

Plymouth 4

a *feet*	*ohm/feet*
40	136
80	166
120	111
160	49
200	69
240	66

Plymouth 5

a *feet*	*ohm/feet*
40	248
80	71
120	73
160	60
200	43
240	43
280	37

Plymouth 6

a *feet*	*ohm/feet*
40	111
80	121
120	107
160	87
200	47
240	81
280	89

Plymouth 7

a *feet*	*ohm/feet*
40	311
80	310
120	148
160	227
200	210
240	187
280	168

Plymouth 8

a *feet*	*ohm/feet*
40	87
80	55
120	46
160	49

Canton 1

a *feet*	*ohm/feet*
20	88
40	118
60	149
75	84
90	93
105	99
120	106

Canton 2

a *feet*	*ohm/feet*
20	159
30	179
40	177
50	179
60	166
70	181
80	194
90	195
100	198
120	212
160	236

Canton 3

a *feet*	*ohm/feet*
40	155
52	165
64	176
76	181
88	192
100	200

ANTHROPOLOGY

THE IROQUOIS IN AMERICAN PREHISTORY *

JAMES B. GRIFFIN

THERE are a number of approaches which can be used in determining the origin and former homeland of a given group of people. The most nearly correct reconstruction will be the one which does least violence to the data known from all of these approaches. All factors must be considered; by constant examination and testing of the various hypotheses which are put forward the less tenable ones will be eliminated and gradually a sound foundation will remain upon which to erect the most reasonable theory of the history of a people and of their culture.

The mythology, folklore, and traditions of a group such as the Iroquois may offer valuable leads, and this approach has been employed by various students of their history. These data, however, should be subjected to the most rigorous examination in order to ascertain the plausibility of origin accounts or migration legends in the light of other known data. Parker, for example, made use of material of this nature in his reconstruction of the origin of the Iroquois.[1] If there are significant references in the unwritten literature of a group to a habitat and an economy differing from those in their historical environment, clues are furnished pointing to areas where such an environment and economy are found.

A study of the ethnology of the cultural group in question will indicate that portion of their workaday life which was well adapted to their habitat and that part which offered but a poor adjustment to the necessity of obtaining a livelihood in their home area. If the tools and the everyday pursuits of a people differ fundamentally from those of neighboring groups, we are justified in seeking a reason. If it can be shown that the economy of a group like that of the Iro-

* The bulk of this paper was read in May, 1941, at the annual meeting of the New York Archaeological Society in Rochester, New York.

[1] Parker, 1916, pp. 482–483. Even if the "tree with the long sword-like leaves" is the "palm tree," it would not suggest a former Iroquoian homeland in the Middle Mississippi area but would point toward Florida.

quois differed radically from that of their neighbors and more closely resembled the type of life practiced in another area, we are warranted in suspecting a genetic connection between the two related methods of making a living. Outstanding features of material culture such as the difference between the utility of the elm-bark canoe of the Iroquois as contrasted with the birch-bark canoe, or the presence of the blowgun among Iroquois, require an explanation in the light of distribution and probable diffusion. On the other hand, how much of the material culture of the Iroquois was shared by the non-Iroquoian tribes around them? Speaking roughly, the degree of cultural homogeneity would reflect the length of contact between two divergent linguistic groups.

In the same way that one can examine the material culture of a group, so can one study the social organization and religion, for example, and determine the degree of resemblance of the Iroquois to the neighboring Algonquian groups, in the social patterns that each tribe followed in its way of life. If there are significant differences, we are justified in seeking reasons for them.

Another method which can be utilized in studying the prehistory of a given group and its relationship to its neighbors is that of linguistics. If a group under consideration speaks a language which is markedly distinct from that of contiguous tribes, we at once look for an area where more closely related dialects are spoken and we seek to determine the degree of likeness of the geographically separated but linguistically related tongues in an effort to obtain some idea of the time in which such a divergence may have been produced. Another linguistic method analyzes the vocabulary of a given group in order to ascertain those words which indicate borrowing from surrounding tribes and, conversely, to identify the foreign words introduced to the native groups by the most recent migrants into the area. Not only may new techniques, tools, or ideas diffuse themselves but the words applied to such instruments or concepts will sometimes be adopted and reveal thereby the parent group from which the new concept was derived.

Still another approach to the understanding of the historical position of a group of people is by the study of their physical type. If they are essentially the same as their neighbors and the same as the physical remains of the precedent archaeological cultures in the area, there is no need to look elsewhere for their origin, although this does not deny that they could have come into the area from a region where the same type is also found. Minor variations from one time period to another may sometimes be explained by a knowledge of the cultural history of the group. Wholesale amalgamations and intermarriage are likely to cause definite shifts in type, especially if these practices have been of long duration.

Historical documents which have been carefully checked for their accuracy and which are substantiated by internal evidence or by complementary data offer most valuable leads toward the discernment of tribal movements during the earliest contact periods, and these in turn may provide clues to the direction of movement of the group during prehistoric times.

The primary historical method to be employed in a reconstruction of the past life of an ethnological unit is that of archaeology. Such a study must begin with the accurate identification of the artifacts and village organization of known tribal units at specific sites. When this has been done comparative studies may be made of contemporaneous and linguistically related or unrelated groups in an effort to determine the degree of homogeneity of the material remains. If at an early historic period the group under consideration possesses an archaeological culture which is completely at variance with that of the surrounding groups in the area, we shall need explanations for these differences. If, on the other hand, there is a relatively small number of divergent elements, our task is not so difficult.

After the culture content of our test group has been ascertained for the earliest historic period, the next step is to search for sites with material of similar type but belonging to the period immediately preceding white contact in the area. Once the artifacts from such sites are studied, it is possible to correlate the early historic with the late prehistoric and to identify the new types of artifacts which were being introduced and something of the direction of the cultural drift which was interrupted by European contact. Such a careful comparative method will ultimately enable archaeologists to reconstruct the prehistoric culture of the ancestors of the group under consideration. This method must be used with caution, however, because tracing the development of a culture element or of a group of elements is not the same as tracing the genetic history of the people who had those traits at any particular moment.

So far as I am aware, Iroquois folklore and tradition have not been exhaustively studied to determine the likeness and the dissimilarity between the different Iroquoian tribal groups, and the resemblance of the Iroquois folklore and mythology to that of their neighbors. The apparent lack of any tradition among the various Iroquois units which explains their presence in the northeast as the result of group movement into the area is certainly significant with regard to the time of arrival of this linguistic stock. As Fenton has pointed out, several elements of Iroquois shamanistic society bespeak intimate acquaintance with their historic habitat, whereas that portion of their religious life which deals with the maize cycle is southern.[2]

The semisedentary villages of the Iroquois based on a maize economy certainly present a subsistence picture different from that of their Algonquian neighbors on the west, the north, and the northeast. The Iroquois agricultural–hunting economy included relatively permanent large villages, with a social structure highlighted by the matrilineal household unit within the clan and by the possession of two moieties. The Iroquois tribal organization was of a high order, and the conceptual framework of the "League" was a still further advance. Much of this sociopolitical organization is at marked variance with the Algonquian tribes to the north and west and, to a somewhat less degree, with the bordering Algonquian groups to the east and south.[3] The southern New England groups had fairly permanent villages with elaborate stockades and rectangular houses, as well as wigwams.[4] The Iroquois village life and house structures also resemble those of the tribes in the Middle Atlantic states.[5]

The recent picture presented by Butler of the aboriginal village pattern in southwestern Pennsylvania at approximately the 1550–1650 period[6] and that of the Fort Ancient group in southern Ohio are also similar.[7] The type of food production among all these groups was very much the same. The presence, however, of the Iroquois as the northernmost dominantly agricultural groups brings up an interesting point concerning the time period at which either the Iroquois brought the corn to the north or the corn was brought to the Iroquois, who successfully cultivated it. According to some of our best authorities, it must have taken several centuries to produce a corn strain which would mature sufficiently rapidly to make possible its

[2] Fenton, 1940, p. 165.
[3] Speck, 1920, p. 66.
[4] Willoughby, 1935, pp. 228–301.
[5] Morgan, 1881, pp. 67, 115–119.
[6] Butler, 1939.
[7] Griffin, 1943.

cultivation as far north as Lake Simcoe and Quebec.[8] If, then, the Huron and related Iroquois were growing corn in the sixteenth century near the border of the Laurentian Upland, a reasonably long period must have elapsed during its adaptation to the shorter growing season. We may also assume that the Iroquoian tribes had known the use of corn for some time, owing to its importance as a basic food supply and to its intimate position in their material culture and their social and religious life.

In a recent analytical study of the Coastal Algonquian groups from northern New England to Virginia Dr. Regina Flannery points out that the tribes of the Middle Atlantic area particularly have a distinctive number of traits which can be attributed to the southeast,[9] and that, though some of them are shared with the Iroquois, there is no reason to believe they were derived from the Iroquois. On the other hand, there are definite traits among the Coastal Algonquian dealing with political organization and warfare which may properly be attributed to Iroquois influence. She notes that this trait-trading was by no means a one-way transaction. Her data present a picture somewhat different from the one now current, since it does not indicate that the Iroquois were as distinctive culturally in the northeast as they are thought to have been. Dr. Flannery's paper was written, however, with particular emphasis on the Algonquian, and a sound comparative historical treatment of Iroquoian ethnohistory would be a distinct aid in an understanding of northeastern Indian cultures.

The political and martial strength of the Iroquois, together with their fortunate but accidental position as a buffer group and a trading front for the Dutch and English, enabled them to maintain their solidarity while other aboriginal tribes were being broken and having their society destroyed by settlers, soldiers, and epidemics. For a time they were a favored nation of British diplomacy, and they occupied this relatively desirable position until the Revolutionary War, when they bore the brunt of such forays into central New York as that of General Sullivan. Their tribes and their culture thus received more sympathetic treatment at the hands of contemporary English observers, and by the time the people of the new American nation had ceased hostilities there were a sufficient number of Iroquois

[8] Jones, Volney H., personal communication.
[9] Flannery, 1939.

available in the United States and Canada to preserve a considerable number of the concepts and part of the content of their life. Some of our outstanding students of American anthropology have devoted much of their time to Iroquoian research,[10] and their work has furnished us with valuable material; but at the same time it emphasizes too much, in my opinion, the superiority of the Iroquois over some of the other groups in the northeast and even over the tribes in the Ohio and upper Mississippi valleys. In a European's eyes, the character of the Indian tribes during the eighteenth century depended on whose side they were fighting. The Iroquois are none too well thought of historically in the Middle West.[11]

I have already mentioned the political and martial strength of the Iroquois, but this seems historically to have been limited to the Five Nations at a particular moment of time and was strongly aided by international developments. On the other hand, there can be little doubt of the uniqueness in the northeast of the Iroquois social structure which was held more or less in common among all the northern Iroquois tribes. This fact alone, without the supporting historical evidence, would suggest a greater age for the social organization as contrasted with the political structure of the Five Nations.

So far as I know, no thorough study of the various Iroquois dialects has been produced with the view of reconstructing a primitive Iroquois tongue which could be compared with Cherokee to arrive at a proto-Cherokee-Iroquois language. It is sufficient to recall here that the Iroquois in the New York–Canadian area had differentiated into a number of distinct dialects, and that the Cherokee in the south had diverged into three. This would not only argue for a reasonably long period for such divergences in the north, but would also ask for a much longer time for the separation of the northern Iroquois from the Cherokee. As for the linguistic relationship between Iroquois and Caddoan, it can only be said here that it is so remote as to be of little or no significance in a reconstruction of the relatively recent migrations and cultural connections of the Iroquoian peoples.[12] It is quite evident, however, that the Iroquois language has no genetic connection with the Algonquian language, and the position of this Iroquois island in a sea of Algonquian requires a historical explanation.

[10] Fenton, 1940, pp. 160–164. [11] Kellogg, 1925.
[12] Griffin, 1943, p. 228.

The physical anthropologists have yet to propose a consistent and well-rounded interpretation of the significance of the physical types in the northeast with regard to known chronological and cultural differences. A beginning in this direction has been made by Georg Neumann of Indiana University, who has adopted the nonlinguistic and noncultural term "Sylvid" to apply to the physical type Hrdlička called "Algonkin" and to Dixon's "Proto-Negroid" and "Proto-Australoid." In New York and in New England there is a definite tendency among the long-headed Sylvid groups, which includes both Iroquois and some Algonquians, toward a lower cranial vault, a lower face, and a broader nose. The Hurons and the Younge site people in eastern Michigan have this same physical type, which has been called the "Younge type." The Sylvid subtype in the northern Mississippi Valley area, called the "Gooden type," has a slightly shorter cranial length. The vault is much higher, the face is longer, and it has a narrow nose. The third subtype of the Sylvid race is the shell-midden people, called the "Indian Knoll type," whose vault proportions have a tendency to approximate those of the western Sylvids, but whose facial proportions more closely approach those of the eastern group. These people differ from both of the other subtypes since they have smaller absolute dimensions.

The evidence for the physical type of the Lamoka Focus, on the basis of published data, is by no means clear. It may be rather closely related to the western Sylvids, such as the Delaware and the people from the Maine shell heaps, and does not offer a progenitor for the eastern Sylvids, to whom the Iroquois and many New England Algonquians belong. On the basis of published evidence, the Vine Valley and Laurentian sites apparently have a continuance of the earlier population, with normal variation or group intrusion represented by a round-headed element which also appears as far north as Newfoundland.[13] In other words, the Iroquois, as well as a strong element of New England Indians, represent a physical type somewhat different from that of their known predecessors in the northeast, and there is little evidence available for wider comparisons. There is also no physical group in the southeast which compares closely with the eastern Sylvid subtype to which the Iroquois belong. That such a group may turn up is, of course, not beyond the realm of possibility.

It is highly important to realize that the Iroquois cannot be

[13] Ritchie, 1943, p. 314.

derived physically from the Indian stock which is most closely connected with the development and spread of the Middle Mississippi Phase. If the Iroquois were in the southeast or the Mississippi Valley, they must have left before the Middle Mississippi period began. In the Fort Ancient Aspect there is a very small proportion of the eastern Sylvid type, but the dominant types belong with the western Sylvids and the Centralids. The latter are usually associated with the Middle Mississippi cultures. Thus on the basis of physical type alone we can hardly look to Fort Ancient to furnish the ancestors of the Iroquois.

We are truly fortunate in having a large body of historical data dealing with the location of the Iroquois in the early historic period.[14] Our earliest records point to the presence of Iroquois parties in 1534 at the mouth of the St. Lawrence, and slightly subsequent ones indicate their strong position around Montreal and Quebec. Our next accurate information is at the beginning of the seventeenth century, when we find that, though the Iroquois in the lower and middle St. Lawrence Valley have evacuated the area and presumably moved south, the Huron, Tobacco, and Neutral nations were occupying their historic seats and were populous and prosperous groups. The Hurons, indeed, referred to their former villages dating back to about 1400.[15] By 1600 the Susquehannock were firmly established in the Susquehanna Valley in Pennsylvania and the Cherokee were, of course, already differentiated in the southern Appalachians. The primary fact ascertainable from this early period is the extent of the Iroquoian occupation, so that a reasonable length of time must be allowed for it to have taken place. The stories related by Perrot [16] telling of the Iroquois having been driven by the Algonquian from their middle St. Lawrence Valley home into New York, which were copied by Colden [17] without giving credit, are substantiated by archaeological findings, as are the historic documents relating to the presence of Iroquois at the mouth of the St. Lawrence. In this regard history, legend, and archaeology coincide.

For many years archaeological excavations in the northeast have indicated that the Iroquois possessed certain distinctive material culture traits which served to differentiate them not only from their contemporary Algonquian neighbors but also from a number of pre-

[14] Fenton, 1940.
[15] Kinietz, 1940, p. 59.
[16] Blair, 1911, I: 42–47.
[17] Colden, 1904, pp. 4–6.

historic culture groups which are found in the same area. In an effort to explain the known data on Iroquois culture, particularly those with reference to archaeology, Parker[18] formulated a hypothesis, published in 1916, suggesting a route by which the Iroquois might have migrated into the northeast. In this paper he took issue with a prevalent concept of a northern origin of all Iroquoian tribes. He advocated the theory that the former home of all the Iroquoian-speaking peoples was in the middle Mississippi Valley, where they were in contact with the Caddoan and Siouan peoples. The Cherokee were the first to start east up the Ohio, where they found the "Mound Builders." These native Ohioans, said Parker, were eventually conquered, and some of their objects were ceremonially destroyed. In this he was following Mills' attempt to explain certain Hopewellian mutilated objects and ceremonial caches which contained intentionally destroyed artifacts.[19] Here Parker evidently meant to call the Fort Ancient people of Mills prehistoric Cherokee. After the assumed Cherokee movement the northern Iroquois arrived in the middle Ohio Valley from its mouth, and a conflict arose between the Cherokee and the proto-Iroquois. The two linguistic relatives then diverged, with the Cherokee going south and the proto-Iroquois going north, where they split into two groups. One of these traveled north around Lake Erie and became the Huron-Mohawk; those who moved east along the south shore of Lake Erie became the Seneca-Erie and Susquehannock. Parker published substantially this same argument in "The Archaeological History of New York."[20]

In 1926 Parker modified his migration hypothesis to take into account additional information which seemed to bear on the problem of the archaeological connections of the Iroquois.[21] In this statement he moved some of the Iroquois up the Mississippi and thence overland toward Lake Erie. In *The Fort Ancient Aspect* I have discussed this hypothesis at some length and suggested that the new interpretation given Parker's data calls for a different formulation.[22]

It must be emphasized that Parker's hypotheses were the result of an endeavor to explain the known data in terms of a consistent and unified migration story and as such have offered a firm foundation upon which to conduct further research. It is the only intelligible

[18] Parker, 1916, pp. 483–484.
[19] Griffin, 1943, pp. 220–221.
[20] Parker, 1922.
[21] Parker, 1926, pp. 17–19.
[22] Griffin, 1943, pp. 223–239.

attempt to offer an explanation for the anomalous position of the Iroquois in the northeast, although others have had the same feeling for a southern origin of the Iroquois.[23]

Leaving the theoretical discussion for a moment, we can examine the archaeological materials from the Iroquoian sites and areas to determine the homogeneity of Iroquoian artifacts, first within the confines of known tribal sites, then between closely related tribes, and finally between historic tribal sites and those which go back into the protohistoric and prehistoric periods. When this has been done — and only when it has been done — can we speak with any assurance of the archaeological groupings into which the various Iroquois tribal units will fall. It will, indeed, be somewhat surprising if Ritchie's present alignment of one tribe per focus stands the test of comparative analysis.[24] Such an analysis needs to be made before we can accurately compare Iroquoian units, and Iroquoian archaeological culture as a whole, with contemporary contiguous nations or earlier cultures in the New York–Ontario and surrounding areas. In the discussion on the relationship of Fort Ancient to Iroquois, which I have mentioned earlier, I made a rough beginning in this direction and attempted to specify some of the resemblances and differences between those two distinctive cultural groups.

Parker, Wintemberg, Skinner, and Harrington have all recognized an archaeological cleavage between the Mohawk-Oneida-Onondaga on the east and north and the Seneca-Erie on the west. This separation, although relatively insignificant when one is viewing northeastern archaeology as a whole, is very important in studying the interrelationships of the Iroquoian units and the connection between those units and surrounding groups. It will be possible to characterize clearly the different focal groupings of the Iroquois and isolate those traits which the Iroquois exclusively possessed. It is my impression that this list of so-called diagnostic traits will be relatively small. In my own opinion such traits, though interesting and valuable, are in some ways rather a side issue in archaeological research, for our proper aim is the recovery of the total complex of the group to be studied, and this total complex must be compared with other artifact assemblages equally complete in order to determine the true connection between groups. The diagnostic trait of today turns out to be the link trait of tomorrow.

[23] Boas, 1910; Speck, 1920, p. 66, n. 23. [24] Ritchie, 1936, p. 4.

A comparison of the available archaeological data indicates that there are indeed many similarities between Fort Ancient and the Iroquois, but there are also important differences which make it impossible to confuse a representative assemblage from sites of both cultures.[25] This is true of all types of the artifacts recovered. Many of the resemblances are of such a general nature that they are of little use in separating Fort Ancient and Iroquois from contemporary contiguous groups, and there are very few, if any, traits which can be said to have been possessed by Fort Ancient and Iroquois alone. Wherever specific Fort Ancient and Mississippi–Ohio Valley traits are found in the Iroquois country they are almost invariably from protohistoric or historic sites. In a number of elements there is as much connection between the Iroquois and their Algonquian neighbors as there is between Fort Ancient and Iroquois. All the archaeological evidence as it is interpreted at present indicates that Fort Ancient was contemporaneous with the Iroquois occupation of between, say, 1500 and 1700. There is a strong probability that the Fort Ancient culture was interrupted by those forces which sent the Iroquois raiding into the Ohio Valley and which concentrated the tribes of that area around either the French trading posts in the Illinois country or the English posts across the Appalachians.

We know from the archaeological evidence that Iroquoian type material is restricted to the area in which the historic Iroquois groups were located, and there is thus no area of which we can say: "Here is the prehistoric material that represents the culture that developed into Iroquois." Those archaeological assemblages formerly interpreted as offering a possible prehistoric Iroquois culture are now believed to be contemporaneous with the Iroquois occupation of the lower Great Lakes.

Many of the Iroquois traits were derived from the southeast,[26] but they did not come *en masse*. A number of them were obtained after the contact period. This does not apply to corn, which is certainly as old as the Hopewellian period and perhaps goes back into the Adena–Vine Valley period. Are there no indications in New York in the artifacts and the pottery of the late Laurentian and Vine Valley cultures that would point toward the traits found in

[25] Griffin, *op. cit.*, 1943.

[26] W. R. Wedel is the authority for the statement that the Iroquois style pottery, mentioned by Holmes (1903, p. 105) as from southwestern Alabama, is actually from New York State.

Iroquoian? Must we look entirely in the southern Ontario area for the developmental phases of Iroquois? Is there any suggestion in the physical type associated with the New York cultures of the middle period which would help in the recognition of early Iroquois? Is the Owasco material earlier than or contemporary with Iroquois? If it is earlier, we should find definite examples of stratigraphy; and if contemporary, we should find definite examples of trade objects and not merely speak of "influences" from one group to the other when they must have been in close geographical proximity.

Let us now examine some of the material and sites which furnish suggestions of Iroquoian origins.

At the Reeve site the dominant surface finish is cord-marked.[27] Some of this continues up to the lip, but the area where decoration was placed on the outer rim was usually smoothed. The most common tempering material is crushed rock, but a minor percentage is crushed shell. This site has a minimum of the Woodland elements in pottery and other artifacts to be found in the Whittlesey Focus. The same thing is true of the Fairport Harbor site and, as indicated by Morgan and Ellis, these two sites have a great deal in common.[28] In them the Woodland tradition is most strongly represented in the pottery of the Fairport Harbor site, even though trade material was found there and was absent at Reeve.

The Tuttle Hill and South Park sites have a high degree of cohesion within the Whittlesey Focus which sets them apart from Reeve and Fairport Harbor.[29] This difference in culture content is entirely in the direction of Woodland, with a significant use of dentate-stamp decoration on the rim. The same situation holds true for the Taylor site in the same area; in a forthcoming report on the pottery from it a more detailed statement will be made regarding these ceramic affiliations.[30] It is suggested that there is discernible in the northern Ohio area a gradual transition from a Woodland pottery complex into an Iroquoian complex and that the transition took place near the close of the Hopewellian chronological period.

In the eastern Mohawk-Onondaga sites the pottery still carries the Woodland traits of check-stamp surface finish and dentate-stamp

[27] Greenman, 1935, p. 25.
[28] Morgan and Ellis, 1943, p. 52.
[29] *Ibid.*, p. 51.
[30] Excavation by E. F. Greenman for the Ohio State Museum. This pottery is being studied in the Ceramic Repository at the University of Michigan.

rim decoration. This is particularly significant, for the center of check-stamp concentration is in the Atlantic drainage of the southeast and the adjacent Gulf coast area. It is also an early surface finish in the area and is associated with the early Woodland cultures of the southeast.

Though Wintemberg noted the resemblance of the culture of the Uren site in southern Ontario to Woodland, he inclined to the view that it was attributable to contact with neighboring Algonquian groups. He considered that the Uren site represented an early prehistoric Neutral group,[31] in which he was probably correct. It is also possible to regard the Uren pottery as a transitional stage between Woodland and Iroquoian. The shape, temper, overall surface treatment by brushing, cord-wrapped paddle, and check stamp, and many elements of the decoration are Woodland in type. The decoration is confined to the rim area, and cord-wrapped stick impressions are found on both the inner and the outer rim surfaces. Linear-punctate lines (called "interrupted lines" by Wintemberg [32]), cord-impressed designs, and rim bosses are other significant Woodland features. On the other hand, a strong proportion of the Uren pottery suggests Iroquoian treatment, and though some of the results resemble patterns occurring in the Whittlesey Focus,[33] others are identical with a grit-tempered pottery type from southwestern New York.

Another site of the focus to which Uren belongs was excavated in 1938 by Philleo Nash. In a brief preliminary report he has emphasized some of the Woodland features.[34]

The dominant pottery type at Uren is almost duplicated at a site in southwestern New York near the Pennsylvania line which was excavated by Ross Pier Wright of Erie.[35] This Westfield site also had a pottery type derived from the Keyser Cord-marked type of the Monongahela Woodland division.[36] It is a shell-tempered ware whose basic jar shape is not essentially different from the Uren and can be regarded as having been derived from Woodland jars. The Westfield shell-tempered pottery is predominantly cord-marked on the body and, when decorated, has an incised rectilinear pattern on

[31] Wintemberg, 1928, p. 50.
[32] *Op. cit.*, p. 20.
[33] Greenman, 1937, p. 350.
[34] Nash, 1939, p. 74–75.
[35] Personal examination of the collection at Mr. Wright's home in Erie and of a large collection of sherds presented by him to the Ceramic Repository for the Eastern United States.
[36] Manson, MacCord, and Griffin, pp. 375–418 in this volume.

the outer rim. The style of decoration is not only related to Iroquoian but is also connected with the Feurt Incised type of the Fort Ancient Aspect.[37] Though there is some evidence of contemporaneity of these two types at Westfield, I believe that the Uren-like type was in use before the shell-tempered ware appeared and that it was not made during the later years of the occupation of the Westfield site. Neither of the Westfield pottery types is typical of the Whittlesey Focus to the west or of the pottery of Ritchie's suggested Seneca Focus.[38]

Quimby has briefly discussed the appearance of Hopewellian pottery in the Saginaw Bay area of eastern Michigan.[39] It is of the Ogden-Fettie-Goodall style of dentate stamping, which is not characteristic of the Ohio Hopewellian pottery. The dentate-stamped designs on the exterior of Iroquoian pottery rims is closely related to this western Hopewell style. Still another possible connection between Iroquoian pottery and that of the earlier Hopewellian horizon can be seen in the small specialized, highly ornamented jars sometimes called "typical Hopewell vessels." The basic shape of many of these jars [40] is not only related to a widespread Woodland cord-marked shape, but is also very much the same as the basic Iroquois shape.[41] One of the essential features of the better Hopewell pottery is the rim area, which is marked by a short vertical upper rim rising from the flaring lower rim. This concept is identical with that on Iroquois vessels. On the rim in both Hopewell and Iroquois there are incised designs. On the former vessels the incisions are almost always fine lines, closely spaced and usually crosshatched. The Iroquois incisions are generally medium-wide, and though they are predominantly oblique, they are rarely crosshatched. The incised Hopewell rim is almost always accompanied by a horizontal row of hemiconical punctates at its lower margin. These punctates are usually oriented horizontally. Many Iroquois rims also have a lower horizontal border of punctates or gashes, but these impressions are usually vertically placed. The most difficult point in this suggested transition or influence is the almost complete absence of body decoration on Iroquois vessels, but the suggested Hopewell prototype normally has the major decoration on that area.

[37] Griffin, 1943, pp. 78–80. [38] Ritchie, 1936, p. 4.

[39] Quimby, 1941, pp. 132–135.

[40] For examples of these shapes see Quimby, 1941; Griffin and Morgan, 1941.

[41] Parker, 1907 and 1922; Bailey, 1938.

This ceramic development is analogous to that which apparently took place in the middle Missouri Valley in the shift from Woodland to Upper Republican. I have elsewhere suggested that the developed rim of Upper Republican was derived from concepts present in the eastern Plains area during the Hopewell-Woodland period.[42] Thus the similarities noted between Upper Republican and Iroquoian may be explained on the basis of both having developed in their respective areas from similar influences which preceded them. This alternative is offered to the explanation that those resemblances are the result of a migration of Caddoan and Iroquoian groups from an unnamed cultural unit somewhere in the middle Mississippi Valley.[43]

Some sixty years ago L. H. Morgan attempted to explain a number of the earthworks of the Hopewell sites in Ohio as substructures for long communal dwellings of the general type prevalent among the Iroquois and eastern Algonquian tribes.[44] He observed that the height and shape of the geometric earthwork enclosures, together with the height of the dwelling, would make of it an admirable defensive structure. He further argued that this defensive principle was common to all house types associated with the type of economy and social pattern that must have been in existence among the "Mound Builders." Some of the possible objections to these theories of Morgan have already been given by Fowke.[45] The latter's sensible suggestions regarding the systematic excavation of these earthworks have never been adequately carried out. The closest approach was the partial excavation of the Turner site, which was completed after Morgan's paper was published and which indicated that portions of the earthworks may have covered dwellings erected upon the natural surface of the ground.[46]

In any event the contention that the type of economy and the social pattern implied by these earthworks were analogous to those among the Iroquois is accepted by all students of the eastern United States aborigines. In spite of the fact that American anthropologists in general have been labeled as belonging to a "historical school," very little attempt has been made in the past to provide a historical (archaeological) explanation of the development of the contemporary Indian cultures.

[42] Griffin, item 12 in Bibliography.
[43] Parker, 1922.
[44] Morgan, 1881, pp. 199–221.
[45] Fowke, 1902, pp. 154–155.
[46] Willoughby and Hooton, 1922, p. 10.

The archaeological stage or cultural period which apparently immediately preceded the development of Iroquois material culture was that which has been known for some time as the Hopewellian Phase. This rather general cultural aggregate was widespread in the Mississippi Valley from the Gulf of Mexico to Minnesota and from the Plains to the Appalachians.[47] It is unlikely that a single tribal group or even a single linguistic stock is responsible for this development and diffusion of the Hopewellian culture complex. Factors contributing to this opinion are the variation in physical type in different areas, the distinctive local foci and even larger groupings which are recognized, and the small size and character of the most distinctive Hopewell artifacts, which make them easy to transport by a few individuals. Indeed, it is probable that the majority of the linguistic stocks in the eastern United States at the period of colonization had ancestors in the area who had participated in the Hopewellian culture.

If the progenitors of the Iroquois did not partake of the Hopewellian complex, they must have been either along the eastern seaboard or already in their historic habitat, where they would have been a northern marginal group. This would help to explain the postulated Hopewellian survivals in some of the Iroquoian material culture traits. These survivals might also be explained as carry-overs from the period when the Iroquois tribes themselves were Hopewellian. The social and economic organization of the Iroquois and the large amount of territory they occupied suggest, as Parker and others have recognized, that the Iroquois probably moved into their historic habitat from the south. The primary error of the earlier hypotheses has been the failure to correlate the time at which this movement might have occurred with the archaeological culture horizon which was in existence at the time the movement could have occurred.

The archaeological program for the future calls for an intensive search for sites in the eastern Great Lakes and St. Lawrence areas which give evidence of earlier and still earlier Iroquois habitations. With a few happy exceptions sites of this type have not been excavated. If such investigations continue to reveal a culture complex which is basically Woodland with additional Hopewellian characteristics, we can then reëxamine with more confidence the various

[47] Griffin, item 12 in Bibliography.

Hopewellian centers in an effort to determine with some exactitude whether any of them provides a possible ancestral area for the Iroquois in that archaeological period or whether we must conclude that the Iroquois were already far to the north by approximately 1200 A.D.

UNIVERSITY OF MICHIGAN

BIBLIOGRAPHY

1. BAILEY, JOHN H. 1938. An Analysis of Iroquoian Ceramic Types. Am. Antiquity, 4: 333–338.

2. BLAIR, EMMA H. (ed.). 1911. The Indian Tribes of the Upper Mississippi Valley and Region of the Great Lakes, 2 vols. The Arthur H. Clark Co., Cleveland.

3. BOAS, FRANZ. 1910. Ethnological Problems in Canada. Journ. Royal Anthropol. Inst. Great Britain and Ireland, 40: 529–539.

4. BUTLER, MARY. 1939. Three Archaeological Sites in Somerset County, Pennsylvania. Bull. Penn. Hist. Comm., No. 753.

5. COLDEN, CADWALLADER. 1904. The History of the Five Nations, Vol. I. A. S. Barnes and Co., New York.

6. FENTON, WILLIAM W. 1940. Problems Arising from the Historic Position of the Iroquois. *In* Essays in Historical Anthropology of North America. Smithsonian Misc. Coll., 100: 159–253.

7. FLANNERY, REGINA. 1939. An Analysis of Coastal Algonquian Culture. Catholic Univ. Am., Anthropol. Ser., No. 7.

8. FOWKE, GERARD. 1902. Archaeological History of Ohio. Fred J. Heer, Columbus.

9. GREENMAN, EMERSON F. 1935. Excavation of the Reeve Village Site, Lake County, Ohio. Ohio State Archaeol. and Hist. Quart., 44: 1–64.

10. —— 1937. Two Prehistoric Villages near Cleveland, Ohio. *Ibid.*, 46: 305–367.

11. GRIFFIN, JAMES B. 1943. The Fort Ancient Aspect: Its Cultural and Chronological Position in Mississippi Valley Archaeology. University of Michigan Press, Ann Arbor.

12. —— Culture Change and Continuity in Eastern United States Archaeology. (*In press.*)

13. —— AND MORGAN, RICHARD G. 1941. Contributions to the Archaeology of the Illinois River Valley. Trans. Am. Philosoph. Soc., N. S. Vol. XXXII, Part I.

14. HOLMES, WILLIAM H. 1903. Aboriginal Pottery of the Eastern United States. Ann. Rep. Bur. Am. Ethnol., Vol. 20.

15. KELLOGG, LOUISE P. 1925. The French Régime in Wisconsin and the Northwest. State Hist. Soc. Wis., Madison.

16. KINIETZ, W. VERNON. 1940. Indians of the Western Great Lakes, 1615–1760. Occas. Contrib. Mus. Anthropol. Univ. Mich., No. 10, University of Michigan Press, Ann Arbor.

17. MANSON, CARL, MACCORD, HOWARD A., AND GRIFFIN, JAMES B. The Culture of the Keyser Farm Site, pp. 375–418 in this volume.

18. MORGAN, LEWIS H. 1881. Houses and House Life of American Aborigines. Contrib. North Am. Ethnol., Vol. IV. Washington.

19. MORGAN, RICHARD GUY, AND ELLIS, H. HOLMES. 1943. The Fairport Harbor Village Site. Ohio State Archaeol. and Hist. Quart., 52 (1) : 1–62.

20. NASH, PHILLEO. 1939. The Pound Village Site, Elgin County, Ontario. Soc. Am. Archaeol. Notebook, pp. 74–75.

21. PARKER, ARTHUR. 1907. An Erie Indian Village and Burial Site. Bull. New York State Mus., No. 117.

22. —— 1916. The Origin of the Iroquois as Suggested by Their Archaeology. Am. Anthropol., 18 : 479–507.

23. —— 1922. The Archaeological History of New York. Bull. New York State Mus., Nos. 235–236.

24. —— 1926. An Analytical History of the Seneca Indians. Researches and Trans. New York State Archaeol. Assn., Lewis H. Morgan Chapter, Vol. VI, Nos. 1–4. Rochester.

25. QUIMBY, GEORGE I., JR. 1941. The Goodall Focus. Ind. Hist. Soc., Prehistory Research Ser., II (2) : 67–161.

26. RITCHIE, WILLIAM A. 1936. A Prehistoric Fortified Village Site at Canandaigua, Ontario County, New York. Res. Rec., Rochester Municipal Mus., No. 3.

27. —— 1943. Recent Advances in New York State and the Northeast. *In* Recent Advances of American Archaeology. Proc. Am. Philosoph. Soc., 86 (2) : 313–315.

28. SPECK, FRANK G. 1920. Decorative Art and Basketry of the Cherokee. Bull. Public Mus. City of Milwaukee, Vol. II, No. 2.

29. WILLOUGHBY, CHARLES C. 1935. Antiquities of the New England Indians. Peabody Mus. Am. Archaeol. and Ethnol. Cambridge, Harvard University Press, Mass.

30. —— AND HOOTON, ERNEST A. 1922. The Turner Group of Earthworks, Hamilton County, Ohio. Papers Peabody Mus. Am. Archaeol. and Ethnol., Vol. VIII, No. 3.

31. WINTEMBERG, WILLIAM J. 1923. Uren Prehistoric Village Site, Ontario. Bull. Nat. Mus. Can., No. 51.

THE CULTURE OF THE KEYSER FARM SITE

CARL MANSON, HOWARD A. MacCORD

AND

JAMES B. GRIFFIN *

University of Michigan

Introduction

IN 1890 Gerard Fowke conducted archaeological investigations in the Shenandoah Valley for the United States National Museum, and in his report [1] he mentioned the existence of several burial caves and mounds in the valley of the South Fork of the Shenandoah River. This early work prompted Manson and MacCord to relocate and further identify these sites, and to search for new ones. They succeeded in finding a heretofore unrecorded burial cave, which was excavated. While trying to correlate this cave with adjacent cultures, they discovered a large village site about one-half mile distant, on the South Fork of the Shenandoah River between Luray and Front Royal, at the foot of Massanutten Mountain (Fig. 1).

Topographic Features

The Shenandoah River, entering the Potomac at Harper's Ferry, West Virginia, drains a large area in western Virginia and eastern West Virginia. About thirty-five miles from its mouth the river divides into the North and South forks. The North Fork runs between the Massanutten and Shenandoah Mountains from Harrisonburg northward to the junction of the forks at Riverton, just north of Front Royal. The South Fork drains the area between Waynesboro

* The sections dealing with the excavation and the artifacts were prepared by Carl Manson and Howard A. MacCord. The one on the pottery was prepared by James B. Griffin, who also edited the manuscript and arranged the illustrations. In addition to the individuals referred to in the footnotes the writers are indebted to Frank M. Setzler, Neil Judd, and Waldo Wedel, of the Department of Anthropology of the United States National Museum, for advice and encouragement during the excavation of the site and the writing of the report.

[1] Fowke, 1894.

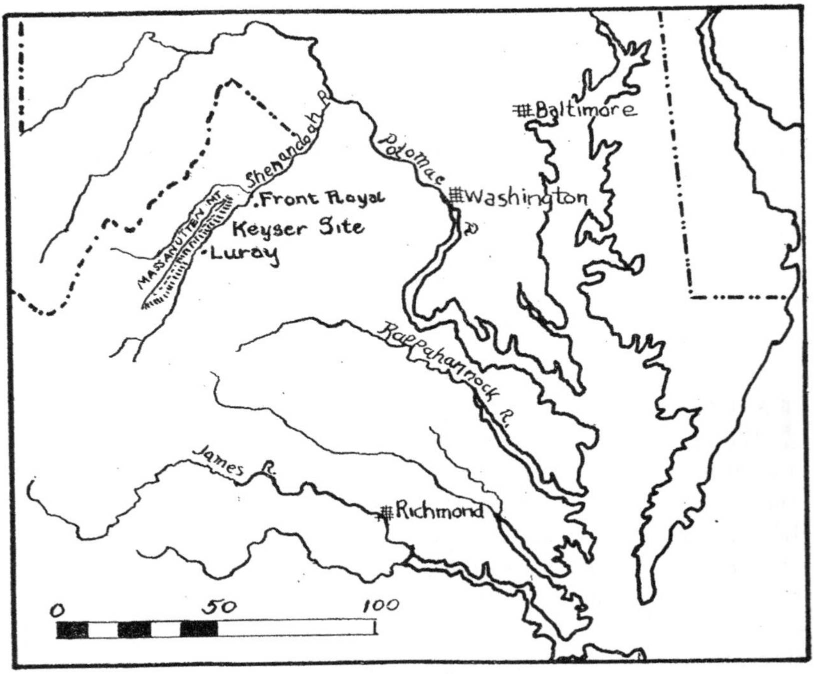

Fig. 1. Location of the Keyser farm site

and Front Royal, meandering between the Massanutten and Blue Ridge Mountains and forming the famous Page Valley.

On practically every stretch of bottom land not silted or entirely eroded away traces of aboriginal occupation may be found. On hillsides there are numerous caves and rock shelters, almost all of which show use by the Indians as either shelters or burial places, while on hilltops and spurs of the mountains there are occasional mounds and rock graves (cairns), most of which have been destroyed by casual investigation.

The Keyser site, which, incidentally, derives its name from the original settler, is located on one of the higher reaches of bottom land on the right (or west) bank, which is subject to flooding only at extremely high water (Pl. I, Figs. 1–2). It is well back from the river, and is separated from it by a ravine and a high sandy ridge. During the unusually high water of 1924 the site was completely buried

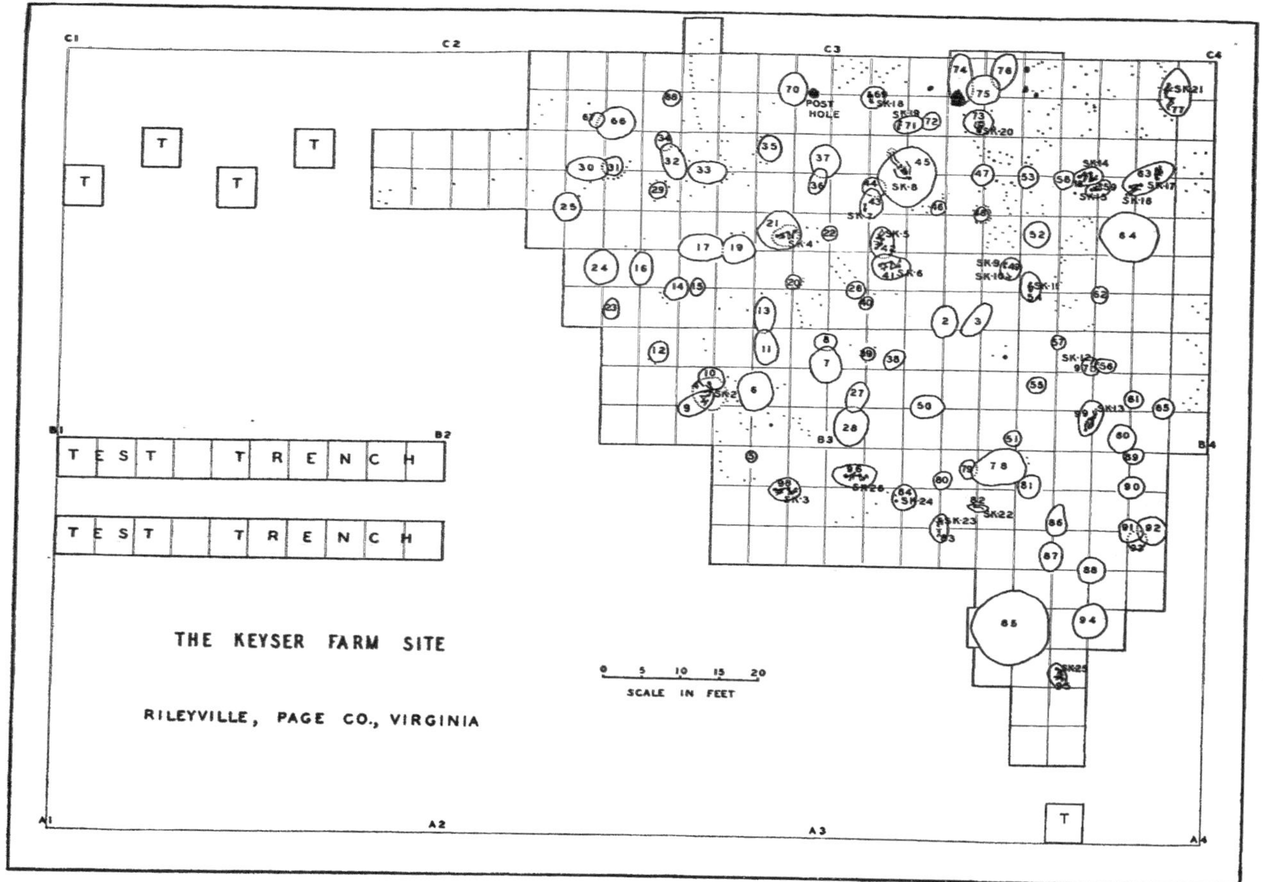

Fig. 2. Horizontal placement diagrams of Keyser farm site

under from four to six feet of silt, which entirely hid every vestige of occupancy. In 1936 the river undid its concealing work of 1924 and left the site in its original condition. In the three years following the latter flood, plowing and the erosive action of the wind and rain further lowered the surface, exposing many pits and burials. It was the profusion of village refuse thereby uncovered that convinced the writers of the value of the site.[2]

At present the site is about 550 feet from the river's edge and is between 13 and 15 feet above normal river level. In the direction opposite the river is a low range of precipitous hills, at the foot of which lies an extremely marshy area, fed by many springs, which issue from the hills at their bases. These springs were undoubtedly the nearest source of potable water for the Indians, and are still in use locally.

Excavation Technique

After having secured permission to excavate the writers proceeded to establish a bench mark and a datum plane, and the area to be explored in one season was divided into fifty-foot squares. These squares, or sections, were subdivided into five-foot squares to facilitate digging (Fig. 2).

At the southern and eastern extremities of the occupied area staggered alternate test trenches were run until pits or post molds were encountered. At this point stripping of every five-foot square began. Topsoil was removed by shoveling to the depth of the plowline. Material found in the topsoil was labeled "surface." The subsoil at the plowline was light yellow sand, and in most places clearly revealed the darker outlines of pits and post molds. Where these features were not met at the plowline, the subsoil was removed to a depth of two feet. Several pits were under a foot or more of sterile sand, but as a rule the outlines emerged very close to the plowline.

Stripping was done in five-foot units except when it was necessary to remove additional squares to expose completely some particular feature. Each pit or other disturbance was entirely uncovered horizontally and triangulated from corner stakes for permanent records before being further developed so as to extract all possible meanings and relationships (Pl. I, Fig. 3). The pits were then cut into from

[2] The fall flood of 1942 severely damaged the site.

one side to ascertain the shape for a vertical section, and to show stratification, if any was present. Wherever possible, the vertical section was made along the east–west line, the pit being entered from the south to ensure proper light for photographing. The data determined were recorded in the field notes, and sketches and photographs were made whenever they were deemed necessary for the record.[3]

Archaeological Features

In ten or more instances layers of refuse from three to eight inches thick were found immediately under the plowline and directly over one or more pits. These layers, although usually found in the subsoil beneath the stratum, were distinguishable by the blackness of the soil, their lenslike character, the presence of artifacts and refuse, and the absence of post molds or outlines of pits. Since no definite outlines could be determined, each layer was carefully troweled through, and the material was kept separate from the underlying pit or pits. The layers showed no line of demarcation between themselves and the pits underneath. The first of these phenomena discovered was not recognized immediately and was thought to be a shallow pit with indistinct outlines, and for this reason was given a number ("Pit 4"). This layer was lenslike and contained a burial, skeleton 2. Later it was found that beneath the layer were the outlines of two very clear-cut pits, numbered 9 and 10 on the plan. In all other examples material recovered from the layers was given a number made up of the numbers of the two underlying pits, e.g. $\frac{\text{P 41–42}}{\text{B 2}}$. In several places there were no pits under the layers, in which case the material was numbered for the square and section in which it occurred, e.g. $\frac{70}{\text{B 3}}$. There are no apparent differences between the material from the refuse layers and the pits under the layers. The exact origin of these layers is not known, but it is surmised that they represent the accumulation of débris and refuse in the depressions and cavities that follow the settling of the organic material of previously filled pits.

[3] The field notes and the majority of the artifacts are now in the collections of the United States National Museum. A representative series of sherds are in the Ceramic Repository for the Eastern United States at the University of Michigan.

The predominant horizontal shape of the pits was circular, and they ranged from 2 to 9 feet in diameter, with the average close to 3 feet. These constituted 64 per cent of the total, whereas elliptical pits comprised 30 per cent. The remaining 6 per cent were pear-shaped or irregular. Considerable variation existed in the shape of the pits as seen in vertical section. They ranged from (1) hemispherical to (2) straight-sided and straight-bottomed to (3) bell-shaped, with every possible combination of the three. The depth varied from 0.4 foot to 5.2 feet; the average was about 3.5 feet. A summary of the pits is given in Table I (pages 380–385).

In many places pits were found which overlapped, a superposition which was very nearly akin to stratification, since it indicated a lapse of time between the periods when the pits were open and being used. It is self-evident that two pits could not occupy the same piece of ground at the same time, so that it is equally self-evident that one pit was dug after an earlier one had been filled. This was further confirmed by the outlines of several of the later pits that were found complete, but the earlier outlines were broken by the excavation for a second pit. Comparison of material from one of these pairs of pits showed no noticeable differences. (For pottery analysis of Pits 11 and 13 see page 402.)

In addition to the feature noted above, there were a number of pits entirely or partly surrounded by post molds. Whether these represented a rooflike structure over the pit or a fencelike row of pickets around it to keep out scavenging animals was not determinable. Similar vertical post molds were found at the Montague site, Somerset County, Pennsylvania,[4] and at Fort Hill, Somerset County, Pennsylvania.[5] This concentric arc of postholes partially surrounded Pits 29, 31, 33, 35, 73, and 77. Pit 48 was entirely surrounded by 15 postholes regularly spaced at intervals of 0.6 foot. This pit was 2 feet in diameter, 2.4 feet in depth, and had a rounded bottom and sloping sides. It contained refuse bone, 35 sherds, several fragments of bone artifacts, and a pipestem fragment. At the bottom was a layer of irregular waterworn pebbles 0.3 foot thick.

In Pit 85, the only one with definite stratification, two distinct layers of cultural material were separated by nearly two feet of sterile sand. This pit was 9 feet in diameter, 4.9 feet deep, and had a straight bottom with sides sloping inward from top to bottom.

[4] Butler, 1939, pp. 12–13. [5] Augustine, 1940, pp. 51–58.

TABLE I

THE PITS AT THE KEYSER FARM SITES

Pit No.	Diameter in feet	Horizontal shape	Depth in feet	Vertical shape	Probable use	Remarks
1	3.0 × 3.0	Round	2.6	Rounded sides and bottom	Undetermined	
2	4.0 × 4.0	Round	3.1	Rounded sides and bottom	Undetermined	
3	5.0 × 3.1	Oval	3.2	Rounded sides and flat bottom	Undetermined	
4	Undetermined	Undetermined	0.4	Lens	Shallow pit or refuse layer that intrudes slightly into Pits 9 and 10	Skeleton 2
5	2.8 × 2.7	Round	1.2	Rounded sides and bottom	Fire	Contained numerous ash lenses and charcoal
6	4.5 × 4.5	Round	2.9	Rounded sides and bottom	Fire	Numerous ash deposits and large amount of fire-burned red sand
7	4.6 × 4.6	Round	3.0	Bell	Storage	Carbonized bark around sides and on bottom
8	2.5 × 3.4	Oval	2.4	Rounded sides and bottom	Undetermined	
9	4.0 × 5.1	Oval	2.0	Rounded sides and flat bottom	Fire	Many ash deposits and charcoal
10	3.1 × 3.1	Round	1.7	Bell; flat bottom	Storage	Bark around edges
11	4.1 × 5.2	Pear	3.7	Rounded sides, narrowing to point at bottom	Storage	Layer of carbonized bark around edges and on bottom of pit

12	3.1 × 3.1	Round	2.8	Rounded sides and bottom	Undetermined	
13	3.2 × 5.0	Oval	3.1	Rounded sides and bottom	Refuse	Abundance of village refuse and ash
14	4.1 × 4.2	Round	2.1	Rounded sides and bottom	Undetermined	
15	2.2 × 3.2	Oval	2.8	Rounded sides and bottom	Undetermined	
16	3.1 × 4.0	Oval	3.1	Bell	Storage	Carbonized bark scattered through pit
17	3.5 × 5.4	Oval	2.9	Rounded sides and bottom	Fire	Much ash and charcoal lenses
18	2.5 × 4.5	Oval	1.8	Rounded sides and flat bottom	Inhumation	Skeleton 14
19	4.1 × 4.2	Round	2.3	Rounded sides and bottom	Fire	Lenses of ash and charcoal
20	2.0 × 2.2	Round	1.7	Rounded sides and bottom	Undetermined	
21	5.0 × 5.5	Round	4.0	Straight sides with pointed bottom	Storage and inhumation	Skeleton 4. Used for storage first (bark layers on bottom); later for burial
22	2.0 × 2.0	Round	2.1	Bell	Storage	Carbonized bark on floor
23	2.5 × 3.0	Oval	1.8	Rounded sides and bottom	Undetermined	
24	4.0 × 4.4	Oval	3.2	Rounded sides and bottom	Storage	Carbonized bark lining
25	4.0 × 4.5	Oval	4.0	Rounded sides and bottom	Fire and storage	Red fire-burned sand on bottom, carbonized bark on sides
26	2.5 × 2.5	Round	1.3	Straight sloping sides with straight bottom	Undetermined	
27	2.5 × 3.0	Oval	2.1	One side straight, other bell-shaped; flat bottom	Storage	Carbonized bark on bottom
28	5.0 × 3.5	Oval	2.7	Straight sides, sloping to rounded bottom	Fire	Much ash and charcoal
29	2.0 × 2.1	Round	0.4	Lens	Undetermined	

TABLE I (*Continued*)

Pit No.	Diameter in feet	Horizontal shape	Depth in feet	Vertical shape	Probable use	Remarks
30	4.5 × 4.0	Oval	3.3	Rounded sides and straight bottom	Fire and storage	Used first as storage pit (carbonized bark on bottom); later as fire pit (red sand over bark)
31	2.4 × 2.6	Round	2.4	Rounded sides and bottom	Undetermined	
32	5.1 × 3.0	Oval	3.0	Rounded sides and bottom	Fire	Much ash and a charcoal layer
33	6.2 × 4.1	Oval	2.7	Rounded sides and bottom	Fire	Red sand (burned) layer and ash
34	2.5 × 2.3	Round	1.1	Lens	Fire	Ash lens and red fire-burned sand
35	3.5 × 3.6	Round	2.3	One side rounded, one side bell-shaped; rounded bottom	Storage	Carbonized bark layers
36	2.5 × 3.5	Oval	1.1	Sides concave and tapering to flat bottom	Fire	Red fire-burned sand on bottom
37	4.5 × 4.5	Round	1.5	Straight sides, sloping to straight bottom	Refuse	Abundant sherds, bone, etc.
38	4.0 × 4.0	Round	0.7	Rounded sides and bottom	Undetermined	
39	3.0 × 3.0	Round	1.6	Rounded sides and bottom	Storage	Bark layer on bottom and sides
40	2.0 × 2.0	Round	1.2	Rounded sides and bottom	Fire	Red fire-burned sand on bottom and much ash
41	3.0 × 4.6	Oval	1.6	Rounded sides and bottom	Undetermined	
42	3.5 × 4.5	Oval	3.2	Rounded sides and bottom	Inhumation	Skeleton 5
43	2.5 × 4.0	Oval	3.0	Rounded sides and bottom	Inhumation	Skeleton 7. Bark layer above burial

44	2.5 × 2.5	Round	2.1	Lens	Undetermined	
45	7.5 × 7.5	Round	3.4	Bell	Inhumation and storage	Skeleton 8. Bark layer on bottom and sides
46	2.0 × 2.0	Round	1.1	Lens	Undetermined	
47	2.5 × 2.5	Round	0.9	One side bell-shaped, other straight; flat bottom	Undetermined	
48	2.0 × 2.0	Round	2.4	Straight sloping sides with rounded bottom	Undetermined	No ashes or evidence of fire; fire-cracked stones on bottom
49	3.5 × 3.5	Round	2.6	Rounded sides and bottom	Inhumation	Skeletons 9–10
50	3.5 × 4.0	Oval	2.6	Bell	Storage	Bark layer on bottom and sides
51	2.5 × 2.5	Round	1.4	Bell	Storage	Bark layer on bottom
52	3.5 × 3.5	Round	0.6	Lens	Undetermined	
53	3.5 × 3.5	Round	2.1	Rounded sides and bottom	Undetermined	
54	3.0 × 4.0	Oval	2.8	Rounded sides and bottom	Inhumation	Skeleton 11
55	2.5 × 2.5	Round	0.6	Lens	Undetermined	
56	2.4 × 2.4	Round	2.5	Rounded sides and bottom	Refuse	Unusual amount of animal bones and village refuse
57	2.0 × 2.0	Round	1.2	Rounded sides and bottom	Undetermined	
58	2.5 × 2.5	Round	1.3	Bell	Storage	Bark layer on bottom
59	2.5 × 3.0	Oval	3.0	One side rounded, other straight; rounded bottom	Inhumation	Skeleton 15
60	3.5 × 3.5	Round	1.3	Lens	Fire	Red fire-burned sand and much ash
61	2.5 × 2.5	Round	1.2	Lens	Undetermined	
62	2.4 × 2.3	Round	2.9	Sides straight and sloping to straight bottom	Refuse	Large amount of animal bones and village refuse
63	3.1 × 6.5	Oval	2.5	Rounded sides and bottom	Inhumation	Skeletons 16–17
64	7.0 × 7.1	Round	5.2	Rounded sides and bottom	Refuse	Unusual amount of refuse

TABLE I (*Concluded*)

Pit No.	Diameter in feet	Horizontal shape	Depth in feet	Vertical shape	Probable use	Remarks
65	2.4 × 2.5	Round	2.7	Rounded sides and bottom	Fire	Fire-burned sand and much ash
66	4.0 × 5.0	Oval	3.1	One side straight, other rounded; flat bottom	Undetermined	
67	2.1 × 2.1	Round	1.7	One side rounded, other straight; rounded bottom	Undetermined	
68	2.5 × 2.5	Round	2.1	Rounded sides and bottom	Undetermined	
69	2.4 × 2.4	Round	3.6	Rounded sides and bottom	Inhumation	Skeleton 18
70	3.5 × 3.4	Round	2.3	One side rounded, other bell-shaped; flat bottom	Storage	Bark layer on bottom and side
71	2.0 × 2.0	Round	2.4	Sides straight and sloping to rounded bottom	Inhumation	Skeleton 19
72	2.2 × 2.2	Round	1.9	Sides straight and sloping to rounded bottom	Refuse	Unusual amount of village refuse
73	3.0 × 3.3	Round	2.7	One side bell-shaped, other rounded; rounded bottom	Refuse and inhumation	Unusual amount of village refuse. Skeleton 20
74	3.1 × 6.8	Oval	4.8	Rounded sides; flat bottom	Undetermined	Rock layer on bottom
75	3.4 × 3.1	Egg	4.0	Rounded sides sloping to flat bottom	Fire	Ash layer at 1.3 level; nest of six perfect beamers in ash layer
76	3.6 × 2.9	Oval	1.1	Lens	Undetermined	Disturbed, but no artifacts
77	5.2 × 3.7	Oval	2.1	Rounded sides and bottom	Inhumation	Skeleton 21
78	6.6 × 5.1	Oval	3.9	Straight sides sloping to rounded bottom	Fire	Ash layer, and large pieces of charcoal
79	2.5 × 2.4	Round	2.3	Rounded sides, flat bottom	Undetermined	

80	2.3 × 2.5	Round	1.8	Rounded sides and bottom	Undetermined	
81	3.1 × 3.2	Round	0.5	Lens	Undetermined	
82	2.8 × 0.9	Irregular	0.4	Lens	Inhumation	Skeleton 22. Shallow layer of stones covering infant burial
83	5.1 × 2.3	Oval	2.9	Sides straight and sloping to flat bottom	Inhumation	Skeleton 23
84	4.1 × 4.2	Round	2.4	Straight sides and bottom	Inhumation	Skeleton 24
85	9.2 × 9.1	Round	3.9	Slightly rounded sides and bottom	Undetermined	
86	4.4 × 4.0	Pear	0.4	Lens	Fire	Ash, charcoal, and red fire-burned sand
87	3.0 × 2.5	Oval	1.2	One side straight and sloping to rounded bottom, other rounded	Fire	Charcoal and fire-burned sand with ash layers
88	3.0 × 2.8	Round	1.1	Lens	Undetermined	
89	2.0 × 2.0	Round	0.7	Lens	Undetermined	
90	3.0 × 3.2	Round	2.0	Both sides straight and sloping to flat bottom	Storage	Heavy layer of bark on bottom and also much organic material
91	3.0 × 2.6	Oval	1.4	Rounded sides and bottom	Undetermined	
92	4.0 × 4.2	Round	2.0	Bell	Storage	Bark and organic material
93	2.1 × 2.1	Round	1.0	Lens	Undetermined	Destroyed by Pit 92
94	4.2 × 4.3	Round	1.8	Rounded sides and bottom	Undetermined	
95	2.0 × 3.1	Oval	0.6	Lens	Inhumation	Shallow burial of skeleton 25
96	5.0 × 3.8	Oval	0.9	Lens	Inhumation	Shallow burial of skeleton 26
97	4.0 × 3.6	Oval	0.9	Lens	Inhumation	Shallow burial of skeleton 3

Immediately below the topsoil was a saucer-shaped layer of heavy black soil with a thick bark layer separating it from the remainder of the pit (Pl. II, Fig. 1). This stratum contained refuse bone, shells, 167 sherds, one broken arrowhead, two fragments of beamers and many chips and spalls, besides much organic material. Below this deposit was a two-foot layer of sand, containing a few flint and jasper chips. The layer was banded, probably indicating a lapse of some time after the deposition of each band. At 3.6 feet below the top of the pit was a 0.3-inch layer of cultural material extending 1.5 feet up the sides of the pit. This stratum contained bone fragments intermingled with 149 sherds, one bone awl, one beamer, and several fragments of other bone artifacts, including parts of an antler celt and a bone chisel. The percentage of grit-tempered pottery was greater from the bottom layer of Pit 85 than from the site as a whole, indicating that possibly the grit-tempered ware was older than the predominant shell-tempered ware.

Where burials occurred in refuse or storage pits they were placed at various depths and orientations, and as a rule were not accompanied by any form of grave goods. Unceremonial disposal of the bodies is evidenced by several finds of skeletons in what would otherwise be typical refuse pits. Where pits were expressly dug for funeral purposes, the bodies were carefully placed on the bottom, accompanied by greater or smaller quantities of grave goods, but with practically no refuse in the fill over the body.

Positive evidence of fire was found in pits, although it was meager. At the most it consisted of a layer of red sand or a layer of ash and charcoal, with calcined bones bedded at intervals. The majority of the pits must have served as storage places for food or other valuables; when they were no longer needed, they were filled with refuse or were allowed to fill up naturally with blown sand and accidental enclosures of cultural objects. This would account for the sparsity of remains in many of the pits.

Hundreds of post molds were scattered over the entire explored area, but in no case could they be interpreted as definite house or stockade outlines, although these features were assiduously looked for. The molds ranged from two to four inches in diameter, although several had a diameter of over one foot. The largest of these was pitlike in character and was two feet in diameter, with nearly parallel sides extending to a depth of over four feet, tapering to a point at six feet six inches. The mold contained the charred remnants of the wood, but no village refuse. No special significance could be attached to this unusually large post mold.

Several groups of postholes forming arcs might be construed as part of a house outline; if they were, the size of the houses was approximately ten by twenty feet, and they were elliptical. There was no evidence to connect the partial outline of these arcs with any other feature, and no trace of a prepared floor was discovered.

Burials

Skeleton 1 was on the surface, where the plow had deposited it. An unsuccessful attempt was made to locate the burial pit, and so the topsoil was sifted. This operation recovered many of the bones of a child, the parietals of which were copper-stained. The only artifacts were numerous shell beads and ten small copper ones.

Skeleton 2, of an adult male, was in a refuse layer over Pits 9 and 10. It was loosely flexed on its right side, head to the west with the forearms together, and with the hands in front of and touching the face. No artifacts were found.

Skeleton 3, of an adult female, was taken from Pit 97, one of the few burial pits. It was loosely flexed on the right side, with the left hand in an unnatural position. The left elbow was placed immediately below the mandible, and the left hand was forced back against the sternum. The complete absence of refuse in the pit and its oval shape indicate that it was dug especially for a burial.

Skeleton 4, in Pit 21, was that of an adult male, buried closely flexed at a depth of 2.0 feet. That this pit was used for refuse is borne out by the haphazard placement of the burial and by the abundance of refuse bones and artifacts under, over, and around the skeleton. The position of the skeleton was as follows: on the right side, knees and elbows touching in front of lumbar vertebrae, the right hand in front of face, left hand touching left knee; feet drawn up very close to sacrum.

Skeleton 5, of an adult male, was in Pit 42, at a depth of 3.2 feet. Orientation was to the east, with the skeleton closely flexed on the left side (Pl. II, Fig. 2). The right hand rested under the chin; the left reposed on top of the right femur.

Skeleton 6, of a young adult female, 18–20 years of age, was loosely flexed on the right side at a depth of 3.2 feet. Orientation was to the southeast. The hands were clasped directly in front of the face, and rested on the floor of the pit (Pl. II, Fig. 2).

Skeleton 7 was that of a two-year-old infant, the bones of which were very fragile and fragmentary. It was closely flexed on its right side, with the head to the east, at a depth of 3.0 feet. The feet were drawn up close against the sacrum, and the arms were folded tightly against the chest. Four *Marginella* beads lay near the neck and inside the mandible; in the dirt over the burial was a small piece of polished chlorite, probably part of a stone tube or ornament.

Skeleton 8, at a depth of 2.7 feet, was that of an adult female (young), who had been interred in an extended position, face up, with arms alongside, and both knees slightly bent to the left. Orientation was to the north.

Skeleton 9 comprised parts of a skull and skeleton of a five-year-old child buried at a depth of 2.6 feet. Orientation was indeterminate owing to the fragmentary condition of the remains. Adjacent to the skull was a crushed but restorable pottery vessel, four inches in diameter and four inches high (Pl. II, Fig. 3).

Skeleton 10 was represented only by a fragmentary skull, that of an infant under two years of age. It was in the same pit with skeleton 9, beside which it lay, but the lack of body bones precluded any determination of position or orientation.

Skeleton 11 (so called) represents the unusual burial of most of the body bones, without head or limbs, of a male subadult, along with several bones of a young child. The only semblance of articulation was in the vertebral column, which was almost complete. All the other bones were jumbled indiscriminately. Scattered among the body bones were several ankle and wrist bones and a number of phalanges. No artifacts accompanied this burial, nor was there anything to indicate the reason for its fragmentary condition. It is possible that the skeleton had lain on the surface long enough for animals to carry off the detached parts of the body. It was found at a depth of 2.5 feet. No orientation, of course, could be determined.

Skeleton 12 was that of a child nearly five years of age, buried at a depth of 1.8 feet. The body had been placed on its back, head to the southwest, legs bowed, with feet touching. The pit, number 97, had been cut into by Pit 56, which had slightly disturbed the burial. The skull was crushed, and the rest of the bones were in a very fragile condition. In the fill directly over the skull were two tubular copper beads, and directly under and in contact with the skull were seven additional specimens. Between the legs and touching the left femur lay four shell pendants derived from the columella of the *Busycon carica.* Also touching the left femur was a shattered but restorable clay pipe.

Skeleton 13, that of an adult male, was buried at a depth of 0.8 foot. The body had been closely flexed on its right side, with head to the east. The knees and feet were together, hands in front of the abdomen. The shallowness of the burial resulted in the crushing of the skull and pelvis by external pressure.

Skeleton 14 was at a depth of 1.8 feet. The body, that of an adult female, had been closely flexed on its right side, with the head to the southeast. The feet and knees were together; the left arm was between the legs, and the right arm was placed across the body.

Skeleton 15 was at a depth of 2.3 feet. The body had been closely flexed on the right side, head to the southeast. The skeleton, that of an adult female, lay almost on its face, with the knees at a lower level than the remainder of the skeleton.

Skeleton 16, that of an adult female, was at a depth of 1.7 feet. The body had been placed on its left side, with the head to the northwest (Pl. III, Fig. 3). The knees were drawn up in front of the sternum, with the heels about six inches from the pelvis. The left hand covered the upper part of the facial region; the right hand lay just below the right knee.

Skeleton 17 was in the same pit with skeleton 16, but at a depth of 2.5 feet. This burial, that of an adult male, was closely flexed on its right side, with the head to the northeast. The pelvis was about 0.6 foot deeper than the skull; the knees and feet were still deeper, at a total depth of 3.5 feet.

Skeleton 18, that of a newborn infant, was buried on its back, with its head to the southwest at a depth of 3 feet.

Skeleton 19, that of a six-year-old child, had been buried with the body loosely flexed on its right side and with the head to the east at a depth of 2.4 feet. Under the head and neck lay a rectangular slab of limestone, 6 by 10 by 2 inches. Directly in front of the face was a large lump of charred wood, 0.4 foot in diameter and about 0.8 foot in length, that was roughly round. It had not formed part of a post, since it lay on its longer side. At the feet of the burial was a roughly elliptical layer of waterworn boulders. The dimensions were 2 by 4.5 feet.

Skeleton 20 (incomplete), at a depth of 2.6 feet, was that of an adult female (Pl. III, Fig. 1). The body, without legs, had been buried on its back, with the face uppermost and the head to the northwest. The right hand lay directly on the right pelvic bone; the left arm was fully flexed, so that the head rested on the left shoulder. There was no trace of the lower limbs, and the sockets of the pelvis were empty. Scattered through the fill, however, were the small bones of the legs — patella, phalanges, tarsals, and metatarsals. None of these were articulated, so that their significance is only conjectural. Accompanying the burial and resting right side up at the right elbow was a fine small pottery jar containing a perfect turtle-shell cup. The left parietals were copper-stained, but no other trace of the metal could be found. A number of perfect bone artifacts in the pit may or may not have had some relation to the burial, but direct association was lacking. There was absolutely no secondary disturbance which might have accounted for the incompleteness of the skeleton.

Skeleton 21, that of an adult male, was loosely flexed at a depth of 2.8 feet in Pit 77 (Pl. III, Fig. 2). The body had been placed with the head to the west, lying on the left side. The hand bones were together in front of and slightly deeper than the face. The clayey nature of the soil at this particular point led to the poor condition of the bones as compared with others in sandy soil as close as ten feet. An arc of postholes adjacent to the pit edge was an unusual feature of this burial.

Skeleton 22, that of a newborn infant, lay at a depth of 0.4 foot. It had been disturbed by the plow, and orientation and position could not be ascertained owing to the fragmentary condition of the bones. Associated with the burial were two stone arrowpoints, and one of antler, one piece of "cut" turkey tibiotarsus, and one gorget-like but unperforated piece of felsite.

Skeleton 23, that of an adult female, was in a fragmentary condition in the topsoil, where the plow had thrust it after disturbing Pit 83. No indication of placement or orientation was discernible.

Skeleton 24 was at a depth of 2.3 feet. The bones, those of a small child, were against the edge of the pit, with the head to the east. Lying around the head and face were twenty-four cylindrical shell beads (of *Busycon carica*) and small shell-disk beads (remains of a string of beads about four feet long). Directly under and in front of the face was a piece of clay pipestem, one and one-half inches long, probably an accidental enclosure.

Skeleton 25, that of an adult male, was at a depth of 0.6 foot. The body had been placed loosely flexed on its right side, with the head to the southwest. It had been disturbed by the plow.

Skeleton 26, that of a male subadult, lay in the top of Pit 96, where the plow had removed the uppermost sections of the skull, pelvis, and feet. Enough remained to show the position of the body, which had been placed closely flexed on its right side, with the head to the southeast. At the head was half of a cylindrical awl or pin. Only the pointed half was found, since the remainder had been carried away by the plow.

The following statement by Dr. T. D. Stewart, Division of Physical Anthropology, United States National Museum, outlines some of the physical relationships of the skeletal material.

The skeletal remains from the Keyser site represent a fairly typical sample of an Indian village population, because they include about 46 per cent subadults — newborns, infants, and adolescents. This high percentage of immature individuals limits the number of adults that can be used for comparative purposes to the remaining 54 per cent of the series — 14 individuals in varying degrees of completeness.

Comparative material from this region is limited at present mainly to that from sites in Fayette and Somerset counties, Pennsylvania, and in Rockbridge County, Virginia. I am in the course of studying the Pennsylvania material, whereas the Virginia material (Hayes' Creek Mound) is known to me only from Hrdlička's measurements.[6] For this reason I can indicate at this time only the general relationships of these series as I see them.

Occipital flattening of moderate grade is apparent in at least two skulls from the Keyser site, and it is impossible to rule out slight occipital flattening as the cause of the roundness of other skulls. I have encountered the same problem in the collections from Pennsylvania. The cranial indices of these two lots compare as follows:

Classes	*Keyser site*	*Pennsylvania*
70–74.9 (dolichocrany)	..	7
75–79.9 (mesocrany)	1	1
80–84.9 (brachycrany)	8	3
85–89.9 (hyperbrachycrany)	4	1
90–94.9 (ultrabrachycrany)	1	1

The Pennsylvania material thus tends more to dolichocrany. The skulls from the Hayes' Creek Mound have an average cranial index of 76.0 (36), and all but

[6] Hrdlička, 1927.

two fall within the dolicho- and mesocranic classes. They are thus even more long-headed.

The three series here compared all agree in being very high-headed, the average mean height index in each case being around 87.

Although the samples are inadequate and as yet incompletely studied, it is my impression that the Keyser remains are closest in type to those from Pennsylvania.

Artifacts

Artifacts were obtained throughout the excavated area, not only in pits, but in the topsoil and on the surface. In some pits none were located, but in the vast majority such remains were common. The materials represented consisted of antler, bone, shell, stone, ceramics, and copper. Practically every article of raw material employed by the Indians was native to the immediate locale, with the exception of the marine shells and possibly the copper. The antler and bone were secured from the local fauna; the majority of the shell material came from the adjacent river beds. All the lithic artifacts were derived from the local igneous and metamorphic deposits characteristic in the two mountain ranges between which Page Valley nestles. The pottery was made almost entirely of the local clays tempered with pulverized mussel shells or comminuted pieces of limestone or quartz. No objects were recovered which indicated white contact.

ANTLER

With a few exceptions the artifacts of antler were not dissimilar to those found on many sites throughout the eastern and northeastern areas. Probably the most unusual feature was the presence of several antler headdresses not previously recorded or described for the area under consideration (Fig. 3).

Four examples of this unique ornament were discovered in the course of the digging. Of these, one was practically complete, and the others were fragmentary. The most nearly perfect specimen consisted of the two antlers and a triangular piece of the frontal portion of the skull of a Virginia deer, cut and modified so as to fit closely the forehead of the wearer. It was probably held in place by a band similar to a carrying strap, which passed around the head; the parietal vascular foramen provides an opening through which a cord could have been passed and tied behind the head. (One specimen is worn around the interior edges of the foramen and supports this hypothesis.) The antler branches were carefully hollowed out and cut away lengthwise to lighten the burden without loss of shape or beauty. The anterior view of the headdress showed the natural surface of the antler since the "channeling" was entirely on the posterior side of the antler beam. On the better specimen the marks of the chisel throughout the length of the channeling disclose the technique by which the hollowing was effected. That this ornament was further embellished is indicated by the traces of red pigment remaining on the less exposed portions along the sides, which suggest that the skin was not worn with the headdress. Two holes drilled

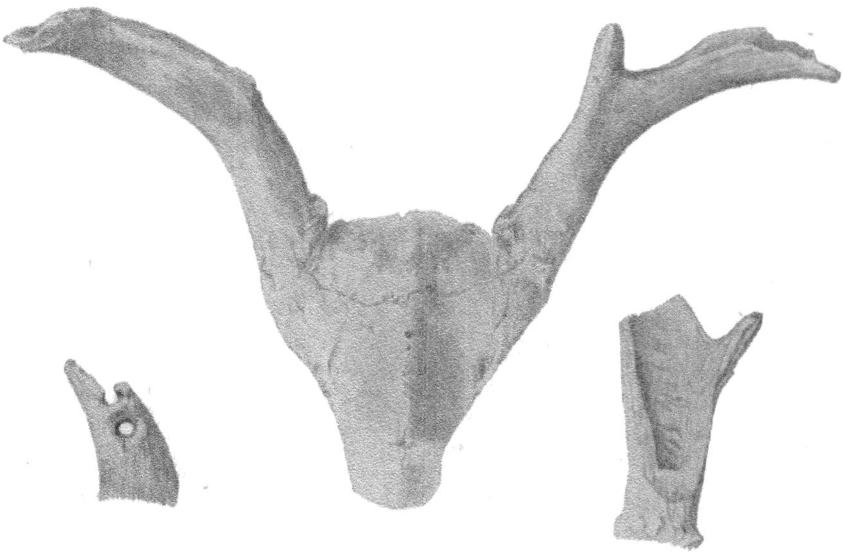

FIG. 3. Antler headdress artifacts (drawing by Dr. Doris Cochran)

near the tip of one antler may have served to repair the broken tip or could have been the means of attaching feathers, pendants, or other decorations to the ends of the individual antlers to make them more imposing. This cannot be ascertained, since the tip of the other antler is missing.

In the next best-preserved headdress the skull portion was in excellent condition, although all of one antler and half of the other were missing. The half recovered showed the chisel marks along the interior of the channeling, but any further decoration had been obliterated by weathering.

The third specimen consisted of six pieces of a hollowed antler branch, worked to a fine polish. Five of the pieces joined, but because of warping their union could not be effected. To judge from the size of the pieces, this headdress was fashioned from elk antler, whereas the other specimens were made from Virginia deer. The chisel marks in the interior of this specimen had been rubbed smooth, but the outside bore a high polish. Two perforations, equidistant from the edge of a fracture, indicated that an attempt had been made to repair the artifact before it was finally discarded. At several points on the edge of the fragments the tooth marks of a small rodent were noticed.

A similar find was made in a shell mound at Belle Glade, Palm Beach County, Florida, by M. W. Stirling.[7] Of this discovery he writes: "An interesting object recovered is what seems to have been a deer headdress consisting of a small portion of the skull with the antlers attached. The antlers have been polished and hollowed out so as to make them light in weight."

The most noticeable difference between the headdress found at the Belle Glade mound and the specimens from the Keyser site is in the position of the channeling. In the Florida specimen the antler is hollowed anteriorly, but those of the Virginia specimens are hollowed posteriorly.

The use of these head ornaments is implied by Arthur C. Parker,[8] who states: "Entire antlers were sometimes used by the Iroquois for the head ornaments of sachems, who, according to the ancient ritual, were 'crowned with deer antlers.'" It is inferred that the source of his information is Iroquois mythology, since no archaeological evidence in New York is known to support it.[9] M. R. Harrington [10] believes that headdresses were undoubtedly used as part of the war or hunting regalia of the Cherokee. But again no substantiating archaeological evidence is adduced. Dr. Frank G. Speck [11] illustrates a similar ornament in actual use by the Penobscots of Maine in their Clown dance. Speaking of the costume worn, he says (p. 297): "A mask made of the headskin and antlers of a deer is also mentioned." Other than the Florida find, the only archaeological specimens of this

[7] Stirling, 1935, p. 374.
[8] Parker, 1922, p. 351.
[9] Private correspondence with W. A. Ritchie, May, 1940.
[10] Harrington, 1922, p. 207.
[11] Speck, 1940, fig. 30.

nature were those excavated by Moorehead in the Hopewell mound, where a burial was discovered crowned with a headdress made of copper in imitation of deer antlers.[12]

Altogether, enough of these unique artifacts are known to indicate that they were of considerable importance to the aboriginal owners, since so much labor was expended on them and since at least one discloses an attempt to repair a break.

Fragments of antler were recovered from almost every pit either as artifacts or as refuse. Two classes of artifacts could be distinguished: those made from the entire antler beam and those made of small pieces of the beam or tine. Among the larger implements were picklike tools, apparently used in the excavation of the pits themselves. Others were chisel-like or gougelike tools, the exact purpose of which is unknown. One peculiar specimen had a definite socket hollowed out on the distal end of one branch, and presumably was utilized as a haft for a smaller tool. Unspecialized branches of antler with a smoothed surface on one side may have been pottery smoothers.

Implements of the second class have much variation and apparently were of considerable importance in the daily life of the people. The most numerous group consisted of the projectile points made from the tip of the tine (Pl. IV, Fig. 1). These were all about one and one-half to two inches in length and were hollowed to a depth of about one-half inch at the base. In many of them the exteriors of the points were carefully polished, but in others the natural surface was left unaltered. The steps in the aboriginal method of manufacture of these weapons were illustrated by rejectage and occasional incomplete forms.

A second important group comprised the chippers or drifts, two-inch long, cylindrical pieces of hard antler, sometimes rubbed to a fine polish, but with occasional rougher examples represented (Pl. IV, Fig. 2).

Fragmentary remains of antler celts, or of some similar implement, suggest that antler was used for purposes other than those mentioned.

The refuse antler reflects the primitive methods employed by the Indians in detaching the antler and reducing it to a workable size. Sometimes it was broken forcibly from the skull of the deer, and at other times it was removed by burning the beam partly through and

[12] Moorehead, 1922, pp. 107–108, 126–127.

then breaking it along the charred area. Incising preliminary to breakage was noticed in only two minor instances. This was the method followed in detaching the tines from the main branches, but not in separating the beams from the skull. The majority of antler pieces were from the Virginia deer, although some were definitely identified as of elk.

BONE

Extensive utilization was made of animal and bird bones in the manufacture of artifacts. Deer bones were the main source of raw material, and for this reason most of the artifacts were of this animal's bones.

The most common bone artifact was the bone beamer or currier, made from the cannon bone of the deer and usually found in fragmentary condition. Fourteen perfect specimens were obtained, however, including a cache of six slightly worn tools. The cache was found under a large sherd of pottery in Pit 75. Most of the seventy-odd broken specimens recovered showed that continuous use had worn down the cutting edge until the bone was no longer thick enough to withstand further strain. Occasionally both halves of a broken tool were thrown into the same pit, in which case they were restorable, but as a rule only one half was found in a pit, and usually even the halves were further fragmented. The beamer has a wide distribution over most of the eastern United States, but it is found in unusual numbers at this site.

Bone awls, the next most common bone artifact, had a great deal of variety in shape and origin (Pl. V, Fig. 4). The sources of the raw material for these tools were as follows:

Turkey joint	10
Turkey splinter	3
Deer joint	3
Deer scapula	3
Deer splinter	6
Cylindrical bones, indeterminate	3
Total	28

Two specimens made from the turkey tibiotarsal bone have notches or tally marks on one side (Pl. V, Fig. 5). Similar marks were also present on one of the deer-scapula awls. No perforated awls were found, although one tibiotarsal with a natural perforation was recovered.

The artifacts listed as cylindrical awls may possibly have been broken hairpins like those illustrated from the Castle Creek site.[13]

Bone bodkins (Pl. V, Figs. 1–2) were represented by two complete and four broken specimens. The source of the material is not known, but the curvature of the implements suggests the rib of a large animal. The length of each of the perfect specimens is seven inches.

Seventeen bone-chipping tools (Pl. V, Figs. 3, 8–15) may be classified by types:

Perforated	7
Perforated by scoring hollow bone	1
Grooved	4
Notched	2
Splinter (unspecialized)	3
Total	17

The length of these chippers ranged from one and one-half to four inches. The working ends are rounded and show the effects of much wear.

The only fishhook found was incomplete; it had been fashioned from a semicylindrical blank formed by scoring and breaking a hollow bone. The lack of fishhooks is not surprising since fishing in the river is better suited to the use of nets or weirs.

One tube, three inches in length (Pl. V, Fig. 6), one bead, and one broken whistle with three openings (Pl. V, Fig. 7) were the only artifacts made from hollow bird bones. The presence of twenty-five pieces of "cut" bone implies, however, a much greater use of this material than was represented.

The carapace of the box tortoise was extensively employed in the manufacture of scoops, cups, and other receptacles. Several examples have perforations near the rim, either for suspension or for the attachment of a handle. The use of these shells as rattles is not indicated by the examples recovered.

Among the rarer bone artifacts were the following: one complete and one fragmentary chisel of deer bone; one perforated bone handle suitable for the insertion of a blade; one fragmentary bone tube, four inches long, split longitudinally; one perforated and worked raccoon baculum; and one worked beaver incisor.

[13] Ritchie, 1934, Pl. XIII.

It is of interest to note that the canine teeth of the bear and the dog had no utilitarian or decorative purposes. Many jaws of these animals were found with all teeth intact, and no attempt had been made to modify or decorate the teeth that did occur separately. Well over one hundred unworked canine teeth were recovered from the excavations.

SHELL

An unusually small number of shell artifacts were found, and they represented but few types. The most common use of shell was in the manufacture of small disk beads, of which there were several hundred (Pl. VI, Fig. 6), recovered mainly in the backfill of burial pits.

Twenty-four tubular beads, one-quarter to three-quarters inch long, were made from the central column of the conch shell (Pl. VI, Fig. 5). Four pendants of the same material, two and three-quarters inches long and having perforations at one end, accompanied skeleton 12 (Pl. VI, Figs. 1–2). This conch was the *Busycon carica,* the nearest source of which was the Atlantic Ocean at the capes of the Chesapeake. Contrast may here be made between these shell pendants and some identical specimens described by Fowke,[14] from the Brumback mound, eight miles to the south, where the pendants had been formed from the *Busycon perversum* of the Gulf of Mexico. One flat pendant, shaped like a bear's claw, was taken on the surface, and a similar broken specimen was obtained in Pit 35 (Pl. VI, Figs. 3–4).

Mussel and snail shells were fairly well preserved, which suggests that the character of the soil was not likely to have been the reason for the paucity of shell artifacts. One fragmentary shell "spoon," four scrapers, and one perforated shell of unknown use were the only artifacts made from the river mussel.

CHIPPED STONE

The most important chipped stone artifacts were projectile points, of which some 120 were recovered (Pl. VII). They range from very crude to medium-fine in workmanship, with a preponderance of the cruder forms. They vary in shape and material as the following list shows:

[14] Fowke, 1894, pp. 49–53.

Material	*Total*	*Percentage*
Quartz	48	40
Quartzite	19	16
Flint	16	13
Jasper	28	23
Rhyolite	5	4
Chalcedony	2	2
Shale (?)	1	1
Chert	1	1
Total	120	100

Shapes	*Total*	*Percentage*
Triangular	104	87
Stemmed	6	5
Unspecialized (crude)	6	5
Drill-like	4	3
Total	120	100

The predominant form is the isosceles triangle, ranging from one-half to two inches from tip to base. All the materials are native to the valley and adjacent mountain ranges.

POLISHED STONE

There were very few complete polished stone artifacts, but celts or fragments of them were fairly common. From the great amount of felsite scattered over the surface and in the general digging it is apparent that considerable celt manufacture was carried on in the area. Since the site was that of a village, the evidence was not so plentiful as one expects to find at a true celt workshop, such as that visited and described by Holmes [15] at the mouth of Pass Creek, three miles west of Luray. Altogether, three perfect celts were collected in addition to several dozen crude and fragmentary specimens.

Stone was used in the manufacture of pipes also (Pl. IX, Fig. 2 D). A portion of the flat stem of a monitor pipe was picked up on the surface. Part of the bowl of another was also a surface find. This is a truncated egg-shaped bowl having a conical cavity that meets the socket for the stem at a right angle. The material is chlorite, and the exterior is decorated with a line of minute incised dots running around the bowl parallel to the lip (Pl. IX, Fig. 2 E). One other small piece of polished chlorite is either part of a pipe bowl or a segment of a stone tube.

[15] Holmes, 1897, pp. 100–103.

The only other articles of polished stone were two fragments of tablet-like objects (Pl. VIII, Figs. 3, 5).

Limonite showing rubbed edges occurred in fragments ranging from one to two and one-half inches square (Pl. VIII, Figs. 1–2). The palette on which the limonite was rubbed is represented by a broken slab of compact sandstone, one side of which still showed the powdered ocher adhering to the rough surface.

Cruder stone implements included hammerstones, both pitted and plain, one net sinker, a cupped stone, three abrasives or sinew stones (Pl. VIII, Fig. 4), and one broken crude discoidal.

COPPER

Copper artifacts were found in two burials only. With skeleton 1 were ten small tubular beads, three sixteenths of an inch in length and one eighth of an inch in diameter (Pl. IX, Fig. 1 B). These had been made by rolling a flat piece of copper into a cylinder producing overlapping edges.

With skeleton 12 were nine tubular beads two inches long and one eighth of an inch in diameter. These also had the characteristic overlapping edges (Pl. IX, Fig. 1 A).

Professor W. C. Root, Department of Chemistry, Bowdoin College, Maine, kindly made a spectroscopic analysis of the copper artifacts. The specimens with burial 1 (Root's no. 1198) were pure copper; those with burial 12 (Root's no. 1199) were copper, with a trace (0.1 per cent) of silver.

Of the copper Professor Root states: "Neither one of these samples had any detectable gold, lead, tin, arsenic, antimony, bismuth, cadmium, zinc, or mercury. I think there can be little doubt the beads were of native American metal."

TRAIT LIST OF THE KEYSER FARM SITE

This trait list is intended to facilitate comparison with similar lists from related sites.

Objects found	*No.*
1. Round pits (top view)	64
2. Overlapping pits	34
3. Refuse layers over pits	10
4. Pits surrounded by post molds	6
5. Burial in refuse pit	1
6. Burials in storage pits	2
7. Burial in refuse layers	1
8. Artifacts with burials	6
9. Pottery as grave goods	2
10. Stone slab for pillow of skeleton	1
11. Flexed burials	15
12. Double burials	2
13. Extended burial, knees bent	1

Objects found	*No.*
14. Fragmentary skeletons in grave	2
15. Beads, copper, rolled	2
16. Bead made from a shell of a snail of the genus *Marginella*	1
17. Bead, shell, large tubular	1
18. Bead, shell, small tubular	1
19. Beads, shell, small disk	7
20. Pendant, columella	1
21. Pendants, shell, imitation bear claw	2
22. Pendant, shell, bipointed	1
23. Spoon, shell, notched edges	1
24. Spoon, shell, plain	1
25. Scrapers, mussel-shell	4
26. Headdresses, antler	4
27. Picklike tools, antler	3
28. Celts, antler	3
29. Chisels, antler	2
30. Drifts, antler	13
31. Arrowpoints, antler	37
32. Arrowpoint, antler, tanged	1
33. Spear points, antler	2
34. Pottery smoothers, antler	3
35. Bead, bone, decorated	1
36. Handle, socketed, bone	1
37. Fishhook, bone, center cut out	1
38. Whistle, bone, three holes	1
39. Chipping tools, bone	9
40. Chipping tools, perforated	8
41. Chipping tool, grooved	1
42. Hairpins, bone	2
43. Pins, bone, double-pointed	2
44. Chisels, bone	2
45. Tubes, bone	2
46. Bodkins, bone	8
47. Beamers, deer, metatarsal	84
48. Baculum (penis bone of raccoon) perforated,	1

Objects found	*No.*
49. Chisel, beaver incisor	1
50. Scoops, turtle-shell, perforated near rim	33
51. Scoops, turtle-shell, plain	46
52. Awls, whole bone	3
53. Awls, splintered bone	9
54. Awl, bone, perforated	1
55. Awls, deer scapula	3
56. Awls, turkey metatarsal, notched	2
57. Pipe, clay, obtuse-angled, decorated	1
58. Pipe, clay, elbow, short stem	1
59. Pipes, clay, all types	4
60. Pipes, clay, with flanged mouthpiece	3
61. Pipes, stone, all types	3
62. Pipe, ovoid stone, decorated	1
63. Pipe, stone, monitor.	1
64. Net sinkers, stone, notched	2
65. Limonite, lumps showing rubbing	18
66. Palette, stone slab for rubbing paint	1
67. Whetstone	1
68. Sinew stones	3
69. Smoothing stones	2
70. Discoidal, plain	1
71. Hammerstones, centrally depressed	4
72. Celts, stone, narrow-polled	10
73. Celt, stone, chipped	1
74. Blade found in cache	1
75. Pendant, stone	1
76. Choppers, chipped stone	2
77. Drills, unstemmed	4
78. Knives, stone, oval	6
79. Arrowpoints, stemmed	6
80. Arrowpoints, triangular	104
81. Potter's clay, in lump	1

The Ceramic Complex

The clay pipes are of the obtuse-angle style, which Holmes demonstrated was the dominant type along the Middle Atlantic area at the time of the English colonization.[16] A complete specimen (Pl. IX, Fig. 2 A) is an excellent example. The bowl section (Pl. IX,

[16] Holmes, 1903, p. 158.

Fig. 2B) has a fine dentate-stamped decoration. Comparable pipes from western Pennsylvania are described and illustrated. The northern and southern limits of pipes of this shape are not clear, although a connection with the shape of late Woodland and Iroquois pipes is obvious, and also with the decoration of the late Woodland pipes of the North Atlantic drainage. The flaring bit on the obtuse-angled elbow pipes is reminiscent of the mouthpiece of some tubular pipes of the Adena-Tchefuncte cultures (Pl. IX, Fig. 2C).

The pottery submitted for analysis from the Keyser site is an interesting mixture of three separate ceramic complexes, which can be recognized immediately because of significant differences in the kind of temper used in their manufacture.

CORD-MARKED BODY SHERDS FROM VARIOUS LOCALITIES SEPARATED BY TYPE OF TEMPER

	Shell	*Grit*	*Limestone*
Pit 11	107	15	..
Pit 13	109	17	8
Pit 85, deep level	57	31	5
Miscellaneous	89	21	5
Total	362	84	18
Grand total			464

The dominant pottery type is clearly indicative of a close relationship with a late prehistoric occupation in southwestern Pennsylvania. The minority types, on the other hand, just as definitely indicate a close connection with the pottery styles found in the Coastal Plain and the Piedmont area of the Middle Atlantic states. Though the evidence obtainable from the pottery itself and from the notes of the excavators is by no means conclusive, there is a suggestion that the minority types were the earlier.

KEYSER CORD-MARKED

The paste of the majority type is composed of a local clay mixed with ground portions of mussel shell. The shell fragments vary considerably in size, from minute flakes of less than half a millimeter to large pieces up to eight millimeters in length. The particles are flaky, and most of them are oriented parallel with the vessel walls. This orientation was probably accomplished during the surface manipulation. The proportion of shell varies considerably, and ranges from approximately 10 per cent up to 30 per cent. The texture of the cross section is almost entirely medium-fine, and in many respects is identical with that of the sherds in the southern foci of the Fort Ancient Aspect. It is also identical with that of the shell-tempered complex in southwest Pennsylvania. The majority of the sherds have an exterior surface hardness of 2.5 and 3. The exterior surface is predominantly dull grayish brown, with a minority of the sherds having a buff to reddish-brown exterior. The interior surface of practically all the sherds is dark gray to black. The core ranges from dark gray to red or buff. Red and buff cores are found principally but not always on sherds with a similarly colored exterior surface.

Surface Finish

The exterior surfaces are all malleated with a cord-wrapped paddle, extending from the lip down the entire body and onto the base of the vessel. On approximately 90 per cent of the rims the cord markings are applied vertically (Pl. X, Fig. 2), but on the remaining sherds the cord impressions are oblique from left to right (Pl. X, Fig. 1). The individual cord impressions are rather indistinct. This is probably the result of a short sliding stroke of the paddle. There was very little subsequent attempt to smooth the exterior surface. The interior surfaces are roughly smoothed.

Decoration

The lip and the immediately adjacent outer rim area received the only decorative modifications, which are as follows:

Keyser Cord-marked Rims Divided according to Lip Treatment and Decoration

Lips with smoothed surface	23
Lips with cord-marked surface	104
Transverse incised grooves on smoothed lip	42
Transverse or oblique cord-wrapped stick impressions	130
Smoothed lips with circular punctates	9
Rims with small lugs	22
Small loop handle	1
Rims with rectilinear narrow line incising and punctates	16
Total	347

Almost one third of the lips bear impressions of a cord-wrapped paddle made either parallel with or transverse to the lip edges (Pl. XI, Fig. 4). Such lips are flat but have the cord-roughened surface. At least half of them project slightly outward owing to the lip treatment. Over a third of the lips have transverse impressions of either the cord-wrapped stick or the narrow edge of the cord-wrapped paddle which was used to work the surface (Pl. XI, Fig. 9). There is considerable variation in the width, depth, and spacing of these indentations. Instead of being transverse a few of these impressions have been placed obliquely. The lip was probably flattened before it was decorated. Less than 10 per cent of the lips have a smoothed surface, which is flattened to rounded. About 12 per cent have a smoothed or cord-marked lip with transverse notches made by a smooth implement. These also vary in size, shape, and spacing from narrow V-shaped cuts to wide U-shaped impressions. There are nine sherds which have a horizontal row of closely spaced circular punctates on the lip surface (Pl. XI, Fig. 7).

About 8 per cent of the Keyser Cord-marked rims have small irregularly shaped solid lugs attached on the outer rim contiguous with the lip (Pl. XI, Fig. 1). The cord-wrapped stick or paddle edge was used to impress the outer surface of the majority of these lugs (Pl. XI, Figs. 5, 8). One such specimen has a deep vertical incised gash (Pl. XI, Fig. 2). Two of the small lugs have each a small horizontal cylindrical hole, which runs underneath the lug at the level of the outer surface of the vessel (Pl. XI, Fig. 6). It is too small to have been of any service in supporting the vessel. The size of the lugs indicates that they were for decorative and not utilitarian purposes. One rim has a small strap or loop handle that is attached at the lip area and extends three centimeters below the lip (Pl. XI, Fig. 3). The lip surface of this sherd has shallow U-shaped transverse notches.

Only a very few sherds have incised lines, and on none of these is the design apparent. The lines, which are rectilinear, are placed obliquely on the outer rim, most often in groups of three. The incisions are narrow, being but two millimeters wide, and are medium-deep to deep. On three sherds irregularly spaced circular punctates were used in connection with incised lines. The lips of these sherds are smooth, and they have plain transverse notches.

Shape

The rim is either vertical and straight or slightly flaring. The lip shape and treatment have been discussed under the head of "decoration." These vessels were wide-mouthed jars with slightly constricted lower rims, slightly expanded bodies, and rounded bases.

In marked contrast to the cord-marked, shell-tempered pottery type discussed above are two grit-tempered types whose paste, surface finish, decoration, and shape clearly set them apart from the dominant ware at this site. There are a number of similarities that will be discussed in a statement of pottery relationships (pp. 409–413). Within the grit-tempered ware a further division can be made on the basis of the type of tempering employed and also of a number of other distinguishing features.

PAGE CORD-MARKED

The smaller series of sherds of the grit-tempered group has large angular particles of limestone as the tempering material. Many of these individual limestone pieces are three and four millimeters in diameter. In the majority of the sherds the limestone fragments nearest the surface have been leached, but in some all the limestone has been dissolved and angular holes have been left. There is a small proportion of limestone. The texture of the cross section is medium-fine to medium-coarse, depending on the degree of prominence of the temper. Otherwise the paste is rather fine grained and relatively compact. The exterior surface has a hardness of 2–2.5 to 2.5, and thus is somewhat softer than the preceding group. At least three fourths of these sherds have buff to reddish-tan exterior surfaces, whereas the rest are dull grayish brown. The interior surfaces are grayish tan to black. The color of the core is predominantly buff to light reddish tan or is gray.

Surface Finish

The exterior surfaces were malleated with a cord-wrapped paddle, but the impressions so made have been partly obliterated. The obliteration is particularly noticeable in the rim area. The interior surface is smooth.

Decoration

Five of the rims have a plain, unmodified exterior; on nineteen there is an added outer rim strip which gives a slightly collared appearance (Pl. XII, Fig. 7). The rim strips vary somewhat in size, conforming in some degree to the size of the vessel. The normal range is from two to three centimeters. Only three of these strips bear no additional decoration beyond the partially obliterated cord-wrapped paddle marks. One sherd is vertically incised at the base of the thickened rim (Pl. XII, Fig. 4); a second has oblique punch marks at the lower edge of the added rim strip. Between the lip of this last sherd and the row of punch marks there are two closely spaced medium-wide and medium-deep U-shaped horizontal incised lines (Pl. XII, Fig. 1). Six of the added rim strips have obliquely placed impressions of a cord-wrapped string, or cylinder, across the lower segment of the rim strip. Four rims show horizontal impressions of cord-wrapped string between the lip and similar impressions that were made vertically or obliquely across the projecting edge of the rim strip. One vessel is represented by three rim fragments and two upper body pieces (Pl. XII, Figs. 2, 5–6). In this vessel there are two parallel rows of cord-wrapped string impressions on the added rim strip and six to seven roughly horizontal rows on the lower rim. The lower edge of the rim strip is cut by short, deep vertical incised gashes. One rim sherd of this type has irregularly placed shallow individual cord impressions on the upper rim and a horizontal row of deep thumbnail gouges 1.5 centimeters below the lip (Pl. XII, Fig. 3). Three of the sherds have cord-wrapped stick impressions vertically or obliquely placed on the inner rim surface.

Shape

The rims are vertical to slightly flaring, with the body somewhat wider than the orifice. No basal sherds were examined.

POTOMAC CREEK CORD-IMPRESSED

The pottery belonging to the most numerous group of the granular tempered sherds has crushed particles of rock for tempering material. The fragments are of various sizes, ranging from less than a millimeter in diameter to three and four millimeters. There is approximately 20 to 30 per cent of grit. Though the exact nature of the rock employed has not been determined, most of it is apparently granitic. The cross-sectional texture is usually medium to medium-coarse. The hardness ranges from 2.5 to 3.5, and the majority of the sherds have a surface hardness of 3. As a rule the exteriors are a grayish black color; a small number of sherds are light tan to buff. There is very little reddish tinge. The interiors are almost all grayish black; the core is dark gray, except on a few of the buff-colored sherds, where the core is also buff.

Surface Finish

Using as evidence not only the decorated rims, but also the undecorated rims and body sherds, one can see that these vessels were originally malleated with a cord-wrapped paddle over the entire exterior surface. On a majority of those sherds that have a rim decoration the rim and the upper body had been well smoothed before the decoration was applied. The cords used in this group for surfacing are rather small and closely braided, and the majority have a clockwise twist. The interior surfaces are roughly smoothed, with fine finger striations showing on the inner rim area. This inner smoothing was not always sufficient to cover the projecting edges of the grit particles completely. There are nine plain cord-marked rims, on six of which the cord impressions were applied obliquely and slant from left to right. The lips of the nine rims are cord-marked and are rounded to flattened.

Decoration

About 90 per cent of the rims on this distinctive pottery type have a decoration on the outer-rim area. It was made either with a single cord, placed against the plastic clay and impressed to form a decorative pattern, or by a technique that is here called "pseudo-cord." The majority of the rims were decorated by the latter technique, and on these sherds the impressions were probably made by a cord wrapped around a flexible element that was either a fiber of some type or another material of similar nature. A numerical listing of the several styles showing variations that may be important in future comparative distributional studies is given below.

POTOMAC CREEK CORD-IMPRESSED RIMS DIVIDED ACCORDING TO DECORATION

Plain cord-marked rims	9
Plain rims with horizontal pseudo-cord impressions	13
Horizontal pseudo-cord impressions, lip notched	8
Lipless rims with horizontal pseudo-cord impressions	19
Plain rims with oblique pseudo-cord impressions	7
Lipless rims with oblique pseudo-cord impressions	2
Patterned arrangement of pseudo-cord impressions	10
Added rim strip with lower edge notched obliquely	3
Added rim strip bearing horizontal pseudo-cord impressions	3
Rims with cord-impressed rim decoration	11
Total	85

By far the most common style was the arrangement of the pseudo-cord impressions in closely spaced, parallel horizontal rows on the outer rim (Pl. XIII, Figs. 2, 4). Depending somewhat on the size of the vessel, this decorated zone extends from two to five millimeters beneath the lip. The notched lips were impressed on the outer lip edge in four examples and transversely across the lip on the remaining four. In three, notching of the latter type is so wide and deep as to present a scalloped appearance (Pl. XIII, Fig. 3). The scalloping was done either with a rather large cord-wrapped dowel or with the edge of the cord-wrapped paddle. One interesting rim with such horizontal pseudo-cord impressions has a small added lug, which was vertically applied on the lip surface and forms a small semicircle projecting above the lip. This has been transversely notched. It resembles in concept the small horizontally projecting lugs on Keyser Cord-marked that were vertically notched.

Seven of the rims have closely spaced parallel oblique pseudo-cord impressions. Five slant from right to left, and the other two from left to right. The lips on two are slightly notched transversely; the others are flattened and cord-marked. These oblique cord impressions begin at the lip level and extend downward to a horizontal pseudo-cord impression which sets off the smoothed lower rim from the decorated upper rim area.

There are ten rims that have a patterned arrangement of pseudo-cord impressions (Pl. XIII, Figs. 1, 5). On them the decorated zone is set off from the lip area and the lower rim by horizontal rows of pseudo-cord impressions. Within this zone the decoration was applied at various angles, forming a herringbone or line-filled triangles or rectangles.

There are six rims with an added strip. Three have right oblique incisions or punctates on the lower margin of this strip. The other three bear horizontal pseudo-cord impressions running parallel around the rim.

The decoration of eleven rims is formed by lines of individual cord impressions in either horizontal or oblique rows. Those placed at an angle are usually set off in the decorated zone by a horizontal cord impression. Two of the lips in this group are notched. One was impressed by a smooth implement and the other by a cord-wrapped stick or the edge of the cord-wrapped paddle. Four of the rims were probably parts of one vessel.

Shape

The rim is either vertical or slightly flaring. The lip is usually approximately the same width as that of the rim and is flattened or rounded, depending upon the amount of modification caused by the application of the cord-wrapped paddle. Most of the lips either are cord-marked or were impressed by the edge of the cord-wrapped paddle, or by the implement used to decorate the rim. Only one good-sized lower body fragment is present in the collection, and it indicates that the vessel base was almost conical. Probably the majority of the vessels of this type had such bases.

POTTERY RELATIONSHIPS

As noted in an earlier paragraph (p. 402), there are three well-defined ceramic types in this collection from the Keyser site. The most common ware is the shell-tempered cord-marked wide-mouthed jar called "Keyser Cord-marked," which has its closest relationships to the northwest and west. A very similar pottery type, from excavations in Fayette and Westmoreland counties in southwestern Pennsylvania, has already been described and illustrated by Engberg.[17] The Ceramic Repository at the University of Michigan has a large collection of sherds from the drainage area of the Monongahela and Youghiogheny rivers. They were collected from various sites by Mr. George Fisher. Some were presented to the University of Michigan by the Pennsylvania Historical Commission, which purchased one of Mr. Fisher's collections; the sherds from documented sites were obtained directly through the courtesy of Mr. Fisher. Subsequent excavation in this same area by the Pennsylvania Historical Commission, under the active direction of Dr. Mary Butler, also uncovered this same ware. Somewhat more distantly related is a shell-tempered type which has been found by Mr. Ross Pier Wright in northwestern Pennsylvania and southwestern New York and by field parties excavating for the Pennsylvania Historical Commission. Another relative of this Keyser Cord-marked type is represented by sherds sent to the Ceramic Repository from numerous sites in the Ohio Valley drainage from northwestern West Virginia by E. W. Fetzer.

There can be little hesitation in assigning the shell-tempered

[17] Engberg, 1930, 1931.

pottery to a group of people who were closely related culturally to what Dr. Butler has called the "Monongahela Woodland Culture." [18] It is becoming increasingly evident that to the east and northeast of the group of sites which have been grouped together into the Fort Ancient Aspect [19] there is a cultural complex that was contemporary with and related to the Fort Ancient and Iroquois divisions. Its manifestation is contained almost entirely within the upper reaches of the Ohio drainage and can be considered the eastern border zone between those late prehistoric and protohistoric divisions now classified in the Mississippi Pattern and those which can be grouped into the Woodland Pattern. The basic shape of the pottery, the overall cord-marked surface, the lack of variety in shape, the relative simplicity of decoration, and the relative absence of utilitarian or ornamental appendages strongly suggest Woodland pottery as the direct genetic background for this type. On the other hand, the shell tempering, the spare use of rectilinear incised and punctate decoration, together with the use of small solid rim lugs and an occasional strap handle, represent the Mississippi ceramic ideas which have been engrafted upon the older Woodland tradition. These Mississippi influences are not so strongly represented on the pottery from the Keyser site as they are on that from southwestern Pennsylvania. Their modifications are analogous to those on the pottery of the Fort Ancient Aspect.

A connection between the Keyser Cord-marked and the Potomac Creek Cord-impressed types at the Keyser site can be recognized in certain features. The fundamental shape is very much the same, but a rounded bottom is typical for the shell-tempered ware and a conoidal bottom for the grit-tempered. Both types have the exterior surface malleated with a cord-wrapped paddle, although the resultant surfaces are somewhat different. On both types the lip surface is sometimes cord-marked or notched with a cord-wrapped dowel or the edge of the cord-wrapped paddle. The differences between these types are equally if not more important. The shell-tempered type probably has a higher percentage of rims that gradually flare outward; the small solid lug on the outer rim is an integral part of the decoration, and the incised and punctate sherds find no counterpart in the grit-tempered type. The cord-marked surface extending to

[18] Butler, 1939.
[19] Griffin, 1943.

the lip and the predominant notched lips provide a distinctive upper rim treatment which is characteristic of Keyser Cord-marked.

The Page Cord-marked type, with its added rim strip, suggests a connection with a limestone-tempered ware found by Fowke at Gala, Botetourt County, Virginia,[20] and by Holmes at a site three and one-half miles north of Luray, Virginia.[21] The ware discovered at these sites is limestone-tempered. The exterior surface of most of it is covered with cord-wrapped paddle, but there is some use of what may have been a plain plaited fabric and a net fabric. The dominant use of an added outer rim strip which has finger notches or incised notches across its lower margin, or a horizontal row of small punctates just below the lip, serves to distinguish this Woodland type from the ones discussed in this paper. The type of rim-strip decoration found at the Keyser site is significantly different, however, from that obtained by Fowke, but that obtained by Holmes is quite similar and also suggests a strong connection with the Potomac Creek Cord-impressed type. The pottery most closely connected with the sherds from Gala is described in Holmes' text as Piedmont Virginia ware, but in the illustration they are called members of the Chesapeake-Potomac Group. Actually both designations would be correct.

The pseudo-cord-impressed and cord-impressed decorated pottery was classed by Holmes as a subdivision of his Middle Atlantic Province. This subdivision extended from the coastal area of the Carolinas as far north as Maine. One of the most important problems of the eastern United States culture history is the classification of the different ceramic types along the seaboard and their arrangement in a chronological sequence in the various subareas. The particular subgroup into which these grit-tempered sherds from the Keyser site can be placed Holmes called "Potomac Creek Ware," the type site being an Algonquian village called Pottowomeck.

Recently excavated sites in the Washington area will undoubtedly provide a sounder basis for a comparative statement on the ceramic connections of this Potomac Creek Cord-impressed type.[22] As a distinctive entity within the very broad Woodland pottery tradition it can confidently be said that this type does not extend very far south of the Rappahannock or very far north of the Susquehanna. Bushnell [23] presents pottery from Jerrys Flats, Richards Ford, and

[20] Fowke, 1894, p. 17.
[21] Holmes, 1903, pl. CXXXIV.
[22] Stewart, 1939 and 1940.
[23] Bushnell, 1935.

Forest Hall, and from an unnamed site near "the large island" in the Rappahannock which belongs to the same Potomac Creek type. He also illustrates pottery from the same and adjacent sites that does not belong to the Potomac Creek Cord-impressed type. In another publication Bushnell [24] figures pottery from the site of Pisaseck which bears little resemblance to the grit-tempered sherds from the Keyser site or to the majority of the pottery he illustrated in his earlier publication from the sites above the falls of the Rappahannock. Just what material collected by him from the area formerly occupied by the Virginia Siouan tribes actually belonged to that linguistic stock is still a matter of controversy. The pottery from the sites along the James reproduced by Bushnell [25] does not closely resemble the Potomac Creek Cord-impressed type or the net or plain plaited fabric (not coiled) pottery that occurs farther north on the coastal area. The Woodland pottery found along the Coastal Plain and the Piedmont area of the Carolinas and Georgia does not indicate a close relationship with the Potomac Creek type. There is only a general resemblance between the Potomac Creek type and the cord-marked grit-tempered or clay-tempered Woodland pottery of the Tennessee, Ohio, and central and lower Mississippi Valley drainages. The ceramic connections are rather to the north along the Coastal Plain and the drainage systems flowing into the Atlantic, until in the New York area there is a westward spread of cord- and pseudo-cord-impressed decoration which carries across southern Canada north of the Lakes and across Michigan and Wisconsin into Minnesota.[26] The techniques employed, but not the actual designs, are the same as those which appear on the Owasco Aspect of New York.[27] The same decorative concepts occur on Woodland ware in the vicinity of New York City [28] and on north into New England.[29] The northern limit is apparently in Nova Scotia.[30] Pottery with a similar decora-

[24] Bushnell, 1937.

[25] Bushnell, 1930.

[26] The recent find of pseudo-cord-decorated pottery as part of the culture associated with an "early" Marksville period mound in Louisiana is significant in estimating the antiquity of this style of decoration in the north and the age of the Marksville period in the lower Mississippi Valley. Ford and Willey, 1940, p. 88, fig. 20 j–k, fig. 28 c; fig. 34.

[27] Ritchie, 1934 and 1936.

[28] Skinner, 1919.

[29] Willoughby, 1935.

[30] Smith and Wintemberg, 1929.

tive technique is widespread throughout Michigan[31] and Wisconsin and into Minnesota. There would be little difficulty, however, in separating distinct ceramic types throughout this distribution if sufficient attention were paid to the material.

It should be noted that the pottery of the Potomac Creek Cord-impressed type has been called "Siouan" by Bushnell and "Algonquian" by Holmes and others.[32] It does not occur on known Siouan sites of the early historic period in southern Virginia and the Carolinas, nor has it been found in the upper Ohio drainage. The related types to the north of Virginia have usually been associated with Algonquian-speaking groups. The majority of related sites in Michigan and Wisconsin have been attributed to Algonquian divisions, but there has been a tendency on the part of some to suggest an Iroquoian connection for the Younge site. The related pottery in Minnesota and northwestern Wisconsin has been confidently assigned to the Dakota Sioux.[33]

POTTERY SUMMARY

The pottery from the Keyser site is composed of two major and one minor ceramic types. The dominant type is a shell-tempered cord-marked wide-mouthed jar called "Keyser Cord-marked," which belongs in the same cultural division as the late prehistoric and protohistoric cultures of southwestern Pennsylvania. The limestone-tempered subgroup of the grit-tempered ware called "Page Cord-impressed" finds close connections in the immediate area, whereas the cord-impressed and pseudo-cord-impressed grit-tempered sherds were called "Potomac Creek Cord-marked," since they are closely related to if not identical with the Potomac Creek type defined by Holmes. Though the evidence is by no means clear, there is a suggestion that the granular tempered wares may be slightly older on the Keyser site. It is roughly estimated that the site was occupied sometime between 1550 and 1650.

Food Remains

Since practically the entire economy of the Indians revolved around the procurement and preparation of foodstuffs, it is of interest to note the results of their efforts as shown by the food remains preserved in the excavations. These may be divided as floral and faunal.

[31] Greenman, 1937. [32] Ferguson, 1937. [33] Wilford, 1941.

The floral portions of the diet were undoubtedly of greater importance than is indicated by the scanty remains. The principal articles of such food were corn, squash, and beans, which were cultivated in the garden patches of the village. These were supplemented by the nuts, seeds, berries, and roots that could be gathered from the surrounding forests, along the river banks, and on mountain slopes.

Carbonized remains of the following foods [34] were found:

Corn	*Zea mays*; abundant
Black walnut	*Juglans nigra* L.; abundant
Big shellbark	*Carya laciniosa* Michx. f.
Sweet pignut (a hickory)	*Carya glabra* Mill.
Squash	(Identification not definite); scarce
Papaw	(Identification not definite); scarce

By far the most plentiful of the remains of foodstuffs were the bushels of animal bones and the pecks of mussel shells, which indicate the great dependence placed on the animal kingdom as a source of food. Among the animals most sought after was the Virginia deer, to judge by the fragments of antler and bone refuse in the pits. All the long bones were cracked to extract the marrow contained in them. Apparently every type of animal in the area was consumed. A list of the mammals identified by the osseous remains follows: [35]

Opossum	*Didelphis virginiana marsupialis*
Elk	*Cervus canadensis*
Deer	*Odocoileus virginianus*
Bear	*Euarctos americanus*
Dog	*Canis familiaris*
Raccoon	*Procyon lotor*
Otter	*Lutra canadensis*
Fox	*Urocyon cinereoargenteus*
Wildcat	*Lynx rufus*
Mountain lion (cougar)	*Felis concolor*
Weasel	*Mustela frenata noveboracensis*
Skunk	*Mephitis mephitis*
Squirrel	*Sciurus carolinensis*
Rabbit	*Sylvilagus floridanus*
Muskrat	*Ondatra zibethica*
Groundhog	*Marmota monax*
Beaver	*Castor canadensis*

Bones of birds, especially those of the turkey, were second in abundance to those of the mammals. The amount of splintered and

[34] Identified by Dr. Paul G. Russell, of the Bureau of Plant Industry.
[35] Identified by Dr. Remington Kellogg, United States National Museum.

fragmentary bone indicates that birds were an important article of diet. Not many species are represented in the collections, but undoubtedly additional ones were considered "fair game." The species identified by bones are as follows: [36]

Wild turkey	*Meleagris gallopavo*
Turkey buzzard	*Cathartes aura*
Bald eagle	*Haliaeetus leucocephalus*
Whistling swan	*Cygnus columbianus*

The amphibians were well represented by two species. One of these is the fresh-water terrapin, or slider. Many bones and fragments of the carapace and plastron were scattered throughout the digging, but there was no indication that these were utilized for ornaments or implements. The second species is the common box tortoise found plentifully in the woods.[37]

That fish were caught and consumed is shown by the presence of one unfinished fishhook, one net sinker, and one fish scale discovered in a pit. Fish abound in the river adjacent to the site, and include bass, carp, perch, catfish, and many smaller species.

Mollusks of various types were eaten and also furnished raw material for artifacts. The river, alternating between a series of rapids and long stretches of still deep water, offered ideal living conditions for a variety of shellfish. Snails of different kinds were also employed to vary the diet, and both land and water species were recovered. The following species, all of which are still found in the region, were identified from the remains: [38]

MUSSELS

Elliptio complanatus (Dillwyn), mainly from the rapids
Anodonta cataracta Say, mainly from the rapids
Lampsilis cariosa (Say), mainly from the deeper water

FLUVIATILE SNAILS

Campeloma decisum Say, from sandy or muddy bottoms

TERRESTRIAL SNAILS

Mesodon thyroidus Say, from sandy or muddy bottoms
Triodopsis albolabris Say, from sandy or muddy bottoms

[36] Identified by Dr. Herbert Friedmann, United States National Museum.
[37] Identified by Dr. Leonhard Stejneger, United States National Museum.
[38] Identified by Dr. J. P. E. Morrison, United States National Museum.

Conclusions

It is difficult at this time to reach any satisfactory conclusion regarding the cultural affiliations of the site. The presence of three pottery types as distinctive as those at Keyser farm site indicates a commingling in the Page Valley of three different cultural units. At this time it is uncertain whether these types were contemporary or whether the Page Cord-marked and possibly the Potomac Creek Cord-impressed were slightly earlier than the Keyser Cord-marked type. With the possible exception of the Monongahela Woodland Aspect the related culture groups in the immediate area have not been adequately described or illustrated in the literature. Fortunately this condition can be remedied in the period after the war by the publication of a number of reports already prepared, and by others on sites in the Washington area which have been excavated.

The Page Cord-marked type is most closely connected to the Shenandoah Valley area, and it is probably the end product of a considerable period of local ceramic development. It is a variant of a pottery series of the widespread Woodland ware. Another type of this series was found at Gala.[39] Limestone-tempered sherds of a related type were collected by Manson and MacCord near Alma, Virginia, about twenty-five miles south of the Keyser site, where they were associated with complicated stamped pottery of the South Appalachian area. These sherds most closely resemble the style of stamping of the late prehistoric period, and they mark the most northerly record of the complicated stamp east of the Appalachians.

The Keyser farm site is not far removed from the Monongahela Woodland sites or some of the Chesapeake area Woodland sites of a late period. The similarity of pipe shapes and decorations between the Monongahela area, the Shenandoah Valley, and the coastal area is evidence of cultural interchange between these areas.

The antler and bone artifacts are not particularly diagnostic. The unusual antler ornaments can be interpreted as local representatives of a rather widespread trait which was present in the eastern United States from at least the Middle Woodland period of Hopewell domination. The notched turkey metatarsal awl is found in considerable numbers in Fort Ancient sites and also in the Dallas

[39] Fowke, 1894, pp. 17–23; also Griffin, 1943, pp. 193–194, and Pl. CXXVII.

Focus of eastern Tennessee.[40] The remainder of the artifacts are at present of relatively little value in indicating closer cultural affiliations. The burial position and the site arrangement can be assigned to the general Woodland culture of the late prehistoric and early historic period along the central Atlantic coast.

LITERATURE CITED

1. Augustine, Edgar E. 1940. Fort Hill. The Pennsylvania Archaeologist, 10 (3) : 51–58.

2. Bushnell, David I., Jr. 1930. The Five Monacan Towns in Virginia, Smithsonian Misc. Coll., Vol. 82, No. 12, Washington.

3. —— 1935. The Manahoac Tribes in Virginia, 1608. *Ibid.*, Vol. 94, No. 8.

4. —— 1937. Indian Sites below the Falls of the Rappahannock, Virginia. *Ibid.*, Vol. 96, No. 4.

5. Butler, Mary. 1939. Three Archaeological Sites in Somerset County, Pennsylvania, Bulletin No. 753. Penn. Hist. Comm., Harrisburg.

6. Engberg, Robert M. 1930. Archaeological Report. Western Penn. Hist. Mag., 13 (2) : 71–103.

7. —— 1931. Algonkian Sites of Westmoreland and Fayette Counties, Pennsylvania. *Ibid.*, 14 (3) : 143–190.

8. Ferguson, Alice L. L. 1937. Burial Area in Moyoane. Journ. Wash. Acad. Sci., 27 (6) : 261–267.

9. Ford, James A., and Willey, Gordon. 1940. Crooks Site, a Marksville Period Burial Mound in La Salle Parish, Louisiana. Anthropological Study No. 3, Department of Conservation, Louisiana Geol. Surv., New Orleans.

10. Fowke, Gerard. 1894. Archeologic Investigations in James and Potomac Valleys. Bureau of American Ethnology, Bulletin 23, Washington.

11. Greenman, Emerson F. 1937. The Younge Site. Occas. Contrib. Mus. Anthropol. Univ. Mich., No. 6.

12. Griffin, James B. 1943. The Fort Ancient Aspect: Its Cultural and Chronological Position in Mississippi Valley Archaeology. University of Michigan Press, Ann Arbor.

13. Harrington, Mark R. 1922. Cherokee and Earlier Remains on Upper Tennessee River. Indian Notes and Monographs, No. 24. Museum of the American Indian, Heye Foundation, New York.

14. Holmes, William H. 1889. Pottery of the Potomac Tide-water Region. Am. Anthropol., 11 (3) : 246–252.

15. —— 1897. Stone Implements of the Potomac-Chesapeake Tidewater Prov-

[40] Prepublication data, by courtesy of the Department of Anthropology of the University of Tennessee.

ince. Annual Report of the Bureau of American Ethnology, 15:13–157. Washington.

16. Holmes, William H. 1903. Aboriginal Pottery of the Eastern United States. *Ibid.*, Vol. 20.

17. Hrdlička, Aleš. 1927. Catalogue of Human Crania in the United States National Museum Collections. (The Algonkin and Related Iroquois; Siouan, Caddoan, Salish and Sahaptin, Shoshonean, and California Indians.) Proc. United States Nat. Mus., Vol. 69, Art. 5.

18. Moorehead, Warren K. 1922. The Hopewell Mound Group of Ohio. Anthropological Series, Field Museum of Natural History, Vol. VI, No. 5. Chicago.

19. Parker, Arthur C. 1922. The Archeological History of New York. New York State Mus. Bull., pp. 235–236. Albany.

20. Ritchie, William A. 1934. An Algonkin-Iroquois Site on Castle Creek, Broome County, New York. Research Records of the Rochester Municipal Museum, No. 2. Rochester.

21. —— 1936. A Prehistoric Fortified Village Site at Canandaigua, Ontario County, New York. *Ibid.*, No. 3.

22. Skinner, Alanson. 1919. Exploration of Aboriginal Sites at Throgs Neck and Clasons Point, New York City. Contributions from the Museum of the American Indian, Heye Foundation, Vol. V, No. 4. New York.

23. Smith, Harlan I., and Wintemberg, William J. 1929. Some Shell-Heaps in Nova Scotia. National Museum of Canada, Bulletin No. 49. Anthropological Series, No. 9. Ottawa.

24. Speck, Frank G. 1940. Penobscot Man. University of Pennsylvania Press, Philadelphia.

25. Stewart, T. Dale. 1939. Excavating the Indian Village of Patawomeke (Potomac). Exploration and Field Work of the Smithsonian Institution in 1938, pp. 87–91. Washington.

26. —— 1940. Further Excavations at the Indian Village Site of Patawomeke (Potomac). *Ibid.*, 1939, pp. 79–82.

27. Stirling, Matthew W. 1935. Smithsonian Archeological Projects Conducted under the Federal Emergency Relief Administration, 1933–34. Annual Report of the Smithsonian Institution for 1934, pp. 371–400.

28. Wilford, Lloyd A. 1941. A Tentative Classification of the Prehistoric Cultures of Minnesota. Am. Antiquity, 6 (3) : 231–250.

29. Willoughby, Charles C. 1935. Antiquities of the New England Indians. Peabody Museum, Harvard University, Cambridge.

PLATE I

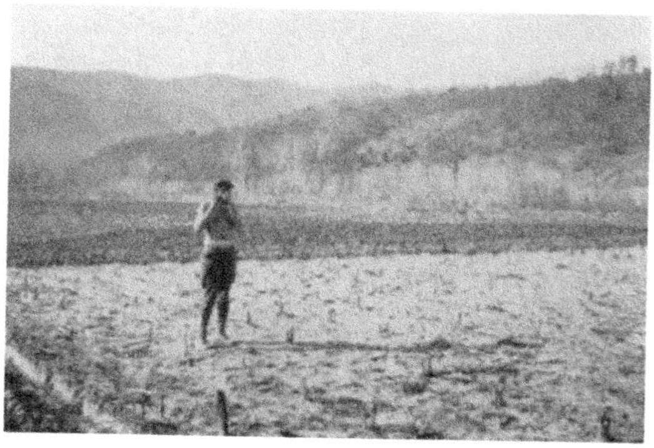

FIG. 1. The Keyser farm site, looking east from stake C 4 (see Fig. 2)

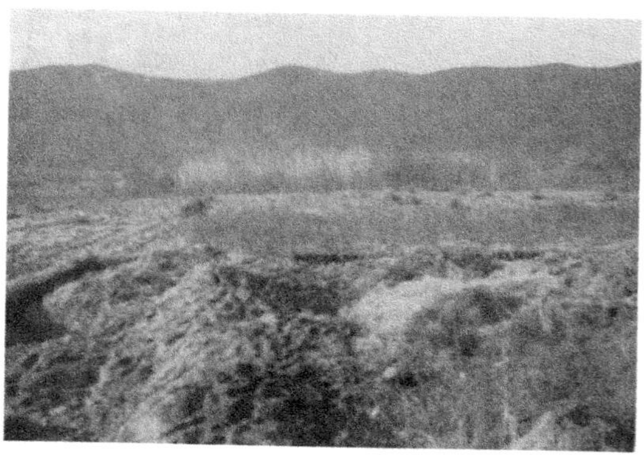

FIG. 2. The Keyser farm site, with Massanutten Mountain in the background

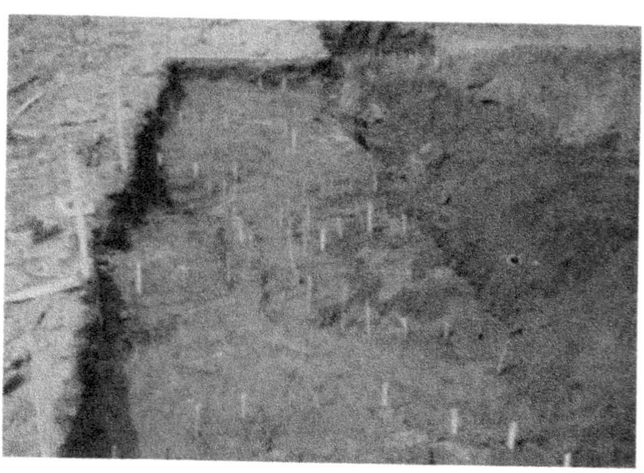

FIG. 3. Pit outlines delineated by stakes after removal of topsoil

PLATE II

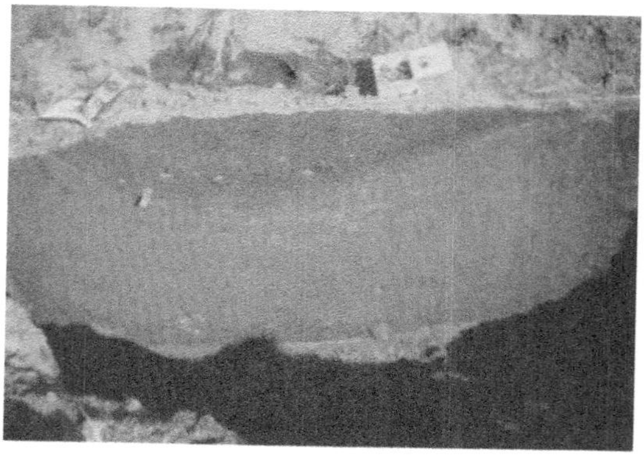

FIG. 1. Profile of Pit 85, with dark upper level clearly differentiated from light lower level

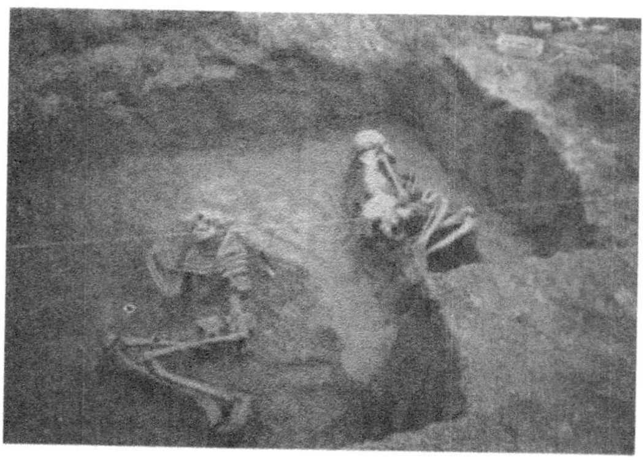

FIG. 2. Burials 5 (right) and 6

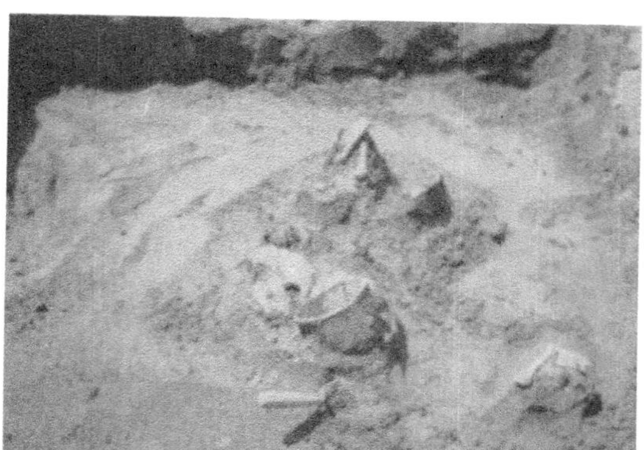

FIG. 3. Burials 9 and 10. The pottery vessel beside the skull is illustrated in Plate X, Figure 1

PLATE III

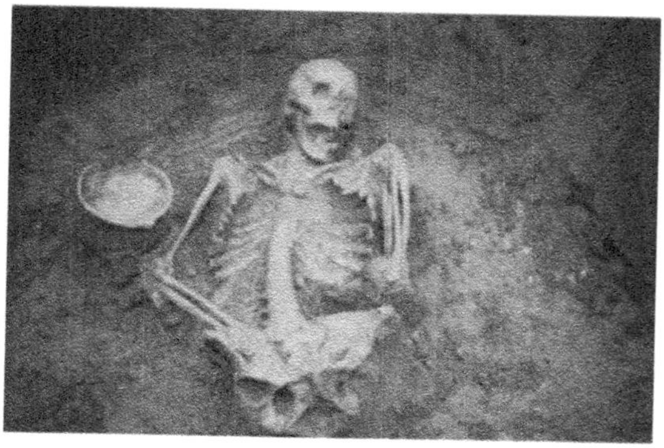

FIG. 1. Burial 20. A female with funerary jar (illustrated in Pl. X, Fig. 2) at right shoulder

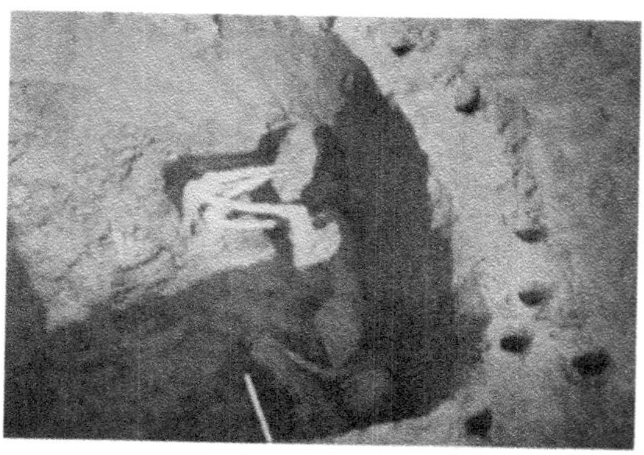

FIG. 2. Burial 21, with posthole outlines bordering Pit 77

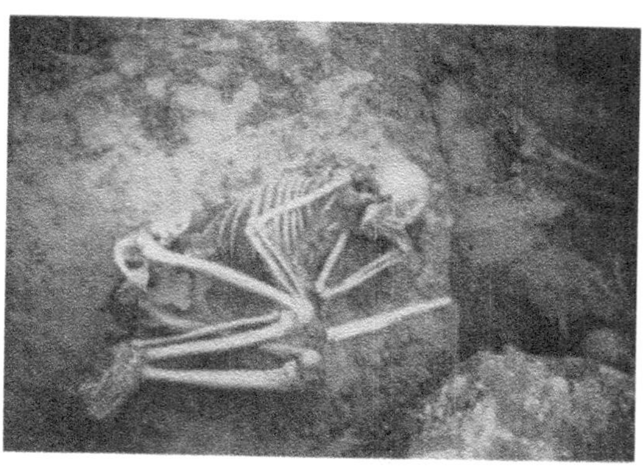

FIG. 3. Burial 16. An adult female

PLATE IV

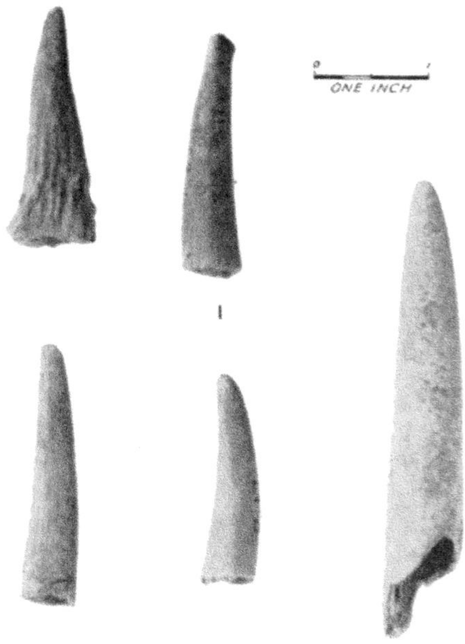

FIG. 1. Antler projectile points

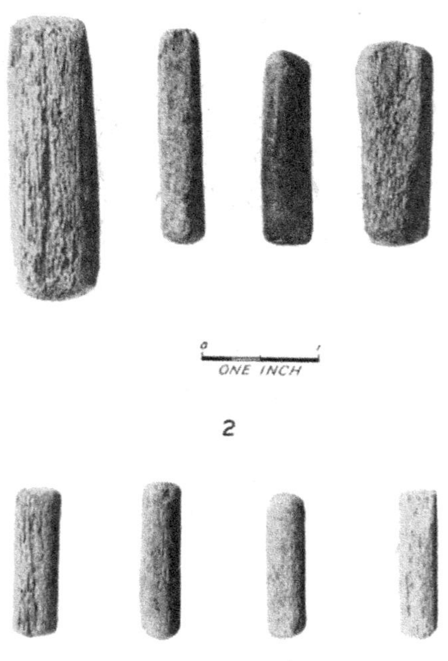

FIG. 2. Antler drifts

United States National Museum negatives 36328–A (Fig. 1) and 36328–D (Fig. 2)

EXPLANATION OF PLATE V

FIGS. 1–2. Bodkins

FIG. 3. Chipper (hollow and notched)

FIG. 4. Scapula awl

FIG. 5. Turkey tibiotarsal awl (notched tally marks)

FIG. 6. Bone tube from Pit 73

FIG. 7. Whistle from Pit 28

FIGS. 8–12. Pierced chippers

FIGS. 13–15. Notched chippers

PLATE V

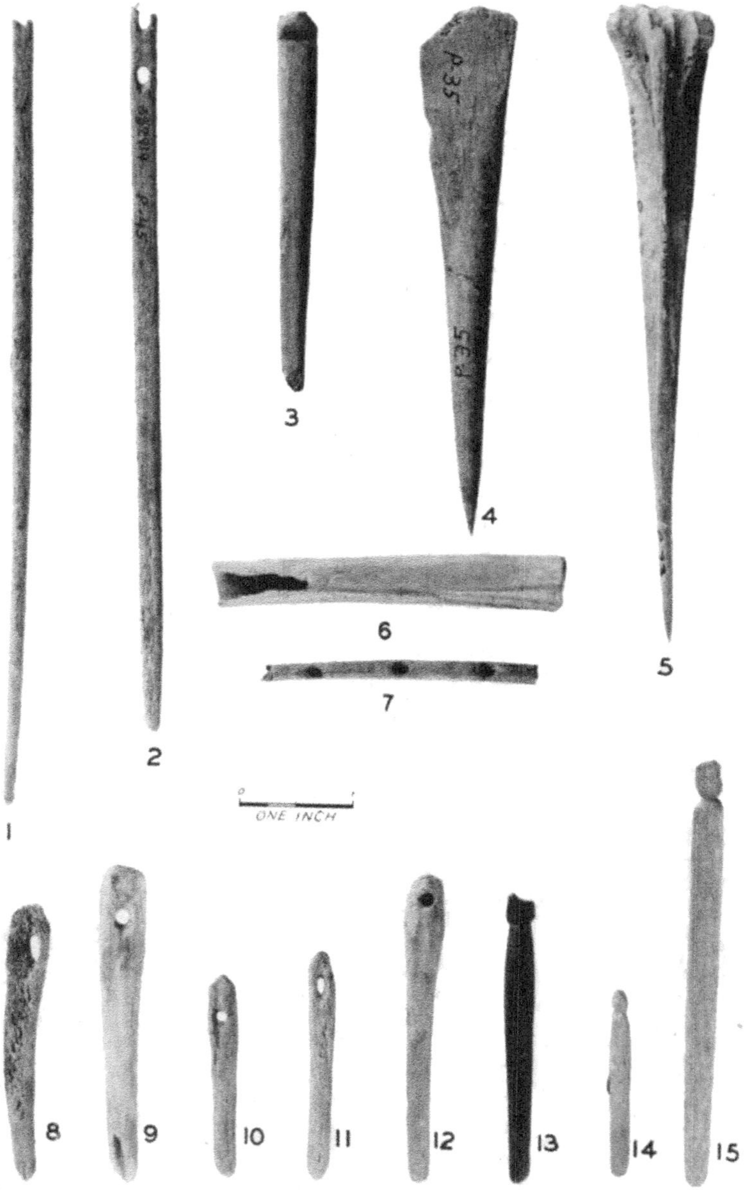

Bone implements (U. S. N. M. neg. 36328-B)

PLATE VI

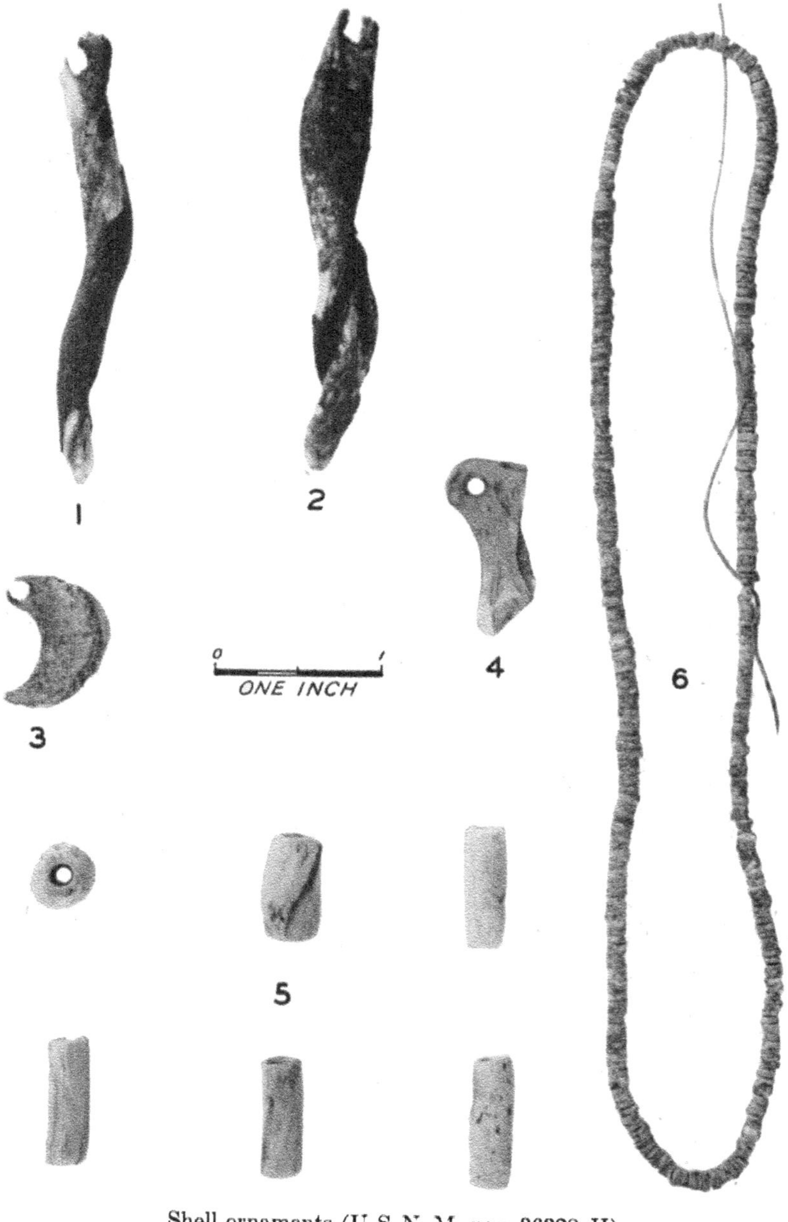

Shell ornaments (U. S. N. M. neg. 36328–H)

Figs. 1–2. Columella pendants with Burial 12

Figs. 3–4. Small pendants (surface finds)

Figs. 5–6. Beads associated with Burial 24

PLATE VII

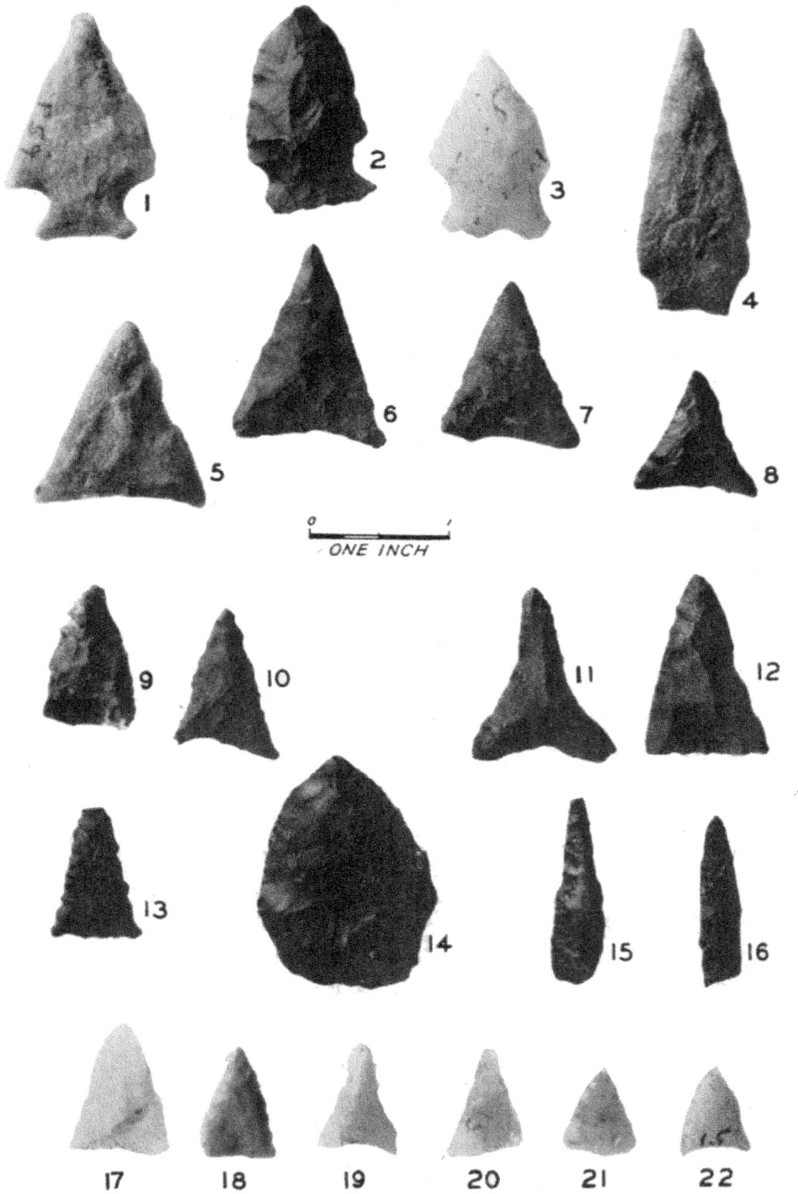

Chipped stone implements (U. S. N. M. neg. 36328–G)

The materials are as follows: chalcedony, 18; chert, 21; flint, 7–9; red jasper, 13–16; quartz, 3, 17, 19, 20, 22; quartzite, 1, 4, 5; shale (?), 6; yellow jasper, 2, 10–12

PLATE VIII

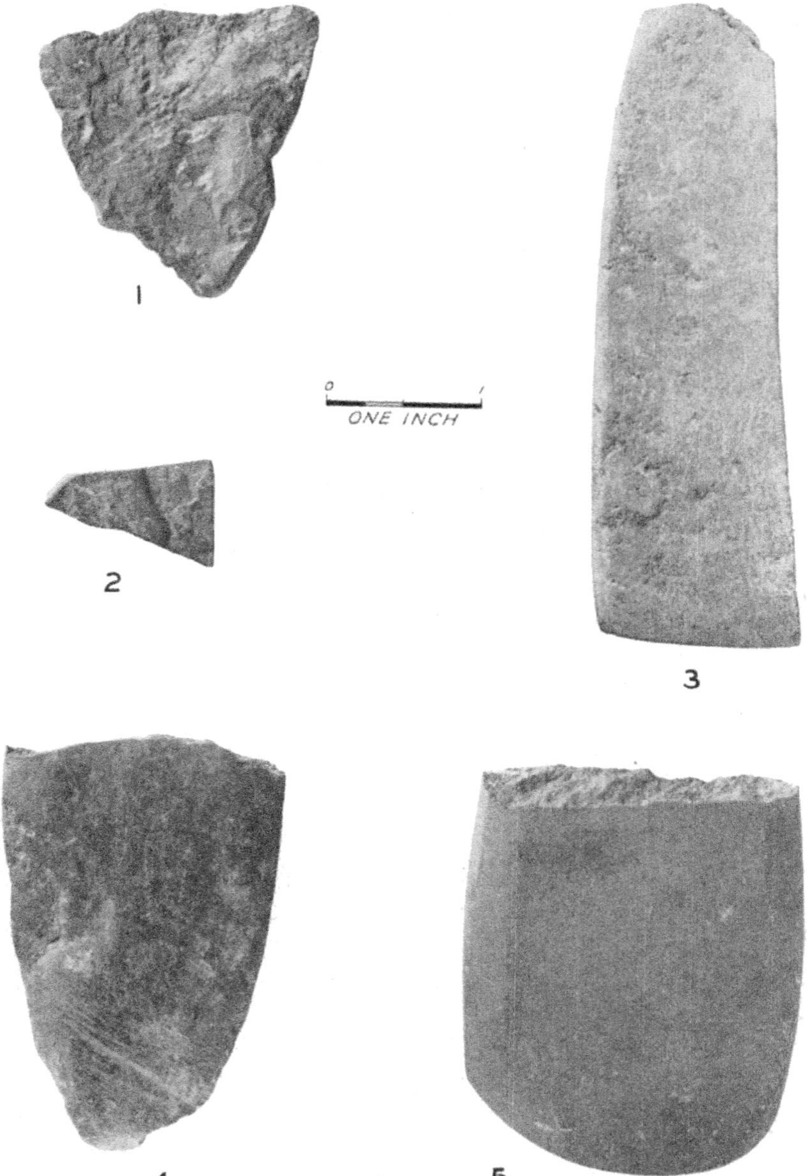

Worked and polished stones (U. S. N. M. neg. 36328–C)

FIGS. 1–2. Rubbed limonite

FIG. 3. Broken stone tablet found with Burial 22

FIG. 4. Abrading stone found with Pit 45

FIG. 5. Unfinished stone tablet

EXPLANATION OF PLATE IX

FIG. 1A. Large copper beads associated with Burial 12

FIG. 1B. Small copper beads found with Burial 1

FIG. 2A. Obtuse-angled elbow pipe of clay with flattened mouthpiece

FIG. 2B. Elbow pipe of clay with fine dentate design on bowl

FIG. 2C. Flattened mouthpiece of clay

FIG. 2D. Fragment of stemless ovoid stone pipe found on surface

FIG. 2E. Fragment of stone pipe bowl found with Burial 7

PLATE IX

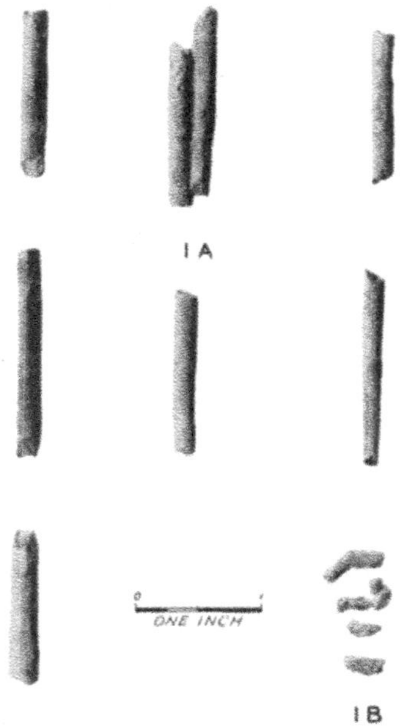

FIG. 1. Copper beads (U. S. N. M. neg. 36328-E)

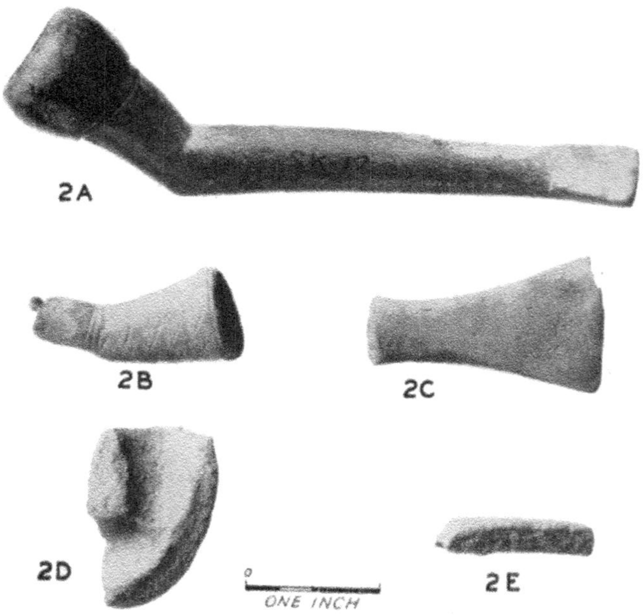

FIG. 2. Pipes and pipe fragments (U. S. N. M. neg. 36328-F)

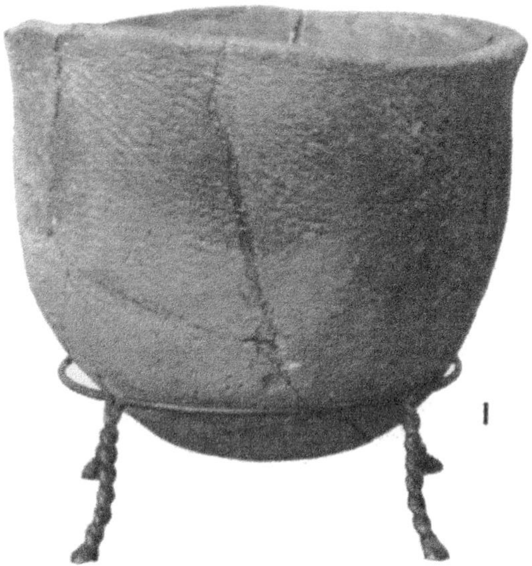

Fig. 1 (Burial 10)

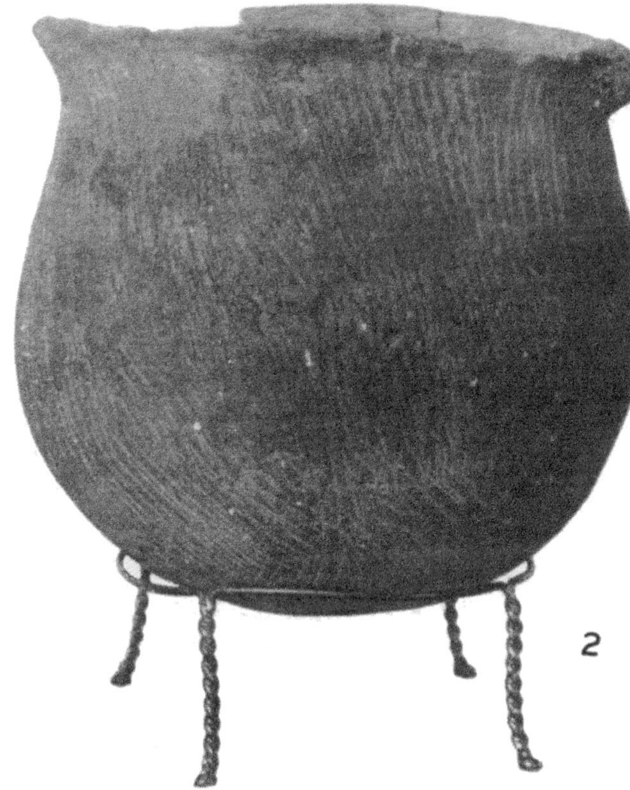

Fig. 2 (Burial 20)

Keyser Cord-marked jars (U. S. N. M. neg. 36328)

EXPLANATION OF PLATE XI

FIG. 1. Small solid lug attached to outer rim at level of lip

FIG. 2. Small solid lug with V-shaped gash

FIG. 3. Small strap handle

FIG. 4. Rim sherd with cord-wrapped paddle impressions on surface of lip

FIG. 5. Small solid lug with cord-wrapped stick impressions

FIG. 6. Small lug with small horizontal perforation and horizontal row of punctates

FIG. 7. Small circular punctates on surface of lip

FIG. 8. Cord-wrapped stick impressions applied vertically on small lug

FIG. 9. Cord-wrapped stick impressions applied transversely on lip surface

PLATE XI

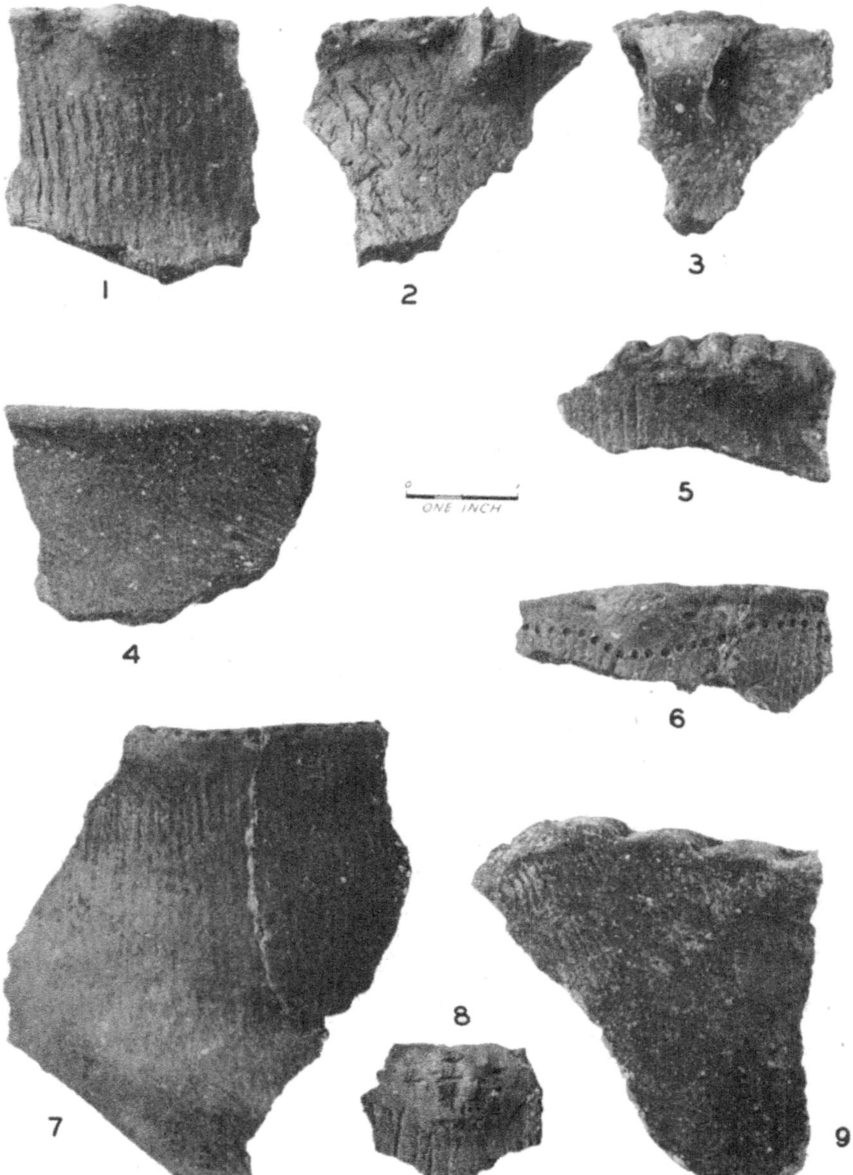

Sherds of Keyser Cord-marked type (U. S. N. M. neg. 36340–B)

EXPLANATION OF PLATE XII

FIG. 1. Oblique punctates at base of added rim strip, with horizontal incised lines on rim

FIGS. 2, 5–6. Cord-wrapped string or cylinder impressions on upper and lower rims

FIG. 3. Individual cord impressions on rim and thumbnail gouges at base of added rim strip

FIG. 4. Short vertical gashes at base of thickened rim

FIG. 7. Undecorated rim with added rim strip

PLATE XII

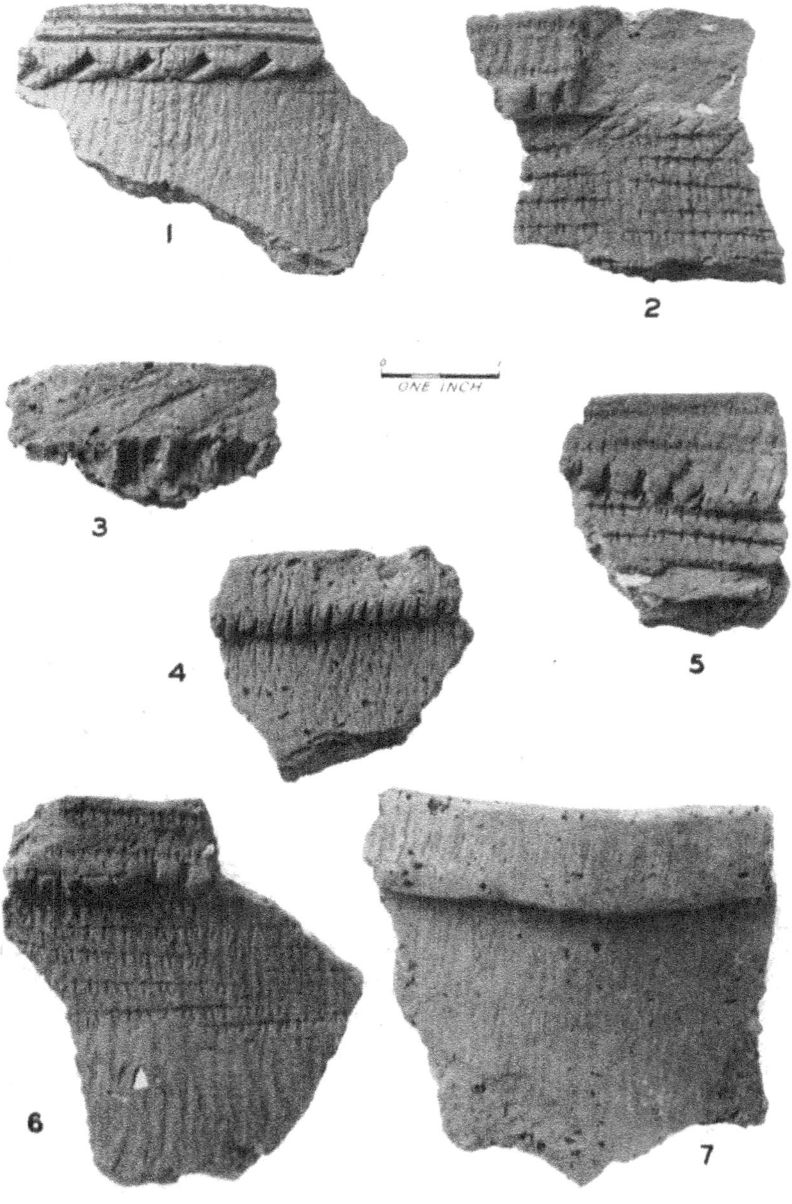

Page Cord-marked sherds (U. S. N. M. neg. 36340)

PLATE XIII

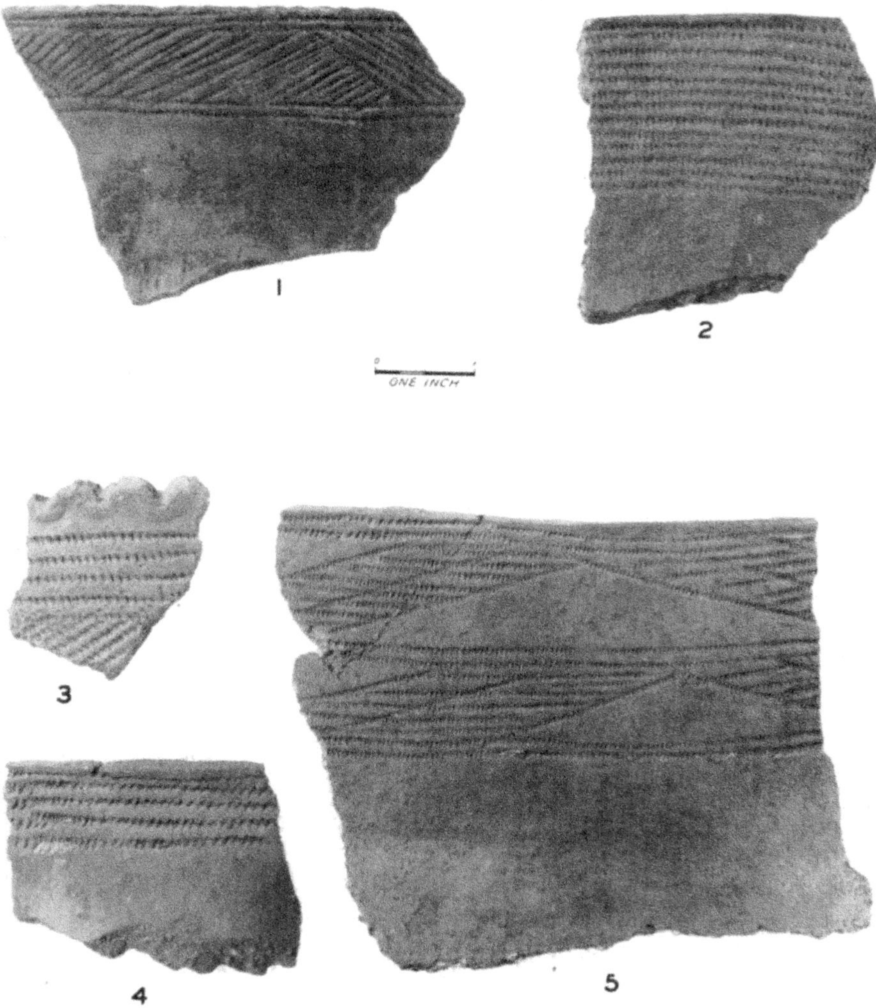

Potomac Creek Cord-impressed rims (U. S. N. M. neg. 36340–A)

These five rims have a decoration of the type called pseudo-cord in the text. A cord wrapped around a narrow pliable element was impressed on the plastic prefired rim

SOME NEW DATA ON THE GOODALL FOCUS

GEORGE I. QUIMBY, Jr.

SOME additional data on the Hopewellian occupancy of southwestern Michigan and northwestern Indiana have been gathered since the publication in 1941 of my article "The Goodall Focus," which was written in 1938.[1] At that time I was unaware of the existence of most of the new material [2] that is presented in this paper.

The term "Goodall Focus," which is applied to one of many foci belonging to various aspects of the Hopewellian phase,[3] designates the archaeological remains of a group of culturally related Indians who lived in southwestern Michigan and northwestern Indiana in prehistoric times. These Indians probably were farmers who were also dependent upon hunting, fishing, and the gathering of wild foods. For ceremonial purposes they constructed dome-shaped burial mounds of earth and had elaborate funeral rituals in which fine ornaments, tools, and clothing were placed in the graves. Tools and ornaments were made of copper, silver, stone, bone, wood, shell, and mica. Garments were manufactured from animal skins and woven cloth. Several varieties of fired clay pottery were used by these Indians, one of which was possibly imported.[4] Nicely executed stone platform pipes provide evidence of their smoking.

Stratigraphic, typological, and distributional studies of other Hopewellian foci show clearly that the Goodall focus is the product of a cultural movement into southwestern Michigan and northwestern Indiana, probably from Illinois by way of the Kankakee

[1] "The Goodall Focus, an Analysis of Ten Hopewellian Components in Michigan and Indiana," *Prehistory Research Series*, Vol. 2 (1941), No. 2. Published by the Indiana Historical Society.

[2] Dr. James B. Griffin of the Museum of Anthropology, University of Michigan, has kindly brought to my attention much of the new material presented in this paper.

[3] The terminology and the system of classification have been described by W. C. McKern, "Midwestern Taxonomic Method as an Aid to Archaeological Culture Study," *American Antiquity*, 4 (1939) : 301–313.

[4] Quimby, *op. cit.* (see note 1), and "Hopewellian Pottery Types in Michigan," *Pap. Mich. Acad. Sci., Arts, and Letters*, 26 (1940) : 489–494. 1941.

River. The individual sites are distributed along drainage systems in such a way that it looks as if there had been a progressive movement northward from one river valley to another, without, however, any appreciable loss of contact with the main culture center or centers in Illinois.

In another paper [5] I have contended that the south-to-north geographic distribution of river valleys containing sites has temporal significance. The St. Joseph, Grand, and Muskegon rivers, named in order from south to north, all flow in a westerly direction into Lake Michigan. The Goodall focus occupancy of the Muskegon Valley is northernmost and therefore represents the latest known period in Goodall focus history. Similarly, the Grand River occupancy is earlier than that of the Muskegon Valley, and the sites in the St. Joseph Valley are earlier than those in the Grand. The southernmost and oldest occupancy in the series is represented by sites in the area drained by the headwaters of the Kankakee. Of course, there must have been some cultural overlapping from one river valley to another and, therefore, chronological overlapping from one period to another, but, by and large, the settlement of each valley seems to have been temporally distinct. Typological studies of the Goodall focus ceramics tend to strengthen this hypothesis.[6] The cultural outline and the conjectural chronology, which have been presented in summary fashion, are intended only for purposes of orientation.

The Goodall site is in northwestern Indiana in the region drained by the headwaters of the Kankakee River. The new material from it is in the collections of the Field Museum (see Pl. I). There are seven small to medium-sized copper celts which are ovate-oblong in outline and narrow-rectangular in section (Figs. F–G, L, S–U, W). One has a slightly flaring bitt (Fig. F); three others are also ovate-oblong in outline, but have flaring bitts and are plano-convex in section (Figs. H, K, V). A fragment of twined fabric adheres to the convex surface of one of these three (Fig. H). There are five double-pointed copper awls, three of which are illustrated (Figs. A–C). Four are short, and one is long. The last, which is the best preserved, clearly shows that the point on one end is sharper and tapers more gradually than the other. Possibly the dull point was jammed

[5] *Idem*, "The Ceramic Sequence within the Goodall Focus," *Pap. Mich. Acad. Sci., Arts, and Letters*, 28 (1942): 543–548. 1943.

[6] *Op. cit.*

into a handle of some kind. Three small fragments of copper appear to be broken sections of awls.

Copper ornaments from the Goodall site are one small spheroidal bead (Fig. I), two small tubular beads made from rolled sheets (Fig. I), one fragmentary hemisphere, which probably was the covering for a wooden button (Fig. J), and two earspools (Fig. O). Each earspool consists of a flat disk joined by an axle to a concavo-convex disk.

Woven cloth is represented by one fragment adhering to a copper celt and by three impressions preserved by copper salts on the surfaces of three celts. The weaving is plain twining of several varieties. The cordage, which is probably bast, is made of two strands twisted clockwise.

The Marantette site is located in the St. Joseph River Valley near the town of Mendon in southwestern Michigan. Material from this site comprises one large, thick tubular bead of copper (Fig. D) and two copper earspools nearly identical with those already described for the Goodall site (Fig. P). All these artifacts are in the Field Museum.

It is interesting to note that copper earspools are found only at sites in the Kankakee and St. Joseph drainages. In view of the geographic–temporal relationships previously postulated, it is probable that earspools are to be associated with the early phases of Goodall focus history and were not used in the late phases represented by the occupancies of the Grand and Muskegon valleys. These spools are unique in that one of the disks in each is flat. The usual Hopewellian variety consists of two disks, which are concavo-convex.

Although hundreds of earspools have been excavated from Ohio Hopewell mounds, there is only one published record of the finding of such spools of the Goodall–Marantette type. A pair came from the Tremper site.[7]

Because most of the cultural similarities of the Goodall focus are with Hopewellian sites in Illinois it seems reasonable to assume that the Goodall–Marantette type of earspool should occur in Illinois also. Unfortunately, such an occurrence cannot be demonstrated because these spools are rarely found in Illinois Hopewellian sites,

[7] Mills, Wm. C., "Exploration of the Tremper Mound," *Certain Mounds and Village Sites in Ohio*, 2: 216. Columbus, 1917.

and the published descriptions of those that have been excavated are not adequate for comparative purposes.

It is unlikely that the Goodall–Marantette type of earspool does not indicate a cultural connection of some kind with other Hopewellian groups. If future evidence demonstrates that this type is lacking in Illinois, I should be inclined to assume that there was contact between some parts of the Goodall focus and some Hopewell groups in Ohio. In my opinion such contact would be of less importance, however, than the contacts with Hopewellian groups in Illinois that have been formulated upon a greater number of cultural similarities.

The Converse site is in the Grand River Valley in the city of Grand Rapids, Michigan. Artifacts from it in the Field Museum include two large ovate-oblong copper celts that are narrow-rectangular in section. One is here illustrated (Fig. E). Additional artifacts are one small copper celt, which is ovate-oblong in outline and narrow-rectangular in section (Fig. R); one long copper needle, square in section at the middle and broken halfway across the eye (Fig. Q); and two medium-sized copper awls, round in section, one of which has an end wound with gut or plant fiber (Figs. M–N).

A visit to the Peabody Museum, Cambridge, Massachusetts, enabled me to ascertain that two silver artifacts which I had thought to be historic trade material associated with intrusive burials[8] were actually prehistoric and therefore should have been included in the original list of culture traits for the Converse site. One that is rather fragmentary appears to be a large sheet-silver arm band. The other is a silver band corrugated on one side.[9] Artifacts of this type, which have been called "conjoined tubes," probably were musical instruments. The complete instrument had reed or bone pipes inserted in the tubes. Such panpipes made of silver, copper, or iron have been found in Michigan, Illinois, Wisconsin, Ohio, Tennessee, Alabama, and Florida.

The Brooks site, located in Newaygo County, Michigan, along the Muskegon River, is the northernmost site in the Goodall focus.

In the Muskegon County Museum there is an additional pottery

[8] "The Goodall Focus," p. 100, note 30 (see note 1 of this article).

[9] This artifact is mentioned by Chas. C. Willoughby, "The Turner Group of Earthworks, Hamilton County, Ohio," *Papers of the Peabody Museum*, 8, No. 3 (1920): 51.

vessel from this site — a small jar about 13 cm. high and 7 cm. in diameter at the mouth. It has a quadrilobate body, a rounded bottom, a constricted shoulder area or "neck," and a slightly flaring rim which terminates in a flattened, insloping lip. The upper rim is thickened and cambered.

The paste of which this vessel was constructed is medium to fine in texture, tempered with small particles of granitic rock, has a surface hardness of 2–2.5, and color ranges in buff and gray. The surface is smooth. Decorative techniques, which are rather poorly executed, consist of incising and punctating. The ornamentation, which covers the exterior from lip to base, is as follows: There is a narrow band of closely spaced crosshatching that is confined to the thickened area of the upper rim. The individual incised lines are narrow, a millimeter or less in width. Beneath this band is a single row of small hemiconical punctates, punched from the left. The remaining smoothed area of the rim and upper shoulder is decorated with two or three rather wide (2 to 3 mm.) shallow incised lines that make a pattern of curvilinear meanders around the vessel. The interlobate areas are smooth and undecorated, but the lobate areas are outlined by narrow shallow incised lines and are filled with short parallel vertical lines of similar type. This vessel appears to be a poor copy of the exceptionally fine Hopewellian ware of the limestone-tempered variety.

In general, the new data do not change the interpretations and conjectures set forth in my first paper, "The Goodall Focus." Traits hitherto unreported for this focus are copper earspools, a copper needle, a silver conjoined tube, and a silver arm band. Other data presented here serve to enrich our knowledge of the culture content of particular sites and also of the frequency of a trait in a given site.

Field Museum of Natural History
Chicago, Illinois

PLATE I

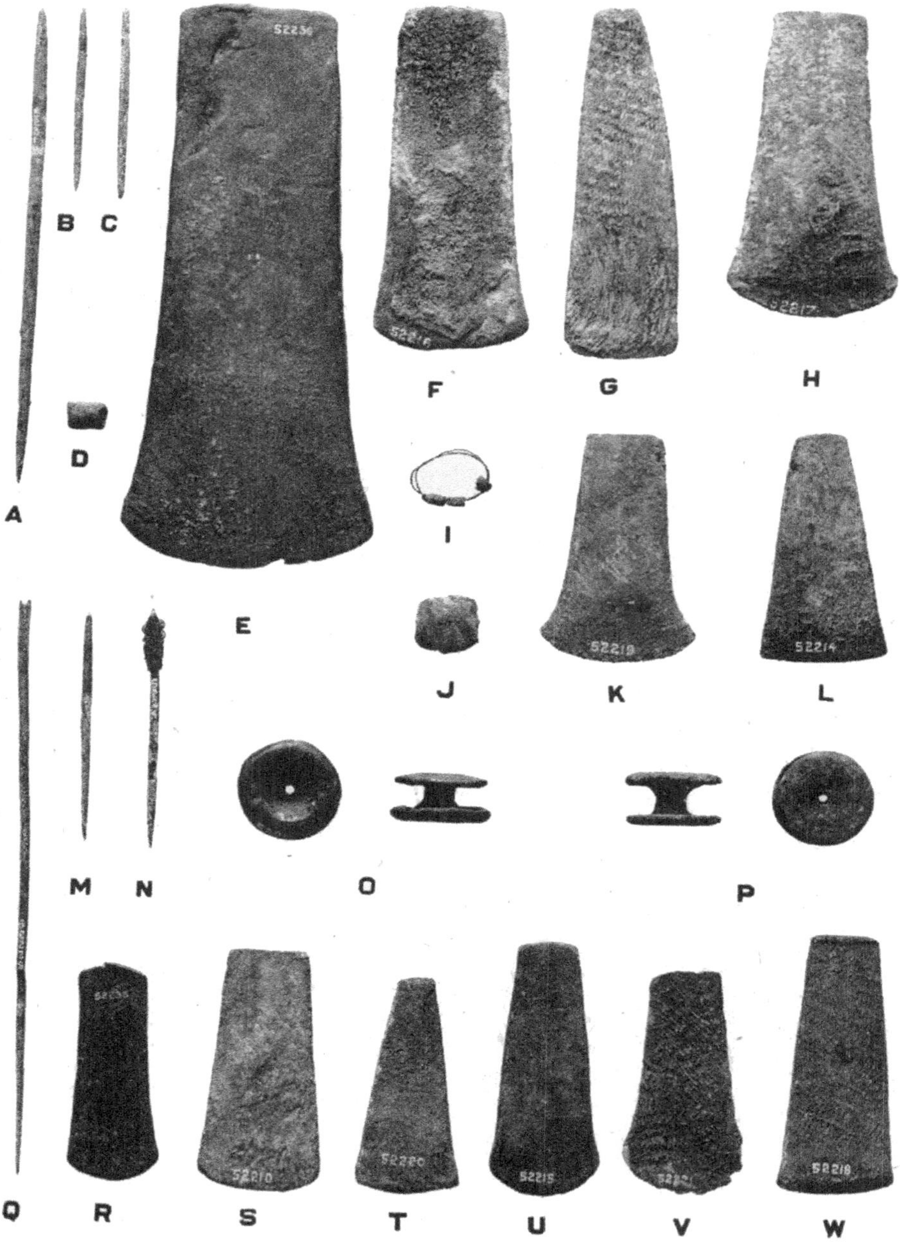

Copper artifacts of the Goodall Focus

Converse mounds: celts, E and R; needle, Q; awls, M and N

Goodall mounds: celts and adzes, F–H, K–L, S–W; awls, A–C; beads, I; covering for button, J; earspools, O

Marantette mound: bead, D; earspools, P

TWO HOPI TALES FROM ORAIBI

MISCHA TITIEV

AMERICAN Indian folklore has been intensively studied by many American anthropologists because of a conviction that "in the tales of a people those incidents of the everyday life that are of importance to them will appear either incidentally or as the basis of a plot. Most of the references to the mode of life of the people will be an accurate reflection of their habits." [1] Although the late Professor Franz Boas has several times demonstrated the possibility of reconstructing a tribe's pattern of culture from its mythology,[2] other scholars have expressed a need for caution in this regard, since "Many aspects of everyday life that are of interest to the people are neglected in the myths, while others are either exaggerated or transformed into their opposites." [3] Nevertheless, even if one grants the likelihood that a number of stories present distorted pictures of a group's life, it is still highly probable that those themes which occur again and again reflect the main interests of a given tribe. This is particularly true of the Pueblo Indians, who live by farming in a semiarid environment and "whose continual reference to the water problem" [4] is a direct expression of a major concern of their lives. The following tale clearly illustrates the attitude of the Hopi, one of the most important Pueblo tribes, toward their supplies of water.

THE STOLEN SPRING [5]

Early one summer, about forty years ago, a Lizard clansman named Na'usi discovered a damp spot while hoeing near his orchard, and on

[1] Boas, F., *Tsimshian Mythology*, in *Annual Report of the Bureau of American Ethnology*, 31 (1916) : 393.

[2] Boas, *op. cit.*, pp. 393–477; *idem*, *Kwakiutl Culture as Reflected in Mythology*, in *Memoirs of the American Folklore Society*, Vol. 28 (1935).

[3] Wittfogel, K. A., and Goldfrank, E. S., "Some Aspects of Pueblo Mythology and Society," *Journal of American Folklore*, 56 (1943) : 17.

[4] *Ibid.*, p. 24.

[5] This tale was narrated in the winter of 1933–34 by a Hopi woman named Hahai'i, late head of the Sun clan at Oraibi. She was the daughter of Homikni, a Lizard clansman with a great reputation as a medicine man, and the mother of Don Talayesva, my interpreter and chief informant.

digging deeper he found a good flow of water just beneath the surface. He told the welcome news to Homikni, his clan brother, and asked him to help develop the spring for watering their orchards. Homikni hurried to the place and on sampling the water found it to taste so good, just like rain water, that he readily agreed. The next day he dug a hole about two feet in diameter and was overjoyed at the steady flow of clear sweet water that resulted.[6] Immediately he filled a water jug (olla), and as he was bringing it home he met Na'usi, who thanked him for having developed the spring and willingly granted him permission to use it.

It happened that the season was particularly dry that year, and it was not long before people began to notice that Homikni's family was fetching plenty of water from some place near the village, whereas they had to carry their supplies a great distance. As a result of the eager questioning of their neighbors, Homikni and his daughter Hahai'i, having filled as many containers as they had, threw the spring open to the use of all. For three days there were steady lines of people going to and fro, and still the spring ran full. In the hope that it might be a permanent source of water it was decided to set aside a day for communal spring building and cleaning. A large force responded to a call for volunteers, and while the men did the heavy work the women brought *somiviki*, *tcukuviki*, and *piki*.[7] Owing to the fact that the spring's capacity was much greater than anyone had guessed the task of building retaining walls proved harder than expected, because the foundations were constantly being undermined. At last, in order to make sure of getting on solid ground, they had to begin again at some distance off. All this took so much time that after forty-eight hours of work it was decided to skip a day so that the men might attend to their farming and other duties. On each of the working days the women brought food to the men, bits of which were "fed" to the spring. When everything was completed the spring was named Masduma ("Grave Soil"), after a point of land near by, and its discoverer put down prayer sticks (*paho*), prayer feathers (*nakwakwosi*), and sacred meal. All were happy to have such a good source of water near the pueblo, and for three

[6] At the start of his work, Homikni deposited four prayer feathers and sacred cornmeal as thanks to the gods for his good luck.

[7] *Somiviki* and *tcukuviki* are foods made on a cornmeal base and wrapped in corn husks; and *piki* is a crisp unleavened bread made with a cornmeal batter and baked in paper-thin sheets.

years it flowed continuously, but in the fourth year it began to fail a little, so that if too many women tried to fill their ollas at one time, the spring would run dry temporarily but would soon recover.

At last it came time for the Snake dance that was to be held in the fifth year following the spring's discovery. On *totokya* (the day preceding the public performance) Hahai'i was hastening to get water with which to make stew for the next day's feast. Clouds were gathering and rain threatened, but she continued on until, just as she came abreast of her father's peach orchard, she noticed a man coming rapidly toward her. He was dressed in ordinary Hopi fashion except for a single eagle plume, which he wore fastened to the top of his head like an *omaunakwa* ("cloud prayer"). On his left shoulder he carried a rolled-up blanket, and in his right hand he held a wing plume of an eagle, and a glass bottle in which Hahai'i made out the movements of a small yellow wormlike creature swimming about in a mixture of grass and water.

The man appeared anxious to avoid detection, but as Hahai'i was squarely in his path he could not escape. He greeted her in the Isleta tongue, whereupon she immediately recognized him as Yayama ("Emerging," a reference to the rising sun), a Chimopovy man of the Pikyas clan who was a medicine man and a member of the Wuwutcim, Snake, and Flute societies at that village. He was in the habit of showing off by greeting people in Isleta before he spoke Hopi, so that Hahai'i had no doubt of his identity.

"Where are you coming from, and what have you got there that seems so important?" questioned Hahai'i, who was made suspicious by his appearance. "Where are you going yourself?" he countered. "You know it's *totokya*[8] today, and you ought to be home watching the dance." "Never mind me," replied Hahai'i. "Answer my question. What have you got there?" Yayama hesitated, and showed signs of alarm before he spoke. "Well," he said at last, "you've recognized me, and you know what I've got here. I came on a secret errand, and I never thought that anyone would meet me here on *totokya*."

By now Hahai'i was convinced of the importance of the en-

[8] On *totokya* there is a sort of rehearsal for the Snake dance, during which a member of the Antelope society dances with a bundle of vegetation dangling from his lips. See H. R. Voth, *The Oraibi Summer Snake Ceremony*, Anthropological Series, Field Columbian Museum, 3 (1903): 321–336.

counter, so she looked carefully at Yayama and then asked, "Why are you dressed like that? What are you doing here?" "Well, there's no hope for me," said the Chimopovy man. "The peopel from my village heard that you folks had discovered a wonderful spring, so I came here to taste it. In this way I found that it is true that someone from Oraibi stole our spring called Chimopovy — no wonder it is getting dry. I came here to take this living creature out and to bring the spring back to our village." [9]

"Did you take that living thing out of the spring?" asked Hahai'i. "Yes," he confessed. "Years ago Kelwushioma, an Oraibi man, came to Chimopovy in secret, and I've heard that he is the one who stole our spring, so I came here to taste the water to see if it's true. Sure enough, it tastes like our water, so this spring doesn't belong to Oraibi but to us." "Well," retorted Hahai'i, "you were sent here to do an errand not for our good but for our harm. What living thing have you got there? Are you taking it out so that our spring will run dry?"

"No," replied Yayama, "I just came here to taste the water and to put some of it into our spring so that it will run better." "But what have you got there?" insisted Hahai'i. "That's a tender living thing," answered the Chimopovy man; "if you touch it with your hand it will die. You've got to lift it out on the edge of this feather." "Why do you put those bits of grass [*patusaka*] [10] into the water?" "That is the food of this living creature," was the reply, "and if I didn't put some in it would die." "You must be one of the bad men [witches]," said Hahai'i. "Yes," admitted Yayama, "I am caught as I told you before, but don't tell this to anyone. Keep it a secret in your own heart." [11]

[9] Every spring and water hole is thought by the Hopi to harbor a *palulokong*, a reptile ranging in size from a worm to a serpent, from which water is supposed to issue. This creature, usually in its large form, plays an important part in Hopi mythology and ceremonialism. For further details consult E. C. Parsons, *Pueblo Indian Religion* (The University of Chicago Publications in Anthropology, 1939), p. 184, note *, *et passim.*

[10] *Patusaka* was described as a grass that grows near water and looks like millet (*lehu*). The narrator said that Kelwushioma, accused of having stolen the spring from Chimopovy, belonged to the Water Coyote clan, but in my Oraibi census he is listed as Lehu. Both clans are in the same phratry.

[11] The Hopi generally hesitate to reveal the identity of suspected witches for fear of their revenge. In addition, sorcerers often try to bribe people to keep silent. See M. Titiev, "Notes on Hopi Witchcraft," *Pap. Mich. Acad. Sci., Arts, and Letters,* 28 (1942) : 549–557. 1943.

"Well," argued Hahai'i, "you warn me not to tell this story, and here you are taking this thing to your home. It seems to me that it must be a *palulokong*, but here you have only a tiny creature. We were glad to have a spring close by, and now you are taking it from us." "Listen to me, my daughter," [12] said Yayama, "I will pay you to keep it secret. Now it looks like rain, so hurry home, and I will repay you by having it rain so hard that it will fill all your cisterns." [13]

With that he hurried off, and Hahai'i saw him stop in the center of her father's orchard, where he put down the plume he had been wearing as a plea for rain. Then Hahai'i went on to the spring. When she arrived the water looked soapy (foamy), and it was rolling and dashing so madly that she was afraid to go down to the edge, but at last she got up enough courage to approach and begin filling her ollas. Whenever the lightning flashed it seemed to skim along the top of the water in broad jagged streaks, and the thunder kept up a continuous roar. Badly frightened, Hahai'i hurriedly filled her water jars and started for home, pausing only to examine the plume which Yayama had left in the orchard.

Before she could get back to the village it began to rain so hard that soon she was soaked to the skin. When she was near her house one of her neighbors called out: "I'm sorry for you. When it's raining so hard why don't you empty your jugs so that you can get home faster? There's plenty of water here."

"I went to the spring to bring this water," Hahai'i shouted back, "and I'm not going to waste my time by dumping it out. This rain won't hurt me." [14]

Although she did not know where he had gone, Hahai'i felt that Yayama had kept his promise to pay for the spring by sending a heavy fall of rain. From that day on she kept close watch of the spring. Later that year it began to fail, and by the next year it had completely dried up. Although Yayama had stolen the entire spring from Oraibi, it had not done the Chimopovy people much good, for

[12] A medicine man has the right to call all people his children.

[13] The fact that rain was already imminent when she set out for the spring is completely discounted by the narrator, who was convinced that the storm which soon broke had actually been sent by Yayama.

[14] Water is such an unqualified blessing to the Hopi that they go to extremes never to waste any, and they have nothing but contempt for those who fear to get wet.

their spring, too, ran dry, probably because Hahai'i had caught the thief in the act.[15]

As a punishment for his misdeeds Yayama died about eight years later. Feeling that she was at last relieved of the need of maintaining silence, Hahai'i went to Chimopovy to talk with the dead man's daughter, who happened to be an albino. From her she learned that in Yayama's last illness the water (perspiration) had poured so abundantly from his body that it had been impossible to keep dry the sheepskins on which he lay. He complained constantly that he could not cool off; his eyes were big and gaping, and his tongue hung out like a dog's. Then, just before he died, the folds of his skin took on the appearance of a *palulokong* design, a sign that his death was being brought about by his theft of the living creature from the Oraibi spring.

The constant concern of the Hopi with water, which is strikingly reaffirmed in the story given above, has long been known and has properly been associated with the basic agricultural pattern of their lives. Less widely recognized is their interest in meat foods and hunting. This may be due to the fact that in recent years the game supply has shrunk so markedly that hunting is greatly restricted. Nevertheless, rabbit hunting is still an important practice,[16] and there is reason to believe that in pre-European times, before the introduction of domesticated livestock, the natives put far more emphasis on success in the chase. Again and again in Hopi folklore one finds expressions that show the great joy of the Hopi when game is plentiful and their sorrow when it is scarce. Moreover, a number of rituals whose primary purpose is to stimulate the growth of crops contain important pleas for success in the hunt. During the performance of the Oraibi Soyal ceremony, for example, special objects called hunting prayer sticks (*makbaho*) are manufactured;[17] the

[15] Witches caught in the act are supposed to have their plans frustrated, and the Hopi believe that they will be punished by death, although it is acknowledged that many of them live on indefinitely. On this point consult M. Titiev, *op. cit.* (see note 11).

[16] For details regarding this aspect of Hopi life see E. Beaglehole, "Hopi Hunting and Hunting Ritual," *Yale University Publications in Anthropology*, 4 (1936) : 3–26; and *Hopi Journal of Alexander M. Stephen* (edited by Elsie Clews Parsons), pp. 1023–1024, *et passim* (Vol. 23 [1936] in *Columbia Contributions to Anthropology*).

[17] Dorsey, G. A., and Voth, H. R., *The Oraibi Soyal Ceremony*, Anthropological Series, Field Columbian Museum, 3 (1901) : 57.

men of the pueblo deposit other offerings at an Antelope shrine with requests for good luck in hunting; [18] and the rites culminate in a three-day rabbit hunt by the members of the Soyal society.[19] Furthermore, one of the most important deities worshiped in the course of Oraibi's Tribal Initiation ceremonies is Tuwapongtumsi, patron goddess of all game animals.[20] From these examples it seems plain that hunting was once an extremely important aspect of Hopi culture, and the following tale supports this contention.

Sitiyo and the Hawk

Once there was a boy named Sitiyo ("Blossom Youth") who lived at Oraibi long, long ago. He was very eager to become a successful hunter, but try as he would he never had the luck to kill even the tiniest of rabbits. Sitiyo had an only sister, of whom he was very fond, and he used to feel heartbroken at the conclusion of every hunt when he thought of all the other families feasting on freshly killed game while his sister went without because of his lack of skill. The poor girl's mouth would water as she watched her neighbors dining on roasted rabbits and other dainties, but long though she did, she never dared ask for some for fear of being repulsed.

During the Soyal ceremony Sitiyo made some special prayer feathers and gave them with fervent prayers to be deposited at a shrine by the head of the Rabbit clan, but throughout the three-day hunt that follows the termination of the Soyal he did not succeed in killing a single animal. By now his kiva mates had lost all patience with poor Sitiyo. "Why is it that you have legs and arms like the rest of us, and yet you never kill anything?" they asked angrily. "If you don't contribute something after the next Soyal, we won't let you join our feast." [21]

[18] Titiev, M., "Old Oraibi," *Papers of the Peabody Museum of American Archaeology and Ethnology*, 22 (1944) : 144–145.

[19] See footnote 17.

[20] At Oraibi Tuwapongtumsi ("Sand-altar-young-woman") is also known as Tuwapongwuhti ("Sand-altar-woman"), and as Tihkuyiwuhti ("Childbirth-water-woman"). Compare E. C. Parsons, as cited in footnote 16, p. 178, and H. R. Voth, *The Traditions of the Hopi*, Anthropological Series, Field Columbian Museum, 8 (1905) : 140.

[21] Kivas are subterranean rooms, generally shared by groups of men who belong to the same secret societies. In the post-Soyal hunts each kiva unit acts independently, the total kill of its members being pooled. Sitiyo's mates were angry not only because he partook of feasts without contributing to them, but also because his lack of skill lowered his group's average and made it look as if the entire kiva consisted of poor hunters.

At last another year passed by, and again the men were out on the hunt that follows the Soyal. It happened to be a particularly good season, and most of the men were enjoying exceptionally fine luck, some of the best hunters getting as many as eight or ten rabbits on the very first day. As each man returned to the kiva, the kiva chief greeted him, saying, "How many rabbits did you touch [kill] today?" and each would joyfully announce a number, but when it came the luckless Sitiyo's turn he was forced to admit complete failure.

"I'm sorry, my son," said the kiva head kindly, "but I've got to warn you. Your fellow members are angry, and if you don't bring in something before the hunt is over, you won't be allowed to take part in our feast."

Still, try as he would, Sitiyo failed to kill a thing during the entire three days, and when the rabbit feasts began in all the kivas he appeared hopefully, carrying a batch of fresh *piki* bread that his sister had prepared for him. However, when the chief of the kiva called the men to partake of the stew that was set on the kiva floor, Sitiyo's name was not mentioned, and he sat alone eating dry *piki* while the others dined heartily on their favorite dish. Some of the meaner fellows went so far as to taunt the forlorn youth as he sat disconsolately by himself, and no one took pity on him or offered him a single mouthful, until at last he lost what little appetite he had and left the kiva, carrying his scarcely tasted bundle of *piki* with him.

Ten days later an announcement was made in the village that another hunt was to be held, this time in the direction of the Little Colorado River.[22] In spite of all his previous failures Sitiyo decided to try his luck once more, so he prepared his weapons, asked his sister to put up a sufficient quantity of baked sweet-corn meal (*tosi*) and *piki*, and set out with the others. Although each man left singly they united into kiva groups when they reached the hunting ground, and early the next morning the chase started. In the evening, as the hunters began to return, the kiva leader greeted them in the usual manner, and when Sitiyo arrived he again had to reply that not a single rabbit had fallen to his lot. "Well, there's no hope for you," answered the chief of the kiva. "When we roast our rabbit meat for supper you won't get any."

Thus it happened that, while everyone else was supping on savory

[22] In former days the Little Colorado River territory was a favorite hunting ground among the Hopi.

flesh, the unhappy Sitiyo had to satisfy his hunger with *piki* and as much rabbit gut as he could force himself to eat.

So it went until noon of the fourth day, when it was decided to bring the hunt to a close. Each kiva head then began to portion the combined kill evenly among all the members of his unit. Sitiyo looked on longingly, hoping against hope that the chief might relent at the last minute and give him one or two rabbits with which he could gladden his sister's heart. First the cottontails were parceled out, and then the jacks were distributed, yet not a single one of either kind did Sitiyo get.

It was with a heavy heart that Sitiyo watched his mates tying up the rabbits that had fallen to their lots, and when the men began to depart for home he could not bring himself to join them empty-handed, but decided to remain a little longer in the hope that his luck might change. As soon as he was alone he made a prayer feather for Tuwapongwuhti, the patron deity of all game animals, praying fervently that she might help him. Then he set out to hunt for the rest of the day, but when night fell he had nothing to show for his efforts. Sadly he prepared some supper, and while it was cooking he wondered what he should do next. As he sat looking at the light that still lingered after the sun had set, he thought he heard something being sung just across the river, and soon he was able to distinguish the words. "Over there, over there," sang a woman who seemed to be grinding corn, "Sitiyo was walking while the afterglow was fading."

Then the singing stopped, and without waiting to taste his supper Sitiyo hurried to investigate. Crossing the river, he headed for a large house from which a light was gleaming. Here he found a ladder, and climbing up he peeped through a partly opened door. Inside he saw a woman busily grinding corn; she was singing the song that had attracted his attention. He was afraid to enter because he did not know what sort of power this person might have, but as soon as she noticed him she invited him to come in, and she fixed a seat for him.

"I knew that you were tired and hungry," she said, "so I sang my song to bring you here." "Yes," admitted Sitiyo readily, "I am tired and hungry and very nearly worn out."

At this the woman bustled about and soon spread before Sitiyo a bowl of hot rabbit stew and a plentiful supply of *piki*. At her invitation he ate and ate until his hunger was completely sated. When he had finished his hostess promptly cleared away the dishes and then spoke. "I took pity on you," she said, "because you were so unlucky during the last four days. Now you and I will sleep together tonight, and tomorrow I will tell you what I have in mind for you."

The woman was young, pretty, and as plump as most men desire, so Sitiyo offered no objections and did not hesitate to accept her offer. Next morning, even before breakfast, his companion said to Sitiyo: "Follow me and I'll show you something." She led him down some steps, and on her opening a door there was heard the sound of a large number of cottontail rabbits running about. "Now, 'my husband,'" said the woman, "we have enjoyed each other, and I'm going to give these animals to you as a reward.[23] They are my children and my pets, but I give them to you."

Then she led him to other doors facing in each of the four cardinal directions, where she gave him jack rabbits, deer, and antelope. "All these other boys," she said, "have become lucky at hunting because they have slept with me. Now I am going to make you a good hunter, so go out and kill all the game you want before you go home. From now on you will be a good hunter." The boy thought for a while and then replied: "Yes, I think I know you. You must be the one living here at Little Colorado, as my ancestors have told me." "Yes," she admitted, "I am the woman whom the Hopi call Tihkuyiwuhti, who owns all the game.[24] But I'm not the only one. When you leave here you must go along the river until you meet an old woman, and she will tell you something else to do."

So Sitiyo took leave of his hostess and followed the river, watching closely for the person he was supposed to meet. Suddenly he felt the need of performing his natural office, but just as he prepared to relieve himself a voice said: "Please don't do that here, my grandson, move off a little way." "I'm sorry," replied Sitiyo, "I don't like to harm anyone, but I did not see you there. Who are you?" "I am the one you are looking for," was the reply. "When you are ready, come back to me." Sitiyo did as he was bid, and then Spider

[23] It is a Hopi convention that all who receive sexual favors must pay for them. Voth has recorded a different tale which contains several of the features in this story. Consult Voth, *op. cit.* (see note 23), pp. 120–121.

[24] For a description of this supernatural personage see *ibid.*, p. 140.

Woman — for it was she whom he had disturbed — led him to her *kiva* [house] and asked him to enter.[25]

"Now," said Spider Woman, "it was I who told Tihkuyiwuhti to call you with her grinding song. After you had enjoyed her she promised to make you into a good hunter like the other men, and sent you to me. It is getting late, and we must hurry. Have you any baked sweet-corn meal left?" "Yes," said Sitiyo, "I still have some." "All right," answered Spider Woman, "make some *qömi* [cakes or loaves baked with sweet-corn meal dough] and give some with *nakwakwosi* [prayer feathers] to Kisa [Cooper's hawk][26] and me."

Sitiyo hurried back to his camping place and quickly obeyed Spider Woman's instructions by making two cakes of *qömi* and depositing them with prayer feathers at the edge of a cliff where Kisa was accustomed to perch as he kept a sharp lookout for game. Then the youth ran off and hid, as he had been told to do, but after a time he returned to the spot and, finding that his offerings had disappeared, he realized that they had been accepted by Kisa.

Next day, still acting on Spider Woman's orders, he left his camp early in the morning, carrying his weapons with him. While he was trying to locate his prey, a shadow fell across the plain, and on looking up he saw a hawk circling about high overhead. Soon he noticed a jack rabbit, but before he could throw his stick, the hawk swooped down and killed it with one jab of its powerful sharp-pointed wing. Rabbit after rabbit was slain in this way, until the happy Sitiyo decided that he had had enough and returned to his camp. At the same time Kisa flew back to the cliff where he lived.

Immediately after breakfast the next morning Sitiyo hurried to seek out Spider Woman again. "Are you arrived?" she asked. "Yes, I'm arrived," he replied.[27] "Now it is time for you to go home," spoke Spider Woman. "Your parents and your sister are lonesome for you, and are afraid that some wild beasts may have

[25] This is the conventional mode of meeting Spider Woman in most of the stories told at Oraibi. She is a powerful figure, grandmother of the Little War Twins and ancestress of all witches. She is both venerated and feared.

[26] Kisa is not only a great hunting deity; he is also said by the Hopi to have taught them the use of the curved throwing sticks which they employ for killing rabbits. Cf. Parsons, *op. cit.* (see note 9), p. 187.

[27] This exchange of questions is the formal way in which the Hopi greet new arrivals.

killed you. Now, my grandson, when you get home tell your sister and mother to divide your game into two parts. With one half let them make a stew for your kiva mates, and with the other half let them make stew for everyone else in the village. When your kiva mates taste this stew it will be a lesson to them for having been so mean to you."

"Also," she continued, "when it comes time for the Soyal ceremony and your people make *nakwakwosi* for every living creature, don't forget to make some for your 'wife' [Tihkuyiwuhti], for me, and for your servant Kisa. We will watch you at Soyal to see if you are thinking of us, and we shall be glad when we get *nakwakwosi* from you."

Sitiyo thanked his benefactress and started back for Oraibi, but on account of his heavy load of rabbits it took him about three days to make the return journey. As he was approaching the village he was met by one of the men from his kiva, who said: "I'm glad to see you coming home. Your folks were worried when you didn't return with the rest of us. Now they will be happy. Just let me help you with your load." Then, taking only a few of the rabbits with him, Sitiyo's friend rushed on ahead, promising to send the boy's father to help with the rest of the burden.

Soon the father came running gladly to meet his son, and on seeing the large number of rabbits he exclaimed: "Thank you for bringing so much game. We were worried for fear wild beasts or enemy tribes had killed you." Together they shared the load and speedily reached the pueblo. Sitiyo greeted his mother and sister and then hurried to his kiva, where the men had been puzzled by his absence and had begun to blame themselves for having left him behind. Now they were overjoyed to see him, and all began to ask how many rabbits he had killed. "Oh, I killed a great many," he answered without making mention of his supernatural helpers, "but I didn't bother to count them."

Then Sitiyo announced the two feasts that he was going to provide the next day, and on the morrow there was enough to satisfy everyone in the entire village. As Spider Woman had foretold, his kiva mates were now ashamed of the way they had treated Sitiyo, and many of them apologized, promising to treat him from then on as a true brother.

So the time passed happily for Sitiyo until the next Soyal. He did not disregard Spider Woman's instructions, but prepared prayer feathers for her as well as for Tihkuyiwuhti and Kisa. From that time on he was one of the most successful hunters at Oraibi, supplying much game for making stew to his parents, sister, and paternal aunts.[28]

University of Michigan

[28] Note how kinship factors are woven into this story. Reference is made to the strong bond between parents and children, between brother and sister, and between a boy and his paternal aunts. Moreover, when Sitiyo's kiva mates repent, they promise to treat him like a true brother. The tale thus reflects quite accurately the importance of these relationships in everyday Hopi life.

"ROHONA" IN PUEBLO CULTURE *

LESLIE A. WHITE

ROHONA is a term encountered frequently in the literature on the Keresan pueblos. We find it also in accounts of the Hopi. The western Keres — at Acoma and Laguna — pronounce the word shrohóna,[1] the eastern Keres, róhona.[2] The Hopi say zrohóna or DZó·hona.[3] We shall use the form rohona throughout this paper to designate this concept regardless of local variations of pronunciation.

At Laguna rohona has been identified as a weasel (3, 1 : 28; 4, p. 276; 8, p. 127), as a female mountain lion (3, 1 : 28; 8, p. 127), as a society (7, p. 145), and as "stone fetish animals used by hunters" (9, p. 177). At Acoma it is said to be a lynx (12, p. 23), or "a big cat with spots — a great hunter" (16). Among the eastern Keres a Cochiti informant describes it as a small animal that helps hunters by entering deer through the ears and then killing them (5, p. 85); at Santo Domingo it is the name of a mask, associated with the Hunters' society (4, p. 134; 13, p. 112). At Santa Ana it is "a powerful hunter," a "big spotted cat — a jaguar" (14, p. 283).

Among the Hopi, according to Stephen, rohona is an "unidentified animal, (?) mountain coyote, in size between fox and coyote, associated with southeast" (11, p. 1325). Beaglehole merely speaks of it as "another animal deity," a patron of the hunters (2, p. 9). Dr. Edward A. Kennard, who kindly made inquiry among the Hopi at the request of the writer, reports that it is a small animal, like a

* A portion of the data in this study was obtained on field trips supported by grants from the Horace H. Rackham School of Graduate Studies, University of Michigan, to which grateful acknowledgment is here made.

[1] Boas spells it co·ho·'nα (3, 2: 42), Stirling, shohóna (12, p. 23), Curtis, sohona, the *s* being "approximately s^r," (4, p. 276), Parsons, shuhuna (8, p. 127; 9, p. 177), Kroeber, cuhuna (7, p. 145).

[2] Curtis (4, pp. 134–135, 152) and White (13, p. 112; 14, pp. 283, 287, 293; 16) both have it rohona. Goldfrank spells it rahona (5, p. 85).

[3] Beaglehole renders it DZó·hona (2, p. 9), Stephen, zrohóna (11, pp. 1071, 1325). It is interesting to note that Beaglehole puts the accent on the first syllable, as do the eastern Keres, whereas Stephen places the stress on the second syllable, as is done at Acoma and Laguna.

coyote in appearance but smaller, that lives in the San Francisco Mountains; it is a great hunter, particularly of deer and antelope, and is a night prowler. Thus, to summarize, we find rohona identified as (1) a weasel or other small animal, (2) a foxlike or coyotelike animal, (3) a lynx, (4) a mountain lion, (5) a jaguar, (6) a kachina (an anthropomorphic rain god), (7) a society, and (8) a fetish. In all instances, however, rohona is closely associated with hunting in one way or another.

Because of this variety of interpretation I once thought that rohona might be a mythical animal rather than a real one, the difficulty of identification being due to vagueness of the conception in the minds of the Indians. Dr. Elsie Clews Parsons, in her exhaustive work, *Pueblo Indian Religion,* also speaks of rohona as an "unidentified, perhaps a mythic animal" (10, p. 187). I now feel reasonably sure, however, that it is a real animal and, furthermore, that it is the jaguar (*Felis onca* Hernandesii).

In the first place, it seems unlikely that rohona is a mythologic animal. Mythical beings are, as a rule, easily recognized as such. Whirlwind Old Man, Sun Youth, the kachinas, and the like, are obviously characters in mythology. The manlike being with flint-feathered wings and the water snake that has a horn on his head and cloud designs on his sides are also clearly imaginary beings. We have occasionally had difficulty in identifying real animals, such as maidyupi, the shrew (*Sorex*), but with the possible exception of rohona, we have never found it difficult to distinguish the real from the imaginary. It seems likely, therefore, that in this instance rohona is a real animal and that the only reason for thinking otherwise has been the difficulty of identification. We now believe we have sufficient evidence to identify it.

Among the Keres rohona has been identified once as a "weasel" and once as a "small animal." The description of the "small animal" reported by Goldfrank fits the shrew (maidyupi), except that it enters the animal through the rectum instead of the ear, as Goldfrank's polite informant told her. It is probable, too, that the weasel-like animal reported by Boas at Laguna was really the maidyupi. All other Keresan informants identify rohona as a large cat — a lynx, a mountain lion, or a jaguar. It is significant to note here that the man who impersonates the rohona kachina at Santo Domingo wears a "cougar skin" on his back (4, p. 134). Quantitatively the evidence in favor of a large cat is considerably greater than that in support of a small weasel-like animal or shrew.

Qualitatively, too, the evidence favors the theory that rohona is a large cat. In myths which tell of the origin of hunting customs (3, 15) rohona is a beast of prey swift enough to overtake, and powerful enough to knock down, a deer or a mountain sheep. A weasel, or a shrew, could not do this. Besides, we have the rohona kachina actually wearing the skin of a "cougar." It seems almost certain, therefore, that rohona is a large cat rather than a weasel or a shrew. It is interesting, and perhaps significant, to note, however, that the Tewa call a "small burrowing animal, the sacred beast of the nadir" — in all probability the shrew — "earth mountain lion" (6, p. 30).

If rohona is a big cat, which member of the cat family is it? It cannot be *Lynx rufus*, commonly called wild cat, bobcat, or lynx, for we have identified this animal repeatedly as gyakyu (13, p. 203; 12, p. 23) or tyatyu (3, 2 : 45). Neither can it be the mountain lion (*Felis concolor*). Not only has the mountain lion been identified over and over again as mokaitc (13, p. 203; 3, 2 : 45), but mokaitc and rohona are repeatedly distinguished from each other in ritual. At Santa Ana the hunter, entering the house of the Caiyaik ("Hunters") society before a ceremony, addresses the Caiyaik as rohona. In reply the society members address him as mokaitc, "mountain lion" (14, p. 287). The same custom is observed at Sia (16). In a Santa Ana hunting song the first stanza addresses mokaitc, the second, rohona (14, p. 293). Hunting fetishes at Laguna are called either mountain lion or rohona (8, p. 127). There is only one other wild species of cat in New Mexico — the jaguar (*Felis onca*).

When I first encountered rohona I did not know that the jaguar, fairly common in Mexico, ever got as far north as the Pueblo region; consequently this identification never occurred to me. It is a fact, however, that *Felis onca* has penetrated the Pueblo country occasionally, *but very rarely*, during the past century (1, p. 283). Thus the Pueblo Indians of New Mexico knew the jaguar, but have seen it so seldom that they have never known it well. The Acoma Indian who told me that rohona was a "big cat with spots" did not know the Spanish or the English name of the animal. Nor had he ever seen one. My Santa Ana informant knew the name as well as the appearance of *Felis onca*. He had never seen the beast either, but said that he had the description from his grandfather, who had seen one. The vagueness of the conception of rohona in the minds of the Indians, and the consequent difficulty of identification seem to be due to their unfamiliarity with this animal.

But what about the Hopi who describe rohona as a foxlike, or a coyotelike, animal? My theory is that the Hopi have borrowed both the word and the idea of a hunting animal from the Keres and have identified it with the swift fox (*Vulpes velox*) or perhaps the desert fox (*Vulpes macrotis*). We feel sure, although we cannot prove it, that rohona is a Keresan word rather than a Hopi one; it *sounds* more like a Keresan word. We know that the Hopi have many Keresan words in songs. The word for "bison" used by the Hopi is Keresan (11, p. 124; 13, p. 203). The bison never ranged near the Hopi pueblos, but the Hopi borrowed the word, and, perhaps, the buffalo dance, from the Keres. It seems likely that they borrowed the word rohona and some associated ideas from the Keres in the same way.

Our theory, which we have suggested tentatively before (14, p. 283; 10, p. 187; 12, p. 23, note 57), if not proved by our present evidence and argument, is, we believe, rather well substantiated.

University of Michigan

LITERATURE CITED

1. Bailey, V. 1931. Mammals of New Mexico. U. S. Department of Agriculture, Bureau of Biological Survey, North American Fauna, No. 53. Washington.

2. Beaglehole, E. 1936. Hopi Hunting and Hunting Ritual. Yale University Publications in Anthropology, No. 4: 1–26. New Haven.

3. Boas, F. 1928. Keresan Texts. Publications of the American Ethnological Society, Vol. VIII, Parts 1–2. New York.

4. Curtis, E. S. 1926. The North American Indian, Vol. XVI. New York.

5. Goldfrank, E. S. 1927. The Social and Ceremonial Organization of Cochiti. Memoirs of the American Anthropological Association, No. 33. Menasha, Wisconsin.

6. Henderson, J., and Harrington, J. P. 1914. Ethnozoology of the Tewa Indians. Smithsonian Institution, Bureau of American Ethnology, Bulletin 56. Washington.

7. Kroeber, A. L. 1917. Zuñi Kin and Clan. Anthropological Papers of the American Museum of Natural History, Vol. XVIII, Part II. New York.

8. PARSONS, E. C. 1920. Notes on Ceremonialism at Laguna. Anthropological Papers of the American Museum of Natural History, Vol. XIX, Part IV. New York.

9. —— 1923. Laguna Genealogies. *Ibid.*, Vol. XIX, Part V.

10. —— 1939. Pueblo Indian Religion, 2 vols. Chicago.

11. STEPHEN, A. M. 1936. Hopi Journal of Alexander M. Stephen, edited by Elsie Clews Parsons. Columbia University Contributions to Anthropology, Vol. XXIII, Parts 1–2. New York.

12. STIRLING, M. W. 1942. Origin Myth of Acoma and Other Records. Smithsonian Institution, Bureau of American Ethnology, Bulletin 135. Washington.

13. WHITE, L. A. 1935. The Pueblo of Santo Domingo, New Mexico. Memoirs of the American Anthropological Association, No. 43. Menasha, Wisconsin.

14. —— 1942. The Pueblo of Santa Ana, New Mexico. *Ibid.*, No. 60.

15. —— (MS). Gotsa Tests the Beasts of Prey.

16. —— (MSS). On Sia and Acoma.

ECONOMICS

AGRICULTURE IN WARTIME

ORION ULREY

THE primary function of agriculture in the national life is to provide an adequate supply of food and other raw materials. In an industrial society farm families have the added responsibility of raising a large proportion of the citizens needed in urban life, since the cities do not produce sufficient children to maintain their population. The urban groups are concerned with obtaining the farm products at reasonable prices. Farmers wish to obtain a per capita income comparable with that received by the city people.

At the present time winning the war and the peace is the primary goal of the nation. During the war the growing of the maximum supply of food is especially important. Since the war began in 1939 the demand for food has been expanded by the steady growth in population, the increase in employment and higher urban incomes, the greater consumption by men in the armed forces than by those in civilian life, the lend-lease shipments, and the building up of a storehouse of food for distribution to the peoples of Europe as the Allies advance. The supply cannot be rapidly expanded because agriculture is an industry of diminishing returns and increasing costs. To obtain a ten per cent expansion in output an even greater percentage increase in costs is necessary.

Agriculture has been handicapped for two decades, relatively to most urban industries, by low prices and income. National farm programs have been developed to assist farmers to improve their economic condition. Prices of foods have been rising rapidly during the past two years; and many citizens are concerned with the relation of the upward movement of retail food prices to the demands of industrial workers for higher wages. A national program to control the cost of living was outlined on April 27, 1942, by the President. The controls over food prices and wages have not been complete because farmers and labor represent the two largest political segments of the nation.

Urban labor wants to keep wage rates in line with the rising cost of living and with the higher wages of management. When food prices rise, organized labor requests an increase in wage rates. Farmers favor rising prices which are beneficial to them, but are disturbed by the high wages in war plants which attract needed labor away from the commercial farms. Farm people believe that the war workers should pay a price for food which would permit farmers to pay wages sufficient to retain the necessary farm labor.

In general, the urban press and consumers have been critical of programs of the commercial farm bloc, which in recent years has been very effective in obtaining national legislation affecting prices. Since the war began many questions have been raised which involve economic relationships of farm and urban people. What are the economic forces which have affected agriculture since the start of the present war? Is August, 1939, a fair date to use to compare increases in farm prices and income with increases in urban wages and income? Have farm prices risen excessively? What is the significance of the parity concept, and are the present parity formulas fair to farmers and urban people? What have been the relationships of prices received by the farmer and paid by the consumer for food? Have retail food costs increased more or less than urban income? How do the per capita farm and urban income compare? What has been the relationship of real farm prices and real industrial wages over a period of years? How does the "Little Steel" formula of adjusting urban wages to rising costs of living affect agriculture? Are the methods of financing the war related to the immediate and long-term price and income position of agriculture? Are rising food costs inflationary? How will an increase in farm prices affect the cost of living? Why are farmers opposing price subsidies? What is the relationship of farm prices and food production? These are some of the questions which this paper will attempt to answer, briefly and incompletely.

UNDERLYING ECONOMIC FORCES AFFECTING AGRICULTURE

Many national and international forces were affecting American agriculture when the present war began, some favorably and others harmfully. The war has brought some changes in these forces.

1. Farm prices and incomes were out of adjustment with urban prices and incomes in August, 1939. The prices of farm products were

70 per cent of parity, which is a relationship of prices received by farmers to prices paid by farmers for commodities bought, farm interest payments and farm taxes, compared with those of August, 1909—July, 1914. Income per capita of farm people was only about one half that of urban people, after all adjustments for such factors as food and fuel obtained from the farm, living costs, earnings off farm, and the proportion of children in families. The unfavorable price and income position of agriculture was due to the maladjustments in the price structure resulting from inflation and deflation of the World War I period, and also to underlying forces of many decades, such as the regional and area surpluses of farm people in relation to available rural resources.

The rise in the price level, especially during 1941–42, has corrected most of the disparities between prices received and prices paid by farmers. The opportunities in war plants have drawn farm people from the overpopulated rural areas. A few years of war will bring a better balance between farm population and resources, especially in the submarginal areas.

2. Farmers have been producing and selling under conditions more competitive than those under which they have been buying. The large-scale corporate businesses have administered the supply to maintain retail prices of many of the things farmers buy. The growth of organizations among almost all types of urban workers, industrial as well as professional, has affected the prices of items included in both the production and living costs of the farmer. The fixed and rigid urban price items were unfavorable to farmers during the period of postwar deflation.

The inflation of the past two years has lifted the prices of farm commodities produced under competitive conditions more than the prices of industrial goods farmers buy which are produced under conditions of imperfect competition or pseudomonopoly.

3. The proportion of American agricultural products exported has been declining since the beginning of the century, except during the period of World War I. Only 5 or 6 per cent of the domestic farm production was exported during the thirties. Cotton, tobacco, wheat, lard, and fruits were the only farm commodities exported in significant quantities. The United States had been importing more food than she has exported for many years. The recovery and improvement of

European agriculture after the last war, the expansion of agricultural production in newer countries using exploitative systems, the postwar deflation, and the consequent nationalistic protective policies were important forces restricting exports of American farm products.

The trade-agreements program and, more recently, the lend-lease program, which provides for shipment of food to Russia, England, China, and the other Allies, have again opened foreign markets for American farm products.

4. The application of technology to agriculture has brought about wide differences in earnings of farm people. Farmers who obtain an advantage because of choice of parents, educational opportunities, mental equipment, or health have applied technical knowledge, expanded output, and reduced the per unit costs. Those who are not so fortunate and who have lagged behind in the application of science have continued to lose in the competitive struggle and have been forced to reduce their level of living. Modern urban industry can use efficiently many of the farm people who are unable to compete successfully in agriculture. The urban war opportunities have encouraged a migration from farms of many who are not equipped to operate and manage commercial enterprises.

5. The industrial wage policy since 1920 has been unfavorable to agriculture. The higher earnings through higher wage rates and reduced hours have increased the costs of commodities farmers buy, especially those not manufactured under decided conditions of decreasing costs. The higher wage rates in the distribution of foods have tended to reduce the proportion of the consumer's dollar obtained by the original producers. The trend toward higher wages, primarily in those industries most completely organized and those which could best apply mass production methods, has not been entirely favorable to agriculture. The demand for food is inelastic for the higher-income groups and elastic for the lower-income groups. A more equitable distribution of urban and labor income would have increased the total purchases and the average prices of food.

The increases in the number of employed workers since the war began and the larger number of workers with higher wages have been primarily responsible for the larger domestic purchases and the rising prices of food.

6. For a few decades many forces have been moving toward agricultural planning. The expansion of technology and the resulting urbanization have made necessary the coördination and integration of all phases of the national economy to mitigate depressions, reduce unemployment, and obtain economic security. The extension of education, in both the technical and the social sciences, has been tending to provide the knowledge and the techniques for intelligent planning. The growth of farm organizations, coöperatives, and other institutions of the rural community has permitted farm people to gain experience and vision.

The deflations of 1920–22 and of 1929–32 encouraged very rapid development of national farm programs, out of which policies and principles have been evolving. Under the Agricultural Adjustment Administration Acts provisions have been made for restricting production and marketings, for purchasing and distributing farm excesses abroad and among the low-income groups within the country, for subsidizing consumption, for establishing minimum price levels by loans, and, by other methods, for increasing farm prices and income. Machinery has been set up to assist in the expansion or the contraction of crops and livestock and to encourage improved production-conservation practices — to obtain a balanced output of farm products in line with domestic and foreign demand, at prices more favorable to farmers than have existed in America at any time except during the peak of prices of World War I.

The Agricultural Adjustment Administration has been used, since the present war needs for food were realized, to reverse the trend from contraction to maximum production. Some believe that assistance is not needed for the expansion of agriculture because higher prices will bring the proper results. Others contend that the regulating machinery can assist in wartime changes and that it must be kept in operation to be ready for the needed adjustments after the war period is ended and at least moderate deflation occurs.

Extreme inflation during the war undoubtedly would be followed by postwar deflation and economic nationalism as raw-material producers seek protection from low-priced foreign supplies. The type of international economic arrangements in which we shall participate probably will depend primarily on the relationship of prices within our domestic economy.

Although most of the emphasis has been on national economic

planning for agriculture, state and local groups have been moving forward in expanding educational opportunities for farm youth and adults, in providing a foundation for improved nutrition and health, and in building a rural culture for a technological age. The war has emphasized the necessity for training in technical skills or vocational education for both rural and urban people. Preparation for the peace and a friendly relationship among the peoples of the world will show the need for better training in the techniques of coöperative action, or in the social sciences and humanities. The findings on the physical condition of the young men being inducted into our armed forces are again calling our attention to the necessity of improved diets, medical attention, and housing. The emphasis on nutrition undoubtedly will have great influence upon our domestic food policies and, in turn, upon all phases of health in the postwar period.

7. For over a century agriculture has been declining in relative importance in the national life. The improvements in farming and the larger output of food per farm worker have permitted migration to cities and the growth of industry and services and the resulting higher standard of living.

At the beginning of the war the 32 million farm people constituted about 24 per cent of the total population. The lower-income one third of the farm families, or the 2 million with gross income of less than $400, sent an average of only $100 to market and sold only 3 per cent of all farm products marketed. The middle-income one third, with gross incomes from $400 to $1,000, provided 13 per cent of all farm products sold. The 3.5 million farm workers who have left agriculture since the war began, two thirds to urban industries and one third to the armed forces, were primarily from the lower- and middle-income farms. From 1939 to 1942 the distribution of the 18 per cent increase in agricultural production among the farm-income groups probably was in proportion to the contribution to sales, discussed above. The long-time trend has been for food production to be concentrated on the commercial farms.

The total farm products needed in our domestic economy probably can be supplied by 3 million farm families, constituting only 12 to 15 per cent of the total population. The urban opportunities during the war period are speeding up the movement away from the country. This trend is favorable in bringing about a better balance between people and resources in our less fertile areas, and thus in increasing the average income of farm families. However, the migration from farms and the decline in proportion of farm to total population raise some very critical national problems. The farms and small towns have produced the children for replacements in our population. On the basis of recent vital statistics about 30 per cent of the nation's people should be on farms to maintain the population at its present level. A country with declining birth rate probably will have many international difficulties, especially if a peaceful world is not established. The speeding up of the migration from farms to cities will also bring more rapidly the need for the beginning of studies leading to national population policies and programs.

The decline in proportion of total population on farms has meant that the prosperity of the commercial farmers has become more and more dependent upon urban prosperity, on the amount of and distribution of urban purchasing power. The movement from submarginal areas and farms is also related to urban opportunities for employment. The expansion of the armed forces and of war production and the increase and better distribution of urban purchasing power since 1939 have been very favorable to the commercial farmers of the nation, except those handicapped by labor shortages.

PARITY POSITION OF FARM PRICES

In August, 1939, the ratio of prices received by farmers to prices paid by farmers, interest, and taxes, was 30 per cent less favorable than when World War I began (Table I). Two years were required, or until August, 1941, for farm prices to reach a ratio of parity to these cost items. The rise in prices in relationship to costs during the first two years of the present war was only a recovery from the depression levels that caused financial disaster and severe sufferings among rural people.

Many urban residents hold the erroneous belief that parity prices give farmers equality in real income with other groups. Parity, as statistically defined, is a relationship of prices and costs, and does not guarantee economic equality.

Wages of labor are the only important farm cost not included in the index of articles farmers buy that is used in computing parity. They rose less rapidly during World War I than did farm prices. They remained relatively lower than the other cost items during 1933–40 and, consequently, were not included in the parity formula. The in-

TABLE I

INDEX NUMBERS OF FARM PRICES, PRICES PAID, RATIO OF PRICES RECEIVED AND PRICES PAID, AND FARM WAGE RATES, UNITED STATES

Year	Prices received by farmers	Prices paid, interest, taxes	Ratio	Wage rates
1910–14	100	100	100	100
1917	175	149	117	141
1920	211	205	103	242
1921	125	166	75	155
1932	65	122	53	96
1939 (Aug.) ..	88	125	70	123
1940	98	126	78	126
1941	122	134	91	154
1942	157	152	103	201
1943 (Feb.) ..	178	160	111	223

crease in wages in 1942 at a more rapid rate than prices received or than the other items of cost resulted in the farmers requesting that wages be included in the parity concept. Farmers realize that wages have been too low for good health and social morale of the workers. Farm income has not been sufficient, however, to permit the paying of adequate wages.

BRIEF ANALYSIS OF THE PARITY FORMULAS

In the original Agricultural Adjustment Act of 1933 the parity index was based only on prices of commodities bought, numbering 174, divided about equally between goods farmers use for family living and those used in production. In 1935 interest and taxes were included in the parity formula except for products, such as potatoes and tobacco, that have a postwar parity base as 1919–29 and 1934–39. If these costs were included, parity prices for these commodities would be lower than they were in 1941–42.

Farm wages have not been included in the index. In 1935 the average wage rate was only 3 per cent higher than in 1910–14. The inclusion of wages would have lowered the parity index for commodities with a prewar base by about 4 per cent. The average wage rate for 1942 was about double the average for 1910–14. If wages had been included in the latter year, the index of farm costs would have been raised 5–15 points, depending on the farm product and also on whether only hired labor or both hired and family labor were included. Products which require considerable labor, such as dairying and raising sugar beets, will be benefited by the inclusion of wages in the parity formula more than livestock feeders and grain farmers.

The Soil Conservation and Domestic Allotment Act of 1936 and the Agricultural Adjustment Act of 1938 provided also for a parity income per capita of farm in relation to nonfarm people, based upon the prewar period. As yet, parity formulas based on farm and nonfarm income per person have not been used.

Many criticisms have been offered on the use of 1910–14 as a base period to compare farm prices and farm costs. There are also several weaknesses of parity prices as computed by the present legal formulas. Farm labor is not included in the index of costs of items farmers buy. The relative prices of some farm products have changed since the prewar base period because of changes in the cost of production and the market-demand situation. The index of cost of items farmers buy is based on national averages, and does not reflect the great differences in expenditure for machinery, feed, fertilizer, and labor among different regions and states, and among the various types of farming. The parity index may not be accurate for any single state.

The ideal parity would be based on the net income per person or per farm rather than on the relationship of price and cost items. One of the immediate war goals is to obtain a proper balance of agricultural production. The percentage of parity necessary to achieve adequate production will vary among the farm products, being relatively higher for the dairy products than for the grains.

FARM FOOD PRICES, RETAIL FOOD PRICES, AND COST OF DISTRIBUTING FOOD

Farmers are favored during periods of inflation when they obtain a larger share of rising retail prices, since the costs of distributing food from the country to the consumer rise slowly. From August, 1939, to December, 1942, farm food prices rose 84 per cent, retail food prices 42 per cent, and the cost of distributing the same food only 8 per cent (Table II). Another study on marketing margin of fifty-eight foods indicates no increase in spread between the prices received by farmers and those paid by consumers during the same period. Although wage rates in processing and distributing foods were from 20 to 50 per cent higher in December, 1942, than they were fifty-two months earlier, many services had been eliminated and marketing efficiencies introduced during the period.

TABLE II

INDEX NUMBERS OF FARM FOOD PRICES, RETAIL FOOD PRICES, COST OF DISTRIBUTION OF FOOD, AND MARKETING MARGIN OF FIFTY-EIGHT FOODS *

Year	Farm food prices	Retail food prices	Cost of distribution	Marketing margin — 58 foods
1910–14	100	100	100	..
1918	201	180	158	..
1920	212	207	203	..
1933	53	97	144	..
1939 (Aug.) ..	92	132	168	100
1940	101	136	167	95
1941	123	148	167	93
1942	157	170	178	99
1942 (Dec.) ..	169	173	180	99

* Data on farm food prices, retail food prices, and cost of distribution of food published currently in *Farm Economics, New York State College of Agriculture;* data on marketing margins in *The Marketing and Transportation Situation, Bureau of Agricultural Economics.*

Disequilibria among farm food prices, retail food prices, and the costs of distributing food resulted from the inflation of World War I and the falling-price periods 1920–22 and 1929–32. The corrections had not been made by rising prices during 1933–39, since in August, 1939, the index of farm food prices was 92, retail food 132, and cost of distribution of food 168 (1910–14 = 100). The changes in the prices from the outbreak of the present war to December, 1942, had brought these prices into a more favorable relationship.

FARMER'S SHARE OF RETAIL VALUE OF FOODS

American farmers were receiving too small a proportion of retail values of food when the war began, on account of the maladjustments in farm prices, retail prices, and cost of distribution, brought about by the inflation during and following World War I. The food raisers were obtaining 41 cents of each dollar spent for food at retail in August, 1939, as compared with 53 cents when the earlier war was started. Farmers got 55 per cent in December, 1942 (Table III).

TABLE III

FARMER'S SHARE OF RETAIL VALUE OF FOODS *

Year	Percentage	Year	Percentage
1913–15	53	1939 (Aug.)	41
1920	53	1940	42
1921	44	1941	48
1929	47	1942	53
1932	33	1942 (Dec.)	55
1935–39	42		

* Data published currently in *The Marketing and Transportation Situation, Bureau of Agricultural Economics.*

URBAN INCOME AND RETAIL FOOD COSTS

In 1942 the retail costs of foods constituted about 22 per cent of the income of the average nonfarm family, as compared with 25 per cent during 1939 and 27 per cent during 1918–20 (Table IV). After buying food in 1942 the urban family's balance of $1,817 to use for other purposes was much higher than that for any previous year. The balance of income remaining after purchase of food increased 55 per cent from 1939 to 1942, while the cost of living rose only 24 per cent.

TABLE IV

NONFARM FAMILY INCOME, RETAIL COST OF ALL FOOD, BALANCE AFTER PURCHASE OF FOODS, FOOD COST AS PERCENTAGE OF INCOME, AND INDEX OF COST OF LIVING *

Year	Nonfarm family income	Retail cost of all food	Balance after purchases of food	Food cost as percentage of income	Index of cost of living 1913 = 100
1913–15	$1,022	$330	$ 692	32	102
1918–20	1,659	616	1,043	37	177
1933	1,108	343	865	31	132
1939	1,558	388	1,170	25	141
1940	1,671	394	1,277	24	142
1941	1,957	430	1,527	22	149
1942	2,322	505	1,817	22	164

* Data published currently in *The Marketing and Transportation Situation, Bureau of Agricultural Economics.*

Urban income has risen more rapidly than the cost of food since the war began. The nonagricultural income payments were 80 per cent higher in December, 1942, than in 1935–39, whereas retail food prices were only 27 per cent higher (Table V). The farmers were obtaining 66 per cent more for the same kinds of food, partly because the marketing margin, the difference between what the farmer received and the consumer paid, did not rise materially.

TABLE V

INDEXES OF RETAIL COSTS OF FIFTY-EIGHT FOODS, NONAGRICULTURAL INCOME, AND PAYMENTS TO FARMERS FOR FIFTY-EIGHT FOODS *

Year	Retail cost of 58 foods	Nonagricultural income	Payments to farmers for 58 foods
1929	125	122	138
1935–39	100	100	100
1940	95	115	94
1941	103	137	116
1942	121	166	152
1942 (Dec.) . . .	127	180	166

* Data published currently in *The Marketing and Transportation Situation, Bureau of Agricultural Economics.*

The data on urban income and retail food costs indicate that food took a small proportion of income in 1941 and 1942 as compared with that of any previous period. The food price increases since 1939 have not been an added burden on the average urban family. However, the low-income families in cities, especially those living on pensions and annuities, have had to reduce their level of living in terms of food and other items. As federal war taxes take a larger proportion of income, urban families will be even more concerned with retail food prices.

COMPARISON OF FARM AND NONFARM INCOME PER CAPITA, 1941

The farm income has ranged from about 20 to 40 per cent of urban income per capita (Table VI). The prewar period is not satisfactory as a parity base for comparing farm and nonfarm income since the former was only 28 per cent of the latter in 1910–14. Although some adjustments are necessary for accurate comparisons, the data do in-

TABLE VI

NET INCOME PER CAPITA OF FARM AND NONFARM PEOPLE *

Period or year	Net income from agriculture per person on farms	Income per person not on farms	Percentage of farm to non-farm income
1910–14, prewar	$134	$488	28
1917–20, war inflation ..	287	737	39
1921–22, postwar deflation	136	717	19
1925–29, prosperity	218	837	26
1931–34, depression	98	488	20
1937–40, recovery	178	667	27
1941, present war	254	826	31
1942, present war	368	981	38

* Data from *Net Farm Income and Income Parity Summary, 1910–14*, mimeograph publication of the Bureau of Agricultural Economics.

dicate wide differences in farm and urban income for the past three decades.

After proper adjustments the income of farm people represented about 63 per cent instead of 31 per cent of nonfarm income per capita in 1941 (Table VII).

Farm people worked in factories, mines, lumber camps, and on government jobs and in 1941 earned from such sources about 1.5 bil-

TABLE VII

ADJUSTMENTS FOR COMPARISON OF FARM AND NONFARM INCOME PER CAPITA, 1941 *

	Farm	Nonfarm	Percentage farm to nonfarm
Income per capita, 1941	$254	$826	31
Income after adjustments			
Earnings other than from farm	322	809	40
Food, fuel, at wholesale instead of retail prices	392	809	48
House rent, and more children on farms	507	809	63

* Adjustments suggested in *Parity, Parity, Parity*, by J. D. Black, pp. 111–113.

lion dollars. The outside payments increased farm income from $254 to $322, and reduced nonfarm income from $826 to $809, on a per capita basis.

The farm family income included about $256 worth of food and fuel from the farm at wholesale prices, which would have cost an urban family about $536 at retail prices. This difference of $70 per capita would have increased farm income from $322 to $392, if the farmer had purchased the food and fuel at retail prices.

In 1941 the annual rent of a farmhouse was about $100 a year, or $25 a person, for the average farm dwelling, which was worth about $1,200. The rent for an urban dweller was about $360 a household, or $100 a person. The differences are due to wide variations in standards and to higher city values, taxes, and cost of labor in construction. Since there is a larger percentage of children and aged people on farms than in cities, we can expect a relatively lower income per person.

After making the adjustments noted, which increased the farm income per capita to about 63 per cent of that of nonfarm people, the difference still remained excessive. Although there are wide variations between the average farm and nonfarm incomes which should be reduced, there are also inequities among farm and among urban people which require national attention. The farmers are interested in obtaining their per capita share of the growth in the output of society. During the past thirty years their buying power has increased, but not so rapidly as that of urban families.

The employment opportunities during the war period are correcting only a minor portion of the inequities in income between and within the farm and nonfarm groups. It is very unfortunate that the higher cash incomes cannot be converted into peace goods and services and thus improve the level of living of rural and urban families.

COMPARISON OF REAL FARM PRICES AND REAL WAGES OF FACTORY WORKERS

Real wages of factory workers have been relatively higher than real farm prices during the past decade, if we consider 1924–29 a fair base for comparison. When the present war began, the index of real farm prices (prices received in ratio to prices paid, interest, and taxes) was 82 (Table VIII). The index of real wages (monthly earnings per employed factory worker in relation to cost of urban living) was 113.

Real farm prices and real wages increased rapidly to indexes of 131 and 158, respectively, in December, 1942.

Farm prices were in a poor strategic position in August, 1939, since the purchasing power position was 30 per cent below the 1910–14 level and 18 per cent below that of 1924–29. Industrial wages were in a favorable position, being 13 per cent above the 1924–29 purchasing-power level. Several months of rising prices passed after the war

TABLE VIII

INDEXES OF REAL FARM PRICES AND REAL WAGES OF FACTORY WORKERS

(1924–29 = 100)

Year	Prices received by farmers	Prices paid, interest, taxes	Real farm prices	Monthly earnings per employed factory worker	Cost of living	Real wages
1924–29 . .	100	100	100	100	100	100
1939	63	74	85	90	80	112
1939 (Aug.)	60	73	82	90	79	113
1940	67	75	89	95	81	117
1941	84	80	105	113	85	133
1942	108	90	120	140	94	149
1942 (Dec.)	122	93	131	153	97	158

began before a parity position for farm prices was reached. The urban press and statisticians are unfair to agriculture when they use August, 1939, as a base for comparing changes in farm prices with cost of living and urban wages.

The period 1910–14 is commonly used by agricultural economists to compare the changes in farm prices with farm cost items, since such a base period shows the disarrangement in the price system brought about by the inflation and deflation of World War I. Likewise the prewar base is not a sound period for comparison of real farm prices and real industrial wages. If lines are drawn to show the trend of real farm prices and real industrial wages for the period 1866–1941, then in 1910–14 real farm prices were 12 per cent above the trend line, and real wages were 29 per cent below the trend line. Farm prices rose more rapidly during 1896–1914 than the level of all prices, whereas urban wages did not increase so rapidly as the cost of living.

There are many limitations in making the comparison given of real farm prices and real wages of factory workers. At least one half or more of the 6 million farmers may be classified as primarily self-sufficient. Higher real farm prices will not materially improve the economic position of these farmers. They need a larger farm business and greater efficiency. The index of wages used does not make allowances for unemployment, nor does it include wages of workers in the distribution fields whose earnings have been relatively low. It is not entirely sound to compare the price position of commercial farmers who take economic risks on capital and who have large investments with the wage position of industrial workers whose jobs do not require the training and managerial ability of the commercial farmers.

In periods of falling and low prices the buying power of employed industrial workers increases rapidly, whereas farmers have no increase or even an actual decrease. The opposite is true during rising prices. During the twenties, with a declining level of prices and wages remaining up, urban workers regained a more nearly fair share of the national income than they had obtained since the beginning of the century. A disproportionately high share of the national production went to urban people during the thirties.

THE "LITTLE STEEL" FORMULA AND AGRICULTURE

The "Little Steel" formula, resulting from a wage dispute between the union and the small steel companies, provides that workers whose wage rates have not increased 15 per cent, the rise in the cost of living between January, 1941, and May, 1942, are entitled to a wage increase of that amount in order to maintain their peacetime living standards. Other workers, under the formula, are not entitled to any further increase, except in cases involving inequities, inequalities, or substandards of living.

The formula is very favorable to labor, since it recognizes as a parity base (relationship of wages to living cost) a month when real factory wages were, to that date, the highest in history. Real monthly earnings per employed factory worker were 28 per cent higher in January, 1941, than during 1924–29, which is a fair period to use to compare real farm prices and real wages. Although the cost of living rose 19 per cent from January, 1941, to December, 1942, real wages rose 24 per cent since the monthly cash earnings increased 48 per cent. The increase in monthly earnings per employed factory worker was due primarily to laborers working longer hours on the higher overtime pay and to more workers being engaged in war plants rather than to increases in basic wage rates.

The high real wages in war plants have tended to be inflationary in effect on retail food prices, which is favorable to farmers. They also have been an important force responsible for drawing from agriculture workers needed for food production, especially from farm areas which are near industrial centers, as are those in Michigan.

Higher wages in war industries have been necessary to move labor from industries producing peace goods. Citizens are not ready for the alternative method of shifting labor — that of conscription and allocation to plants and farms where needed.

FARM SITUATION IN RELATION TO FINANCING OF THE WAR

One of the weak spots in our national war economy is our general plan of financing the war. For the present financial year (ending June, 1943) about one fourth of the war expenditures are being met by taxes and three fourths by bond sales. Bonds have been purchased primarily by banks rather than out of current earnings and savings of individuals. The national income has been rising and the supply of consumer goods and services declining. Excess of purchasing power in the hands of individuals in relation to available things to purchase and the payment of taxes is a force tending toward inflation. The administrators of the price control programs will have difficulty keeping prices from skyrocketing upward unless the excess of national income is withdrawn by taxes or by bond sales to individuals.

Farmers, along with the other economic groups, have not realized that the present generation bears almost all the physical cost of the war, that there are dangers of inflation from our present plan of financing, and that a greater equality of sacrifice can be achieved by a plan of financing the war almost entirely by taxes. They should remember the evils of deflation, and especially the harmful effects upon them that resulted from the rising and falling prices during and following World War I. A heavier tax program must fall, however, upon the middle-income groups, which include farm spokesmen, the higher income one third of the farmers as well as the small business men, the labor leaders, and other very important people in the life of almost every community.

Farmers have been very critical of the profits of the war plants, the wages of management, and especially of the wages of the war workers. In order to prevent inflation, to supply the government with needed funds, to secure a better equality of sacrifice, and to bring an improved relationship between rural and urban income, a progressive program for farm organizations would be not only to advocate but to exert legislative pressure to raise 20–25 billions more a year from taxes to finance the war. A more equitable relationship between farm and urban income, after taxes are paid, would tend to keep more workers on farms. Such a tax program would benefit agriculture in the long run, financially by preventing harmful inflation and deflation, and politically by winning the respect of thinking city people. Farmers may need urban friends in the future to obtain essential legislation.

ARE RISING FOOD PRICES INFLATIONARY?

Considerable attention has been given since the middle of 1942 to rising food prices by the urban press. Retail prices rose 43 per cent from August, 1939, to December, 1942, which seems like an excessive increase. When the war began, food prices were too low, however, for the welfare of agriculture. Farmers subsidized urban consumers with cheap food during the twenties and thirties. Of course, food seems high to low-income people, but their difficulty is income that is too low and not food that is too expensive. There are also some savings in the costs of distributing food which might be shared by the farmers and the consumers.

Organized labor groups have emphasized the rise in prices of food in their requests for upward adjustments in wages. Food, which is a regular necessity, is purchased daily and consequently urban consumers are more aware of rises in food prices than of any other items in their family budget.

The retail cost of all foods increased from about 15 billion dollars a year during 1935–39 to 25 billion in 1942, but the income to nonfarm persons rose from 61 to 106 billion (Table IX). The balance available after the purchase of food increased by 36 billion. It seems logical that the increase of 45 billion in urban purchasing power from 1935–39 to 1942 is more inflationary in effect than the increase in gross cash farm income of 7 billion dollars.

The farmers contend that the methods employed in determining war contracts and in paying factory labor are much more inflationary than the higher food prices. The policy of governmental purchase of war goods on a basis of cost plus a profit adds materially to the cost

TABLE IX

INCOMES OF FARM AND NONFARM POPULATION, RETAIL COST OF FOOD, AND BALANCE OF NONFARM INCOME AFTER FOOD PURCHASES *

Year	Total cash farm income	Total cash farm income from food	Income of nonfarm population	Expenditure for food at retail	Balance of nonfarm income after food purchases
	Million dollars				
1935–39 av. .	7,973	6,030	60,899	16,000	44,899
1940	8,340	6,395	72,651	17,000	55,651
1941	11,168	8,548	84,806	19,000	65,806
1942	15,442	11,772	106,400	25,000	81,400

* Data from mimeograph publications of the Bureau of Agricultural Economics.

of the war and to the amount of purchasing power which may be used to bid up prices unless the money is withdrawn by taxes or by purchase of bonds. The paying of wages on a time-and-a-half basis for all hours over forty a week also increases urban purchasing power at a time when the volume of available consumer goods is being reduced. A factory laborer who works eight extra hours receives an income 30 per cent higher.

Farmers are also affected by the higher retail prices of food since it is the largest item they purchase, among both production and living expenses. They bought about 1.6 billion dollars of food during 1935–39 and about 2 billion in 1942.

HOW WILL AN INCREASE IN FARM PRICES AFFECT THE COST OF LIVING?

A 10 per cent increase in the farm price of foods will increase the cost of living of the average urban family less than 2 per cent. An increase in the farm price of food of 10 per cent would have raised the retail price only 5.5 per cent in December, 1942, if distribution costs, which constituted 45 per cent of retail food prices, were not changed (Table X).

In 1941 purchases of food constituted 33 per cent of the average

TABLE X

EFFECT OF A 10 PER CENT RISE IN FARM FOOD PRICES ON RETAIL FOOD PRICES, DECEMBER, 1942

	Original prices	Increase of 10 per cent in farm price	Increase in retail price
Farm price of food	$1.00	$1.10	..
Distribution cost, 45 per cent	.82	.82	..
Retail price of food	$1.82	$1.92	5.5 per cent

expenditures of nonfarm families and single consumers (Table XI). The range was from 41 per cent for families with incomes under $500 a year to 23 per cent for those with incomes over $5,000. An increase of 10 per cent in the prices farmers received for food would raise the average cost of living 1.82 per cent (33 per cent of 5.5 per cent). The cost of living of the lower-income consumers would be increased most since they spend a higher percentage of their total expenditures for food.

TABLE XI

EXPENDITURES OF NONFARM FAMILIES FOR FOOD, PERCENTAGE FOR FOOD IN 1941, AND THE EFFECT OF AN INCREASE IN FARM PRICES ON THE COST OF LIVING *

Net money income classes	Total family living expenditures	Expenditures for food	Percentage for food	Effect of a 10 per cent increase of farm price of food on cost of living
Less than $500	$ 368	$ 150	41	2.41 per cent
$ 500–1,000	753	285	38	2.05
1,000–1,500	1,199	418	35	1.89
1,500–2,000	1,636	542	32	1.73
2,000–3,000	2,264	715	31	1.67
3,000–5,000	3,180	940	30	1.62
5,000 and over	7,199	1,673	23	1.24
Median consumer unit	1,552	518	33	1.82

* Data from study of *Family Spending and Saving in Wartime*, made by the Bureau of Home Economics and the Bureau of Labor Statistics.

WHY ARE FARMERS OPPOSING PRICE SUBSIDIES?

During 1942 and early 1943 the Federal Administration favored and experimented with a subsidy program to prevent a rise in the retail price of foods. The Commodity Credit Corporation bought milk from farmers in the New York City and Duluth markets at one price and resold it to milk dealers at a lower price, absorbing the loss of about $800,000 a day in order to prevent an increase in retail prices. Processors of vegetable oils and small packers of pork also received some assistance. Wheat obtained under governmental loans has been released to farmers at less than the loan rates. The Secretary of Agriculture has suggested a system of price incentives to obtain increased production of needed foods, and has requested Congress to provide $100,000,000 for the program. During 1933–42 the government payments to agriculture amounted to over 5 billion dollars, or to about 6 per cent of the gross cash farm income.

A subsidy system has been used very extensively in England to expand farm output, and in Canada to increase output, to shift production to needed foods, and also to assist in controlling the cost of living and, consequently, the entire price level.

Several of the American farm organizations have objected to bonuses, subsidies, and price incentives. The Federal Congress is assisting the farm leaders by not providing appropriations. The urban press and consumer groups are very critical of the position taken by the farm leaders.

The farmers contend that they have been subsidizing urban consumers with cheap food for over twenty years, that food was too low when the war began, and that urban income, which is the highest on record, is sufficient to permit the payment of fair prices for food. They are critical of the high wages in the war plants which are attracting many of their best farm workers. They wish to exclude benefit and conservation payments from the parity formula in calculating price ceilings. They fear a subsidy and bonus plan because of the danger of state domination and control. They say that rising food prices are a minor factor in causing inflation as compared with the great increase of urban purchasing power. They ask why wages should not be effectively controlled under the price-control acts. They insist that a subsidy program which would give the farmers their fair share of the national income would be far too costly. They state that consumers would obtain the subsidy since retail prices would be lower than the competitive supply-and-demand prices. Farmers are not opposed to benefit payments during periods of low-farm income or to subsidy of consumption when urban employment opportunities are limited. They believe that farm prices at even 110 per cent of the present parity formulas are not adequate to enable farmers to obtain an equitable share of the national income.

The statistics of prices and income between farm and urban groups indicate that the farmers have been mainly correct in their recent stand on parity and subsidy, yet there is danger that upward movement in retail prices of food will be one of the forces responsible for further advances in many prices. If the farmers would also insist on a program of heavy taxation which would reach the urban factory managers and workers, they could achieve their aim of reducing the inequities between rural and urban income. The government certainly needs the support of farmers of programs to raise more of the war expenditures by taxes.

FARM PRICES AND INCOME RELATED TO 1943 FOOD PRODUCTION

Farmers have been asked to produce about 5 per cent more food in 1943 than they did during the past year, when the weather was very favorable — and with less labor, machinery, and fertilizer. A continuation of about 90 per cent of the nation's food supply is assured for several years regardless of low-price ceilings and high urban wages. The marginal 10 per cent is the most expensive part of our food supply, but is very important during the war period. To prevent the movement of farmers, their families, and agricultural workers to urban industry and thereby assure the nation of an abundant food supply farm incomes should be more nearly in line with urban incomes. From the standpoint of national welfare farm-price ceilings should not be fixed at levels that will result in less than full production of all essential foods. Farmers want neither inflation nor deflation, but prices that will obtain for them an equitable share of the national income and that will permit them to produce the maximum quantities of foods.

Michigan State College
East Lansing, Michigan

FOLKLORE

A GROUP OF PERSIAN FOLK SONGS

ANNA RATZESBERGER

THE folk songs here presented were collected during a visit to Persia. Some of them have a special interest, however, for members of the International Institute at Flint, for they may be heard among the 250 Assyrian families who left Iran in the 1920's and settled in this city.

Music, according to Occidental theory, is composed of three elements: melody, rhythm, and harmony. Yet in much Persian folk music one element may be so much more important than another that at times the song is reduced to a monotonous chant, and rhythm becomes the only noticeable element. This is illustrated in the song of the brick tossers as they build a wall in the holy city of Meshed. One workman has the solo part and is answered by an antiphon chanted by the rest of the men in line. In this way the timing of the motion of tossing bricks is maintained in a perfect rhythm, no accident occurs through a brick's missing an outstretched hand and hitting a workman, and the bricks speedily reach the top of the wall.

CHANT OF THE BRICK TOSSERS OF MESHED

Solo: May all the enemies of the Prophet
Accursed be until the resurrection!
To Him and His Family

Chorus: SALAVAT!

ALLAH! ALLAH!

Solo: My head and my soul
A sacrifice to Ali!

Chorus: YA'ALI!

Similar songs are sung for other forms of group work, such as paving the city streets. On reaching the words *Ya'Ali!* or *Ya'allah!* they straighten up from work for a moment to rest their backs.

Work songs of a more melodious nature are heard among the

sailors on the lagoons of the Caspian Sea and in the Persian Gulf after the wind has dropped in the late afternoon and the sails wilt and the men must pull an oar. There we have the sea chanteys of Persia.

In the rice fields of Ghilan, the narrow plain below sea level which forms the southern rim of the Caspian, women make merry as they weed the fields for the final time. On the last day of weeding they sing love songs to their husbands — the men not being present to enjoy the music, however — for at this time they transfer to the men the responsibility for harvesting the crop.

One woman acts as soloist and gives the narrative; the others chant the refrain in a manner similar to our spirituals. "Ghambar Ali," a particularly jolly one, is melodious and, a most unusual thing, has a bit of harmony in the refrain. In the first five stanzas the soloist is the wife asking her husband to buy this and that pretty thing, while the chorus, representing the husband, answers "*Insh'allah!*" ("If God wills!"). Then the melody changes and Ghambar Ali is represented in the stanzas enumerating the things he has bought for his wife, but the chorus gives her answer, "*Insh'allah!*"

GHAMBAR ALI

She: Ghambar Ali um Ghambar
Go to the bazaar, Ghambar. INSH'ALLAH! (*Chorus*)
Ghambar Ali um Ghambar
Buy me a blouse, Ghambar. INSH'ALLAH!
Ghambar Ali um Ghambar
Buy me a coin, Ghambar. INSH'ALLAH!
Ghambar Ali um Ghambar
Buy me a kerchief, Ghambar. INSH'ALLAH!
Ghambar Ali um Ghambar
Buy me some sweets, Ghambar. INSH'ALLAH!
He: What have I not done for you? INSH'ALLAH!
I went and brought figs
From the tree for you. INSH'ALLAH!
I brought you two boxes
Of fruit. INSH'ALLAH!

(*Verses are improvised indefinitely.*)

SONG OF THE RICE WORKER

You were young
When you walked thro' the field
In your leather *chimush* [1]
Bound with cord.

[1] Handmade shoes, like moccasins.

Perhaps you think
You're the son of the Sabs Imam.
Now don't wed another,
But stay with me only.

You bought me for only
A hundred and fifty *tomans;*[2]
And, though it was cheap,
You have treated me well.

Now for a little while
Sit in the house;
Your presence eases
My heart of its burden.

In these songs the melody usually has a range of five notes or fewer. Phrases are repeated with great monotony. Frequently a song ends on the fourth or the fifth of the scale instead of the tonic. Because the Persian scale is divided into seventeen parts it is impossible to reproduce the melody on a piano tuned to the Occidental scale. Such playing can give only a distortion of the real melody.

Children's singing games have the same pattern. "Wiggly, Waggly, Falling Leaves" is typical of this group. In this game the children stand one in front of the other, lock arms, and sway backward and forward to the rhythm of the song.

WIGGLY, WAGGLY, FALLING LEAVES

Wiggly, waggly, falling leaves!
Your little mother, Lady Zenab,
Has braids as long as a bow;
Yea, longer than any bow,
And blacker than lust.
I do not want a husband;
I do not want a husband.
Instead I want a turquoise comb.
Ha, I found! Va, I found!
In the pool I silver found.
The silver became my salt dish;
Sir Haj'Ali became my sacrifice.

The wedding song is sung to greet the veiled bride as she enters the room where all the female guests are awaiting her. With the first word everyone begins to clap rhythmically. There is no other accompaniment to the song.

[2] A coin equal, at par, to one dollar.

WEDDING SONG

Bada, bada, bada!
If God wills, blest be this union!
Wedding like a king,
Great is the rejoicing.

Bada, bada, bada!
If God wills, blest be this union!
The bride is tall. Oh, yes!
The street is small. Oh, yes!

Bada, bada, bada!
If God wills, blest be this union!
We came, and again we came.
From the bridegroom's home we came.

May the belovèd be blest,
If God wills, blest be this union!

Most songs are unaccompanied. When someone sings a popular folk song in the teahouse, he may be accompanied on the piano or the violin or the *tar*, which is only a reinforcement of the melody in octaves or single tones, or the accompaniment may be merely the rhythmic clapping of hands. Frequently dances are performed only to the rhythm of clapping hands, snapping fingers, or castanets.

Probably the most beautiful song I found in Persia was a lullaby which a Parsee woman sang to her babe. Although some of the folk songs are highly irregular in meter, almost like recitatives at times, this song was remarkably metrical. Strict adherence to a fixed meter is no doubt due to the nature of the song — a lullaby that gives the effect of rocking a cradle.

Any harmonic accompaniment would be foreign to these melodies. Therefore in transcribing this song I kept the accompaniment as simple as possible and used only the two tones of a perfect fifth in the folk song proper in order to emphasize the rhythmic element. The unusual intervals in the melody cannot be written exactly. Where the note *f* appears I have indicated that the tone is in reality somewhere between *f* and *f*♯ by marking ♯$\frac{1}{4}$ above the *f*. Actually, the interval is between a quarter and a third.

It is interesting to note that in the last stanza of this little lullaby there appears a metaphor that recalls a similar idea in one of our English childhood verses.

PARSEE FLOWER LULLABY

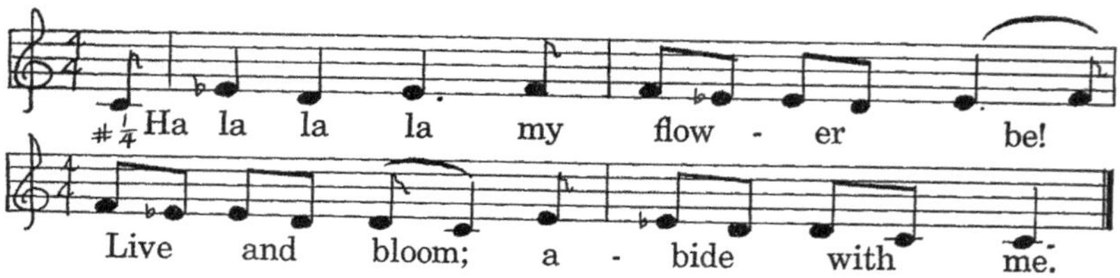

Ha la la la my flower be!
Live and bloom; abide with me.

Ha la la la ye la la yee
My little one is sad, is lonely.

Ha la la la my flower of flowers,
Daddy's gone. What loneliness is ours!

Ha la la la my poppy flower, sleep!
Daddy's gone. May God him keep!

Ha la la la sweet clover bloom,
Daddy's gone. My heart is filled with gloom.

Ha la la la my little caraway seed,
In mother's arms is rest indeed.

INTERNATIONAL INSTITUTE
FLINT, MICHIGAN

HISTORY AND POLITICAL SCIENCE

THE TRANSFORMATION OF RURAL LIFE IN MICHIGAN SINCE 1865 *

WILLIS F. DUNBAR

BY THE year 1865 southern Michigan was no longer a frontier region. The land was largely cleared and under cultivation. One had neighbors close by, too close in some instances. There was no danger of want or privation. Schools and churches had been established. A few miles away was a village in which supplies not produced on the farm could be procured. Not very far distant was a sizable city. The farmer was still to a high degree self-sustaining, however; hardware, salt, sugar, wagons, shoes, and some clothing had to be obtained from outside, but for most foods the farmer relied on his own acres. For social recreation he also was wont to depend on his own ingenuity. His house contained, of course, no bathroom or running water, no radio, no refrigerator, telephone, or vacuum sweeper. His sons and daughters might go to high school, though this was unlikely, and it was even more exceptional for one of his children to enroll in a college. The weather was the most important topic of conversation; family affairs ran a close second, with politics (save during a particularly "hot" campaign) a poor third. What happened outside the United States was quite beyond the farmer's sphere of interest. It did not occur to him to question the social and economic order in which he lived, much less to spend time pondering on religious matters. He would have said, if pressed, that he believed in God and in the church (and usually he meant the Protestant Church), but he would have acknowledged that in the matter of church attendance he was a little lax. Prices which his salable products commanded were something he might grumble about, but on the whole took to be, like death and taxes, entirely outside his control. At any rate, unless a man was lazy, he always could make a living.

The mass of gadgets and machines that are part of the standard

* This article has been published, without documentation, in the *Michigan Alumnus: Quarterly Review*, 49 (1943): 352–359.

equipment of the farmstead of 1943 constitute the surface manifestation of the profound change which has come about in rural living. Of much greater importance is the transformed character of the farmer himself and the members of his family. He is less self-reliant, more dependent on others. He is something of a scientist and a full-fledged business man. By virtue of necessity he has learned how to coöperate with his neighbors in marketing and purchasing. He is better informed about national and international affairs. He is less bound by custom and tradition. He has fewer dealings with the business men of the neighboring villages. He is more class-conscious, less of a peasant, more of a bourgeois. We may understand this transformation better by examining the various factors which brought it about.

Nothing has had more to do with changing the pattern of rural life than the improvement of transportation facilities. Until the close of the Civil War railroad building in Michigan lagged, first, because the lakes were barriers to through east-west lines and, secondly, because the character of the northern portion of the Lower Peninsula had not invited settlement and hence north-south railroad lines. The increase of railroad mileage from 779 in 1860 to 8,195 at the turn of the century, though caused to a large extent by the demands of the lumbering and mining industries, greatly facilitated the marketing of farm products and was the first step in bringing the farmer into closer contact with the city.[1] As steam railroad construction tended to taper off after 1900, promoters turned their attention to the building of electric interurban lines. By 1918 almost a thousand miles of main line track was in operation.[2] The interurban cars offered the first challenge to the public-be-damned policies of the steam railroads. They operated on frequent schedules, they stopped at every crossroad, and they offered lower rates. Needless to say, they were a great boon to the farmer.

But it was the coming of the automobile, the bus, and the truck in the years following World War I that brought about the greatest change in the pattern of rural life. Eggs, dairy products, and other perishables could be marketed more quickly and in better condition. It put the farm family into closer contact with cities and their cultural

[1] *Thirty-fourth Annual Report of the Michigan Commissioner of Railroads for 1906* (Lansing, 1907).

[2] *First Annual Report of the Michigan Public Utilities Commission for the Year Ending December 31, 1919* (Lansing, 1921), p. 193.

advantages. It established a new pattern of courtship for young people. It made the county or township consolidated school practical and thereby started the decline of the one-room country school. It encouraged the movie habit. It made possible for the family a trip in winter or between summer and fall work. It dealt a telling blow to the provincialism of the farmer and his traditionally restricted and narrow point of view. There were scores of changes in the way things were done on the farm. For example, one of the onerous tasks performed by the rural housewife or the children was the churning of butter. In the year 1910 over 130,000 pounds of butter was churned for sale on Michigan farms. By 1940, although milk production had nearly doubled, the amount of butter churned was less than one quarter the figure of thirty years before.[3] The reason is quite obvious: the truck or the family car was taking the cream to the city creameries and saving the farmer's family many hours of monotonous toil. In 1920 only 4 per cent of the livestock delivered at the Detroit stockyards came by truck; twelve years later 75 per cent came that way.[4] Only 40 per cent of the farms of Michigan had an automobile in 1920; by 1930 almost 79 per cent had one or more, and in 1940 almost 82 per cent.[5] The farmer tended to retain his car longer than the city dweller, it would seem, for in 1940 the date of the average car on Michigan farms was 1933.[6]

Closely allied with improved transportation was speedier communication. The farmer of 1865 received mail only when he could call at the nearest postoffice for it or when some obliging neighbor would do so for him. Few farmers therefore subscribed to a daily paper, and by no means a majority even got a weekly publication. Rural free delivery was first tried out in Michigan in 1896. The village of Climax in Kalamazoo County was selected as the point of departure for the first carriers. One of the first two carriers used a horse and buggy; the other, a bicycle.[7] By 1899 fifteen routes were in operation, and the service was thereafter gradually extended to reach practically every farm in the state.[8] This innovation did a great deal

[3] *Sixteenth Census of the United States, Agriculture* (Washington, 1942), I: 775.

[4] Fuller, George N. (ed.), *Michigan: A Centennial History of the State and Its People* (4 vols., Chicago, 1939), I: 498.

[5] *Sixteenth Census, Agriculture,* I: 781.

[6] *Ibid.,* I: 781.

[7] Brown, J. H., "How We Got the R. F. D.," *Michigan History Magazine,* 6 (1922): 442–459.

[8] Fuller, *op. cit.,* II: 235.

more than simply speed up the delivery of personal correspondence. The metropolitan newspaper found its way into the rural home. The huge annual catalogues of Sears Roebuck and Montgomery Ward became a standard item on the reading table and served to whet the appetite for a range of merchandise far exceeding that available in the country stores. The inauguration of parcel-post deliveries about 1910 was another milestone in rural betterment.

Then came the telephone. The expiration of the Bell patent in 1893 made possible the organization of hundreds of local companies, many of them coöperatives. It was some time, however, before telephone service was extended beyond the cities. During the first decade of the twentieth century private companies or local coöperatives built their first lines into rural areas.[9] The average farm telephone was on a party line, usually including several neighboring farmsteads, and became by virtue of this fact an invaluable aid to community visiting and gossip. By 1920 there were telephones on almost half the farms of Michigan. But rising costs of service, the coming of the automobile, and the installation of radios during the 1920–40 period resulted in a sharp decrease in the number of rural telephones, so that in 1940 only about 27 per cent of the farms in the state had a telephone.[10]

By the later twenties radio sets were becoming so inexpensive that even the poorest farmer could acquire one. They introduced fine music, lectures, and news as well as jazz (later "swing") and Hollywood chitchat, into the rural home. From a more practical angle, the radio became invaluable as a means of keeping the farmer posted on market prices, weather reports, and scientific aids to agriculture. It is significant that WKAR, operated by Michigan State College and especially pointed to the interests of rural listeners, is one of the few noncommercial stations that survived the early days of broadcasting. The radio took the place not only of the telephone on many farms but also of the daily newspaper.

A third transformation that has taken place in rural living is the greatly increased use of machinery in the work of field and home. The intricate combines employed extensively on the western prairies today are lineal descendants of a machine devised near Climax, Michigan, and put into use in 1845 by John Hascall and Hiram

[9] Fuller, *op. cit.*, II: 240.
[10] *Sixteenth Census, Agriculture,* I: 781.

Moore.[11] But Michigan was not destined to be a great wheat-producing state. For the Michigan farmer, the Deere plow with its steel moldboard, which appeared in the fifties, the Oliver plow, which was fabricated of chilled steel and was put on the market about 1870, and the Deering binder, used widely after 1879 to bind the sheaves of wheat with twine, were more important.[12] Then came the threshing machine, operated by horsepower, steam, or gasoline, making its seasonal rounds of a community of farms and doing away with the hard work of threshing the grain with flails on the barn floor. Threshing time became a midsummer event on the farm. The men toiled like titans in the boiling sun while the women labored around the kitchen stove for hours on end to provide the best of cookery for the famished menfolk at dinner and supper times. The hay-loader was another "must" item for the farmer by the early twentieth century. And then came silos, cream separators, manure spreaders, spraying equipment, and dozens of other aids that the progressive farmer soon regarded as indispensable. During the twenties and thirties tractors attained wide popularity, and by 1940 one third of the farms in Michigan had one or more of them.[13] In 1900 Michigan farmers were using machinery and implements to the value of $29,000,000 in their fields, sheds, and barns. Forty years later the figure had risen to $120,000,000.[14]

The farmer's wife shared with her spouse the conveniences which the machine age contributed to the farmstead. The automobile, telephone, and radio made the farm home a much less lonely place. To her the coming of electricity was an especial blessing. This great servant not only lighted the house and barn but operated a radio, a cream separator, a vacuum sweeper, a poultry incubator, and numerous other contrivances. At first electric current was generated by plants operated by gasoline engines, but during the last twenty years power lines have been developed to reach a far larger number of rural homes. As recently as 1920 only 8 per cent of the farmhouses of Michigan were lighted by electricity, whereas the census of 1940 reveals that no less than 71 per cent enjoyed the use of electric power in that year.[15]

[11] *Michigan Pioneer and Historical Collections,* XXVII (Lansing, 1907): 527, 543, 544.

[12] Hacker, Louis M., and Kendrick, Benjamin B., *The United States since 1865* (New York, 1932), pp. 175–177.

[13] *Sixteenth Census, Agriculture,* I: 781. [14] *Ibid.,* I: 786. [15] *Ibid.,* I: 781.

These and other changes, as well as trial and error, brought about a transformation in the type of products which came from the fields of Michigan farms. Until almost 1900 wheat was the leading grain crop of the state, and as late as 1890 Michigan ranked eighth in the list of states producing this grain.[16] But the competition of the prairie states and the prairie provinces of Canada was too great, and Michigan now has less than a third of the wheat acreage it had in 1880.[17] The cattle raised on Michigan farms at the close of the Civil War were of the dual-purpose type, acceptable both for milk production and for meat. The mighty increase of cattle on the western plains subordinated the meat-producing cattle business of Michigan to small proportions. More and more farmers turned to the raising of dairy cattle. Synchronized with this change was the increased acreage planted to corn and oats as wheat production shrank. Michigan today ranks high among the states in the production of milk, cream, butter, cheese, ice cream, and condensed milk. In 1934 the income from dairy products was ten times that from wheat.[18]

Wool and wheat were Michigan's most important farm products prior to the Civil War. Between 1860 and 1890 this state ranked either third or fourth in wool producing. But sheep raising, along with wheat growing, declined as the century drew to a close. The number of sheep in 1900 was hardly more than half that in 1880.[19] Hogs, along with dairy cattle, had become numerous by 1900. In 1892 the number of swine was estimated at 301,000; in 1920 well over a million were being raised, though since that year the number has declined by almost one half.[20] The raising of driving and work horses once was an important business, but motor vehicles have displaced old Dobbin. Today the number of horses on Michigan farms is only about one half what it was in 1920.[21] This change affected the growing of forage. Timothy hay gave way to red clover and alfalfa. The latter crop, called "lucerne" in the early days, was first sown by a French farmer near Monroe in 1845, but for many years remained almost unknown. With the change from wheat-sheep-horse production to dairy-cattle-hogs-corn-rye-oats production, alfalfa at

16 *Eleventh Census, Statistics on Agriculture* (Washington, 1895), p. 335.

17 *Sixteenth Census, Agriculture,* I: 776.

18 Fuller, *op. cit.,* I: 493.

19 *Thirteenth Census, Abstract, with Supplement for Michigan* (Washington, 1913), p. 329.

20 *Sixteenth Census, Agriculture,* I: 775; Fuller, *op. cit.,* I: 485.

21 *Sixteenth Census, Agriculture,* I: 775.

length came into its own. The amount of this crop grown multiplied more than twelve times between 1920 and 1934.[22]

The trends so far indicated were fairly general throughout the state. There has developed a definite movement toward specialized production in certain restricted areas since 1865. For instance, in the Upper Peninsula and in certain counties in the northern part of the Lower Peninsula potatoes have become the big money crop. The extensive growth of navy beans in the Saginaw valley makes Michigan a rival of New York in the production of this crop. The Dingley tariff law of 1897 provided protection for domestic sugar and encouraged the growth of the beet-sugar business. No fewer than eighteen sugar refineries were built in Michigan during the years 1898 and 1899.[23] The "Thumb" region is the center of the beet-sugar industry. In the southwestern part of the state fruit growing became a major farm occupation. Within the "fruit belt" are several subregions specializing in grapes, apples, peaches, or berries. Cherry raising has been developed extensively in the Grand Traverse region. Michigan now produces more fruit than any other Middle Western state and has a larger acreage devoted to small fruits than any other state in the Union. Certain areas specialize in the production of celery and peppermint oil. Thus there was much less "general farming" in 1940 than in 1865. The farmer became less interested in supplying his family's needs from the products of the farm; instead, he endeavored to grow and market some money crop from the proceeds of which he could buy food, clothing, machinery, and other things which he and his family wanted.

Another phase of this transformation has made farming less of an art, more of a science. Reluctantly (for he is innately conservative) the farmer has been forced by competition and social pressure to rely increasingly on the scientist for gadgets and advice. Michigan State College has played a primary rôle in this phase of rural change. One of the oldest state agricultural colleges in the nation, its services to farmers have been constantly expanded since its opening in 1857. In 1875 it undertook extension work, carrying valuable information directly to the farmer. The distribution of regular bulletins describing agricultural experiments started in 1885. Federal aid for the

[22] Fuller, *op. cit.*, I: 482.

[23] Utley, Henry M., and Cutcheon, Byron M., *Michigan as a Province, Territory, and State* (4 vols., New York, 1906), IV: 275.

establishment of an experiment station was utilized beginning in 1889. Subsequently there were many other kinds of federal grants for agricultural education. It was the Agricultural College that made available and stimulated interest in new and better grains such as "American Banner" wheat, "Rosen" rye, "Markton" oats, and "Spartan" barley. Better breeds of farm animals and new strains of fruits and vegetables have come in steady succession from the College experiment stations.[24] The farmer has been encouraged to keep systematic records and to carry on his work by business methods. Science has come to the farm and has to a large extent displaced the superstitious hocus-pocus that clustered for centuries around the art of agriculture.

The farmer has become less individualistic, less eccentric. He has lost some of his suspicion of "newfangled" ideas. He is less self-reliant. The farmer of 1865 had plenty of capacity for teamwork, for he gave and received help cheerfully and effectively at harvest time, during sickness or trouble of any sort, in house raisings and the like. But when it came to selling his own products or buying to supply his own wants, he wanted to make the deal himself. Vital in the breakdown of this attitude was the growth of great corporations such as the railroads and manufacturers of farm machinery and the increasing cost of middleman services. The farmer learned to coöperate because necessity drove him to it. Then he found that coöperation with his neighbors had its pleasanter aspects. The Grange came to Michigan in 1872. Three years later there were six hundred local units. A bit later the Gleaners started with a similar program. The Grange led in the fight for cheaper railroad rates and was somewhat involved in politics during the eighties. But perhaps the most significant work of both Grange and Gleaners was their promotion of coöperative marketing and buying. Elevators, canneries, and retail stores were established in many towns and villages. Through them the farmer learned to coöperate with his neighbors in the solution of common problems and for mutual benefit.[25]

The Grange and the Gleaners were only two among many farmers' organizations that came into being during this period. Examples of others are the Michigan State Horticultural Society (1870), the Michigan State Association of Ginseng Growers (1904), the Michigan

[24] Beal, W. J., *History of Michigan Agricultural College* (Lansing, 1915).
[25] Chase, Lew A., *Rural Michigan* (New York, 1922), pp. 338 ff.

Milk Producers' Association (1916), the Michigan State Poultry Improvement Association (1925), and a host of others. The Michigan State Farm Bureau was formed in 1919 as a central organization to look after the farmers' interests in matters like transportation, taxation, and legislation. Fire and life insurance became available after 1928 through the Bureau, although there were other farmers' coöperatives in the insurance business, too.[26] In these and other associations the farmer learned to work coöperatively to a greater degree than ever before.

One of the problems that the farmer had to contend with throughout this period was the steady loss of manpower. The fertile lands of Nebraska, Minnesota, the Dakotas, and the Canadian Northwest lured many of his sons away. At the same time others were enticed by the high wages paid in the rapidly growing cities such as Detroit and Chicago. This movement to the cities was accelerated tremendously in Michigan by the mushroom-like growth of the automotive industry. Until 1910 Michigan had a relatively low proportion of urban population. In that year the percentage of the population of the state living in incorporated towns and cities of over 2,500 population was 47.2 as compared with an average of 52.7 for the states of the Old Northwest. Twenty years later (1930) the percentage of urban population in Michigan rose sharply to 68.2 as compared with an average of 66.4 for the states of the Old Northwest and a national average of 56.2.[27] Thus it is evident that from 1910 onward urbanization proceeded more rapidly in Michigan than was generally true throughout the country.

The gains from immigration and the natural population increase enabled the rural areas to gain or at least to hold their own for many years. The number of farms increased from 62,422 in 1860 to 206,906 in 1910.[28] Thereafter there was a decline to 169,372 in 1930. Then the depression started a back-to-the-farm movement, and by 1935 the total number listed was 196,517. Returning prosperity renewed the movement to the city, so that in 1940 only 187,589 farms were operating.[29] The average size of farms has not changed very much; in 1860 it was one hundred and twelve acres and in 1940 ninety-six

[26] Fuller, *op. cit.*, II: 611 ff.

[27] *Sixteenth Census, Population* (Washington, 1942), Vol. I, *passim.*

[28] *Eighth Census, Agriculture* (Washington, 1864), p. 204; *Thirteenth Census, Abstract, with Supplement for Michigan*, p. 268.

[29] *Sixteenth Census, Agriculture*, I: 774.

acres.[30] The value of farm land was greatly enhanced by the prosperity of the World War period. The average Michigan farm, including buildings, was worth about $2,800 in 1900; in 1920 the average value had increased to $7,313. Thereafter there was a steady decrease, so that the average value in 1940 was only $4,865.[31]

A trend in agriculture that has disturbed many observers during the past fifty years is the increase in tenancy. In Michigan this has not been a serious problem. Nearly three fourths of the farms in Michigan today are operated by full owners. The number occupied by tenants increased slowly after 1880 until 1920, when it started to decline. During the hard times of the thirties some farmers lost their lands, but by 1940 the percentage of tenancy was actually less than it had been a score of years previously.[32]

In spite of all these changes there were some constants. Among these was the unchanging rhythm of the seasons which still governed the pattern of life and labor on the farm. The laws of nature were neither modified nor repealed. The weather continued, as always, dependably undependable and the perennial subject of conversation. The rural counties of Michigan remained consistently Republican in politics through thick and thin. The ability and disposition to coöperate on a voluntary basis were as much a part of the farmer's nature as rugged individualism, but he lost none of his deep suspicion and dislike of anything resembling regimentation or control from above. Technology in its many forms had changed the work patterns of the farm family and had left more time for other things. But technology did not bring about in the country, as it had in the cities of Michigan, the gradual elimination of the little fellow. Three out of every four farmers still owned the lands they tilled and were accountable to no one save their God and their government. In these ways, at least, the transformation of rural life was not complete.

Kalamazoo College
Kalamazoo, Michigan

[30] *Eighth Census, Agriculture,* pp. 76, 204; *Sixteenth Census, Agriculture,* I: 774.

[31] *Thirteenth Census,* VI, *Agriculture* (Washington, 1913), p. 768; *Sixteenth Census, Agriculture,* I: 775.

[32] *Thirteenth Census,* VI, *Agriculture,* p. 769; *Sixteenth Census, Agriculture,* I: 774.

THE CONTEMPORARY BRITISH OPPOSITION TO THE STAMP ACT, 1764–65

FRED J. ERICSON *

ONE of the persistent half myths in the history of the American Revolution is the belief that there was little contemporary British opposition to the Stamp Act.[1] It is based on the uncritical use of such statements as Horace Walpole's "one slight day on the American taxes,"[2] Edmund Burke's 1774 reference to a "languid debate,"[3] and a misinterpreted comparison to a hotly contested turnpike bill.[4] If the usual interpretation drawn from these sources were correct, much weight would be added to the theory that British colonial policy was characterized by continuity in the years before the Revolution. However, it will be shown in this paper that the usual treatment is based on faulty evidence and that many influential British leaders opposed the innovation that they recognized the new tax to be.

The customary story has been secured from self-justifying statements of British politicians on both sides. The claims of the Grenvillites that there was "very little opposition" to the Stamp Act were colored by the desire to prove that the "supposed Patriots" had not protested when they had the chance.[5] The Old Whigs also came to

* The writer wishes to express his appreciation to Dr. John Alden, of Bowling Green State University, and to the editor for checking the references and making valuable suggestions.

[1] See, for example, C. W. Alvord, *The Mississippi Valley in British Politics* (Cleveland, 1916), I: 227 f. On the other hand, the correct interpretation is given by W. T. Laprade, "The Stamp Act in British Politics," *American Historical Review*, 35 (1929–30): 735–757.

[2] Walpole to Hertford, February 12, 1765, *The Letters of Horace Wapole Fourth Earl of Orford*, ed. Mrs. Paget Toynbee, VI (Oxford, 1904): 187. See also C. H. Van Tyne, *The Founding of the American Republic*, Vol. I: *The Causes of the War of Independence* (Boston, 1922), p. 142.

[3] *Parliamentary History*, XVII: 1253.

[4] *New-York Mercury*, May 27, 1765. See also C. A. and M. R. Beard, *The Rise of American Civilization*, I (New York, 1927): 207.

[5] *A Short History of the Conduct of the Present Ministry* (London, 1766), pp. 4–7.

have the impelling desire to agree that they had not incited colonial disobedience. Burke's oft-quoted denial contained this justification: "Far from any thing inflammatory, I never heard a more languid debate in this House."[6] These partisan accusations and apologies should not be taken too seriously; they cannot cover the whole story of propaganda, committees, and other forms of opposition made to taxing America.

After a half century of demands — from Keith to M'Culloh[7] — the stamp-tax principle was approved in the 1764 "Sugar Act" resolution. According to Almon, this was passed in a "thin House, late at night, and just at rising without any debate."[8] The statement is probably misleading. To be sure, the official Grenville explanation for postponing the application of the tax for one year was British "Tenderness" to America and the desire to "enquire" about the suitability of the proposed method.[9] But there was little contemporary doubt that Grenville was forced to postpone action against his wishes and that considerable opposition was expressed in Parliament.

A rather complete story in the colonial newspapers in early May, 1764,[10] related that "it had been previously debated in the parliament, whether the[y] had power to lay a tax on colonies which had no representatives in parliament"; this matter was determined in the "affirmative." On March 9 Grenville had given a "long harrangue" on the finances and had proposed the Sugar Act provisions. The story continued:

> . . . Besides this, an internal tax was proposed, a stamp duty, etc. but many members opposing it, this was defered till next session . . . .

---

[6] *Parliamentary History,* XVII: 1253. John Eardley-Wilmot likewise denied in 1779 the "least opposition in Parliament" to the Stamp Act and admitted that his "only purpose" in going over "beaten ground" was to show that America's resistance was "voluntary and spontaneous on her part, and not excited by any violence in either branch of the Legislature . . . ." *A Short Defence of the Opposition* . . . (London, 1779), pp. 6–7.

[7] A most interesting prophecy of the troubles that would ensue after peace and the effort to make America contribute to imperial expense is contained in the pamphlet *The Interest of Great Britain in the Approaching Congress Considered* . . . (London, 1761), pp. 40–43.

[8] Almon, John, *Prior Documents* (London, 1777), p. 5.

[9] [Grenville, George,] *The Regulations Lately Made concerning the Colonies, and the Taxes Imposed upon Them, Considered* (London, 1765), p. 101.

[10] *New-York Mercury,* May 7, 1764; *Massachusetts Gazette and Boston News-Letter,* May 10, 1764; *Pennsylvania Gazette,* May 10, 1764; *Newport Mercury,* May 14, 1764. Some of these accounts are incomplete.

That Mr. Jackson, agent for Connecticut, (a Member of the house) exerted himself nobly, and that it was chiefly owing to him that the stamp act did not take place; likewise, that Mr. Allen of Philadelphia, was indefatigable in remonstrating to many of the Members, with whom he was acquainted, on the illegality of an internal Tax, and had considerable Influence in preventing it.

In June several American papers [11] carried a less detailed account which said the act would "inevitably have passed this Sessions" had not William Allen exerted influence through important personages. In a private letter from Eliphalet Dyer of Connecticut to Jared Ingersoll, April 14, 1764, credit was again given to Richard Jackson. Dyer said there was a determination in Britain to raise money from America and that a stamp duty would probably be attempted. Grenville had spoken strongly for British power and the right to tax the colonies, "but much has been said by Agents and others from the Continent," although they would not be able to avert the "Impending blow." [12]

Later in the year Boston-born John Huske entered the list as a claimant for the honor of postponing the Stamp Act. Writing to the Boston merchants, August 14, 1764, he said that "Altho' my private attempt to delay the introduction of the stamp duty, was ineffectual, yet my public one succeeded . . . ." His greatest difficulty was in removing prejudice caused by "erroneous and alarming doctrine" expressed by some Americans on the constitutionality of taxation. He further declared that Grenville "agreed reluctantly" to the delay and, "being irritated," was resolved to convince the colonies "next session" that they were "as subject to an inland tax" as to all others they had paid.[13] William Knox was also undeservedly given some credit in the postponement by his fellow Grenvillite, William Henry Lyttelton, governor of Jamaica, who said in a letter of July 22, 1764, that the immediate imposition of the tax might have been "inefficatious" in some colonies "and in others have been productive of greater discontents than the object was worth." [14] Jared Ingersoll, also, who was later to be hanged in effigy as a Stamp man, warned Whately, July 6, 1764, that if the colonial legislatures did not levy

[11] *Newport Mercury*, June 11, 1764; *New-York Mercury*, June 4, 1764; *Boston Gazette and Country Journal*, June 11, 1764, etc.

[12] Dexter, Franklin B., ed., "A Selection from the Correspondence and Miscellaneous Papers of Jared Ingersoll," *Papers of the New Haven Colony Historical Society*, IX (1918): 290–291.

[13] *New-York Mercury*, November 5, 1764.

[14] William Knox Papers in the William L. Clements Library, Ann Arbor, Michigan (hereafter referred to as "WLCL"), Vol. I, fol. 13.

the proposed taxes, they would "go down with the People like chopt Hay." [15]

Further evidence of parliamentary controversy over internal taxation comes from Thomas Hutchinson's famous *History*. In it, he said that "many members of the house of commons, in the last session of parliament [1763–64 if Hutchinson was correct] had declared against the stamp duty, while it was mere matter of conversation." [16] Thomas Pownall, in his first edition of *The Administration of the Colonies*, which appeared not later than March, 1764, [17] intimated serious doubts as to the constitutionality of internal taxation. He said:

> . . . But besides the difficulties, respecting the rights of the crown, which may attend this scheme [an internal land tax], it is a point that ought very deliberately and dispassionately to be weighed, how far even the supreme government of the mother country can, consistent with general liberty, proceed in laying taxes on its colonies, where the consent of the people cannot be, in any constitutional way taken.

This doubt was never expressed in later editions of Pownall's book, although a somewhat milder hint that excises and stamp duties were "precluded" from British action because the colonies granted revenues to the Crown was allowed to stand.[18]

What was Grenville's course between the passage of the Sugar Act and the introduction of the Stamp Act? William Knox in 1769 and Israel Mauduit in 1775 told how Grenville had asked the agents in 1764 to inform their colonies that, if they could suggest a substitute for the stamp tax which would be of "equal efficacy, he would adopt it." Mauduit said that Grenville had stated to the agents that he was not "set upon this Tax" and would accept any practical alternative.[19] There is good evidence that Grenville's professions were not sincere. Jenkinson wrote to him, July 2, 1764, crediting him with the postponement of the stamp tax, but at once taking away the value of the credit by a damaging revelation:

[15] *M^r. Ingersoll's Letters Relating to the Stamp-Act* (New Haven, 1766), p. 10.

[16] *The History of the Colony and Province of Massachusetts-Bay*, ed. L. S. Mayo (Cambridge, Mass., 1936), III: 84.

[17] It was reviewed in the *Gentleman's Magazine* for March, 1764, pp. 103–108.

[18] Pp. 67–68; Second Edition (1765), p. 92.

[19] [Knox, William,] *The Controversy between Great Britain and Her Colonies Reviewed . . .* (London, 1769), p. 199; Mauduit, Israel, *Mr. Grenville's Offer to the Colony Assemblies to Raise the Supply Themselves, instead of Having It Done by a Parliamentary Stamp Act* [1775], p. 2.

In the last session of Parliament you assigned as a reason for not going on with the Stamp Act, that you waited only for further information on that subject. This having been said, should not Government appear to take some step for that purpose? I mentioned this to you soon after the Parliament was up. I remember your objections to it; but I think the information may be procured in a manner to obviate those objections, and without it we may perhaps be accused of neglect.[20]

This entirely accurate prediction was supplemented by an even stronger warning from the Earl of Mansfield, who was to be staunch in defending Parliament's supremacy. On December 24, 1764, he wrote to Grenville:

I have thought of the observation you made yesterday from looking into the charters of some of the Charter Governments.

Though the question certainly does not want this, or any other authority, yet it will be a striking alteration to ignorant people, and an unanswerable argument ad homines; and, therefore, I wish you would employ somebody to look with this view into the origin of their power to tax themselves and raise money at all.[21]

As to whether Richard Jackson, William Allen, Huske, or all of them as well as other persons in Parliament should have credit for postponing the Stamp Act there would seem to be no final agreement. However, the willing tenderness which Grenville and his friends assumed to themselves can be seriously discounted on the basis of the evidence here cited, especially Jenkinson's admission. That opposition had been expressed and that the agents had secured some British pressure on the colonial side is undeniable. Grenville had even been warned by an important legalist that it would be an "unanswerable argument ad homines" to alter the methods of taxation. Indeed, about a month before the reappearance of the issue in the winter of 1764–65 the famous *Letter concerning Libels*, which dealt with the Wilkes question, referred to the 1764 intimation of a stamp tax for the colonies as a "momentous resolution." [22]

The record of the course of the Stamp Act in its two-month passage through Parliament in February and March, 1765, must be

[20] *The Grenville Papers: Being the Correspondence of Richard Grenville, Earl Temple, K. G., and the Right Hon. George Grenville, Their Friends and Contemporaries*, II: 373. Edited, with Critical Notes, by William James Smith. London, 1852–53.

[21] *Ibid.*, p. 478.

[22] *A Postscript to the Letter on Libels, Warrants, . . .* (London, dated Jan. 24, 1765), pp. 6 f. This postscript was also incorporated with the *Letter concerning Libels, Warrants . . .*, the famous Wilkes case pamphlet, at least as early as the third edition (London, 1765), *Postscript*, pp. 6–7.

drawn from letters and from reports that appeared chiefly in the American press, since the *Parliamentary History* was fragmentary. The various accounts were contradictory, but agreed that considerable opposition was made within and without Parliament. It will be asked how the agitation over an important bill could be so quickly forgotten. There are several reasons that may be offered. The opposition to the Stamp Act can be confused with the rather divided stand of the Old Whig group. The Whigs were divided among themselves; the ever-fretful Newcastle was anxious to alienate neither the West Indian nor the North American interests.[23] It is not surprising that they did not make a strong showing on the Stamp Act and that Onslow could report to Newcastle, March 19, 1765, that Pitt was "not without his complaints of the American tax being not sufficiently objected to this year." [24] Further, the Whigs were willing to allow the story of their opposition to go by default in order to establish an excuse for so soon repealing the Stamp Act. This they secured on the plea that Grenville had hurried the bill through before the members had "opportunity to inform themselves on the head." [25] Thus Grafton said in 1766 that the law would not have passed had the American position been known.[26] What Parliament would have done had it been aware of the full weight of colonial protest is not ascertainable with finality. But it was so well acquainted informally with American opinion that he would have been a dull politician indeed who was not cognizant of it.

Another possibility of misjudging the weight of anti-Stamp Tax

[23] Sutherland, L. Stuart, "Edmund Burke and the First Rockingham Ministry," *English Historical Review*, 47 (1932) : 54–55, 58; Winstanley, D. A., *Personal and Party Government; a Chapter in the Political History of the Early Years of George III, 1760–1766* (Cambridge: The University Press, 1910), pp. 214, 218.

[24] Winstanley, *op. cit.*, pp. 218–219, footnote, citing Add. MS 329 ff., fol. 69.

[25] *Considerations on the American Stamp Act, and on the Conduct of the Minister Who Planned It* (London, 1766), p. 27.

[26] "Debates on the Declaratory Act and the Repeal of the Stamp Act, 1766," *American Historical Review*, 17 (1911–12) : 580. The American Dr. Ruston said, in a letter of March 3, 1766, that Grenville secured the passage of the act "merely by browbeating our Agents, and suppressing Remonstrances which were ordered by his Majesty to be laid before the House . . . ." WLCL, manuscript letter, Dr. Thomas Ruston to Job Ruston, March 3, 1766. Another American, Benjamin Gerrish, wrote to John Hancock, March 1, 1766, that it was thought in Britain that Grenville would be impeached for suppressing secret information the former year. WLCL, manuscript letter, Benjamin Gerrish to John Hancock, March 1, 1766.

pressure lies in confusing the paucity of constitutional opposition with opposition based on policy and expediency. As Governor Bernard told the Massachusetts legislature, "It is said that the gentlemen who opposed this act in the House of Commons, did not dispute the authority of parliament to make such a law, but argued upon the inexpediency of it at this time, and the inability of the colonies to bear such an imposition." [27] Indeed, there is some evidence that one argument for passing the Stamp Act was to punish the "Refractoriness" of the colonies, as Edward Sedgwick of the Grenvillites said, and to establish the right of taxation "by a new execution of it, and in the strongest instance, an Internal Tax, that of the Stamp Duty." [28] Franklin likewise testified that "the Tide was too strong against us. The nation was provoked by American Claims of Independence, and all parties joined in resolving by this act to settle the point. We might as well have hindered the sun's setting." [29]

An example of how contemporary evidence has been misjudged is the customary conclusion drawn from the American complaint: "There has not been so much opposition as to a common Turnpike Bill." [30] This was literally true; for the statement as first used probably referred to no hypothetical measure. The Earl of Sandwich had a turnpike bill before Parliament during the same time that the Stamp Act was in passage. He wrote to the Earl of Denbigh, March 18, 1765, saying: "My turnpike bill was warmly contested in the House of Commons and became a mere party affair . . . ." It had carried by 125 to 66 in committee; he hoped that Denbigh would be on hand, for the vote in the Lords would be close.[31]

The specific evidence of controversy is of several kinds. Especially significant is the contradictory course of the Grenvillites in claiming that there was no opposition to the Stamp Act and that the American disobedience was stimulated by British politicians. In the debates of January, 1766, Grenville made his well-known assertion that none had denied the right of Parliament to tax America. But, perhaps in an unguarded moment while trying to prove another point, he also

[27] Almon, John, *Prior Documents*, p. 17.

[28] *Royal Historical Manuscripts Commission, Eglinton . . . Weston Underwood*, p. 382.

[29] Franklin to Charles Thomson, July 11, 1765, *The Writings of Benjamin Franklin*, ed. A. H. Smith (New York, 1906), IV: 390.

[30] *New-York Mercury*, May 27, 1765; Beard, *op. cit.*, I: 207.

[31] *Royal Historical Manuscripts Commission, Denbigh*, V: 294.

said: "The seditious spirit of the colonies owes its birth to the factions in this House. . . . We were told we trod on tender ground; we were bid to expect disobedience." [32] There was, of course, no contradiction in the two statements if they are carefully considered; few had opposed the legality of taxation, but the Whigs had warned of resistance.

Evidence of opposition came also from — of all persons — Edmund Burke. In the *Annual Register* for 1765 he wrote that, although the previous resolution had gone through "smoothly," the "final laying it on in the present [session] was attended with no small debates" both as to right and as to expediency.[33] Another contemporary testimony was in a review or excerpt in the *Gentleman's Magazine* of April, 1765, from a pamphlet that made an apparent allusion to the Stamp Act as the bill "that has been so warmly agitated without doors [that is outside Parliament]." [34]

The agents for the colonies naturally claimed to have made all possible efforts to stop the tax. Franklin said he had taken "every step" in his power; [35] Professor Crane, after careful study, agrees with his contention.[36] The best known of the agents' activities was their meeting with Grenville for two hours, during which he is supposed to have silenced the Americans by showing they could offer no practical substitute contribution to the empire. Actually, Franklin proposed a tax on American paper money; for obvious political reasons he suppressed this fact and only revealed it privately to Galloway in 1766.[37] Jared Ingersoll testified that at this meeting Richard Jackson had told Grenville that spending American money without check would mean that "the Assemblies in the colonies would be subverted." [38] Ingersoll also told how Jackson had produced copies of the acts which had given members to Chester and Durham before

[32] *Parliamentary History,* XVI: 102.

[33] Laprade, *op. cit.*, p. 745 and n.; *Annual Register for 1765* (1766), p. 34.

[34] Review of *The Political Balance* . . . (London, 1765), in *The Gentleman's Magazine,* April, 1765, pp. 188–189.

[35] Franklin to David Hall, February 14, 1765, *The Writings of Benjamin Franklin,* IV: 363.

[36] Crane, V. W., "Benjamin Franklin and the Stamp Act," *Publications of the Colonial Society of Massachusetts,* 32 (1933–37): 57; also *idem, Benjamin Franklin, Englishman and American* (Baltimore, 1936), *passim.*

[37] Franklin to Joseph Galloway (Oct. 11, 1766), WLCL, Franklin-Galloway Letters; see also the two references to Crane in the preceding note.

[38] Ingersoll to Fitch, February 11, 1765, *Mr. Ingersoll's Letters Relating to the Stamp-Act,* p. 18.

they were taxed.[39] Even more significant is Ingersoll's story in the fall of 1765 about how the pamphlets published in America "were dispersed among the Members, but every one seemed to think the Reasonings were not conclusive." [40] At all events, the members probably needed no petitions to inform them of the American attitude.

The agents did not work without the aid of the time-honored "friends of America" — the British merchants. At the close of a letter on the Stamp Act, February 11, 1765, Ingersoll said the merchants were "alarmed at these Things; they have had a Meeting with the Agents, and are about to Petition Parliament upon the Acts that respect the Trade of North-America." [41] On March 6 he mentioned a petition which Fuller, a West Indian planter, tried to present on behalf of the London merchants — vainly, since it was against a money bill.[42] In his fall letters Ingersoll recounted how Barlow Trecothick, who was reared in Boston, had opposed the Stamp Act "with all his Might." [43] Again he said that the merchants of London appointed a committee "to make all the opposition they could to the Stamp Bill" and that Trecothick was its chairman. The committee collaborated with the agents; "they were frequently together and several times before the minister, upon the Stamp and other Bills, that related to America . . . ." [44]

The early accounts that reached America may have conditioned historical interpretation of the parliamentary story. According to an item in the Boston papers in early April, "not a Man spoke who did not declare his Opinion that America ought to be taxed . . ." and that no petition was offered against the tax.[45] A mild, and pro-Grenville, report also appeared in several American papers in late April and early May. It mentioned Grenville, Townshend, and Barre, and insisted it would have been dangerous to show Parliament the "warm and unbecoming Expressions" of the American addresses.[46] Another item, copied from other papers by the *Newport Mercury*, May 6, 1765,

[39] *Ibid.*, p. 13.

[40] Ingersoll to Connecticut General Assembly, September 18, 1765, *ibid.*, p. 31.

[41] *Ibid.*, pp. 19–20.

[42] *Ibid.*, pp. 20–22.

[43] To Connecticut General Assembly, September 18, 1765, *ibid.*, p. 31.

[44] Letter to *Connecticut Gazette*, reprinted in the *New-York Mercury*, September 23, 1765; see also Dexter, *op. cit.*, pp. 331–334.

[45] *Massachusetts Gazette and Boston News-Letter*, April 4, 1765; *Boston Evening-Post*, April 8, 1765.

[46] *Newport Mercury*, April 29, 1765; *Boston Evening-Post*, May 6, 1765; *Massachusetts Gazette and Boston News-Letter*, May 9, 1765.

emphasized the virtual representation theory in Parliament and declared: "The power of parliament was asserted, and so universally agreed to, that no petition disputing it will be received." [47] The division on a test vote was reported as 245 to 49. These earlier items corroborate the view that there was a rather full discussion as early as the first days of February.

During May, 1765, however, the reports emphasized the number of speakers who opposed the passage of the bill. On May 13 the *Newport Mercury* cited the *New York Gazette* of May 6 for an item which complained that earlier accounts had been "extremely erroneous." It was not true that none had opposed taxation. Rather, "the speakers in favour of the colonies were more numerous than their opposers, much better speakers, and incomparably superior in point of argument, tho' outnumbered by the ministerial party, and dependants upon court favour." Barre's speech was also favorably reported, and it was said that a voice vote had given only "about 40 voices" for the bill at one stage. The account added: "Without doors we hear, every person who was at all qualified to form any judgment of the matter, seem'd unanimous in favour of the colonies. And it was hoped they would assert their rights, by resolves . . ." and other means to show they would not be enslaved. This was one of the earliest hints from Britain that any resistance was expected or desired.

More details were given in an article of May 27, 1765, in the *New-York Mercury* and in the *Boston Gazette and Country Journal*, June 3, 1765. This purported to be based on an extract from a letter of February 16. It emphasized Conway's insistence that the colonial position be heard. Of a speech by Charles Yorke favoring the government the statement was made that "He answered all the speakers against the bill . . . ." The supposed extract concluded:

> In short, from the opinion of the most sensible unbiassed Members of Parliament, I am induced to think it one of the best speeches ever made there. The Bill was read the first time without a Syllable being said, nor do I expect any more will. I think matters have been ill managed. There has not been so much opposition as to a common Turnpike Bill.

It will be noted that equally significant with this pessimistic denial of opposition is the assertion that Yorke answered "all the speakers against the bill." [48]

[47] Also in the *Massachusetts Gazette and Boston News-Letter*, May 9, 1765.

[48] For corroboration that Charles Yorke supported the Stamp Act see George Harris, *The Life of Lord Chancellor Hardwicke* . . . (London, 1847), III: 444.

The British press in the meanwhile was not silent on the question of taxing America. Its opinions have been satisfactorily treated in an article by William T. Laprade [49] and need not be repeated here since they tend only to amplify the reports that reached America.

The fullest contemporary account of the Stamp Act controversy was that given in the letters by Jared Ingersoll. On February 11, 1765, he wrote to Governor Fitch to tell of the "principal Attention" which was being given the Stamp Act. The arguments of both sides were summarized. On the one hand, those in favor of the bill said the colonies were under British authority and that they had escaped internal taxation only because of "Infancy and Inability." But Ingersoll also gave the opposition arguments:

> On the other Hand, Those who oppose the Bill say, It is true the Parliament have a supreme unlimited Authority over every Part and Branch of the King's Dominions, and as well over Ireland as any other Place; yet, we believe a British Parliament will never think it prudent to Tax Ireland: --- [The theory of representation of all subjects is admitted but its effects in Britain and America] must be infinitely different in the Article of Taxation. . . . we think, say they, That it will be only to lay a Foundation of great Jealousy and continual Uneasiness, and that to no Purpose, as we already, by the Regulations upon their Trade, draw from the Americans all that they can spare; at least, they say, This Step should not take Place until, or, unless the Americans are allowed to send Members to Parliament . . . .[50]

It is possible that Ingersoll here merged into his account some arguments from other than Parliamentary sources. However, he gave details of occurrences in Parliament. He cited Grenville's speech, and continued: "The Argument was taken up by several who opposed the Bill, viz. by Alderman Beckford, who and who only, seemed to deny the Authority of Parliament, by Col. Barre, Mr. Jackson, Sir William Meredith, and some others." [51] If to this list of opponents of the bill are added others named by various observers already specified, it will be seen that the usual "two or three" members allowed by most historians must be expanded to at least eight, even if the "some others" represent none additional.

In his letter to Fitch of March 6, 1765, Ingersoll mentioned "some farther Debates" in the Commons, especially in the second-reading stage.[52] He also referred to Charles Yorke's "very long speech" favoring the bill and said the Virginia petition question

[49] *Op. cit.*, pp. 745–747 and notes.
[50] *M^r. Ingersoll's Letters Relating to the Stamp-Act*, pp. 13–14.
[51] *Ibid.*, p. 15.
[52] *Ibid.*, p. 20.

"drew on a pretty warm Debate." The rule against petitions on money bills was invoked, but Conway vehemently denied Parliament's authority. Ingersoll went on:

> . . . The Hardships and Inconveniences were also again urged, and placed in various Lights, by our other Friends in the House. — And here, I find myself obliged to say, that, except the Gentlemen interested in the West-Indias, and a few Members that happen to be particularly connected with some of the Colonies, and a few of the Heads of the Minority, who are sure to thwart and oppose the Ministry, in every Measure, of what Nature or Kind soever; I say, except these few Persons so circumstanced, there are scarce any People here, either within Doors or without, but what approve the Measures now taking, with regard America.[53]

Interpreted in its most unfavorable light, this account confirms the support by "our other Friends" in Commons; its catalogue of economic and political enemies of the Stamp Act consists of a considerable group when the opposition could muster little strength on any question. But the pessimist of March 6 should not be taken too literally; Ingersoll was then probably preparing to justify his accepting a Stamp office. By September 18, 1765, it was again necessary for him to show what efforts had been made to stop the tax. So the account of the hard efforts of agents and British friends once more appears. He again mentioned those who made up the forty-odd members opposed and added: "yet there Opposition to it was not on Account of it's being Unconstitutional, but because they tho't the Measure imprudent, and perhaps, burthensome. — These Things are no Secrets, but well known to Hundreds of Persons besides myself." [54]

Another way to gauge the amount of controversy over the Stamp Act is to examine the admissions made in progovernmental pamphlets that essayed to defend it. Because of distance any pamphlet published in Britain before May, 1765, was written without any knowledge of how Americans actually received the news of the bill's passage. Only such pamphlets will be considered here. The so-called "Grenville pamphlet," *The Regulations Lately Made concerning the Colonies . . .* (London, 1765), was one of the more prominent of the contemporary apologies. It dealt with the whole colonial policy of the administration and defended the Stamp Act against the charges that it violated American rights and that it differed fundamentally from

[53] *M*r. *Ingersoll's Letters Relating to the Stamp-Act*, p. 22.
[54] *Ibid.*, p. 31.

the duties long imposed upon trade.[55] It thus covered much of the ground to be trodden in the next few years.

William Knox, a staunch supporter of Grenville, wrote his first American pamphlet early in 1765.[56] This was *The Claim of the Colonies to an Exemption from Internal Taxes Imposed by Authority of Parliament . . . Examined . . .* (London). In it Knox readily accepted the defensive, saying he refused to aid the movement against the tax because of his "thorough conviction" of British authority. But he admitted:

> This opinion I know is not the general opinion of the people in America; at the same time I do not believe the contrary is so generally adopted, as people on this side of the water suppose it to be from the writings they have seen of some political zealots.[57]

The pamphlet as a whole made a thorough defense of the constitutionality of taxation of America, but Knox acknowledged that Parliament should be "more tender" to America because of differences in circumstances.[58] As to practicability, it was urged that the old requisition method had failed. But one of the arguments against it was a confession of British malpractice. The fate of the 4.5 per cent duty in the Windward Islands that had been granted for fortifications but never used for them should "make every colony cautious of granting money, which neither parliament not its particular legislature has a right to enquire into the expenditure of . . . ." [59] As if by malice aforethought, Knox also stated the other example of British practice in revenue matters; this was the point, afterward soft pedaled, that Virginia had settled a permanent revenue for its civil government and was thenceforward ruled by a deputy.[60] His reason for mentioning these practices was simply to conclude that parliamentary taxation was preferable to them. That Americans and their friends would use the same two cases to show the danger of any taxation by distant authority did not seem to occur to him. Knox also made an argument for colonial representation that would be used for the next ten years by the colonial side. It was that the nonrepresented in Britain lived in the same areas with the members of Parliament and that this was a "security" for the British nonrepre-

[55] Pp. 100 ff.
[56] It was reviewed in the *Critical Review* for March, 1765.
[57] Pp. 1–2.
[58] *Ibid.*, pp. 16–17.
[59] *Ibid.*, p. 20.
[60] *Ibid.*, p. 20 n.

sented which did not operate for the colonies. Rather, in colonial taxation, the parliamentary member's "own property must be eased thereby."[61] Knox favored an interval between the proposal and the imposition of taxes on colonies, saying, in good American style, that otherwise: "At every pinch or occasional exigence, when the landed gentlemen are to be eased, or the trading people flattered, a new tax upon the colonies will be the means proposed . . . ."[62] On the whole, his admissions of dubious British practice and the dangers of unrestricted taxation are more important than his trite assertions of legality. In years to come Knox was to leave British faults unexplored in his controversial writings.

Another pamphlet published during the height of the agitation was Soame Jenyns' *The Objections to the Taxation of Our American Colonies . . . Briefly Considered* (London, 1765). It contained an interesting admission of the reason why it was written. Jenyns said the right and expediency of parliamentary taxation were "so indisputably clear" that he would not have bothered to speak of them "had not many Arguments been lately flung out both in Papers and Conversation, which with Insolence equal to their Absurdity deny them both."[63] He gave the usual Grenvillite justification for the Stamp Act and answered the characteristic American arguments.

The piece of contemporary evidence that went farthest to show opposition to the Stamp Act on principles of policy was an anonymous document in the Shelburne Papers in a hand which has been thought to be that of Sergeant John Glyn.[64] It was clearly written before the colonial reaction to the tax could be known. Internal taxation was termed a "new Measure and the Adoption and Avowal of a new principle"; it was important because, if one inland tax was

[61] Knox, p. 28 of the work cited in the text on the preceding page.

[62] *Ibid.*, pp. 37–38.

[63] Pp. 3–4. The pamphlet was listed in the *Gentleman's Magazine*, February, 1765.

[64] Undated and untitled document: ["On the right and expediency of taxing America," 1765], WLCL, Shelburne Papers, 85: 71 ff. The present writer feels that the identification of the handwriting as Sergeant Glyn's is open to question. A reference in the document to a resolution of a previous year which the colonies ignored has led some to think it was written in 1766–67 and referred to the Declaratory Act. However, the 1764 Stamp duty resolution is the far better resolution to connect this reference with; it was meant to be acted upon and the Declaratory Act was not. Also the whole tenor of the document in question shows that nothing was as yet known of American reaction except by extremely accurate conjecture.

proper, so would be a hundred. The British "Ancestors" had never laid inland taxes on Ireland, even though it was a conquered country. British pleas for support from America were derided, for it was said "in the Language of modern Politics, this weak parent is with a strong hand to compel its vigorous Offspring to give it the necessary assistance." [65]

Internal taxation, it was asserted, raised the question of the "Relation" of Britain and her colonies. If they were "mere Dependencies," they should have such "Special Rights" as "an Exemption from internal Taxes." But if they were "constituent Parts" of Britain they should not be exempted from a "single Tax," should have representatives, and should enjoy the "general Equity" of trade equality with Britons. Economically, taxation was described as "indeed a System so Absurd and Chimerical that nothing but the present attempt to tax America internally can justify the bringing it into Supposition." [66] It was further asserted that a hypothetical agreement had once been made between Britain and her colonies. She was to give them protection and they must respect her monopoly. They were told in effect: "But for the rest, you shall manage your internal Governments according to the Charters and Powers given or to be given you." British trade profits would take the place of taxes. This statement of a hypothetical agreement was followed by the observation that it contained nothing against Britain's "true Interest." [67]

A long discussion followed on the economics of taxation. To take American specie would be nugatory and unjust because the colonists needed it to support their own governments. "Besides, as all the Money of America does ultimately rest in Great Brittain, how strangely impolitic must it be to arrest it in the very Act of Circulating brittish Commodities in order to remit a poor barren Sum into the Exchequer" rather than to obtain the same money in the form of trade profits. Further, "this new System" would defraud the very British manufacturers who would now become unemployed. The writer said in ridicule: "It is wonderful that Gentlemen should think of getting more Money from a Country than it has, or of enriching the Exchequer by impoverishing of Great Brittain." The accurate prophecy was made that, if Americans were compelled to pay taxes,

[65] *Ibid.*, pp. 71 and 71 verso. [66] *Ibid.*, pp. 71 v. and 73.
[67] *Ibid.*, pp. 73 v., 74, 74 v.

they would be forced to enter into "Associations for the Encouragement of their own Manufactures and the Rejection of the English." The conclusion on economic grounds was that a "paltry Sum" for the Exchequer was not worth the loss of British prosperity.[68] Such a line of argument was to be repeated many times in the years 1765–75.

The writer in question discussed in a most interesting way the matter of how the Stamp Act would be received in America. It had been supposed the colonies would be "intirely passive under this Imposition." This would be true as to the islands; but North America was "not of a Temper to acquiesce under such Injuries." Nothing could be expected but resistance, perhaps independence; it was further said that the "most favourable Event" would be that assemblies, juries, judges, and lawyers would regard the tax as of "no Effect." He continued:

> In which Case our Ministers made wiser by events may suffer it to sleep in Oblivion, but I believe rather than pay this Tax the Americans would enter into Treaties with France . . . . I am afraid a right Measure has not been taken of the Temper of the Americans otherwise such a momentous Step would not have been made. I am inclined to think that their Neglect of the Resolution of last Year [1764] has proceeded rather from a Confidence of their own Strength than from any Spirit of Complaisance or Submission.
>
> If Divisions and Clamours ensue what may be the Consequence and who shall recede? Not the Americans, if I judge rightly of their Temper; and it will but ill suit with the Dignity of Great Brittain to repeal a Solemn deliberate Act in Complaisance to the Clamours nay Threats of the Colonists . . . . America is grown too large to be trifled with or made the Subject of new Experiments in politics . . . [the reformers were important but peter's pence had important results in the Reformation] and in like manner if we will exact from America Tributes and peter pence not founded in justice and the Nature and ends of our Connection, it must happen that America will as soon as she is able throw off so disgraceful a Subjection and ever after bear a particular Enmity to a Country who may have the Vanity & presumption to retain and avow such ridiculous Claims upon her.[69]

The evidence presented in this paper does not prove opposition to the Stamp Act in 1764–65 equal to that made in 1765–66. Neither the Americans nor their merchant allies in Britain were as skilled in propaganda as they were soon to be. It is possible that the opposition by riot, boycott, and political pressure was so striking after the summer of 1765 that the more normal protests of the former winters sank into the background in men's memories. Both Grenvillites and Old Whigs were willing to allow the facts to be suppressed. Historians

[68] Pp. 74 v. – 75 v. of the document cited in note 64.

[69] *Ibid.*, pp. 76–77.

have too frequently quoted a few statements tending to prove that there was little discussion. However, the most significant agitation was probably that made in committee meetings, private interviews, and activities outside Parliament. This type of pressure forced the postponement of the tax for one year; during that time Grenville was warned by Mansfield that the new taxation would be a "striking alteration" and provide an "unanswerable argument ad homines."

By reconstructing the story of the winter of 1764–65 it is possible to see that the agents exerted many kinds of pressure and enlisted many political and economic forces on the colonial side. They had held meetings, seen Grenville, proposed alternative forms of taxation, and distributed pamphlets; Franklin was probably most influential with tongue and pen. Especially important was the aid secured from such men as Trecothick and Fuller — who were to be key men in the movement for the repeal of the act they had sought to prevent. Even the deprecatory lists of opposing forces showed a respectable number of political and economic leaders on America's side. It seems clear that the Old Whig party did not come out in full force against the Act; but this does not prove the lack of activity on the part of the "friends of America."

Evidence from all types of controversialists confirms the fact that numerous speeches were made on both sides. The "two or three" speakers against the bill must be enlarged to at least eight on the basis of definitely named speakers. There was, to be sure, little questioning of Parliament's legal right to tax colonies; two speakers, at most, took that position. The significant opposition was that made on the grounds of policy and expediency, not constitutionality. In the pamphlets written to defend the Stamp Act the whole field was well covered in 1764–65; an anonymous paper also referred to the "new principle" of internal taxation and gave an accurate prophecy of America's reaction to "Peter pence." The evidence, on the whole, supports the contention that a large number of influential Britons felt that a significant innovation in colonial policy was being made and that they raised vigorous opposition to it.

MICHIGAN STATE NORMAL COLLEGE
YPSILANTI, MICHIGAN

MICHIGAN'S FIRST SUPREME COURT ELECTIONS, 1850–51 *

CLARK F. NORTON

PRIOR to the admission of Michigan to the Union in 1837 there were only two states that had provided for the popular election of their superior court judges.[1] Consequently, when the Michigan constitutional convention of 1835 adopted the appointive method for the selection of supreme court justices, it was acting in accord with a practice almost universal among the states. Each of the ten men who sat on the highest bench of Michigan before 1850 received his position by virtue of gubernatorial nomination and senatorial approval, without direct expression of the people's will on his candidacy. Nearly a decade elapsed after Michigan attained statehood before New York in 1846 became the third state to democratize its method of choosing members for its court of review. Immediately thereafter the movement for popular election of judges spread rapidly, no fewer than eleven old states [2] and five new ones [3] incorporating this method of selection into their constitutions by 1857. Michigan was in the forefront of this nation-wide movement.[4]

* This article is based upon material appearing in the writer's doctoral dissertation, "A History of the Supreme Court of the State of Michigan, 1836–1857" (unpublished, 1940), pp. 352–357, 539–562.

[1] The two states and the dates when they adopted this system are Georgia, 1777, and Mississippi, 1832. For statistics and a discussion concerning the methods used to select highest state judges see the following works: Carpenter, W. S., *Judicial Tenure in the United States* (New Haven, 1918), pp. 4–5; Bryce, J., *The American Commonwealth* (New York, 1910), I: 510–511; Dealey, J. Q., *Growth of American State Constitutions* (New York, 1915), pp. 38–39; McCarthy, Sister M. Barbara, *The Widening Scope of American Constitutions* (Washington, 1928), p. 13; Schouler, J., *Constitutional Studies, State and Federal* (New York, 1897), p. 65. For a good account of judicial appointments in New York State see "The Selection of Judges in New York," *Proceedings, New York State Bar Assn.*, LVI (1933): 206–208.

[2] Mich. (1849–50), Ky. (1850), Mo. (1850), Pa. (1850), Va. (1850), Ind. (1851), Md. (1851), Ohio (1851), La. (1852), Tenn. (1853), Iowa (1857). Carpenter, *op. cit.*, p. 181, footnote.

[3] Wis. (1848), Cal. (1849), Kan. (1855), Minn. (1857), Ore. (1857). *Ibid.*

[4] Places of publication not indicated in the titles of Michigan newspapers and periodicals are as follows: *American Citizen*, Jackson; *Free Press*, Detroit;

In 1845 there was widespread agitation for judicial reform in Michigan, much of it instigated and led by Dr. Samuel Denton and Mr. John Allen, both of whom were elected in that year to the state senate from Washtenaw County. One of the several changes which they sponsored was an amendment to the state constitution providing for the election of all state officials — executive, legislative, and judicial.[5] This proposal received no support in 1845, but during the next five years the subject became of increasing prominence and significance. It appears that at first the supposedly conservative Whigs were more in favor of popular election for members of the supreme court than were the Democrats. Because every appointment to the state supreme court had been made by a Democratic governor and because every appointee to the bench had come from Democratic ranks, the zeal of the Whigs for this reform may have been founded in part upon the hope that under a different process of selection such positions would be more readily obtained by members of their party. With prospects of immediate success in the gubernatorial race not great, the Whigs may have believed that there would be more chance to elect their favorites to judgeships from limited districts, or even from the state as a whole, than to wait until some governor should appoint a Whig to the supreme court.[6]

Whatever the cause, among the Democrats only those belonging to the more radical faction were solidly behind the movement from the beginning;[7] many recognized Democratic sources, on the other hand, voiced direct opposition.[8] In his annual messages to the legis-

---

Grand River Eagle, Grand Rapids; *Michigan Argus*, Ann Arbor; *Michigan Expositor*, Adrian; *Michigan State Journal*, Ann Arbor; *Oakland Gazette*, Pontiac; *True Democrat*, Ann Arbor; *Washtenaw Whig*, Ann Arbor.

5 *Journal of the Senate of the State of Michigan* (hereafter cited as *Senate Journ.*), 1845, p. 63. Under the constitution of 1835 practically all important officers were appointive except the governor, lieutenant governor, and members of the legislature. In another report Senators Denton and Allen, as members of a committee to consider a bill for improving the administration of justice, similarly advocated popular election of judges. *Documents of the Senate of the State of Michigan* (hereafter cited as *Senate Docs.*), 1845, Doc. No. 13, p. 19.

6 Although this is primarily an assumption not substantiated by direct proof, a somewhat similar idea may be inferred from an address made by Austin Blair in the house of representatives in 1846. In arguing for the election of all state judges Blair stated that all members of the supreme court not only were Democrats but also were mere politicians. *Michigan State Journal*, April 18, 1846.

7 For instance, see the *True Democrat*, Dec. 26, 1845, and Sept. 3, 1846.

8 The *Michigan Argus*, Aug. 26, 1845, opposed the election of judges on the ground that the judiciary should not be mixed with political campaigns. The

latures of 1846 and 1847 Governor Alpheus Felch recommended in a general way, it is true, that the choice of a larger number of office holders be committed to the people, but it seems clear that he had in mind mainly the local officials and not the superior court judges.[9] By 1847, however, when two of the leading Democratic journals of the state announced their support of the popular-election principle as applied to justices of the supreme court, it was evident that the views of the two parties were coalescing in this respect.[10] Another indication in 1847 that final action on this question would not long be postponed was the narrow defeat in the legislature of a joint resolution proposing several constitutional amendments, one of which provided for an elective supreme court.[11]

Epaphroditus Ransom, a former member of the supreme court who became governor on January 1, 1848, advised in his first message to the legislature that the constitution be so modified as to authorize the election of all county officers. He added further that it was "worthy of inquiry, whether all *State* officers also, may not, with advantage to the public interest, be selected by the direct action of the electors." [12] Ransom said that previously he had been opposed to the election of superior court judges, but that his opinion on that subject had been shaken recently because of an experiment made in another state (New York).[13] The legislature responded by adopting a joint resolution for a constitutional amendment which stipulated that the justices of the supreme court, together with such executive officers as the auditor-general, state treasurer, secretary of state,

Hillsdale County Democratic Convention in 1846 adopted a resolution expressing a similar view. *Ibid.*, Sept. 9, 1846.

[9] *Joint Documents of the Legislature of the State of Michigan* (hereafter cited as *Joint Docs.*), 1846, p. 7; *ibid.*, 1847, p. 16.

[10] *Free Press*, Jan. 14, 1847; *Kalamazoo Gazette*, Jan. 8, March 12, Nov. 26, 1847.

[11] The joint resolution for these amendments passed the senate nearly unanimously (15–1), but was rejected by the house, 28–30. See the *Senate Journ.*, 1847, pp. 27, 31, 61, 108, 143, 154, 189, 190, 191, 225, 248, 371, 470, 476, 477, 478, 494, 519; *Journal of the House of Representatives of the State of Michigan* (hereafter cited as *House Journ.*), 1847, pp. 283, 284, 331, 332, 375, 384, 385–389, 407–410. The senate committee which reported the joint resolution made the following statement: "The mode of appointment under our organic law does not comport with the spirit and genius of our republican institutions. It is a relic of a monarchical age which freemen should throw away." *Senate Docs.*, 1847, Doc. No. 9, pp. 4–5.

[12] *Joint Docs.*, 1848, p. 24.

[13] *Ibid.*

attorney-general, superintendent of public instruction, and the various prosecuting attorneys should by law be made elective by the people. An additional provision was included which declared that the members of the supreme court should be ineligible to hold any other than a judicial office during the term for which they had been elected and for one year thereafter.[14] This latter provision resulted from the dissatisfaction felt over the fact that on more than one occasion judges of the superior courts had used, or attempted to use, their judicial office as a steppingstone to a higher political office.[15]

In accordance with the method of changing the state's fundamental law as set forth in the Constitution of 1835[16] the proposed amendment was referred to the legislature of the next year, and the secretary of state was ordered to publish it for three months previous to the next general election.[17] The idea seems to have been quite well received,[18] although one journal counseled delay until sufficient time should have passed to evaluate the effect of the elective experiment in New York.[19] At the beginning of the 1849 session Governor Ransom indicated that, while he was in favor of the elective method for the selection of all public officers, he thought that it might not be expedient to propose constitutional amendments at that time, inasmuch as many people believed there should be a revision of the entire constitution.[20] However, the legislature proceeded (in conformity with the constitution) to approve the resolution for an amendment by the necessary two-thirds vote of each house and to submit it to the people for their approval at the next general election.[21]

By 1849 public opinion had become fairly well united upon the

[14] *Acts of the Legislature of the State of Michigan* (hereafter cited as *Mich. Acts*), 1848, Joint Resol. No. 37, p. 458.

[15] Examples of this were: Chancellor Farnsworth's running for governor in 1839; Justice Felch's election as governor in 1845; Justice Ransom's active campaign for the United States senatorship in 1847; Justice Ransom's election as governor in 1847.

[16] Art. XIII, sec. 1.

[17] *Mich. Acts*, 1848, p. 458. The house adopted the resolution by a vote of 45–5. See *House Journ.*, 1848, pp. 211, 243, 264, 271, 272, 284, 286, 287, 596, 597, 614, 626; *Senate Journ.*, 1848, pp. 231, 232, 280, 353, 377, 379, 524.

[18] *True Democrat*, March 28 and April 4, 1848; *Kalamazoo Gazette*, June 23 and Oct. 20, 1848.

[19] *Grand River Eagle*, Feb. 25, 1848.

[20] *Joint Docs.*, 1849, pp. 13–14.

[21] *Mich. Acts*, 1849, Joint Resol. No. 12, pp. 367–368.

desirability of popularly elected state officials, including court personnel. Apparently it had ceased to be a controversial issue in partisan politics. For example, a mass meeting of Whigs at Jackson [22] and Democratic conventions in Jackson and Hillsdale counties [23] all approved the measure. Newspapers of both parties advocated adopting the amendment.[24] In view of this widespread support it is not surprising that the amendment was ratified by an overwhelming majority at the general election held on the first Tuesday in November, 1849.[25]

Some doubt existed, however, concerning the most appropriate time and method of putting this amendment into effect. The question was complicated by the fact that a proposal to revise the constitution as a whole had been ratified by the people at the same general election in 1849. Despite the mandate in the amendment ordering the legislature to provide by law for the election in 1850 of state officers and judges, certain legislators and other leaders believed it would be more advisable to defer the matter until the constitutional convention (which was scheduled to convene shortly) should have considered the subject.[26] Governor John S. Barry, who was beginning a third term as chief executive, was among those favoring such a plan. With some logic the governor contended that, inasmuch as the terms of a majority of the supreme court justices would expire during July, 1850, it would be wise to postpone the election until later; Barry suggested that, if the legislature adopted this plan, the vacancies which would occur in the meantime could be filled by temporary appointments.[27] Because the tenure of the incumbent justices had not been altered by the amendment only three of the five places on the court would become vacant in 1850.[28] Nevertheless, the legisla-

[22] *Washtenaw Whig*, June 20, 1849.

[23] *Michigan Argus*, Sept. 12, 1849.

[24] *Grand River Eagle*, Aug. 3, 1849; *Free Press*, Sept. 11, 1849; *Michigan Argus*, July 13 and Sept. 12, 1849.

[25] The amendment was supported in every county of the state. The total vote was: for — 38,117; against — 728. *Mich. Acts*, 1850, pp. 473–474.

[26] Several debates on this question were held in the legislative session of 1850. They are reported in the *Free Press*, Jan. 24, 25, and 31, Feb. 5, 6, 13, 14, and 19, 1850.

[27] *Joint Docs.*, 1850, pp. 4–6.

[28] The terms of Justices Warner Wing, Sanford N. Green, and George Miles would expire, but those of Chief Justice Charles W. Whipple and Justice Mundy would not. For a discussion of this point see the *Grand Rapids Enquirer*, Jan. 23, 1850.

ture, unwilling to evade its constitutional obligation or to delay the popular election of judges until any similar action taken by the forthcoming convention could become effective, passed an act directing that three justices of the supreme court (and also numerous other state officers) should be chosen at the general election to be held in November, 1850. Meanwhile, it was made the duty of the governor, if any of the offices within the scope of the act became vacant, to appoint successors, who would then occupy the positions only until the election had been held.[29] Thus the way was prepared for the first popular election of superior court judges in Michigan.

The task of selecting candidates for the bench was made more difficult by the death on August 24, 1850, of Justice George Miles.[30] Miles was the first member of the state supreme court to die in office and one of the three members whose terms expired prior to the election. Convening at Marshall on September 19, 1850, the state Democratic convention proceeded to nominate with little opposition Sanford N. Green and Warner Wing to succeed themselves on the supreme court; in place of Miles the delegates, by a vote of seventy-seven to thirty-nine, selected Abner Pratt of Marshall in preference to David Johnson of Jackson.[31] Five days later the Whigs of Michigan (assembled at Jackson) chose Henry Chipman of Wayne County, Alexander Tiffany of Lenawee County, and Charles Draper of Oakland County to oppose the Democratic nominees.[32] Tiffany declined

[29] *Mich. Acts*, 1850, pp. 18–19. Long discussions were held in the house upon the question of when the new officers would take their seats. The bill was passed, 36–21. *House Journ.*, 1850, pp. 140, 149, 150, 163, 174–180, 189, 198, 201–206, 209, 210, 241. In the Senate Isaac P. Christiancy led the opposition to the bill. *Senate Journ.*, 1850, pp. 127, 136, 137, 143, 144, 145, 205, 211, 212, 228, 230–232, 238–246, 258, 261, 262.

[30] *Michigan Argus*, Aug. 28, 1850; *Free Press*, Aug. 26, 1850. Miles had been appointed to the bench in 1846. *Ibid.*, Oct. 5, 1846.

[31] For the proceedings of the Democratic convention and comments on its sessions and nominees see the following Michigan newspapers: *Free Press*, Sept. 21, 1850; *Michigan Argus*, Sept. 25, 1850; *Coldwater Sentinel*, Sept. 27, 1850. Daniel Goodwin, a former justice of the supreme court, and Randolph Manning, a former chancellor, received one vote and six votes, respectively, during the convention's ballot for judicial candidates.

[32] Reports on the Whig convention and comments on the nominations are in the following issues of papers: *American Citizen*, Sept. 25, Oct. 2, 1850; *Coldwater Sentinel*, Sept. 27, 1850; *Grand River Eagle*, Sept. 30, 1850; *Pontiac Jacksonian*, Oct. 2, 1850; *Michigan Argus*, Oct. 2, 1850; *Oakland Gazette*, Oct. 5, 1850; *Washtenaw Whig*, Oct. 9, 1850. One Democratic newspaper said of Draper: "We don't want to laugh, for Charley is a good and clever fellow, but we do say that

to run, however, and three weeks later Samuel H. Kimball of Jackson County was named in his place.[33]

Shortly after the nominations were made Governor Barry, acting upon the authority which had been given him by the act of February, 1850,[34] appointed Abner Pratt, one of the Democratic candidates, to the vacancy on the bench caused by the death of Justice Miles.[35] Although this action was excused by one Democratic newspaper on the ground that its purpose was to give Pratt "a chance to get his hand in before the first of January," [36] it appears to have been primarily a move of political strategy; at least we know that Pratt did not attend the only term of the supreme court held between the time of his appointment and the first session in 1851.[37] The net result of the move was to extend to all three nominees of the governor's party whatever advantage there might be in holding office before the election occurred.

From the beginning the campaign was conducted along strict party lines, and the election returns demonstrate conclusively the partisan character of the contest. Since the members of the court were selected on a state-wide vote and not from limited districts or circuits, it is interesting to compare the total ballots cast for the justices with the total cast for the other state officials who were elected at the same time; such comparison shows that with few exceptions the people of the state voted a straight party ticket without splitting their ballots in favor of one candidate or another. All the Democratic nominees, judicial as well as administrative, received totals which were between 32,000 and 33,000, whereas all the Whig candidates received totals between 27,000 and 28,000. In most instances only two or three hundred votes separated the various candi-

---

he is just a little unqualified for that office. Only think how lean and small our judge is, and what an enormous quantity of turkey and goose and sauce and soup will be required to swell his meager size to due proportion. This duty, connected with the necessary labor of a review of his law books, will be too much for him . . . ." *Pontiac Jacksonian*, Oct. 2, 1850.

[33] *Michigan Expositor*, Oct. 8 and 29, 1850; *American Citizen*, Oct. 16, 1850; *Washtenaw Whig*, Oct. 23, 1850.

[34] *Mich. Acts*, 1850, No. 23, pp. 18–19.

[35] *Free Press*, Oct. 7, 1850; *Michigan Argus*, Oct. 9, 1850; *Kalamazoo Gazette*, Oct. 11, 1850.

[36] *Coldwater Sentinel*, Oct. 11, 1850.

[37] The court met for four days in October at Pontiac. Supreme Court, Fourth Circuit, *Journal*, pp. 125–133.

dates within each party.[38] A similar distribution occurred within the individual counties, where small differences frequently existed in the number of votes polled by all contestants on the Democratic slate or by all on the Whig slate. Although the Democrats were victorious by narrow margins in several counties, such as Jackson,[39] Kent,[40] and Kalamazoo,[41] in Wayne[42] and Oakland[43] they beat the Whigs by several hundred votes. In Lenawee County[44] the Whigs managed to edge out the Democrats by a majority of about fifty in a total of more than four thousand ballots. It seems justifiable to conclude that the first superior court judges ever chosen by popular vote in Michigan were elected primarily on the basis of party membership rather than because of individual merit or ability.

The year 1850 was marked by the drafting and the ratification of a new state constitution. As might be expected, the principle of popularly elected state officers, which had been the chief feature of the amendment adopted during the previous year, was retained by the new document. One important change was made, however, in the method of selecting supreme court justices; the practice of election from the state at large was replaced by a system whereby one member of the court was to be chosen from each of eight judicial districts into which the state was divided. As previously, the justices meeting together in bank were to comprise the supreme court, and individually each justice was to preside in the circuit court of the district from which he was elected.[45] In accordance with the revised constitution the legislature in 1851 adopted an act providing for the election of circuit judges (and of University regents) on the first Monday in April of that year.[46] Changing the date of the election from fall to spring was commended because it would tend to remove the election of judges from the great party strife that always accom-

[38] The official state canvass was as follows for the Democrats: Wing, 32,606; Green, 32,533; Pratt, 32,265; for the Whigs: Chipman, 27,231; Draper, 27,360; Kimball, 27,589. See the *Michigan Argus*, Dec. 25, 1850; *Oakland Gazette*, Dec. 28, 1850; *Grand River Eagle*, Jan. 3, 1851.

[39] *American Citizen*, Dec. 4, 1850.

[40] *Grand River Eagle*, Nov. 12, 1850.

[41] *Kalamazoo Gazette*, Nov. 22, 1850.

[42] *Free Press*, Nov. 18, 1850.

[43] *Pontiac Jacksonian*, Nov. 20, 1850.

[44] *Michigan Expositor*, Dec. 3, 1850.

[45] Constitution of 1850, Art. VI, secs. 2, 6; schedule, sec. 29. For the counties composing each circuit see Norton, *op. cit.*, Appendix, Table IV.

[46] *Mich. Acts*, 1851, pp. 20–23.

panied the general elections in November, and would thereby encourage greater attention by the voters to the qualifications of the candidates, irrespective of party.[47]

For several months prior to the scheduled election its importance was emphasized frequently by many newspapers of the state. Stressing the experimental nature of the innovation, the editors of both parties warned that the elective system would be on trial and urged that the best men available should be chosen for the supreme court. It was commonly asserted that the designation of incompetent judges at the approaching election would result in agitation for return to the appointive method.[48] Of special interest are the many contemporary expressions of respect and esteem for the supreme court as an institution.[49] Nevertheless, when it came to the actual nomination and election of judges, it appears that much of this altruistic and benevolent attitude was lost in the political whirl. Party organs were quick to recognize that political machinery would be needed in each new judicial circuit to aid in selecting candidates and in conducting the campaign.[50] An indication of how political leaders responded to this need can be found in the following charge made by an editor who condemned particularly the election of judges by circuits instead of by a general state-wide election:

> In the primary meetings and conventions for nominating candidates, the same appliances have been called into action, the same huxtering and trafficking have been resorted to which are common to the most petty demagogue who seeks everlasting glory by a nomination as a candidate for representative in the Legislature; and the result in many cases has been the nomination of judicial candidates possessing the same order and amount of qualifications and fitness.[51]

Although this description was no doubt tinged with the personal feelings of the writer, the existence of considerable partisan activity in

[47] *Grand River Eagle*, Jan. 10, 1851.

[48] *Ibid.; American Citizen*, Jan. 15, March 26, 1851; *Michigan Expositor*, Jan. 28, 1851; *Pontiac Jacksonian*, Feb. 5, 1851; *Free Press*, Feb. 25, 1851; *Michigan Argus*, March 26, 1851.

[49] For instance, the *Pontiac Jacksonian* said on February 5, 1851: "Our Constitution makes no higher, no more important office than that of judge of the Supreme Court, and no branch of our government exercises a greater influence upon the commonweal, than the judicial. Its importance, therefore, can scarcely be over estimated."

[50] *American Citizen*, Jan. 15, 1851; *Michigan Expositor*, Jan. 28, 1851; *Kalamazoo Gazette*, Feb. 7, 1851. The last newspaper proposed that the Democratic state representatives name a committee in each circuit whose duty it would be to call a nominating convention.

[51] *Pontiac Jacksonian*, March 19, 1851.

most of the circuits during the election can be demonstrated by ample evidence.

In the first circuit Warner Wing, who had been reëlected to the bench in 1850, was chosen by the Democrats as their candidate,[52] and Alexander R. Tiffany, who had declined a similar nomination in 1850, was named by the Whigs.[53] Wing was bitterly attacked by William Draper, an attorney of Oakland County. Draper claimed that Justice Wing, when presiding in the Oakland circuit court, had refused to listen to the full presentation of his argument in a certain case, and that in the supreme court Wing had accused him of falsehood. In Draper's estimation Wing lacked certain traits of character that were necessary for a good judge.[54] Apparently his opinion was not shared by the people of the first circuit, who returned a majority of more than two thousand votes for Wing.[55]

Chief Justice Whipple was unanimously nominated by the Democrats of the second circuit.[56] His opponent, Samuel J. M. Hammond, was described as a "rabid abolitionist." [57] It does not appear that the chief justice, who, after twelve years on the bench, was highly respected and well liked in his circuit, had any difficulty in winning the election.[58]

A complex state of affairs developed in the third circuit, which was composed entirely of Wayne County. The Democrats nominated Andrew T. McReynolds, who was said to have "very sparing qualifications" and who reputedly was offered the judgeship "as a reward for services in the Mexican war." [59] Rufus Hosmer, editor of the *Detroit Advertiser*, had been originally selected by the Whigs of that circuit, but he withdrew in favor of Samuel T. Douglass,[60] reporter of the supreme court from 1845 to May, 1850. Although Douglass had always been regarded as a Democrat in politics, he ran as a so-

52 *Coldwater Sentinel*, March 7, 1851.

53 *Michigan Expositor*, March 18 and 25, 1851.

54 *Oakland Gazette*, March 8, 1851.

55 *Coldwater Sentinel*, April 25, 1851.

56 *Ibid.*, March 28, 1851. This journal said that Whipple had been very satisfactory to the bar and to the people: "Possessed, in an eminent degree of the necessary legal attainments, and combining those attainments with a moral character above reproach, he has gained many warm friends throughout the district."

57 *Ibid.*

58 *Free Press*, April 11, 1851. Exact statistics not located.

59 *Washtenaw Whig*, April 2, 1851.

60 *Free Press*, March 28, 1851.

called "independent" candidate, drawing support from dissatisfied Democrats and Whigs alike.[61]

The movement in favor of Douglass was vigorously opposed by factions of both parties. Some of the Whigs held a second convention at Dearborn; failing to agree at this meeting on another candidate, they adopted a resolution stating that they were not bound to vote for either of the "Democrats" (McReynolds or Douglass) nominated.[62] In leading the fight against Douglass the *Free Press* insisted that it was the duty of all good Democrats to vote for McReynolds.[63] At a mass meeting of Democrats, which had been convoked by a petition bearing approximately one thousand signatures (twenty of which were those of local attorneys) and which met in Detroit on April 2, resolutions were adopted approving the candidacy of McReynolds and censuring Douglass as follows:

> Resolved, That Samuel T. Douglass, in allowing his name to be used, first, to remove and disgrace the whig candidate, and second, to defeat the democratic candidate, and destroy the democratic party, has, like the man in olden times, turned his hands against every man, and he will on the day of election find the hand of every honorable man turned against him.
>
> Resolved, That for anyone to participate in a democratic convention, is a virtual pledge to abide its results. And that Mr. Douglass in violating this rule has destroyed the only satisfactory evidence we had, that he is possessed of that high sense of honor, and scrupulous integrity, without which no man is qualified for any office, particular[ly] one of a high judicial character.[64]

At the polls, however, Douglass defeated McReynolds by nearly one thousand votes,[65] although he failed to capture the city of Detroit.[66]

There appears to have been a considerable amount of political intrigue in the fourth circuit, where David Johnson of Jackson[67] and Olney Hawkins of Ann Arbor[68] were nominated by the Demo-

[61] *Ibid.*, March 28 and 29, 1851; *Michigan Argus*, April 2, 1851. The *Free Press* (March 29, 1851) said that a few dissatisfied elements were trying to sell out both parties by electing a man (Douglass) "who cannot place himself fairly and honorably on either ticket, but must needs 'sail under false colors,' entrapping *here a Whig* and *there a Democrat* by the *stale* and *wornout* cry of being an 'Independent Candidate.'"

[62] *Free Press*, April 5, 1851. See also the *Michigan Argus*, April 2, 1851, and the *Coldwater Sentinel*, April 4, 1851.

[63] *Free Press*, April 2 and 3, 1851.

[64] *Ibid.*, April 4, 1851.

[65] *Washtenaw Whig*, April 16, 1851.

[66] *Free Press*, April 9, 1851.

[67] *Michigan Argus*, March 12, 1851.

[68] *American Citizen*, March 5, 1851.

crats and the Whigs respectively. Both men had been prominent attorneys for many years in their communities and were competent and experienced lawyers; nevertheless, in the campaign each party assailed the ability and integrity of its adversary's candidate.[69] Perhaps of significance is the fact that this conflict more than that in any other circuit reflected the national cleavage over problems connected with slavery. The Whigs publicly accused Johnson of seeking Free-Soil support, and charged that he had gone so far as to hire pretended Free-Soilers to circulate a story to the effect that Hawkins favored the new fugitive slave law.[70]

Another phase of the political activity in the fourth circuit was much less publicized; in fact, probably only a few persons knew about it until the trial of the railroad conspirators occurred more than two months after the election. If the words of the principal witness in that trial can be accepted without qualification, the accused men (who were charged with committing various criminal acts against the Michigan Central Railroad Company) actively supported the candidacy of Hawkins because they had been promised, or at least had expected, more judicial leniency in the event that Hawkins instead of Johnson was elected. The direct testimony of the witness on this matter was as follows:

> . . . he [Corwin] said Fitch had seen Hawkins, a candidate for judge, and he had made some pledges by which Hawkins would befriend them in case he was elected; also that they were doing all they could to defeat Johnson . . . . Hawkin's prospects were a good deal talked of; they said they could elect him, and that he stood pledged to them in passing counterfeit money or anything else, as Fitch had informed them . . . he [Fitch] told me on my return home, I must do all I could for the election of Hawkins — to defeat Johnson, as H. was all right.[71]

Although the existence of such an agreement cannot be proved or disproved by the evidence available, at least one competent contemporary observer, James V. Campbell, was somewhat skeptical of Hawkins' participation in such a scheme.[72]

[69] *Washtenaw Whig*, March 5 and 12, 1851; *American Citizen*, March 5, 12, and 26, April 2, 1851; *Michigan Argus*, March 12, 1851.

[70] *American Citizen*, April 2, 1851. Also bound with this copy of the newspaper (in the Jackson Public Library) is a Whig handbill attacking Johnson's honesty in several court cases.

[71] Testimony of Henry Phelps, given on June 25, 1851, printed in *Report of the Great Conspiracy Case: The People of the State of Michigan versus Abel F. Fitch and Others* (Detroit, 1851), pp. 137–138.

[72] J. V. Campbell to his wife, June 26, 1851, Campbell Papers, Michigan Historical Collections, University of Michigan. Campbell wrote: "He [Phelps]

In the election Hawkins carried Washtenaw County by nearly four hundred ballots,[73] but Johnson received sufficient votes in Jackson and Ingham counties to gain the judgeship. Johnson was particularly strong in his home city of Jackson.[74] The Whigs of Washtenaw County were chagrined at the result; according to their version, Hawkins had been defeated because the Whigs of Jackson County had deserted him in favor of a local Democrat.[75] On the other hand, it was asserted that certain Whigs from Washtenaw County had electioneered against Hawkins in Jackson![76]

Prior to the election in the fifth circuit there was an alleged attempt at gerrymandering by the legislature. Bills were passed by the house and senate to transfer Barry County from the eighth to the fifth circuit,[77] supposedly because the Democrats feared that a Whig might be elected in the fifth.[78] The bill was laid on the table, however, after passage and never became law.[79] This action was explained on the ground that the Democrats could not command the two-thirds majority necessary to put an act into effect in less than ninety days after the end of the session, and without immediate effect the transfer would not serve their purposes.[80]

George Woodruff, county judge of Calhoun County, was nominated without difficulty by the Whigs of the fifth circuit.[81] On the other hand, the Democratic convention of this circuit met for four days at Kalamazoo without being able to select a candidate; for

---

has made some very awkward disclosures, and among others, that the conspirators labored hard to elect Olney Hawkins as Judge, and urged their associates to do so on the ground that Hawkins had given them to understand he would be a safer Judge than Johnson. I have heard this story before. It is certain that he [Hawkins] and Mr. Higher law Foster got the whole town of Leoni [center of conspirators' activity], either unanimously or lacking but a very trifling number of votes. It seems rather hard to believe Hawkins said so — although there is no doubt Fitch so represented."

[73] The result in Washtenaw County was: Hawkins, 2,342; Johnson, 1,969. *Michigan Argus*, April 16, 1851.

[74] In Jackson the vote was: Johnson, 577; Hawkins, 191. *American Citizen*, April 9, 1851.

[75] *Washtenaw Whig*, April 16, 1851.

[76] *American Citizen*, April 23, 1851.

[77] *Senate Journ.*, 1851, pp. 34, 54, 62, 64; *House Journ.*, 1851, pp. 93, 110, 111, 112.

[78] *Michigan Expositor*, Feb. 25, 1851.

[79] *House Journ.*, 1851, pp. 111–112.

[80] *Michigan Expositor*, Feb. 25, 1851.

[81] *Kalamazoo Gazette*, March 28, April 4, 1851.

more than fifty ballots Abner Pratt, who had been elected to the supreme court in 1850, was tied with Samuel Clark of Kalamazoo at eight votes apiece. After the convention adjourned the delegates preferring Pratt met by themselves and nominated him. A little later Clark also was nominated by his friends, but he withdrew, leaving the field clear for Pratt.[82] Although Woodruff defeated Pratt in Kalamazoo County by two hundred votes,[83] the latter was elected by virtue of his strength in the other counties of the circuit (Allegan, Calhoun, Eaton, and Van Buren).[84]

In the sixth circuit Justice Sanford M. Green, Origen D. Richardson, and James B. Hunt were announced in February as candidates for the Democratic nomination.[85] The main contest, however, developed between Richardson and Joseph T. Copeland of St. Clair County. After two hundred and thirty ballots Copeland was nominated.[86] The successful candidate was charged with being a representative of the "railroad clique." [87] Because he was a state senator Copeland was declared by the Whigs to be ineligible for the office under the constitutional provisions which said that the members of the legislature should continue in office until their successors were elected and qualified (Schedule, sec. 7), and that no member of the legislature should receive any other appointment in the state during the time for which he was elected (Art. IV, sec. 18).[88] Moses Wisner, an attorney of Pontiac, was nominated by the Whigs,[89] but Copeland was elected.[90]

John S. Goodrich (Whig) and Ellsworth Walkley (Democrat), both from Genesee County, were the nominees for the judgeship in the seventh circuit.[91] Available records reveal no information about the election there except that Goodrich was victorious.[92]

[82] *Coldwater Sentinel*, April 4, 1851; *Kalamazoo Gazette*, March 28, April 4, 1851. The *Gazette* supported Clark until he withdrew.

[83] *Ibid.*, April 25, 1851.

[84] *Free Press*, April 11, 1851; *Washtenaw Whig*, April 16, 1851. Exact statistics not located.

[85] *Pontiac Jacksonian*, Feb. 19, 1851.

[86] *Ibid.*, March 12, 1851; *Coldwater Sentinel*, March 14, 1851.

[87] *Oakland Gazette*, March 8, 1851.

[88] *Ibid.*, March 8 and 22, 1851. The *Gazette* said that two of the present members of the supreme court had informally expressed an opinion that Copeland was ineligible for the second reason mentioned above.

[89] *Ibid.*, Feb. 15, 1851.

[90] *Free Press*, April 11, 1851. Exact statistics not located.

[91] *Oakland Gazette*, March 8, 1851. Walkley had been county judge of Genesee County.

[92] *Free Press*, April 11, 1851.

Unforeseen circumstances intervened in the eighth circuit to produce much excitement. In February Justice Edward Mundy, who had been appointed to the supreme court in 1848 and had served continuously since then, had consented to accept the Democratic nomination if it were offered to him.[93] When there seemed to be little doubt that he would be chosen, an illness which had forced him to be absent from the January term of the supreme court[94] caused his death on March 15.[95] Thus he was the second member of the state court to die while in office; his demise followed that of Justice Miles by little more than six months.

In place of Mundy the Democrats of the eighth circuit named as their candidate Alexander F. Bell of Ionia, a man who had been seeking the judgeship even prior to Mundy's illness and death.[96] This choice was condemned not only by the Whigs but also by many Democrats. According to the leading newspapers of both parties, Bell was a most unsuitable candidate; he was accused of political knavery and corruption, and was said to have none of the requisite qualities.[97] The Whigs unanimously selected George Martin, who had been county judge of Kent County since 1849.[98] In the election Bell received sizable majorities in Clinton and Ionia counties, but suffered defeat in the circuit as a whole by virtue of Martin's great strength in Kent and Ottawa counties; Martin also won in Barry County by a slim margin.[99] Commenting upon the result, one Democratic journal in the southwestern part of the state frankly acknowledged that Bell was morally and legally unfit and deserved to lose.[100]

[93] *Grand Rapids Enquirer*, Feb. 12, 1851.

[94] Supreme Court, First Circuit, *Journal*, II: 132–174. The *Free Press* reported on January 13 that all hope for his life had been lost, but on January 31 it said that he was recovering.

[95] *Grand River Eagle*, March 14, 1851. For other accounts of his death see the *Washtenaw Whig*, March 19, 1851; *Free Press*, March 20, 1851; *American Citizen*, March 26, 1851.

[96] *Grand River Eagle*, Feb. 14 and 21, 1851.

[97] *Ibid.*, March 21 and 28, 1851; *Grand Rapids Enquirer*, March 26, 1851. The latter, a Democratic organ, refused to support Bell. The *Washtenaw Whig*, April 2, 1851, said that he was "as unfit for office as he can be." When Bell had been appointed Register of the Land Office at Ionia in 1845 several Democrats of Grand Rapids protested, stating that his character was openly bad and his conduct disgraceful (July 1, 1845, Lyon Letters, W. L. Clements Library, University of Michigan).

[98] *Grand River Eagle*, March 21, 1851.

[99] *Grand Rapids Enquirer*, April 23, 1851. For the whole circuit Martin had a majority of 1,075 votes.

[100] *Coldwater Sentinel*, April 18, 1851.

On June 25, 1851, Governor Barry (acting by authority of a special law of 1851 which gave the executive power to fill by appointment any vacancy in the supreme court before January 1, 1852) [101] named Martin to complete the rest of the unexpired term of Justice Mundy, and with little delay the senate unanimously confirmed the nomination.[102] This was the first time that any governor of Michigan had appointed a member of the opposing party to the highest state court. The significance of this action is greatly reduced, however, by the fact that the results of the recent election made it almost mandatory for Barry to select Martin, despite his Whig affiliation.[103]

With the exception of Martin, who, by virtue of his special appointment, took his place on the bench at the term of the supreme court for October, 1851,[104] the newly chosen justices were scheduled to take office on January 1, 1852. However, for the third time within fourteen months death intervened in the history of supreme court personnel. Justice-elect John S. Goodrich of the seventh circuit, the youngest of the newly elected members (he was only thirty-six years of age at the time of his election), died on October 15, 1851, two and one-half months before he would have commenced his judicial duties.[105] The problem of filling the vacancy thus created was different from the situation existing after the death of either Justice Miles or Justice Mundy. It will be recalled that Miles and Mundy had become members of the supreme court while it was operating under the constitution of 1835, whereas Goodrich had been elected under the revised constitution of 1850. In the case of Miles the person appointed to succeed him already had been nominated for the court

[101] *Mich. Acts*, 1851, pp. 149–150.

[102] *Senate Executive Journ.*, 1851 (Extra Session), p. 156.

[103] *Grand River Eagle*, July 1, 1851. Justice James V. Campbell made a completely erroneous statement when he said that, under the constitutional amendment of 1849, ". . . Judge Martin was the only member of that court who was elected and not appointed." "Judicial History of Michigan," in *The Semi-Centennial of the Admission of the State of Michigan into the Union* [1886], p. 128. Three members were elected under that amendment, none of whom was Martin. As explained above, Martin was elected under the constitution of 1850, but was appointed to the court before his elective term began.

[104] Supreme Court, Fourth Circuit, *Journal*, p. 139. Although there were three days in this term, Martin was present at only one session. It is possible that he may have attended the July, 1851, term at Jackson, but no evidence to that effect has been found.

[105] *Free Press*, Oct. 17, 1851. Typhoid fever was reported to have been the cause of his death.

and was shortly afterward elected a member thereof by the people. After Mundy's death the governor waited until the approaching election had taken place and then appointed the candidate who had won in the new circuit corresponding most closely with Mundy's former circuit. According to the constitution of 1850, in the event of a vacancy in the supreme court the governor was empowered to designate a temporary appointee, who would hold office only until a permanent successor should have been chosen by popular election (Art. VI, sec. 14). Since Goodrich had not yet taken office, there was no need for an interim appointment, and so Governor Barry called a special election to be held in the seventh circuit on December 29, 1851.[106]

Democratic aspirants for the post were William Fenton, F. C. Whipple, and Justice Sanford M. Green.[107] Green was the only supreme court incumbent who had not been reëlected the previous April. Despite the fact that he was a resident of the sixth and not of the seventh judicial circuit, Green was nominated by the Democrats of the latter area to oppose Artemus Thayer, the Whig candidate.[108] In protest a mass meeting of Shiawassee County citizens named Josiah Turner, judge of the Livingston County court, to run on an Independent ticket, but Turner telegraphed from Detroit his refusal to participate and urged his friends to support Green.[109] This may have had some effect on the election, although it is probable that Green would have won anyway, in view of the fact that he polled a majority of more than five hundred votes.[110] As a result, Green received an assignment to a full six-year term only two days before he would have been forced to leave the bench.

Thus ended the final episode in the series of Michigan's first supreme court elections. There can be no doubt that all these elections in 1850 and 1851 were mainly partisan in nature, conducted with a minimum of attention to the candidates' legal attainments or judicial qualifications. Loyalty to party led voters in general to cast their ballots for the nominee of their own faction, and, because of

[106] *Coldwater Sentinel*, Nov. 28, 1851.

[107] *Free Press*, Dec. 18, 1851.

[108] *Michigan Argus*, Dec. 24, 1851.

[109] *Free Press*, Dec. 30, 1851. In 1857 Turner became a member of the supreme court for a few months.

[110] *Ibid.*, Dec. 31, 1851, and Jan. 3, 1852; *Coldwater Sentinel*, Jan. 2, 1852; *Washtenaw Whig*, Jan. 7, 1852.

their numerical superiority, the Democrats won most of the positions. Nevertheless, a study of the men chosen and of their subsequent work on the bench shows that a majority were competent justices and that a few became outstanding in their field. The periods involved are not long enough to justify making a comparison of the relative merits or disadvantages of the appointive and elective methods of selecting judges as they were employed in Michigan in the decades immediately before and after the constitutional amendment of 1849. On the other hand, there is enough evidence to prove that both systems, though often resulting in the choice of capable and honorable personnel, demonstrated the weaknesses which characterize all partisan judicial selections, whether by popular election or by executive choice.

University of Michigan

LANGUAGE AND LITERATURE

MELANCHOLIC VILLAINY IN THE ELIZABETHAN DRAMA *

LAWRENCE BABB

IN AN early scene of *Much Ado about Nothing* Don John is charac terized as a person "of a very melancholy disposition." [1] This remark carried a significance for the Elizabethan audience which the modern reader misses. A melancholy man, it was believed, was likely to be a shrewd and malevolent fellow, and the Elizabethan playgoer consequently suspected Don John's criminal bias some time before it was revealed in action. The popular association of melancholy with villainy was based, directly or indirectly, upon ideas which appear in scientific works dealing with the psychology of the four humors. It is my purpose to explore the scientific and semiscientific literature of the period for the origins of this association and to illustrate its influence in the drama.[2]

I

In the language of the Elizabethan an individual is said to be melancholy when a melancholic humor, due either to natural disposition or to accident, is so abundant in his system that it predominates over the other humors. According to Renaissance psychology, the passions characteristic of a man — and therefore his personality and behavior — are largely determined by his dominant humor. An abundance of melancholy commonly engenders fear and sadness. But there are other melancholy passions: melancholy men are "suspicious . . . enuious, and ielous," [3] "distrustful . . . envious, ma-

* A part of the material in this study was gathered while the author was a fellow in the Folger Shakespeare Library.

[1] II, i, 6. I am using W. J. Craig's "Oxford Shakespeare."

[2] It should be noted that the melancholy villain was not the only melancholy type character which the Elizabethan audience was prepared to recognize. There were various less sinister types.

[3] Bright, Timothy, *A Treatise of Melancholie* (London, 1586; Facsimile Text Society Reprints, 1940), p. 124. All passages from which I quote in the first section of this paper refer, expressly or presumably, to the "natural" melancholy

licious . . . repining, discontent . . . *injuriarum tenaces,* prone to revenge." [4]

The passions of the melancholy man, moreover, are stubbornly persistent. For melancholy is a heavy and viscid humor,[5] so thick and adhesive that physicians have great difficulty in qualifying or evacuating it, and the mental conditions which it engenders are accordingly highly tenacious. Melancholy men, therefore, are "stubborne, intractable, obstinate"; [6] "constant in their determination." [7] When the melancholic is "throughly heat" with passion, he "retaineth the feruency thereof farre longer time then anie other complexion: and more feruently boyleth therewith." [8] Melancholics harbor "hatred long in their breasts." They are "hardly incensed with anger," but if they become angry, it is "long ere this passion bee appeased and mitigated." [9] Rancor and malice thrive in the melancholic mind: "if Serpents breed in standing waters, evill thoughts maintaine themselves in plodding humors." [10] When black bile

> . . . exceeds Natures bounds, [it] is most fit to move us to any wickednesse. For men of this constitution conceive grievous and sharp passions, and that last long, for the contumacy of the humour, that will hardly melt and be dissolved. Whence it followes, that evill thoughts and apprehensions stay long in their minds, which sometimes break forth into action.[11]

There is, furthermore, the traditional association of melancholy with the malign planet Saturn. Like the melancholy humor, Saturn is "by Nature Cold and dry," and Saturn is said to be "Author of Melancholly." [12] Persons born under Saturn's influence are of melancholy temperament. They are "false, envious, and full of debate . . . heavy, thoughtful, and malicious . . . [Saturnists] shall

---

humor or its effects. According to some authors, "unnatural" varieties of the humor may also cause evil inclinations.

[4] Burton, Robert, *The Anatomy of Melancholy,* ed. Shilleto (London, 1926–27), I: 451.

[5] Lemnius, Levinus, *The Touchstone of Complexions,* tr. Thomas Newton (London, 1576), fol. 136v.

[6] *Ibid.,* fol. 146r.

[7] Walkington, Thomas, *The Optick Glasse of Humors* (London, 1639), p. 131. This work was first published in 1607.

[8] Bright, *op. cit.,* p. 130.

[9] Walkington, *op. cit.,* pp. 130–131.

[10] Du Bosc, Jacques, *The Compleat Woman,* tr. N. N. (London, 1639), p. 33.

[11] Lemnius, Levinus, *The Secret Miracles of Nature* (London, 1658), p. 63. This work was first published in 1559–64.

[12] Coley, Henry, *Clavis Astrologiae Elimata* (London, 1676), p. 37.

be full of law and vengeance, and will never forgive till they be revenged . . . and all evils shall grow in them." [13] Indeed, Saturn is

> A *Catelline*, to mortall *temperament:*
> That would blow vp the *Capitol* of *man*
> With *enuious influence;* melencholy, wan,
> And much resembling, a deepe plodding *pate,*
> Whose *sallow iawbones,* sinke with *wasting hate.*[14]

Although melancholy men are taciturn persons who seem gloomily apathetic, their minds are continually busy. In the "exercise of their wittes . . . they be indefatigable." [15] This constant mental industry increases their power to do evil. The melancholy man "oftentimes . . . by his contemplative faculty, by his assiduity of sad and serious meditation, is a brocher of dangerous Matchiavellisme, an inventor of stratagems, quirkes, and policies." [16] Saturn's melancholic influence inspires

> *Prodigious thoughts,* and *deepe-fetcht treacheries,*
> Beating the *skul* with *sullen phantasies.*[17]

There is general belief among the learned, moreover, that melancholy may under some conditions produce a high degree of mental ability.[18] The melancholy man may be "of a deep reach, excellent apprehension, judicious, wise, & witty." [19] Renaissance writers credit melancholy minds with many achievements of worth and dignity. Yet this intellectual ability may be perverted to evil ends. A melancholy saturnist may be "most ingenious," but

> Wittie in nothing, but things desperate;
> To glut reuenge, with studious memorie
> Of shallow *wrongs,* or some slight *iniurie.*[20]

The melancholy man, then, is subject to evil passions which are tenaciously persistent and is under the malign influence of Saturn.

[13] *The Kalendar & Compost of Shepherds,* ed. Heseltine (London, 1930), pp. 141–142. The first English edition of this work was published *c.* 1518.

[14] Anton, Robert, *The Philosophers Satyrs* (London, 1616), p. 9.

[15] Bright, *op. cit.,* p. 130.

[16] Walkington, *op. cit.,* p. 129.

[17] Anton, *op. cit.,* p. 9.

[18] This belief was due to the influence of a passage in the Aristotelian *Problems,* 30, 1. I have dealt with the subject elsewhere: "The Background of 'Il Penseroso,'" *Studies in Philology,* 37 (1940) : 257–273.

[19] Burton, *op. cit.,* I : 451.

[20] Anton, *op. cit.,* p. 9.

When one adds a morose mental assiduity and a high degree of astuteness, the result is a very dangerous person, amply endowed with the qualifications of a villain. An English moralist writes that melancholy men "are more wickedly bente" than men of any other complexion, that they are "the causes of mischiefe euery where." [21]

II

These villainous connotations of melancholy, traceable to learned works, at times appear clearly in the lines of Elizabethan and early Stuart plays. To illustrate, Aaron, in *Titus Andronicus,* is plotting iniquities. Saturn, he says, "is dominator over" his desires:

> What signifies my deadly-standing eye,
> My silence and my cloudy melancholy . . .? [22]

Shakespeare's King John, on the point of instructing Hubert to murder young Arthur, makes a show of hesitation. He tells Hubert that he would speak plainly, however,

> if that surly spirit, melancholy,
> Had bak'd thy blood and made it heavy-thick.[23]

Lady Macbeth, as she determines to murder Duncan, calls upon the "spirits That tend on mortal thoughts" to "make thick my blood." [24] Thick blood is blood with an intermixture of melancholy. A character in Webster's *A Cure for a Cuckold,* in planning treachery, declares that "My melancholly and the devil shall fashion't." [25] A villain in Shirley's *The Politician* acknowledges dependency on "witty melancholy." [26] In Nabbes' *Microcosmus* Melancholy (personified) boasts that he "could hatch a conspiracy . . . should cause posterity attribute all Matchiavillianisme to Melancholy." [27]

Several of the malicious characters of the drama are definitely labeled as persons of melancholy temperament. There is, of course, Shakespeare's Don John. His tool Conrade was "born under Saturn." [28] In Fletcher's *The Faithful Shepherdess* there appears a

[21] Rogers, Thomas, *A Philosophicall Discourse, Entituled, the Anatomie of the Minde* (London, 1576), fol. 79.

[22] II, iii, 31–33.

[23] *King John,* III, iii, 42–43.

[24] *Macbeth,* I, v, 41–44.

[25] *Works,* ed. Lucas (London, 1927), III: 76.

[26] *Works,* ed. Gifford-Dyce (London, 1833), V: 100.

[27] *Works,* ed. Bullen (London, 1887), II: 181.

[28] *Much Ado about Nothing,* I, iii, 12.

"Melancholy Swain" [29] whose evil practices cause distressing complications among the rustic characters. He is "More sullen Discontent than *Saturns* Brow." [30] The Cardinal in Webster's *The Duchess of Malfi*, archvillain of the play, is "a mellancholly Churchman" who "laies [wicked] plots." [31] His iniquitous brother Ferdinand becomes the victim of a melancholy madness.[32] The villain of Jonson's *The Sad Shepherd* is the witch Maudlin, a "fearfull, and melancholique" creature.[33]

In some cases the melancholy villain is merged with another melancholic type common on the Elizabethan stage, the melancholy malcontent. Since this compound type character, the malcontent villain, deserves special consideration, I shall digress for a moment to characterize the malcontent.[34] The melancholy malcontent was a social type that actually walked the streets of London. As he is depicted in Elizabethan satire and drama, he is a man who has a quarrel with the world because it has not recognized and rewarded the mental endowments and acquirements which he believes he has. His characteristic state of mind is often called "discontent," and the stock adjective applied to him is "discontented." His melancholy is supposed to signify astuteness and profundity of mind. It has been deepened presumably by disappointment and frustration. He advertises his intellectual superiority and his tragic situation by various melancholic mannerisms: black garments, an appearance of dishevelment and personal neglect, a pose of profound meditation with arms folded and hat pulled over the eyes, a sour and moody expression, unsociability, and taciturnity. In spite of the last trait he has occasional spurts of caustic volubility, when he rails at the sins and follies of a world which has failed to treat him kindly. He is often represented as wretchedly poor.

Many of the satirists regard the malcontent as a ridiculous fraud

[29] *Works of Beaumont and Fletcher*, ed. Glover-Waller (Cambridge, 1905–12), II: 419.

[30] *Ibid.*, p. 380.

[31] *Works*, II: 41.

[32] *Ibid.*, pp. 106–107.

[33] *Ben Jonson*, ed. Herford-Simpson (Oxford, 1925–41), VII: 40. See also p. 39.

[34] The malcontent type has been well defined. See especially E. E. Stoll, "Shakespere, Marston, and the Malcontent Type," *Modern Philology*, 3 (1906): 281–303; Zera S. Fink, "Jaques and the Malcontent Traveler," *Philological Quarterly*, 14 (1935): 237–252.

and overwhelm him with gay mockery. Others, however, take him more seriously, regarding his melancholy as real and his brooding discontent as dangerous. His sense of injured merit makes him "the sparke that kindles the Commonwealth."[35] Thomas Lodge draws a portrait of a poverty-stricken malcontent who is a disciple of Machiavelli and who hates his country and its governors through "meere innated and corrupt villanie; and vaine desire of Innouation. . . . [He] delighteth in nought els but traiterous and deuillish stratagems [;] his daily companion in walke, bed, and bord, is rebellion and disobedience."[36] Another writer sees in the melancholy malcontent the perversion of good abilities:

> Wit's a disease that fit employment wants;
> Therefore we see those happiest in best parts,
> And fortunes under-born unto their merits,
> Grow to a sullen envy, hate, and scorn
> Of their superiors; and at last, like winds,
> Break forth into rebellious civil wars
> Or private treasons: none so apt for these
> As melancholy wits, fetter'd with need.[37]

In the malcontent, then, the evil potentialities of the melancholy temperament are especially dangerous, for he is shrewd and ambitious but frustrated, poverty-stricken, and embittered. A great man who needed a tool for criminal service might wisely look for just such a person. A malcontent, one would think, could be bribed, either with money or with promise of preferment, to perform the greatest iniquities. His bitterness and his melancholy have smothered all his scruples; his needs are urgent; he has little to lose and everything to gain.

At least two villains of the drama are melancholy malcontents. Both of them, to use a phrase of E. E. Stoll's,[38] are tool-villains, that is, cat's-paws for villains of higher social rank. One of these is Lazarotto of *The First Part of Jeronimo*. In this play Lorenzo, son of the Duke of Castile, wishes to hire a criminal tool. He engages Lazarotto because he is

[35] Earle, John, *Micro-Cosmographie*, ed. Arber (Westminster, 1895), p. 28.

[36] *Wits Miserie* (*Works*, Hunterian edition [1883], Vol. IV), pp. 23–24.

[37] Field, Nathaniel, *A Woman Is a Weathercock*, *Old English Plays*, ed. Dodsley-Hazlitt (London, 1874–76), XI: 63.

[38] *John Webster* (Cambridge, Mass., 1905), pp. 96, 104, 118, etc. Stoll alludes to the connection between tool-villainy and melancholy (see footnotes on pp. 125, 126).

> A melancholy, discontented courtier,
> Whose famisht iawes look like the chap of death;
> Vpon whose eie browes hangs damnation;
> Whose hands are washt in rape, and murders bould.[39]

Lazarotto, however, proves deficient in melancholic shrewdness, and his career in crime is brief and ignominious.

Another example is Bosola, malcontent tool-villain in Webster's *The Duchess of Malfi.* Bosola, probably somewhat melancholy by nature, has been rendered bitterly malcontent by the Cardinal's failure to reward him for past services, apparently criminal services.[40] He is a witty and acid railer, "The onely Court-Gall." [41] His "foule mellancholly," says Antonio,

> Will poyson all his goodnesse . . .
> . . . . want of action
> Breeds all blacke male-contents.[42]

The Cardinal and Ferdinand, the two major villains of the play, see in Bosola the ideal instrument for their cruel practices upon their sister. He accepts the offer of this ignoble employment with no illusions regarding the foulness of it and is the brothers' agent through most of the play.

Sometimes a major villain of the drama, deceived by the appearance of discontented melancholy, makes a grave mistake in choosing his tool. Mendoza, in Marston's *The Malcontent*, is thus misled by the discontented behavior of "Malevole" (Duke Altofronto in disguise). Mendoza needs an instrument to commit murder and act as pander. He chooses Malevole apparently for no better reason than the fact that he is (or seems to be) a poverty-stricken malcontent.[43] Malevole makes good use of his position of trust in thwarting Mendoza and in pursuing his own righteous ends.

Chapman's Bussy D'Ambois is a soldier to whom Fortune has been unkind,

> A man of spirit beyond the reach of fear,
> Who (discontent with his neglected worth)
> Neglects the light, and loves obscure abodes
> . . . .
> None loathes the world so much, nor loves to scoff it.[44]

---

[39] *The Works of Thomas Kyd*, ed. Boas (Oxford, 1901), p. 301.
[40] *Works*, II: 38. [41] *Ibid.*, p. 37. [42] *Ibid.*, p. 39.
[43] *Plays*, ed. Wood (Edinburgh, 1934–39), I: 180–183.
[44] *Bussy D'Ambois, Tragedies*, ed. Parrott (London, 1910), p. 6. It is characteristic of the melancholy man that he loves solitude and darkness. See Bright, *op. cit.*, p. 124; Burton, *op. cit.*, I: 445.

He is very poor; his clothing is threadbare.[45] All in all, he has such promise of value as a tool that Monsieur seeks him out and takes him into his service. Bussy proves to be a very unmanageable instrument, but undoubtedly, during the first act, the Elizabethan audience was as much deceived by appearances as was Monsieur.

In Tourneur's *The Revenger's Tragedy* a villain of high rank twice makes this kind of mistake. Lussurioso feels the need of a tool, preferably

> some strange digested fellow . . .
> Of ill-contented nature, either disgracst
> In former times, or by new groomes displacst.[46]

In other words, he wants a malcontent. Vendice, who has undertaken the revenge murder of Lussurioso and his entire family, presents himself disguised as "Piato," acts the rôle of malcontent, is hired, uses his position to carry out designs of his own, and disappears. Later Lussurioso once more needs a tool. This time Vendice appears in his own person. An accomplice has predisposed Lussurioso in his favor by characterizing him as a man "full of want and discontent," one "in whom much melancholy dwels." [47] Vendice once more plays the rôle of malcontent — a rôle apparently not altogether foreign to his real character —, and Lussurioso once more makes the mistake of engaging and trusting the man who is seeking his life.[48]

III

When the Elizabethan playgoer learned that Don John was a melancholy man, he was wholly prepared for Don John's treacheries; he found Bosola's discontented melancholy a sufficient reason for Bosola's villainies; Mendoza's blunder in hiring Malevole seemed to him quite natural. The foregoing study has attempted to reconstruct the preconceptions which made such characters and such situations seem plausible to the Elizabethan audience.

Although the evidence is meager, I should like to add some conjectures regarding the appearance of the actor who represented the melancholy villain on the Elizabethan stage. We know that melancholy men usually wore black clothing, and the melancholy stage vil-

45 *Bussy D'Ambois, Tragedies*, p. 7. Cf. *The Revenge of Bussy D'Ambois, Tragedies*, p. 87.

46 *Works*, ed. Nicoll (London, 1930), p. 81.

47 *Ibid.*, pp. 130, 131.

48 See especially pp. 133–134.

lain was probably dressed accordingly. The actor doubtless assumed a gait and manner like those of the malcontent whom Lodge describes: "his lookes suspitious and heauie, his left hand continually on his dagger: if he walke Poules, he sculks in the backe Isles, and of all things loueth no societies." [49] The melancholy man's face is rather clearly depicted in nondramatic works: He is "of colour blacke and swart . . . leane, and spare of flesh: which causeth hollownes of eye, and vnchearefulnes of countenance." [50] His habitual expression is "grim and frowninge." [51] It seems hardly likely that an actor could contrive to look like this without the benefit of modern make-up. That the Elizabethan actor at least tried to do so is indicated by the lines in *The First Part of Jeronimo* (previously quoted) describing Lazarotto's lean jaws and beetling eyebrows. A comment of Beatrice's on Don John's appearance is also suggestive: "How tartly that gentleman looks! I never can see him but I am heart-burned an hour after." [52]

MICHIGAN STATE COLLEGE
EAST LANSING, MICHIGAN

[49] *Wits Miserie*, p. 23.
[50] Bright, *op. cit.*, p. 124.
[51] Lemnius, *Touchstone*, fol. 146 (see note 5).
[52] *Much Ado about Nothing*, II, i, 3–5.

WAS MONTESQUIEU SENT TO LONDON BY LOUIS XV IN 1748?

ANDRÉ B. DELATTRE

IT IS well known that Montesquieu spent some time in England during the three years, 1728 to 1731, in which he toured Europe. New material allows us to raise the question whether he was not sent to London by Louis XV twenty years later in order to take part in the negotiations in 1748 which were to put an end to the War of Succession of Austria.

A few unpublished letters of Montesquieu to a correspondent whose name and identity remain doubtful are among the Tronchin papers at the Bibliothèque Publique et Universitaire de Genève, and together with those letters is an answer of the addressee, who writes as follows to Montesquieu from Geneva, sometime in the year 1748: "Monsieur, Si je n'ay pas répondu plus tôt à la lettre obligente que vous m'avez fait l'honneur de m'écrire le 31 janvier dernier, ce n'a été que parce que le bruit couroit icy [à Genève], peu de jours après l'avoir reçue, que vous aviez fait un voyage à Londres, pour y travailler au grand ouvrage de la paix, et que cela me fut confirmé par un amy de Hollande, qui me marqua que vous étiez effectivement à Londres, où le Roy vous avoit envoyé pour donner du succez à ses intentions pacifiques." [1]

Not very much is known concerning the Geneva correspondent. All the information about him has to be drawn from the ten letters which Montesquieu addressed to him between 1746 and 1754, and which are among the Tronchin manuscripts in the library of Geneva.[2] We learn from them that his name is either Grenouillo or Grenouillac; that in 1745 he had become involved in "*assemblées*" (whether of a political or religious character, it is not clear) in the province of Guyenne; that he fled to Geneva and remained there until 1754 at least; that he was a lawyer, and a relative of Montesquieu. In

[1] Fonds Tronchin, A. 99, folio 8.

[2] They are to be published by the author this year in the *Romanic Review*.

Geneva he was in close relations with Pierre Mussard and with Jean-Louis Saladin, both of whom had served the Republic as diplomats or at least as envoys. Perhaps Grenouillo (or Grenouillac) obtained his information through Jean-Louis Saladin, who often traveled between Paris and Geneva, frequented the salon of Mme de Tencin in Paris, and had become acquainted through her with her great friend Montesquieu. It is also worth noting that between 1731 and 1734 Saladin had been the minister in Paris of George II of England, in his capacity as prince of Hanover.[3]

There is evidence that on various occasions Montesquieu displayed a definite interest in the life of a diplomat. His correspondence testifies that, after he had sold his charge at the Parliament of Bordeaux, he tried repeatedly, between 1728 and 1731, to obtain from the Secretary of Foreign Affairs a post at one of the major European courts. Again a few years later, in April, 1734, his friend le père Castel insists that "un penseur [comme lui], un approfondisseur, un voyant clair, un combinateur (*sic*) politique d'histoire de mœurs, de lois"[4] should be used in some practical capacity by the State. Although no other mention has been made of a diplomatic mission of Montesquieu to London in 1748, nor even of discussions about entrusting him with such a mission, the document cited brings up a question which cannot be ignored. If Montesquieu was sent to London by Louis XV, it was in February or March, 1748, when preliminary negotiations began with London and Vienna. From indications provided by his published correspondence we know that Montesquieu was in Paris on February 3, and also on March 28, 1748. There is no clue whatever to his activities between those two dates. The London negotiations were followed by the Congress of Aix-la-Chapelle, in which Austria, Prussia, Holland, England, and France were represented. We know that Montesquieu was not among the French envoys at Aix-la-Chapelle. Did the ministers of Louis XV try, however, to make use of the popularity in England of the author of *Les Lettres persanes* by sending him to London for the preliminary negotiations? For an answer we have only conjectures to propose, along with groping explorations of possibly connected circumstances.

[3] Cf. J. Senebier, *Histoire littéraire de Genève* (3 vols.; Genève: Barde, 1786), III: 280–285.

[4] *Correspondance de Montesquieu*, published by F. Gébelin and A. Morize (Paris: Champion, 1914), II: 312.

Of all the major French writers of the eighteenth century Montesquieu is the one whose life is least well known. Whether he was actually sent to England or whether there merely was a plan afoot to make him the French envoy, may remain an undecided question in the history of the very obscure and nonetheless very important relations between him, Mme de Tencin, and her brother, Cardinal de Tencin.

Mme de Tencin was far from restricting her activities exclusively to the field of literature. For a full quarter of a century she was the foremost "*intrigante*" in Paris, continually concerned with the internal and foreign policies of the French court, and with the repercussions by which she could further her own ambitions and the political career of her brother. Her relations with the English court and the English diplomatic service were among her strongest assets. Her salon was visited by Lord Chesterfield, by Bolingbroke, and by other distinguished English travelers. Around 1723, in the days of l'abbé Dubois, her political influence had been built on her friendship with Luke Schaub, the envoy in Paris of the King of England.[5] Her brother, Cardinal de Tencin, was a Minister of State and maneuvered to become the Prime Minister. Opposed as he was by the influence of Mme de Pompadour, his successes in foreign politics were his best weapons against her. He was the avowed friend of Charles-Édouard, the Stuart Pretender; nonetheless, or perhaps because of that fact, he was on excellent terms with George II. According to d'Argenson, Tencin rendered King George a very important and very well paid service, in November, 1747, when he weakened by Machiavellian advice the position of the Pretender.[6] We must also observe that Cardinal de Tencin seems to have been particularly aware of the possible diplomatic usefulness of great literary names. In 1757, when he endeavored to come back to power during the Seven Years' War by engineering peace negotiations with Frederick II, it was Voltaire he used as an intermediary.

A letter Mme de Tencin wrote to Montesquieu on November 14, 1748, provides another fragment of evidence: "Vous vous êtes tiré avec honneur de votre négociation avec Maran. Il est bien dommage

[5] Luke Schaub is the envoy about whom le duc de Saint-Simon got so indignant: ". . . Schaub, . . . ce drôle si intrigant, si rusé, si délié, si anglais, si autrichien, si ennemi de la France, si confident du ministère de Londres . . ." (*Mémoires* [Paris: Hachette, 1884], XIII: 38).

[6] *Journal et Mémoires du Marquis d'Argenson* (Paris: Renoir, 1863), V: 98.

que vous n'ayez été chargé de celle de la paix; en vérité, je ne crois pas qu'elle ait été plus difficile."[7] These words, if unsupported by any other text, could very well be taken simply as a humorous compliment on the diplomatic talents employed by Montesquieu in his difficulties with his friend Maran. However, if we bear in mind the rumor echoed in Grenouillo's letter, it seems much more probable that they were meant quite literally, and expressed a regret by the sister of the chief Minister of State, that Montesquieu was given no part in the preparation of the peace treaty.

That Montesquieu was the candidate of a political faction for the negotiations seems therefore very probable. That he was not one of the French envoys to Aix-la-Chapelle is beyond a doubt. That he went to London as a negotiator is possible, but is not certain, in spite of the rumors current in Geneva and in spite of the Dutch eyewitness mentioned by Grenouillo; yet there is no ground to dismiss that possibility entirely. To offer a last supposition, one may wonder whether the disgust of Montesquieu at being a pawn in the struggle for power between Mme de Pompadour and the Tencin team did not play a part in the failure of the mission to materialize.

WAYNE UNIVERSITY
DETROIT, MICHIGAN

[7] F. Gébelin–A. Morize, *op. cit.*, II: 70.

THE MONTREAL SCHOOL AND THE FUTURE OF CANADIAN LETTERS

EDWARD B. HAM

DESPITE sundry books and articles, in English as well as in French, it is evident that the present-day widening of literary horizons in the United States has scarcely begun, even yet, to reveal French Canada on this side of the border. It has long been the Anglo-American habit either to ignore Quebec literature entirely or else to dismiss it as the vaporings of delayed adolescence among retarded colonials. Likewise, the modest poetry which serves as the point of departure for this paper has frequently been waved aside, even by occasional French Canadians, as little more than pallid facsimile of European originals.

It is not my intention here to file a special plea for any author or group of authors, but, rather, to outline certain justifications in terms of historical facts and future hopes. It should be remembered, however, that writers on French Canada who have bothered to learn something of their subject have invariably come to view it with sympathetic discernment. For example, one casual *amateur des lettres* from France itself (where admirers of French Canada are few and far between) has said of Albert Lozeau, possibly the only Canadian versifier whom he knows, that "one could name greater poets, (but) I know of none more delicate, more appealing, or more noble." [1]

The literary school of Montreal is important above all for the new directions which its membership gave to French-Canadian literary aspiration, and which still promise to dominate the remainder of the twentieth century. Prior to the decade of the nineties the literature of Quebec was almost entirely (one might even say "overwhelmingly") patriotic. The novels of Philippe Aubert de Gaspé and Antoine Gérin-Lajoie, still regarded as Canadian classics, had set the tone; writers strove to foster a just pride in racial and Christian achievement. Thus, the poets of the nineteenth century —

[1] *Bulletin des études françaises*, 7 (Montreal, 1942) : 176.

Crémazie, Lemay, Fréchette, Chapman — were content to follow indigenous themes and to look for non-Canadian verse models, in general, only among the French romantics. But French-Canadian men of letters, as a group, have never been forgetful of their shortcomings. Even Crémazie, the founder of the School of Quebec, was forthright enough to say that "in our country our taste in poetry is not very delicate; mix in a few rhymes like *gloire–victoire*, *aïeux–glorieux*, *France–espérance;* season these rhymes with a few sonorous words like 'our religion,' 'our native land,' 'our language,' 'our laws,' 'blood of our fathers'; cook the whole mixture over the patriotic flame, and serve hot: everyone will call it magnificent."

The School of Quebec has long since ceased to exist as such, but its spirit continues to animate numerous writers with virtually undiminished vigor. The creation of the School of Montreal in 1895 marks a vast expansion in the range of Canadian culture, though it has never been the intention of its members to stifle expression of the patriotic sentiment.

Inasmuch as changing literary trends ordinarily reflect the changing social and economic situation, it is pertinent to recall the principal transformations in French-Canadian life during the past century. In recent decades the old agricultural economy has become disorganized and has yielded to forces which are primarily industrial. Moreover, such external factors as radio, cinema, popular magazines, and athletics have drastically revised the attitude of even preëminently rural communities. Anglo-American influences in general are naturally omnipresent.

In other words, the leaders interested in preserving French ethnic identity are at last recognizing with a more and more compelling realism how futile it is to condemn the universal urge toward greater social mobility. French Canada was for many years sufficiently sheltered geographically and economically to ignore this undeniable characteristic of all peoples in all lands. But leaders and writers alike are coming to face the ethnic problem in its proper light; that is to say, the problem is not how to *resist* the drive for increased social mobility, but, rather, how to *adjust* cultural survival in French Canada to actual conditions (present and future).

The School of Montreal has been one of the component forces shaping the cultural trends of today. The direction and strength of the resultant of all such forces cannot yet be defined with precision.

A few examples will emphasize the diversity of components which serious students must evaluate. It must be emphasized in this connection that literature cannot be filed away in some compartment conveniently separated from politics and sociology. The would-be judge of poetry in Quebec must understand, for instance, the policy of broader collaboration with Anglo-Canadians which political leaders like Adélard Godbout and publicists like Émile Vaillancourt are urging as a means of raising the popular cultural level in the province. At the same time more frankly nationalistic thinkers (such as Léopold Richer and Omer Héroux of the Montreal *Devoir*) clamor for greater political autonomy as the key to ethnic independence in Quebec. On the economic front Edouard Montpetit, Esdras Minville, Victor Barbeau, and others are striving to consolidate a self-respecting French position in industry, as well as in linguistic attainment. For years Abbé Lionel Groulx has been working toward modernization of an educational system which must nonetheless remain Catholic and French.

In scholarship it is important to appreciate the range of humanistic interest reflected in such French-Canadian periodicals as *Le Canada français, La Revue trimestrielle canadienne, La Revue dominicaine, Culture,* and the recently established *Bulletin des études françaises.* The *Bulletin des recherches historiques* and the *Cahier des dix* have long been maintaining a good level of historical research. In history and folkways writers like Benjamin Sulte, Thomas Chapais, Marius Barbeau, Adjutor Rivard, Aegidius Fauteux, Gustave Lanctot have proved themselves worthy legatees of the vigorous tradition begun a century ago by François-Xavier Garneau. It is admittedly not enough merely to recite names of men and books to prove the growth of French-Canadian scholarship, but the foregoing list will provide an adequate hint to the observer who would measure the steady expansion in French-Canadian intellectual interests outside the purely literary domain.

The problem for Quebec today, in all forms of culture, is nowhere more eloquently revealed than in a recent appeal addressed to the younger generation by the conservative Abbé Groulx (*Action nationale,* October, 1940):

Qu'elle se replie sur nos hérédités héroïques, sur ses croyances, sur Dieu. Dans le potentiel de notre passé, dans la fierté, dans la joie de se savoir français, d'appartenir à une culture faite d'ordre spirituel, d'harmonieuses convenances avec l'esprit humain; dans la certitude de posséder, par sa foi catholique, une saine philosophie politique, une saine philosophie économique et sociale, dans l'ensemble de ces sentiments et de ces forces, il y a toutes les conditions d'une cure d'air natal; il y a de quoi tout refaire.

Abbé Groulx is on solid ground in searching for a solution in which faith is reinforced with renewed pride of ancestry. But though the way doubtless lies in faith and pride of race, it is implicit in this very quotation that French-Canadian writers are still bewildered and groping, and that there has not yet been offered any complete working specific for the future.

In literature, poets like Robert Choquette, Alfred DesRochers, and Clément Marchand, and novelists like Panneton-Ringuet and Grignon-Valdombre have recognized the necessity for a comprehensive maturity of outlook which Aubert de Gaspé and Gérin-Lajoie in their day apparently never suspected. Even the literature *à thèse nationalisante* is undergoing basic changes today; these were ably discussed by Professor Antoine J. Jobin a year ago before the Michigan Academy in a paper stressing the increasingly militant spirit of such literature in the presence of Canadian industrialization and American influence.[2]

That Quebec literature is very much at the crossroads is nowhere clearer, perhaps, than in the two recent novels of Léo P. Desrosiers, *Les Engagés du grand portage* (1938) and *Les Opiniâtres* (1941). Both point up the courage and strength of the early French settlers, but neither story is a *roman à thèse*. *Les Engagés* chronicles the occasionally perverse side of their tenacity, while *Les Opiniâtres* reproduces with real brilliance their struggles and hopes. In both books the principal characters are humbler even than the Moisan family in *Trente arpents*, and have no conception that they are working for the establishment of any kind of tradition. It is only indirectly that *Les Opiniâtres* pleads for the preservation in our time of that for which the early colonists sacrificed themselves. Desrosiers seems passionately interested in survival of the race, but he is too much a student of history to be drawn (like Savard in the forceful *Menaud maître-draveur*) into any exclusively bucolic school of Canadian literature.

Abbreviated as my remarks have been up to this point, they

[2] "Present Tendencies in French-Canadian Nationalistic Literature," *Pap. Mich. Acad. Sci., Arts, and Letters*, 28 (1942) : 647–656. 1943 See also Mr. Jobin's article in the *French Review*, 16 (1943) : 312–318.

represent a minimum survey of the many considerations underscoring the achievement of the School of Montreal. When these youthful poets of the nineties began their Friday meetings in the Château de Ramezay (now a historical museum), they had only their own enthusiasm and the somewhat dubious blessing of the older Fréchette to determine their future course.

The Montreal poets knew from the start that their very reason for being demanded new interests, new horizons, new aspirations. The little côterie of the Château de Ramezay introduced to Canada the French Parnassians, and even Baudelaire and the Symbolists. Regionalism was to conform to the canons of a purer artistry, and poetic themes were to be restricted only by the essential necessities of sound literary form.

Such objectives seem reasonable enough on this side of the Canadian border, but even since before the formal organization of the School of Montreal it has had influential and highly articulate opponents. Napoléon Legendre was inveighing against "realists and decadents" as early as 1890. In 1919 a magazine called *Le Nigog*, as unobjectionably liberal and humanistic as Pierre Baillargeon's *Amérique française* today, was attacked with equal vigor by the outright regionalists. Even now there are those who remain literally loyal to the catch phrase: "Traiter des sujets canadiens et les traiter d'une façon canadienne." Writers like Blanche Lamontagne, Félix A. Savard, and Frère Marie-Victorin have been unswervingly devoted to themes of the *petite patrie* and *terroir*. The School of Montreal has not had easy sailing.[3]

How well did the first Montreal poets fulfill their self-imposed mission? To answer this question it is not necessary to repeat the substance of useful historical chapters by Mgr Camille Roy, Jean Charbonneau, Mgr Émile Chartier, Jane Mason Turnbull, and others.[4] The problem here involves only the new directions given by the School of Montreal to French-Canadian literature.

[3] For further samples of adverse criticism see Ian Forbes Fraser, *The Spirit of French Canada* (New York, 1939), pp. 103–105.

[4] In *Essential Traits of French-Canadian Poetry* (Toronto, 1938), pp. 116–164, Miss Jane M. Turnbull gives the only important account of the School of Montreal available in English. The two principal books on the subject are by Jean Charbonneau, *L'École littéraire de Montréal — ses origines, ses animateurs, ses influences* (Montreal, 1935), and Germain Beaulieu, *Nos immortels* (Montreal, 1931). An excellent review of verse forms and themes is provided by Mgr

Of the more than thirty members before the turn of the century the best known are Émile Nelligan, Charles Gill, Albert Ferland, and Jean Charbonneau. (From their names, obviously, like William Chapman before them, neither Gill nor Nelligan had French-Canadian fathers; the former is of Scottish descent, the latter of Irish.) In his eloquent preface to the *Soirées du Château de Ramezay* (Montreal, 1900) Gill sums up the purposes of the *École littéraire* as follows:

> Les fleurs sacrées des bords de la Seine que nous voulons cultiver ici, ont à souffrir de la neige et des grands vents; pourtant, si elles sont chétives, l'espèce en est bonne . . . elles s'acclimateront . . . nous verrons à ce qu'elles ne meurent pas.

Émile Nelligan, one of the tragic figures in literary history, is deservedly regarded by many as the most talented of French-Canadian poets.[5] Yet all his verse was written before his nineteenth birthday (1901); during the last forty years of his life he was insane. Steeped in late nineteenth-century French poetry, he particularly admired Verlaine and Baudelaire. Moreover, there is scarcely a line of his verse which would betray Canadian origin. The measure of his sensitiveness and originality can best be felt, perhaps, in *Le Vaisseau d'or*, *Potiche*, *Deux portraits de ma mère*, *La Romance du vin*, and in the three following poems, which deserve to be given here in full:

UN RÊVE DE WATTEAU

Quand les pastours, le soir des crépuscules roux,
Menant leurs grands boucs noirs aux râles d'or des flûtes,
Vers le hameau natal, de par delà les buttes
S'en revenaient le long des champs piqués de houx,

Bohêmes écoliers, âmes vierges de luttes,
Pleines de candeur blanche et de jours sans courroux,
En rupture d'étude, aux bois jonchés de brous
Nous allions gouailleurs prêtant l'oreille aux chutes

Des ruisseaux dans le val que longeait en jappant
Le petit chien berger des calmes fils de Pan
Dont le pipeau qui pleure appelle tout au loin . . .

Chartier, *Au Canada français: la vie de l'esprit, 1760–1925* (Montreal, 1941), pp. 145–166. Mgr Chartier has also placed at my disposal his manuscript concerning research in the field, to be published by the Carnegie Foundation in Volume VII of its *Critical Bibliography of French Literature*.

[5] Cf. *Émile Nelligan et son œuvre*, edited by Louis Dantin and Thomas M. Lamarche (Montreal, 1932).

Puis las, nous nous couchions, frissonnants jusqu'aux moelles
Cependant que parfois dans nos palais de foin
Nous déjeunions d'aurore et nous soupions d'étoiles.

SOIR D'HIVER

Ah! comme la neige a neigé!
Ma vitre est un jardin de givre.
Ah! comme la neige a neigé!
Qu'est-ce que le spasme de vivre
A la douleur que j'ai, que j'ai!

Tous les étangs gisent gelés,
Mon âme est noire: où vis-je? où vais-je?
Tous ses espoirs gisent gelés;
Je suis la Nouvelle Norvège
D'où les blonds ciels s'en sont allés.

Pleurez, oiseaux de février,
Au sinistre frisson des choses,
Pleurez, oiseaux de février,
Pleurez mes pleurs, pleurez mes roses,
Aux branches du genévrier.

Ah! comme la neige a neigé!
Ma vitre est un jardin de givre.
Ah! comme la neige a neigé!
Qu'est-ce que le spasme de vivre
A tout l'ennui que j'ai, que j'ai! . . .

L'IDIOTE AUX CLOCHES

Elle a voulu trouver les cloches
Du Jeudi-Saint sur les chemins;
Elle a saigné ses pieds aux roches
A les chercher dans les soirs maints,
    Ah! lon lan laire!

Elle a meurtri ses pieds aux roches;
On lui disait: "Fouille tes poches.
— Nenni, sont vers les cieux romains:
Je veux trouver les cloches, cloches,
    Je veux trouver les cloches
Et je les aurai dans mes mains."
    Ah! lon lan laire et lon lan la!

Or vers les heures vespérales
Elle allait solitaire aux bois.
Elle rêvait des cathédrales
Et des cloches dans les beffrois,
    Ah! lon lan laire!

Elle rêvait des cathédrales,
Puis tout à coup, en de fous râles,
S'élevait tout au loin sa voix:
"Je veux trouver les cloches, cloches,
    Je veux trouver les cloches
Et je les aurai dans mes mains."
    Ah! lon lan laire et lon lan la!

Une aube triste, aux routes croches,
On la trouva dans un fossé.
Dans la nuit du retour des cloches
L'idiote avait trépassé,
    Ah! lon lan laire!

Dans la nuit du retour des cloches,
A leurs métalliques approches
Son rêve d'or fut exaucé:
Un ange mit les cloches, cloches
    Lui mit toutes les cloches,
Là-haut, lui mit toutes aux mains,
    Ah! lon lan laire et lon lan la!

In the light of these verses it would be banal to say again that Nelligan is a new voice in French-Canadian poetry, were it not for the boundless ignorance of his work on this side of the border. The important fact here is simply that, more than any other single writer, Nelligan gave assurance to French Canadians of the present century that their literature could reach out to all themes and moods compatible with true artistic inspiration.

While Nelligan is significant in his own right, Charles Gill (1871–1918) owes his reputation more to the influence he exercised than to the poetic quality of his muse. Charbonneau [6] describes Nelligan as "grand, mince, les cheveux en broussaille, majestueux, un pli d'amertume à la commissure des lèvres, les yeux perdus dans l'infini; il n'avait pas l'air de tenir au monde matériel." Gill, on the other hand, is represented by his associates as a sort of ivory-tower Hercules, impractical, unaffectedly bohemian, who "évoluait en marge de la société en ce sens qu'il se complaisait uniquement à vivre sa vie intérieure." [7]

It is probably not uncharitable to regard Gill either as a minor painter who wrote verse or as a minor poet who taught drawing. His one published volume is the unfinished *Cap Éternité,* based on impressions gathered while he was painting landscapes in the Sague-

[6] *Op. cit.* (see note 4), p. 119.
[7] Charbonneau, *op. cit.*, p. 109.

nay country. The setting of Cape Eternity evoked a modest symbolism, subjectively conceived, and supplemented by his enthusiasm for Lamartine's *Jocelyn* and the *Poèmes barbares* of Leconte de Lisle. Sensitive to the limitations of the School of Quebec, but too casual and undisciplined in his work, Charles Gill was at least an attractive and respected leader who opened new vistas to devotees of the patriotic theme.[8]

Of the principal writers of Montreal Albert Ferland hews most closely to the line of nationalistic verse. His reputation rests almost exclusively on the four pleasant *plaquettes* which make up *Le Canada chanté* (1908–10), and certainly not on the mediocre earlier verse preserved in the *Soirées du Château de Ramezay*. Reacting against his own inept beginnings, as well as against the more turgid heroics of the School of Quebec, Ferland is reminiscent of an amiable latter-day Malherbe in Canadian letters. Without the gift for great poetry, he knew how to value natural simplicity, and consequently learned to avoid the ostentation of involutions and pedantic references. In so far as his influence has remained effective, his preoccupation with straightforward style has helped to save nationalistic writers from the pompous excesses current in the period of nineteenth-century Canadian adolescence.

As critic and versifier, no member of the School of Montreal has been more articulate than Jean Charbonneau, and yet it is not easy to evaluate his exact accomplishment for the future of Quebec literature. In poetry he was too obsessed with notions of grandiose synthesis, as in *L'Âge de sang*[9] (1921), to do much more than discourage (by his example) run-of-the-mine writers from undertaking epic structures — in itself no small contribution. In his informative volumes of literary history and criticism Charbonneau has come up sharply against the problem which concerns us here. His book on the School of Montreal contains a forty-one-page analysis of the influences operating upon and deriving from the eighty-odd volumes published by the different members. But he ends on the not too world-shaking conclusion, somewhat reluctantly advanced, that it is appropriate to search the future. What group, if any, asks Char-

[8] Cf. Mgr Olivier Maurault, "Charles Gill, peintre et poète," *Revue canadienne*, N. S., 24 (1919) : 18–31, 180–197.

[9] Cf. Louis Dantin, *Poètes de l'Amérique française* (Montreal, 1928), pp. 85–103.

bonneau, will ever replace the School of Montreal? And as recently as 1935 his only answer is a statement that this is an obscure problem, which cannot be solved in these times of uncertainty and depression! But in any event Charbonneau has explicitly formulated the issue upon which the attention of French-Canadian men of letters must henceforth be focused.

The variety of inspiration fostered by the School of Montreal is well illustrated in the "windowpane pieces" [10] of Albert Lozeau and in the studied exoticism of Paul Morin. Confined to his room for nine years, Lozeau found an outlet in a poetry of simple themes (*L'Âme solitaire*, 1907), religious emotion, and eventually occasional Canadian topics (*Lauriers et feuilles d'érable*, 1916). He gave to French Canada a new poetry of personal humility, again at variance with the often haughty strophes of the School of Quebec.

Paul Morin's *Paon d'émail* (1911) and *Poèmes de cendre et d'or* (1922) should be set off against the work of Albert Ferland, since it would not be possible to find two more opposite men within the same literary school. The most widely traveled and the most widely informed of Canadian poets, Morin abandons indigenous inheritance entirely, and fairly overwhelms his readers with a fanfare of foreign lore and tradition. One is surprised, however, that no Canadian writer had turned to exoticism before the third decade of the School of Montreal.

The few samples just discussed indicate the achievements of the Montreal poets, the diversity which they have been introducing to Canadian literature over the past half century. The work of Robert Choquette and Alfred DesRochers, two leading poets of the last two decades, shows how the pioneering of the School of Montreal is already being crystallized in terms of the immediate future. Both Choquette and DesRochers possess the art of poetizing personal experience, of varying their subjects judiciously (except for *Metropolitan Museum*?), of tempering their style with maturity and *mesure*.[11]

To return again to Charbonneau's unanswered question, What does the contribution of the School of Montreal indicate for the Canadian literary future, against the background of Canadian life as

[10] Turnbull, *op. cit.* (see note 4), p. 142.

[11] For intelligent and more detailed commentary on both poets see Fraser, *op. cit.*, pp. 139–142, 167–172, 179–185.

a whole? A general formula was advanced by Mgr Chartier over twenty years ago, when he wrote that "nationaliser, c'est développer les sujets même étrangers à l'aide de nos propres façons de penser, de nos propres expressions parfois." A few prognostics, based on the developmental tendencies of French-Canadian literature, may not come amiss, even in 1943.

It may be taken as certain that writings of primarily religious inspiration will last as long as the Quebec Church. Writings in the spirit of Lozeau or Savard will spring from religious faith, but the *terroir* will be less and less often the sole center of interest.

Writers will presumably follow the lead of Desrosiers, Ringuet, and Grignon in the sense that fictional characters will be less self-consciously poetic about their feeling for soil and other traditions. Ferland's *La Patrie au poète* is significant in this regard.

The industrialization of Quebec is bound to color the literature, and if urban culture does not lose its French imprint, nationalistic writings may even extend to novels of manners, which have been virtually nonexistent in French Canada. Given the Quebec temperament, however, one should not look for any immediate large-scale trend toward the social sophistication which characterizes the modern literature of France. As a tentative example of the current transition in this direction, *La Chesnaie* (1942) by Rex Desmarchais is a meritorious beginning.

In the wake of the School of Montreal more and more writers will seek to enrich their literature with non-Canadian themes, themes which need be neither negative nor unsalutary factors in the struggle for ethnic survival. Success in non-Canadian themes would manifestly enhance the future prestige of French-Canadian letters as a whole.

Quite apart from the School of Montreal, but in harmony with its essential character, French-Canadian writing is likely to become more concretely responsive to social and economic changes induced by the present war, in particular, concerning the rôle which one hopes will fall to Canada with respect to France itself.

The activities of the School of Montreal, during its period of formal organization, were as unassuming as they have been positive; in fact, unlike their Pléiade and Parnassian forerunners, the members never treated themselves to the luxury of a group manifesto. More than any other single group the poets of Montreal have laid the foundation for Quebec's eventual literary coming of age. Despite opposition to their work, the basis for a new tradition has been firmly established; but responsibility for attaining maturity in the twentieth century lies squarely upon French-Canadian writers themselves.

University of Michigan

"EDUCATION OF THE HEART"

OBSERVATIONS ON THE EIGHTEENTH-CENTURY ENGLISH SENTIMENTAL MOVEMENT

JAMES H. WARNER

RECENT study has convinced me that historians of English thought and literature have failed to do justice to the Sentimental Movement of the eighteenth century. Ernest Bernbaum and C. A. Moore are notable exceptions,[1] but Irving Babbitt and Leslie Stephen represent the prevailing attitude of scorn or neglect.[2]

This unfavorable opinion is reflected in the degeneration of the word "sentimental." The origin is the Latin *sentire*, "to feel." Then came the French *sentement*, used by Chaucer to indicate personal experience or feeling. The suffix "-al" was then added, with the result that Lady Bradshaigh inquired in 1749, "What is the meaning of the word *sentimental*, so much in vogue among the polite? Everything pleasant and agreeable is comprehended in that word."[3] It gradually degenerated, however, until today *Webster's New International Dictionary* confirms its meaning as "excess of sensibility, or (sometimes) an affectation of sentiment." This unfortunate word history has probably been, in turn, an important factor in the treatment accorded the whole sentimental movement.

Before proceeding further with this defense let us define our terms. Shaftesbury and Hutcheson have been properly called the philosophical fathers of eighteenth-century English sentimentalism. I therefore employ their chief tenets as criteria, namely, (1) an intuitive moral sense, rather than original sin, (2) a God concept of

[1] Bernbaum, Ernest, *Drama of Sensibility* (New York, 1915); Moore, C. A., "Shaftesbury and the Ethical Poets of England, 1700–60," *Publ. Mod. Lang. Assn.*, 31 (1916): 264–325.

[2] Babbitt, Irving, *Rousseau and Romanticism* (New York, 1919); Stephen, Leslie, *English Thought in the Eighteenth Century* (London, 1876). Nearly all literary histories have based their findings on the modern meaning of the word "sentimental." They have thus limited the term largely to the excesses of the movement. For details see pp. 553, 556, 557, 559.

[3] See the *Oxford Dictionary*.

beneficence and love rather than justice or power, and (3) an emphasis on emotions rather than reason. I shall apply the word "sentimental" to all writings which stress these points. Like all literary movements, this reaction against the classical separated into numerous and often conflicting currents. Usually, however, it is not difficult to distinguish the sentimentalist from the classicist. Furthermore, I shall not confine myself to aesthetic or literary criteria. On the contrary, I shall emphasize social, philosophical, and educational results. Limitations of space forbid presentation of all the evidence. I shall therefore give only examples of each of the main types of influence, with emphasis on the more neglected.

My first point is the unfair restriction of the movement to its weakest branches, the weeping comedy and the highly emotional types of fiction. There is no foundation for this, either in the word "sentimental" or in the works of Shaftesbury. On the contrary, he specifically cites poetry as the ideal vehicle of the emotions.[4] In spite of the fact that Thomson's *Seasons* has long been recognized as reflecting all the major tenets of Shaftesbury, it is rarely classified as sentimental. If Thomson, Collins, Blake, and Burns had been included in the sentimental category, as our definition of the movement would justify, the reputation of this movement would not have suffered so unfairly. Instead, various indefinite terms, such as "romantic" and "pre-romantic," have been applied to them.[5]

Turning to the field of aesthetics, we note at once the resemblance between "sentimental" and modern conceptions of impressionistic art. The individual, internal emphasis of sentimentalism naturally encouraged revolt against mere external realism. Archibald Alison, in his *Essays on Taste* (1790), praises Corneille for going beyond realism to "exalt and elevate the imagination" and "to adorn the events he represented with all that eloquence and poetry could afford." [6] Here, as in much of Alison's work, we find the union of the sentimental and the "sublime." More specifically, the Edgeworths

[4] Shaftesbury, A. A. C., *Characteristics of Men, Manners, Opinions, Times* (ed. 5, 3 vols., London, 1732), III : 308.

[5] Professor Ernest Bernbaum's illuminating Introduction to his edition of *English Poets of the Eighteenth Century* (New York, 1918) is an almost unique exception. Among general histories of English literature the one by Émile Legouis and Louis Cazamian, *History of English Literature* (New York, 1930), most nearly approaches adequate treatment. One section is devoted to "Poetry of Sentiment."

[6] I (1825) : 5.

approved of Sir Joshua Reynolds' subordination of details to one dominant motive.[7]

Religion also felt the beneficial effect of the sentimental movement. It is not difficult to perceive the connection between Shaftesbury's insistence on a joyous religion and on a beneficent deity with the later evangelical emphasis on a god of love.[8] All of this was quite different from, and more emotional than, the classical conceptions. This contrast is exemplified in the hymns of Isaac Watts and Charles Wesley. John Wesley was so impressed with the sentimental novel, *The Fool of Quality* (1765), that he edited the work and made it required reading for his converts. Mrs. Thomas Day enables us to see the blending of the sentimental and the evangelical:

> Heaven wept that men might smile,
> Heaven bled that men might never die.
>
> . . . What heart but glows at thoughts like these? . . . Endeavour as much as thou canst to copy the bright pattern of goodness which Jesus so illustriously exhibited. . . . Difficult as it is, strive to imitate his diffusive benevolence, uncircumscribed by narrow distinctions.[9]

Of course, many critics would question that the evangelical constituted a religious advance. But the sentimental emphasis on joyous religion was exhibited, particularly by the Methodists, with the result that Thomas Sheridan noted with amazement the emotional pleasure they obtained from their services.[10] And few will doubt that the new emphasis on a god of beneficent love was desirable.

In the field of philosophy, sentimentalism constituted a strong and healthful reaction against the doctrine of self-love emphasized by Hobbes and Mandeville. According to Frances Hutcheson, "An ultimate desire for the happiness of others [is] as easily conceivable and as certainly implanted in the human breast, though perhaps not so strongly, as self-love."[11] Samuel Richardson exclaimed that "The God of Nature intended not human nature for a vile and contemptible thing" and that it is "surely both delightful and instructive to dwell sometimes on this bright side of things."[12] Henry Brooke, author of

[7] *Practical Education* (1798), II: 618.

[8] Shaftesbury, *op. cit.*, I: 37, 39; II: 273.

[9] "Reflections upon . . . Good Friday," *Selected Miscellaneous Productions* (London, 1805), p. 157.

[10] *A Course of Lectures on Elocution* (Providence, Rhode Island, 1796), p. 155.

[11] *Essay on the Nature and Conduct of the Passions* (London, 1728), p. xi.

[12] "Concluding Note by the Editor," *The History of Sir Charles Grandison* (7 vols., London, 1902), VII: 296.

the sentimental *Fool of Quality* (1765), contrasts disparate matter and matter submissive to the formative influences of the Potter (compare "Rabbi ben Ezra"). Only when matter thus abandons its useless selfishness does it fulfill its natural function of benefiting the whole.[13]

But it is in political thought that the influence of the sentimental movement has been most underestimated. For example, Edward Dowden maintained that a school of reason, exemplified by Paine and Godwin, were the chief supporters of the French Revolution in English literature.[14] J. H. Harder likewise maintains that "English sentimentality in the eighteenth century was domestic and humanitarian rather than lyrical and revolutionary."[15] However, the "reason" to which Paine and Godwin appealed was far from the reason concerned with inductive processes. The former, frequently called "right reason," was a special sort of insight which enabled its possessor to pierce the outer veil of nature and to peer into nature's innermost secrets. It was more closely allied to the intuition of the sentimentalists than to the reason of the classicists. It was also closely related to the "pure reason" of Kant's *Kritik der reinen Vernunft* (1781). Enfield's popular *Speaker* (1795) gives us a clear definition n an explanation of "good sense":

> . . . Were I to explain what I mean by good sense, I should call it right reason: but right reason that arises not from formal and logical deductions, but from a sort of intuitive faculty of the soul, which distinguishes by immediate perception: a kind of innate sagacity, that in many of its properties seems very much to resemble instinct.[16]

Mary Wollstonecraft, in her *Historical and Moral View of the French Revolution* (1794), pictures this reason as a beautiful goddess: "Reason has, at last, shown her captivating face, beaming with benevolence; and it will be impossible for the dark hand of despotism again to obscure its radiance."[17]

Fairchild similarly finds that "at the close of the eighteenth century, rationalism had long forgotten empiricism" and that it is impossible "to make a hard and fast division between the revolu-

[13] *The Fool of Quality* (5 vols., London, 1765–70), I: 127.

[14] *The French Revolution and English Literature* (New York, 1897), pp. 8, 127.

[15] *Observations on Some Tendencies of Sentiment and Ethics in Minor Poetry and Essay in the Eighteenth Century until 1777* (University of Amsterdam, 1933).

[16] William Enfield, *The Speaker: or Miscellaneous Pieces Selected from the Best English Writers with a View to Facilitate the Improvement of Youth, First American Edition* (Boston, 1795), p. 80.

[17] P. 19 (London, 1794).

tionary cult of reason and the revolutionary cult of emotion."[18] In his work on prose fiction Wright adds that "too often, Bage, Holcroft, and Godwin have been thought of as pure intellectuals who would establish a Utopian order ruled entirely by the dictates of reason." Godwin's anarchical state, Wright notes, presupposes citizens endowed mainly with good impulses and emotions.[19] I would quarrel, however, with Fairchild's "Conclusion," in which "Romanticism" is presented as the "Villain" of the work and "Right Reason" as the "Hero." These two actors would find themselves too closely related to engage in active combat.[20]

Similarly Carl Becker, in his exposition of the background of the American Declaration of Independence, should have given more credit to sentimental philosophy.[21] The very foundation concepts of freedom and equality were therein based on sentimental intuition, insight, or "right reason": "We hold these truths to be self-evident." Thomas Jefferson, prime author of the Declaration, once exalted the impulses of the "heart" over those of the "head":

> . . . If our country, when pressed with wrongs at the point of a bayonet had been governed by its heads instead of its hearts, where should we have been now? Hanging on a gallows as high as Haman's. You [the heads] began to calculate and to compare wealth and numbers: we [the hearts] threw up a few pulsations of our blood; we supplied enthusiasm; we put our existence to the hazard when the hazard seemed against us, and we saved our country.[22]

But undoubtedly the greatest benefit springing from eighteenth-century English sentimentalism lay in the realm of social ideals and practice. The ideal tutor of *Sandford and Merton* (1783–90) exemplifies the increased emphasis on giving to the poor. He insists that his problem child, Tommy Merton, practice the virtue of generosity at every opportunity: "Nothing could equal the joy which appeared on the poor boy's countenance at receiving the present excepting what Tommy felt the first time at doing a generous act."[23] The

[18] Fairchild, Hoxie, *The Noble Savage* (New York, 1928), pp. 140–141.

[19] Wright, Walter F., *Sensibility in English Prose Fiction, 1760–1814* (Urbana, 1937), p. 127. Wright cites Allen Gregory, *The French Revolution and the English Novel* (New York, 1915), as an example of incorrect interpretation.

[20] *Op. cit.* (see note 18), p. 507.

[21] *The Declaration of Independence, a Study in the History of Political Ideas* (New York, 1922).

[22] Letter to Mrs. Conway, October 12, 1786, *Writings of Thomas Jefferson*, ed. A. E. Bergh (Washington, D.C.), V: 444.

[23] *The History of Sandford and Merton. Intended for the Use of Children* (London, 1803), p. 37.

"gentleman" of the classicists differed considerably from the ideal of the sentimentalists. According to Henry Brooke, the qualifications of the latter were (1) generosity to the poor, (2) delicacy to women, (3) deference to others, (4) honor, (5) love, rather than envy, and (6) independence of fashion.[24]

Lack of space forbids the presentation of the tremendous part the sentimental movement played in support of the numerous social reforms of the day, including abolishment of slavery, reduction of the number of capital crimes, dispensing with public hangings and the infamous "ride to Tyburn," prisons, treatment of the insane, abolition of imprisonment for debt, foundling asylums, and kindness to animals. The trend toward beneficence found vent in countless examples. A rather famous one is James Thomson's tribute to the Jail Committee headed by General Oglethorpe in 1729:

> And here can I forget the generous band,
> Who touched with human woe, redressive searched
> Into the horrors of the gloomy jail?
> Unpitied, and unheard, where misery moans;
> Where sickness pines; where thirst and hunger burn,
> And poor misfortune feels the lash of vice.
> While in the land of liberty, the land
> Whose every street and public meeting glow
> With open freedom, little tyrants raged;
> Snatched the lean morsel from the starving mouth;
> Tore from cold wintry limbs the tatter'd weed;
> Even robbed them of the last of comforts, sleep;
> The free-born Briton to the dungeon chained,
> Or, as the lust of cruelty prevailed,
> At pleasure mark'd him with inglorious stripes.[25]

The hero of that highly popular sentimental novel *Sandford and Merton* protested vigorously against the "cruel" sport of bullbaiting.[26] Cowper attacked slavery in Book Two of *The Task*, and Thomas Day satirized Americans for proclaiming the virtues of independence simultaneously with enslaving Negroes. Day's sentimentalism here merged with primitivism:

> . . . I would endeavour to impress upon your mind the feelings of a being [the slave] fully as sensible, and perhaps more innocent, than you or I, which is thus torn in an instant from every thing that makes life agreeable; from country, friends, and parents . . . which, possest of feelings more exquisite

---

[24] *The Fool of Quality* (5 vols., London, 1767), II: 196 ff.

[25] *Winter*, ll. 358–373.

[26] *Op. cit.* (see note 23), p. 218.

> than European hearts can conceive, is separated from all it loves. . . . I shudder at the horrors I describe. . . . If there be an object truly ridiculous in nature, it is an American patriot, signing resolutions of independency with one hand, and with the other brandishing a whip over his affrighted slaves.[27]

And yet A. E. Dobbs' extended volume, *Education and Social Movements, 1700–1850* (1919),[28] gives no credit to Shaftesbury or to the sentimental movement.

Lastly, we have the marked influence of sentimentalism on eighteenth-century education. The tendency of the sentimentalist to listen to the voice within resulted in Rousseau's famous concept of education as an unfolding of inner propensities rather than as an injection from without. Of course, numerous impractical experiments resulted, but critics quite generally agree that this movement was a sound and necessary reaction against the mechanical, memorizing, and purely classical education then in vogue.

Sentimental tendencies in the pedagogy of the last decade of the century were especially healthful. Leslie Stephen's charge of effeminacy may have applied to some sentimental tales, but not to the educational novels of the 1790's.[29] Their heroes often fought with their fists in defense of the poor and the oppressed. The heart was not left to its own devices. It was educated and warned against excesses. *Sandford and Merton* glorified manual labor, particularly out-of-doors work on a farm. Liberty and independence of fashion were extolled. A healthful type of Christianity was emphasized, one which called for imitation of the humble and benevolent deeds of Jesus rather than faith in the supernatural. The section in Enfield's *Speaker* entitled "Pathetic Pieces" contained much that was far from pathetic in the modern sense. Such vigorous appeals to patriotism as that of Henry V in Shakespeare's play of the same name were frequent.[30] But still more important was the part played by sentimental "enlargement of the heart" in creating the first juvenile

[27] "Fragment of an Original Letter on the Slavery of the Negroes," *Four Tracts* (London, 1785), pp. 30–33.

[28] London, 1919. For a brief acknowledgment of the social influence of the sentimental movement see Stuart Sherman, "Laurence Sterne," *Critical Woodcuts* (New York, 1926), p. 292.

[29] *Op. cit.* (see note 2), II: 442. Stephen preferred an imperialistic type of religion which would dominate the world: "The true sentimentalist accepts all that appears to be graceful, tender, and pretty in the Gospels and turns away from the sterner and more masculine teaching which enables a religion to rule the world."

[30] *Op. cit.* (see note 16), pp. 251–320.

literature in England and in establishing separate religious exercises for children. For the first time men realized that children are not miniature adults.

A writer for the *British Magazine* (1761) summarizes the moral and philosophical basis of sentimental education in these words:

> The heart, cultivated by precept and warned by example, improves in sensibility, which is the foundation of taste. By distinguishing the influence and scope of morality, and cherishing the ideas of benevolence, [the heart] acquires a habit of sympathy which tenderly feels responsive, like the vibrations of unisons, to every touch of moral beauty. . . . Hence it is that a man of social heart . . . is waked to the most pathetic emotions by every uncommon instance of generosity and greatness of soul.

The same writer concludes by suggesting methods for the "education of the heart," a popular phrase among the sentimentalists:

> . . . Historical knowledge indeed becomes necessary, . . . but, as the formation of the heart is of first consequence, and should precede the cultivation of the understanding, striking instances of superior virtue ought to be culled for the perusal of the young pupil, who will read them with eagerness and revolve them with pleasure. Thus the young mind becomes enamoured of moral beauty, and the passions are lifted on the side of humanity.[31]

To us this phraseology sounds affected. But are not both the method and the goal of this education essentially sound? Now, as never before, we need to lift the passions "on the side of humanity" to the support of some form of world organization.

I have attempted to indicate briefly and sketchily some of the ways in which the sentimental movement in eighteenth-century England benefited the nation. These results appeared in the fields of literature, aesthetics, philosophy, politics, religion, education, and social reform. It seems unfortunate that the word "sentimental" has been limited to the inevitable but undesirable excrescences. In any case, I hope that I have assisted in demonstrating that this movement does not deserve the neglect and disfavor that historians have usually accorded it.

University of Michigan

[31] III: 493.

THE SELF-AWARENESS OF THE RENAISSANCE AS A CRITERION OF THE RENAISSANCE

HERBERT WEISINGER

EVER since Burckhardt [1] first asserted the primacy of the Italian Renaissance scholars have been trying to fix the chronological limits of the Renaissance as a whole. For a time after Burckhardt the main concern of scholars in the field was the nature of the break between the Middle Ages and the Renaissance. This interest resulted largely from the influence of nationalism on scholarship, for the primacy of Italy was vigorously disputed and the claims of other nations were advanced. As scholars searched the history of the Middle Ages in their particular countries for the earliest signs of characteristic Renaissance activity it soon was apparent that the line of demarcation became more and more indistinct and therefore incapable of exact

[1] Jakob Christoph Burckhardt, 1818–97, Swiss historian of the Romantic-Nationalistic school of historians, whose *Die Kultur der Renaissance in Italien* (Basel, 1860), translated from the fifteenth German edition by S. G. C. Middlemore (Second Edition, London, 1890), brought the Renaissance problem to the attention of scholars by advancing a well-worked-out and provocative theory to the effect that the Renaissance was caused by the emergence of the spirit of individualism. Since the appearance of Burckhardt's work scholars have attempted to show that the Renaissance did not appear first in Italy, as Burckhardt claimed it did, that it was not so sharp a break from the Middle Ages as he asserted, and that the concept of individualism is not capable of strict application as he used it, but Burckhardt's thesis still remains a challenging one. Scholarship on the Renaissance problem is vast and controversial, and one might even compile a bibliography of bibliographies on the subject, but for the reader who wishes to survey the field of post-Burckhardtian studies of the Renaissance problem the following brief sketches are recommended: Roland H. Bainton, "Changing Ideas and Ideals in the Sixteenth Century," *The Journal of Modern History*, 8 (1936) : 417–443; Douglas Bush, *The Renaissance and English Humanism* (University of Toronto Press, 1939), Chapter I; Robert H. Fife, "The Renaissance in a Changing World," *Germanic Review*, 9 (1934) : 73–95; E. F. Jacob, "Changing Views of the Renaissance," *History*, 16 (1931) : 214–229; Norman E. Nelson, "Individualism as a Criterion of the Renaissance," *Journal of English and Germanic Philology*, 32 (1933) : 316–334; and A. S. Turberville, "Changing Views of the Renaissance," *History*, 16 (1932) : 289–297.

delineation. Consequently, at the present time, it is fairly well agreed that if there was a Renaissance — which some scholars, notably Professor Lynn Thorndike, deny altogether — it came not as a sharp and sudden break from medieval civilization, but as the result of a long-drawn-out and continuous series of small and almost imperceptible movements originating deep in the Middle Ages and ranging over many areas. Thus the previously held notion of the Renaissance as a separate and clearly defined period has been broken down.

Since scholarship on the Renaissance problem is primarily concerned with fitting the Renaissance into previously determined theories of history, with the result that the same materials read in the light of different hypotheses yield divergent conclusions, it is unlikely that there will be anything approaching agreement on the solution to the problem. However, to suggest another possible approach to the question, this paper is intended to call attention to an extensive body of materials written in the Renaissance itself but hitherto neglected by scholars — materials which shed considerable light on the Renaissance problem. For convenience the term "Renaissance" is used as though there were agreement as to its signification, that is, for the purpose of this paper it is necessary to assume that something happened in the course of history in the fourteenth, fifteenth, and sixteenth centuries which merits consideration. If, then, we examine what certain writers in those centuries thought they were doing that distinguished them from their predecessors, we shall have the benefit of contemporary observation and discussion of the phenomenon or, more properly, the phenomena of the Renaissance.

A survey of the evidence shows that the men of the Renaissance were acutely aware of the fact that they were participating in activities which set them apart from the men of the preceding centuries. They did not hesitate to name names or to fix dates; they were cognizant that in some way their age was different, and they tried to account for that difference. Furthermore, one gets the impression that the writers of the Renaissance who undertook, more or less consciously, to characterize the Renaissance were describing contemporary phenomena; almost always descriptions of any aspect of the Renaissance are referred to events which really took place or to specific persons; there is a strong feeling of contemporaneity which leaves the impression that to the writers of the Renaissance at least the Renaissance was a real thing which actually happened to them, for they centered their attention on activities which they themselves had either engaged in or observed, such as changes in style of painting and sculpture, the recovery of lost texts, the building of libraries, the revision of curricula of universities, the effects of voyages to new lands and of new inventions. Finally, they were well aware that the Renaissance could not be encompassed by one simple generalization; they recognized that it was made up of several simultaneous movements which interacted on each other. The revival of the fine arts was discriminated from the revival of ancient learning; the Middle Ages were distinguished from the modern era; the effects of nationalism were estimated; the relationship between the Reformation and the Renaissance was established; the influence of the rise of science and of the idea of scientific progress was studied; and, at the same time, it was felt that the Renaissance was the synthesis of these movements, that it made up a whole of which these movements were the parts. In short, the idea of the Renaissance is a Renaissance invention.

I have collected some fifty datings of the Renaissance made in the Renaissance from 1395 to 1599. In these datings the Renaissance is said to be taking place "in our times," and though this phrase may apply to any period from the end of the fourteenth century to the end of the sixteenth, nevertheless it refers to a set of events taking place contemporaneously, that is, at the time of the writing. According to contemporaries, then, the Renaissance was in continuous existence for over two centuries, for almost as soon as we take the turn into the seventeenth century, the Renaissance is spoken of as an event already in the past. The establishment of the Renaissance at a date not earlier than the fourteenth century and not later than the sixteenth is accomplished by the designation of certain individuals as originators of different aspects of the Renaissance. Of the men who were considered most responsible for initiating the revival of ancient learning Petrarch is by far the most frequently mentioned, with Dante and Boccaccio close behind, and the Greek émigré scholars, headed by Chrysoloras, often praised for their services in reviving the study of Greek (the myth of dating the beginning of the Renaissance at the fall of Constantinople is of much later origin than the Renaissance), while Bruni, Poggio, Ambrogio, Valla, Erasmus, Reuchlin, and More are credited with having restored pure Latinity in their respective countries.

The course of the revival of the fine arts was traced as early as 1382 or thereabouts when Filippo Villani wrote that the art of painting reached its height with the Greeks, then declined for many years until Cimabue and later Giotto restored it to its natural style; and this epitome was repeated with variations by Palmieri, Ghiberti, Alberti, Valla, Erasmus, Gyraldus, Vespasiano, Palladio, Bartoli, and Da Vinci, so that, when Vasari came to write his *Lives*, there was considerable information concerning the history of the fine arts available to him which he in turn amplified and regularized in a systematic account on a high historical and philosophical level. Likewise, the humanists from Petrarch on immediately recognized how important the revival of ancient learning was, and they were keenly aware of the significance of their own work. There is no point in merely listing Renaissance accounts of the revival of ancient learning; I have found about fifty, and they range from brief passing mention, as though this revival were taken for granted, to the extended studies of Biondo, Bruni, Dolet, Le Roy, Gyraldus, and Scaliger, and they are European in their origin and scope.

I have already mentioned the names of individuals who were considered to be innovators of various aspects of the Renaissance. But this method was only one of many by which the Renaissance was accounted for in the Renaissance. The fact of the matter is that Renaissance historiography was sophisticated enough not to be content with attributing historical change only to Providence, chance, or individuals; it was seen that the movements of history were the result of forces larger than individuals and less inscrutable and more understandable than Providence. Consequently, we find the Renaissance considered from a number of different angles; as a reaction against the past, as a product of certain other contemporary movements, as a cause in turn of these movements, and as a part of larger and even universal historical patterns, and subsumed under the idea of decline, the doctrine of uniformitarianism, the cyclical theory, the climate theory, the theory of the plenitude of nature, and the idea of progress.

The Renaissance reaction against the Middle Ages was compounded of many factors, the chief of which were the change in taste in the fine arts, the trend toward simplification and uniformitarianism in philosophy, and the aversion to the Germanic tribes, the monks, the Church, the Arabs, and the Turks, each of which was held to have destroyed classical culture which lay dormant during the Middle Ages until revived by the humanists. This revival instituted "a Renaissance" and created an epoch different from the Middle Ages. Though most of the humanists could see no good in the Middle Ages, there were some who were not altogether devoid of any appreciation of them, and if the Renaissance was the first to disparage the Middle Ages, it was also the first to defend them.

The desire to award the palm for learning to one's own country was a strongly motivating force behind the work of the humanists. The primacy of Italy or, more exactly, of Florence, was handsomely acknowledged by the French, German, and English humanists, but they also sought to compete with the Italians and even to vanquish them at their own game. Tissard, Pasquier, Du Bellay, and Amyot in France; Agricola and Reuchlin in Germany; and Sidney, Daniel, Harvey, Edwards, Best, and others in England: each scholar wished his own nation to be thought the leader of the Renaissance, and this ambition helped bring about a Renaissance within its borders. By the time the Renaissance came to England the Italian Renaissance was in its decline, a fact noted by both the English and the Italians; and Polydore Vergil, Aldus Manutius, Aeneas Sylvius, and Jacopo Sadoleto admit that England is now the center of learning and culture.

The relations between the Reformation and the Renaissance are too complex for more than cursory mention here, but out of the struggle between the two came an increased awareness of the individuality of each. The split between Erasmus and Luther, that is, the split between the Renaissance and the Reformation, came over the use to which the new learning should be put. In line with the main tradition of the Italian humanists Erasmus insisted that the study of good letters was an end in itself, though, like Valla, he thought it could be used to clear away errors and confusion in theological matters. On the other hand, Luther, and with him Calvin and Melanchthon, themselves humanists of great repute, believed that the reforms made necessary by the new learning should be fully carried out even if this step meant the destruction of the Church as an institution.

So far the Renaissance has been considered in terms of a revival, of a return of a way of life and thinking which had disappeared for almost a thousand years but which was now back; this is for the most part the humanist attitude, and it is primarily a looking backward to the greatness of classical civilization. The humanists knew they were different from the men of the Middle Ages, but they thought they were like the ancients. Another group of men, however, turned their faces forward, because they believed that the Renaissance was different in kind both from the ancient world and from the Middle Ages. In the estimation of these men what distinguished the modern period from all those which had preceded it was the rise of science, which to them meant the discoveries and the new information revealed through them, the invention of instruments the ancients had not known, the effects of these inventions, and, finally, the application of science toward more discoveries and inventions, so that the outlook for the future was not one of sameness, as in the humanist sense, but one of continuous change and of change for the better. The doctrine of decay was attacked, and in its stead were placed the idea of progress, scientific method, perfectibility, and the plenitude of nature. The rise of science introduced a new element into the Renaissance idea of the Renaissance; men became aware of their era not because it was like another period which had died away and was reborn but because it was different from any other era on the basis of its own distinguishing accomplishments. This notion of uniqueness, the result of the rise of science, was converted in the seventeenth century into the concept of modernity; the modern world takes its rise in the Renaissance by virtue of the rise of science, and the idea of a chronologically limited Renaissance is merged with the idea of a continuing modern era.

It seems to me that there are three possible interpretations of the body of evidence which I have considered. The first interpretation looks on the idea of the Renaissance with considerable suspicion. It argues that since the idea of the Renaissance is itself a Renaissance invention, it is merely another means by which the humanists established themselves. They had a living to make and a business to foster, and they used the notion of the Renaissance as a promotion scheme to sell themselves and their wares. Merely because they said there was a Renaissance is no reason to assume that one actually took place; the assertion of difference is not necessarily the guarantee of that difference, especially under such suspect circumstances. And it is indeed possible to take some of the Renaissance accounts of revivals in various fields, especially in the sciences, and to show how completely unfounded were the contemporary claims to primacy and originality.

The second interpretation would accept the Renaissance idea of the Renaissance as a literal account of what actually happened. This view is based on the principle that contemporary evidence is primary evidence. The contemporary observer is in a position to describe what he sees happening about him, though he is of course limited by being too close to the object, by the danger of partisanship, and by not having all the evidence at his disposal. But even when we have made all the necessary allowances for these limitations, we have left a picture of the times which is likely to be more correct than one drawn from any other sources.

These two interpretations, however, I am inclined to reject for a third which sidesteps the difficulties inherent in them. Where the first is overskeptical, the second is overcredulous. As a consequence, I am led to the conviction that, even if we nowadays have discovered that the Renaissance was not altogether what it thought it was, the fact that it did think of itself as a Renaissance is an objective criterion of the Renaissance. The self-awareness of the Renaissance remains a fact, regardless of what we think of it, which constitutes a distinguishing mark of the Renaissance. The assertion of difference is in this case a substantiation of difference, and those who deny the existence of the Renaissance must be prepared also to deny its self-consciousness. While other criteria of the Renaissance have collapsed because they have reflected changing conceptions of historical method, and while what one generation of scholars has accepted the next has rejected, the criterion of self-awareness remains constant. Its validity depends not on our judgment of it but simply on our recognition of it. I therefore come to the conclusion that there was a Renaissance.

Michigan State College
East Lansing, Michigan

PHILOSOPHY

OUR BELIEF IN REASON

WILLIAM K. FRANKENA

WE LIBERALS, humanists, and believers in democracy have written a good deal of literature recently on the present cultural, intellectual, or spiritual crisis in which civilization finds itself. A reader of this literature is certain to observe, among other things, that we are proclaiming an adherence to the banners of reason in opposition to the antirationalism or anti-intellectualism which is rampant in Europe. We say, with R. B. Perry, that rationalism is one of the ingredients of democracy,[1] and that we have "faith in the beneficence of truth, and the cultivation of man's intellectual faculties by education and liberty." [2] We say, with W. T. Stace, that the philosophy of democracy embodies "the principle of the primacy of reason." [3] We say, with Dean M. Ten Hoor, that "the official philosophers of German National Socialism are wrong in their contention that man is basically a non-rational animal." [4] In all these ways and in many others we declare a living faith in reason.

The reader of this literature to which I am referring is also sure to notice, however, that our faith in reason is not that of earlier generations. The Age of Reason is over, we say; this is an Age of Unreason, or nearly so. Gone is the confidence in reason and in the rationality of man which characterized the outlook of the Greeks, and of the intellectuals of the seventeenth, the eighteenth, and even the nineteenth centuries — gone, not only from the minds of the Nazi-Fascist theorists, but also from the minds of most of the liberal thinkers among us. This, too, we proclaim.[5]

[1] *Shall Not Perish from the Earth* (1940), p. 94.

[2] "A Friendship of Virtue," *Soviet Russia Today*, 11, No. 8 (1942) : 31.

[3] *The Destiny of Western Man* (1942), p. 163.

[4] "The Crisis in Civilization," *Association of American Colleges Bulletin*, 33 (1942) : 381.

[5] See, e.g., C. L. Becker, *New Liberties for Old* (1941), pp. 82–83, 131; Ten Hoor, *loc. cit.* Our state of mind has been described by many writers, such as J. Ortega y Gasset, *Toward a Philosophy of History* (1941), pp. 169–183; I. Edman, *Candle in the Dark* (1939), pp. 9–37; M. Horkheimer, "The End of Reason," *Studies in Philosophy and Social Science*, 9 (1941) : 367; J. Huizinga, *In the Shadow of Tomorrow* (1936), *passim;* H. Kohn, *Force or Reason* (1937), Chap. II.

Inevitably a question must arise in the mind of our reader. We liberal intellectuals are opposing ourselves to both the contemporary irrationalism of Italy and Germany and the past rationalism of our ancestors. What, then, is our position? Clearly, it is not merely another form of irrationalism which we arbitrarily prefer to that of Germany and Italy, though some of our writers do make it appear to be so. Clearly, also, it does involve a denial of some of the cherished but "naïve" (for so we style them) beliefs of our predecessors in the western tradition. But just what is it that we are believing and what is it that we are denying? What is the content, positive and negative, of our belief in reason? This is the question with which I am concerned. The literature does not contain a satisfactory answer to this question, though it does contain all the parts of the answer. We must, however, determine systematically what it is that we believe or disbelieve, if we are to get our democratic theory straight. It is to help in this task that this paper has been written.

It should be clear that I am concerned here only with the formulation of our beliefs and disbeliefs about reason, not with their causes, reasons, or validity. I must also explain that, when I say that "we" believe or do not believe something, I do not mean either that I believe or do not believe it or that all defenders of democracy and western civilization believe or do not believe it, but only that the average liberal intellectual of today believes or does not believe it. I think that there is a set of beliefs and disbeliefs about reason which the average liberal thinker of today is holding and that it is worth while to know what this is. But even if there is no such set of beliefs, or if it is not important that we should know what it is, my presenting a set of beliefs with the claim that they are what "we" are believing may still stimulate us to make up our minds individually what it is that we do believe. And this is what I am really interested in.

Since we oppose ourselves, on the one hand, to the extreme rationalism of the seventeenth and eighteenth centuries, and, on the other, to the complete irrationalism of Germany and Italy, I shall adopt the following procedure. First, I shall formulate the articles of faith involved in a thoroughgoing rationalism such as was approximated in the centuries in question. Then I shall indicate the opposing views of the Nazi-Fascist irrationalists. Finally, I shall state what seem to me to be the beliefs and disbeliefs about reason which we ourselves tend to be holding.

In order to make clear the position of the "compleat" rationalist I shall put down a classified list of all the propositions which express some sort of belief in reason. All these beliefs may be called rationalistic in the broadest sense of this term. They are of four kinds: metaphysical, epistemological, psychological, and ethical.

(1) Under metaphysical beliefs in reason the first to be mentioned is (*a*) the view that reason is the sole or basic or ultimate reality, maintained by Hegel and, possibly, by Leibniz. The other beliefs that come under this heading are all forms of the view that the world is rational, as is the first, but, unlike it, they do not imply that the world is mental in character. The most extreme of them is (*b*) the doctrine that the world is somehow a logically concatenated system. The seventeenth-century rationalists held this doctrine. Less extreme, perhaps, is (*c*) the belief formulated by A. N. Whitehead in the words, "Faith in reason . . . is the faith that at the basis of things we shall not find mere arbitrary mystery." [6] Still less extreme is (*d*) the claim that the world and its contents behave according to regular laws. Even Hume held the world to be rational in this sense. Least extreme is (*e*) the view that the laws of logic apply to reality, which many empiricists accept.

(2) Epistemological propositions involving a belief in reason are of several kinds. In general, we may say that epistemological rationalism is a reliance on reason, pure or empirical, in all questions of opinion or truth, normative or descriptive, as over against a reliance on intuition (of the irrational kind), faith, or authority. But several propositions must be distinguished. There is, to begin with, (*a*) the doctrine maintained by Plato, Descartes, Kant, and others that some of our ideas or concepts are discovered or contributed by reason in essential independence of experience. Then there are three propositions about the possibility of rational knowledge of the world, human and nonhuman. The first of these is (*b*) the belief that we can know the world and what is in it, at least in part, by reason. This belief takes two forms, depending on what is meant by reason. The stronger of these is the view, held by Descartes, that pure reason can attain to knowledge of the world without the aid of experience. The weaker claim is that empirical reason or science can make discoveries about the world, and is advanced by the ordinary scientist. The second proposition about the possibility of rational knowledge of the world is

[6] *Science and the Modern World* (1925), p. 27.

(*c*) the view that reason not only can know the world but takes primacy over other ways of knowing, such as perception and intuition, if these exist. This view likewise has two forms, depending on which is referred to, pure or empirical reason. The third and, in a sense, the most extreme proposition on the present point is (*d*) the conviction that reason and reason only can bring us knowledge, and this, again, takes two forms. Plato and Descartes held that pure reason alone can give us knowledge. Scientific empiricists usually hold that empirical reason or science alone can do so.

A fifth doctrine of an epistemological sort has to do, not with the relation of reason to knowledge, but with the relation of reason to belief or faith. It is (*e*) to the effect that we are not to believe anything except on rational grounds and then only with the degree of belief which is warranted by these grounds.[7] This is "the sentiment of rationality" against which James inveighs in *The Will to Believe.* It also has two forms. Descartes in "Meditation IV," forgetting that he elsewhere appeals to faith, exhorts us to believe only what is clear and distinct to pure reason. Scientific empiricists, on the other hand, maintain that we may assent only to that which is warranted by scientific evidence and only with the degree of assent which is proportionate to the probability established by that evidence.

The last epistemological belief of a rationalistic sort is of yet a different color. This is (*f*) a belief in the epistemological value of objectivity in rational inquiries — the belief, that is, that objectivity is a necessary and helpful condition of knowledge.

(3) Psychological beliefs in reason play a more frequent part in recent literature. All of them are, of course, to the effect that man is rational. The first is (*a*) the proposition that all men have reason, at least in some degree or as a potentiality, to which it is often added that reason is what distinguishes man from (other) animals. A belief in this proposition was proclaimed by Descartes in the opening sentence of his *Discourse on Method,* but it was the common property of centuries of western thought. To say, however, that all men possess reason is not to say that they are rational in any important sense. They may still be, in Adler's words, not rational animals, but rationalizing brutes.[8] Reason may be universally present and yet be a mere

[7] Cf. A. M. Frye and A. W. Levi, *Rational Belief* (1941), p. iv.

[8] Adler, M. J., "This Pre-War Generation," *Harper's Magazine,* 181 (1940): 529.

slave of the will or the passions. Thus the next proposition to be listed here is (*b*) that reason is not just rationalization, but is, in principle at least, capable of something else, namely, impartial cognition — it does not, or need not, merely invent reasons for what we do or believe on instinct. This was certainly a belief shared by all the rationalists of history. Closely related to it is (*c*) the belief that the rational faculties of all men have a generic character such that in matters of cognition they operate according to the same laws, and that there can, therefore, be an international or intercultural science.

Perhaps no one has ever claimed that man is rational to the extent of being perfectly reasonable in all his actions and persuasions. But another claim has been made by Plato, by the seventeenth-century rationalists, and, according to Dean Ten Hoor,[9] by modern educational theory, viz., (*d*) that man is dominated by reason in the sense that his actions and opinions would be wholly rational if it were not for his ignorance. On this view, virtue is knowledge; we are rational in so far as we have rational insight, and it is only ignorance which is responsible for our irrational actions, prejudices, and methods of settling disputes. James Mill is described by his son as coming very close to holding this view.[10] But what he and other eighteenth- and nineteenth-century thinkers were probably believing is a different proposition, that is, (*e*) that men are largely or on the whole rational in their conduct and convictions. The modern idea of progress involved this belief. The last doctrine which needs mention here is still less extreme. It is (*f*) the belief that reason can be primary or, at any rate, very influential in human life, that men can be largely rational in their actions, persuasions, and dealings with one another, even if they are not. This, of course, rationalists have always believed. It is, I suppose, what is meant when it is said that man is "basically" rational.

(4) Ethical beliefs in reason are also very prominent in the literature, although they are not usually distinguished very carefully from psychological beliefs in reason, with which, of course, they are intimately related. I shall list first a belief which has to do with ethics but which is strictly more epistemological than ethical. This is (*a*) the

[9] *Op. cit.*, pp. 379–381.

[10] Mill, J. S., *Autobiography* (Oxford, 1924), p. 89. P. F. Drucker regards (*d*) as an essential part of *all* rationalist liberalism. Cf. *The Future of Industrial Man* (1942), pp. 214–215.

view that ethics and value theory are rational or scientific in character, that questions of right, good, ends, and ideals are susceptible of rational or scientific treatment, even the basic ones. On this view, ethical judgments, like others, are true or false, and their truth or falsity is, in principle at any rate, determinable by rational or scientific means, for example, by rational intuition, by deduction, or by induction. With it is usually associated the conviction that judgments of obligation and of value have universal validity. Many rationalists and empiricists have held this view. The age-old belief in natural law as a basis of ethical and political theory is but one of its forms.

My remaining assertions are more properly called ethical, since they either are or contain judgments of obligation or of value. They all figure in the literature, and I shall state them essentially as they appear there, at the cost of considerable overlapping. They are: (*b*) the belief in the intrinsic value of rational insight, knowledge, science, truth, or intellectual development, which we say began with the Greeks; (*c*) the belief in the ethical primacy of reason, held by the Greeks and by the Age of Reason, according to which reason should rule our lives, whether it does or not; (*d*) the view that the life of reason, the life in which reason is cultivated and in which it rules the other parts of the soul (in Aristotle's terms, the life of intellectual and moral virtue), is the good, happy, or satisfactory life; (*e*) the faith in the extrinsic value or beneficence to humanity, individually or collectively, of reason, science, knowledge, or truth — in the desirability and necessity "of utilizing the full resources of intelligence for the welfare and happiness of mankind";[11] (*f*) the conviction that the individual, society, or both have an obligation to cultivate or foster the cultivation of reason and science; (*g*) the persuasion that every individual has a right to know and to think freely; (*h*) the belief in the obligatoriness or desirability of objective, impartial, or critical reflection on the part of some or of all members of society, either from the point of view of the individual or from

[11] Rader, M., *No Compromise* (1939), p. 86. This faith is also involved in the modern idea of progress. It takes not fewer than three forms. The most extreme of these is the belief that reason, knowledge, or science is a *sufficient* condition of human well-being — that its results are certain to be beneficial. Less extreme is the view that the results of cultivating and applying reason or science are on the whole good, or at any rate better than those that would otherwise be obtained. Least extreme is the opinion that rational or scientific knowledge is a *necessary* condition of human betterment.

that of society; (*i*) the view that individuals, groups, and governments are morally obliged to use rational persuasion, not force or irrational propaganda, in their dealings with one another, whether in matters of opinion or of interest; and, lastly, (*j*) the doctrine of the dignity or sanctity of reason or of rational beings. This doctrine is perhaps the most vague of all the ethical beliefs in reason. It may mean that man is intrinsically better than the other animals, that reason is intrinsically better than will and appetite, that reason by its very nature should have primacy in our lives, that the exercise of reason is intrinsically good or obligatory, or that rational beings should be treated as ends and not just as means.

Well, these are the various propositions which may be said to express a belief in reason. Now I can define the position of the "compleat" rationalist. It consists in the affirmation of all these propositions. Possibly no one has ever maintained this position. But this is the position which was approached by the Greeks, by the seventeenth-century rationalists, and by the thinkers of the Age of Reason. Hence it will serve to indicate the sort of rationalism which we say is no longer possible for us.

Having sketched, approximately, the articles of faith of the sort of rationalism to which we oppose ourselves on the one side, I shall indicate the contradictory tenets of the Nazi-Fascist ideology to which we oppose ourselves on the other side.[12] The complete irrationalist is, of course, one who denies all these beliefs in reason. It will be apparent from the following summary that the Nazi-Fascist theorists come very near being complete irrationalists in this sense.

(1) In metaphysics they deny the rationality of the world. According to Rosenberg, "nature and its happenings have nothing to do with reason or logical requirements." [13] For them Mythos is king, having driven out Logos.

(2) In epistemology they disparage reason both as a way of knowing and as a ground of belief, in favor of intuition, instinct, feeling, faith, and authority. They also deny the value, epistemological and otherwise, and even the possibility, of objectivity and impartiality in the activity of our intellects. This Johan Huizinga has called "the disavowal of the intellectual principle" and "the

[12] In this summary of these tenets I am indebted most to M. Rader, *op. cit.*, and A. Kolnai, *The War against the West* (1938).

[13] Quoted by Rader, *op. cit.*, p. 20.

anti-noetic creed of life."[14] Instead, the Nazi-Fascist theorists exalt subjectivity, and proclaim the essentially biased and racially conditioned character of all intellectual activity. The test of truth, they say, is not the individual reason, but the racial consciousness or instinct, which varies from race to race. It follows that there can be no international or intercultural science, and they insist on this.[15]

(3) In psychology these theorists affirm that man is not rational. Reason is not generic; the Negro and the Jew do not have the same sort of intellect that the German has, and "2 + 2 = 4" is somehow differently tinged in their minds. Reason is entirely subservient to either the individual's egoism or the racial blood and will — they do not decide which. Certainly man's actions and his more profound beliefs are determined, not by his intellect, but by his feelings, instinct, character, or will.

(4) In ethics they maintain that ethical principles and value judgments are not rational or scientific in character. Sometimes it is Hitler's dictum, sometimes the racial soul or blood, which is the standard of right and wrong and of good and evil — and, of course, these are supposed to come to the same thing. Sometimes it is conduciveness to the welfare of the race which is said to be the standard of right and wrong, but then it is added that this cannot be determined by rational means. Reason, our theorists continue, is not something noble; the best part of man is not his intellect but his life, blood, or will. Rational knowledge is not valuable for its own sake. Nor is it an extrinsic good; the rule of reason in life spells perdition for both the individual and society. It is the racial instinct, not the intellect, which must be cultivated and brought to bear as the director of our deeds. Finally, men need not be persuaded by the use of reason; they may be propagandized, forced, or eliminated, the right being to the strong.

The two systems over against which we set ourselves are now before us. By looking at these two systems on the one hand, and at

[14] *Op. cit.*, pp. 99, 169, *et passim.*

[15] Presumably the Germans are applying reason and scientific method to the physical world just as rigorously as we are, whatever we may think of their militaristic purposes for doing so. In fact, we often cite the Nazi use of science and technology as a case of *wertfreie Wissenschaft* gone berserk. But here their practice would seem to be inconsistent with their theory. In any case they have given up being scientific in their social studies, in both theory and practice.

the views expressed in our literature on the other, one can, I believe, construct a set of beliefs and disbeliefs which we may fairly be said, on the average, to be holding. This is the procedure which underlies what follows.

(1) Metaphysics does not play any considerable rôle in the literature on the present crisis, except in the case of Catholic writers. But it is, no doubt, safe to say that we do not accept the first two of the metaphysical beliefs in reason which I have listed. We do not believe either that reality is reason or that it is a logically concatenated system. On the other hand, we probably do, most of us, subscribe to the last two of the foregoing metaphysical propositions of the rationalist, namely, that the laws of logic apply to the world and that the world operates according to "natural" laws also, at least in large part, and is susceptible to scientific study, even the world of man. Some of us are, indeed, quoting Whitehead's fine statement of the remaining metaphysical belief in reason, "that at the basis of things we shall not find mere arbitrary mystery,"[16] nevertheless, this belief, so far as I can see, is one which most of us do not really share.

(2) I take it that in epistemology most of us are, in a broad sense, empiricists. We do not believe in pure reason even to the extent to which Kant did. We do not think that our intellects can furnish us with any a priori or nonempirical concepts in terms of which we may think about the world, or with any a priori knowledge of the world, actual or real, sensible or intelligible. But we do believe in empirical reason and science; we regard them as providing us with reliable knowledge of both nature and man. In fact, we contrast our attitude on this point with that of the Nazis, who, we say, have given up relying on scientific method in the human disciplines, their racial theories being a case in point. On the question whether or not empirical science alone gives us knowledge we are by no means unanimous. Still, the majority of us would probably answer in the affirmative. Our position is the same, I think, on the question whether or not we should give assent only when and in so far as we have scientific grounds for assenting or could provide them if asked. Most of us, but not all, would reject James' doctrine of the will or right to believe and thereby give a positive reply to the question — which is not to say that we do not all honor James' doctrine in practice.

[16] Cf. E. A. Mowrer, *Germany Puts the Clock Back* (Rev. Ed., 1939), p. 28; Kolnai, *op. cit.*, p. 273.

The epistemological issue which is most prominent in recent discussion is that of the epistemological value of objectivity in the investigation of the world and of man. Here the Nazi-Fascist position is clear. They deny the epistemological value of objectivity and impartiality, as well as their psychological possibility. Objective reason and science, even if there could be such things, would be misleading, they say; it is as if a colony of ants were to degenerate and assume a mixed odor pervaded by the smell of parasites, and thus lose their instinctive faculty of discriminating between the parasites and their own kin.[17] Since we decry this exaltation of subjectivity, even if it be subjectivity relative to the race and not to the individual, our own position is also clear. We are still affirming the epistemological value of objectivity, that is, we are believing that objectivity is a help rather than a hindrance in the pursuit of knowledge and truth. It is true that many of us are questioning the possibility of complete objectivity in the social sciences,[18] and that we are even wondering whether science should be *wertfrei*.[19] But when we do so, we are still thinking that subjectivity is epistemologically regrettable, even if it is psychologically inevitable and, in some fashion, ethically desirable.

(3) Our task is more difficult in the case of psychological beliefs about reason, yet, even here, what we are holding both as compared with the Age of Reason and as compared with Nazism and Fascism seems on the whole to be clear. We believe, to judge by our complaints against the Nazis, that there is an international science, that mathematics and science are capable of being grasped in the same way by Teuton, Jew, and Negro, given the same degree of intelligence, that science is not conditioned by blood or race in any way which affects the universal validity of its findings.[20] We likewise hold that a general science or psychology of human nature is possible which applies alike to all men in its main outlines (variations in human beings are taken care of in the statement of its results).[21] Thus we are asserting that there is a generic human nature, not just an Aryan nature and a non-Aryan nature, and that reason is one of the common ingredients in this generic human nature. We are subscribing

[17] Cf. Kolnai, *op. cit.*, p. 29.

[18] Cf., e.g., G. Sabine, "Social Studies and Objectivity," *University of California Publications in Philosophy*, 16 (1941).

[19] Cf. R. Anshen, *Science and Man* (1942), *passim*.

[20] Cf. G. De Laguna "Cultural Relativism and Science," *Philosophical Review*, 51 (1942): 140–166.

[21] Cf. Stace, *op. cit.* (see note 3), pp. 70–74.

to the rationalistic faith that all men have, in some degree or as a potency, a faculty of reason which operates on the same laws in everyone. The fact that we generally regard it as one of the presuppositions of democracy is evidence of our acceptance of this belief.[22]

We should contend, next, I believe, that reason is not mere rationalization. Granting that much reasoning is rationalization (and we like to say that the reasoning of the Nazi-Fascist theorists *is* rationalization), we should insist that a large part of our inductive reasoning, whether consciously done or not, is not pure rationalization. For instance, my jumping out of the way of a car rests on a piece of inductive reasoning, consciously or unconsciously performed, and this reasoning, though it subserves my will to live, is not just a rationalization of something which my will impels me to do or to believe. Again, granting that desire and not reason furnishes the motive power of all our activity, we should still maintain that reason is not a complete slave of the passions. If it were, it would not tell me that I must jump when I see a beautiful pattern of sense-data bearing down on me.[23]

It may be objected that the pragmatists, of whom there are many among us, *are* really holding that reason is rationalization.[24] But this, I think, is a mistake. Even if "reason is an instrument developed to serve the interests of the organism," as the pragmatist holds,[25] it does not follow that all reasoning is rationalization. It may still be genuinely cognitive of objective fact. My illustration of a moment ago shows this. As C. L. Becker points out, the biological function of reason may be precisely "to discriminate . . . the dependable fact from the deceptive illusion, in order that the organism may pursue the better rather than the less good interest," [26] and this is not mere rationalization — alleging reasons for what we are moved wholly by irrational forces to think or to do anyway.

[22] Cf. R. B. Perry, *On All Fronts* (1941), p. 117; C. L. Becker, *Modern Democracy* (1941), pp. 14–15; W. Cerf, "Philosophy and This War," *Philosophy of Science*, 9 (1942): 181.

[23] Cf. Stace, *op. cit.*, pp. 265–285, for a longer discussion of this point.

[24] W. T. Stace and M. J. Adler seem to interpret pragmatism thus. See Stace, *op. cit.*, pp. 195–197; Adler, *op. cit.*, pp. 527, 529, 532. But the pragmatists may agree with everything that Stace establishes against Schopenhauer in the discussion referred to in the previous note. Cf. J. Dewey, *Human Nature and Conduct* (1922), Part III, Sect. 111.

[25] Cf. C. L. Becker, *New Liberties for Old*, pp. 32, 93, 133, 137, 147–148.

[26] *Ibid.*, pp. 137, 147–148.

We obviously do not believe that man is wholly or perfectly rational, as the ideal man of so many rationalists of the past would be. The opinions of men are not always rationally founded, and their actions are not always determined by a consideration of what is best in view of their ends, actual or proper, individual or social. But none of our rationalistic ancestors believed this either. Yet we differ from them; man, we say, is much less rational than they thought he was.[27] We must, then, be maintaining that men are not even as rational as their knowledge permits them to be, and that other things besides ignorance (for example, passion) keep them from being wholly reasonable in their persuasions, conduct, and conflicts with one another. It is true that our educational system seems to imply that we believe that knowledge is virtue.[28] It is also true that many people are still arguing that *the* trouble with mankind and *the* reason why man is not better off is that the social sciences have lagged behind the physical sciences — a contention which implies that knowledge is enough. Nevertheless, it strikes me that the opposite conviction is now in the ascendancy. More and more we are insisting that besides light, knowledge, science (even social science) man needs sweetness, love, good will, moral purpose.[29]

In fact, if I read our state of mind correctly, we should not even go so far as to assert that men generally are largely rational in their actions and opinions.[30] History, politics, and psychology have been our teachers here. We should hold, however, that some men are largely rational, and that some peoples and ages are more nearly rational than others. Thus we believe, I take it, that man is "basically" rational in the sense that reason *can* come to control our lives and our societies in certain important respects, and that there is some ground for hoping that we may sometime come to listen to reason more than we now do. Even psychoanalysts among us express this belief.[31]

[27] Cf. Becker, *op. cit.* (see note 22), p. 27; Edman, *op. cit.* (see note 5), pp. 35–36; G. R. Morrow, "The Philosophical Presuppositions of Democracy," *Ethics*, 52 (1941–42) : 307.

[28] Cf. Ten Hoor, *op. cit.*, p. 380.

[29] *Ibid., passim;* B. Russell, *What I Believe* (1925), Chap. II; Becker, *op. cit.* (see note 25), pp. xvi–xvii; Stace, *op. cit.*, Chap. XII; Perry, *Our Side Is Right* (1942), pp. 27–29; A. D. Ritchie, "Science as a Creative Power," in *What We Defend*, ed. by E. F. Jacobs (1942).

[30] But cf. Morrow, *op. cit.*, p. 305.

[31] Cf. F. Alexander, *Our Age of Unreason* (1942).

(4) We appear, therefore, to be resolutely opposed to the Nazi-Fascist view of human nature. We object to it as emphasizing the animal in man at the expense of his rationality. As Kolnai has said, its "central dogma . . . is that human nature is nature in the first place, and only additionally human." [32] But it is in ethics that our opposition to Nazi-Fascist irrationalism and our adherence to the creed of rationalism is most complete. Some of us would subscribe to all the ethical propositions of rationalism, at least in their less extreme forms. It is going too far, however, to say that "we" accept them all. Take the first. Many of us do not believe in the possibility of any rational or scientific treatment of basic or ultimate judgments of value and of obligation. Most of us deny that there are any natural laws, in the ethical sense, which may be discovered by reason, and that there are any ethical propositions which are self-evident.[33] But the relativists, positivists, and *Existenz* philosophers among us go much farther. They contend that our basic ethical standards and value judgments are not true or false in any sense, being simply irrational preferences, prejudices, or postulates.[34] A scientific study of ethical judgments as phenomena is still possible on this view, but the judgments themselves are not cognitive in character and cannot be adjudicated by scientific or rational means. On this point, then, it would seem that many and perhaps most of us are irrationalists — as Perry, Stace, Adler, and others have pointed out.[35]

I find it somewhat difficult to determine what the majority of us believe about some of the subsequent propositions in my list. I suppose that we should agree, even those who are pragmatists, that experiences of knowledge and rational insight are good in themselves. Possibly we should also allow that the life of reason is the good life, but certainly many of us would add, "for them as likes it," by way of expressing a doubt that the good life is the same for everyone. Even so, we do seem, on the whole, to be agreed that a large measure of rational control is necessary for the good life, of both the individual

[32] *Op. cit.*, p. 195; cf. pp. 202, 206.

[33] Cf. Morrow, *loc. cit.;* Becker, *op. cit.* (see note 25), p. 93; S. Hook, "The Philosophical Presuppositions of Democracy," *Ethics*, 52 (1941–42) : 285–286. Rader, *op. cit.*, pp. 65–66, 94; A. Cobban, *The Crisis of Civilization* (1941), Chaps. VIII–IX; and W. Lippmann, *The Good Society* (1937), pp. 369–390, are exceptions.

[34] Cf. Russell, *loc. cit.;* Cerf, *op. cit.;* Becker (see note 25), *op. cit.*, p. xvi.

[35] Perry, *op. cit.*, pp. 10–16; *idem, Shall Not Perish from the Earth*, pp. 21–23; Stace, *op. cit.*, pp. viii, 12–16; Adler, *op. cit.*, *passim.*

and society, though we should also insist on the importance of feeling. That reason *ought* to have a considerable degree of primacy in the life of the individual and in that of society is a view we certainly appear to be holding, especially when we are thinking not only of the welfare of the individual but also of that of society. We should, however, commonly add that love, too, must have a certain primacy in our lives.[36]

This brings us to another article in our creed. We do believe in the extrinsic value or beneficence of reason and science, for both the individual and mankind. The destructive applications of science in the hands of militaristic peoples give us pause, but, on the whole, we regard them as effects of the misuse of science, and not as essentially bound up with science.[37] In fact, as I have mentioned, some of us still believe that we have only to multiply and spread knowledge, adding social science to physical science, and all will be well. It is true, if I was correct in what I said previously (p. 582, § 3), that we tend to insist that love and good will, as well as knowledge, are necessary for social well-being, but even those of us who do so insist are nonetheless thinking that knowledge is one of the conditions of progress. It is also true that some of us are suggesting that the past indifference of science to considerations of value has been one of the sources of our present troubles, and that science should no longer go on being indifferent to such considerations. Yet even these people seem to be holding that the *pursuit* and *application* of science must be motivated by moral purpose, and not to be denying that science should be cultivated, or that it is a condition of the improvement of human life. We seem, in short, to be agreed in believing in "the advancement of learning," but many of us are maintaining that good will must be promoted also, while others are of the opinion that learning should be advanced anyway.

But we do not assert only that learning should be advanced. As believers in democracy we believe in the cultivation of reason and knowledge, as far as possible, by *every* individual, both for his own sake and for that of society. This involves a belief in the universal right and ultimate beneficence of freedom of thought and of critical reflection. Some of us hold that certain restrictions must be put on this freedom of thought and expression, at least in an emergency such

[36] Cf., e.g., Stace, *op. cit.*, Chap. XII.
[37] Cf. Adler, *op. cit.*, p. 527.

as the present one;[38] others deny this.[39] But we all insist that such restrictions must be kept to a minimum.

If there is any ethical belief in reason to which we are cleaving, it is the belief in the obligation of individuals, groups, and governments to maximize the use of rational persuasion in their dealings with each other, and to minimize the use of irrational propaganda and force.[40] We almost define a democratic society as a society in which the use of rational persuasion pervades all relationships.[41] Whitehead has even described civilization as the victory of persuasion over force.[42] There is, indeed, much talk in educational and other quarters of "indoctrination for democracy" and of the use by democratic societies of democratic propaganda for either internal or external consumption.[43] But what is envisaged is the use of methods of persuasion which enlist the emotions as well as the intellect, not the use of methods which violate the principle here in question by failing to respect the intellect.

I have already stated that, as against the Nazi-Fascist theorists, we believe in the epistemological value and the psychological possibility of objectivity in science and in the activities of reason. It must now be added that we also believe in the ethical desirability of such objectivity or impartiality. We are holding, with Russell,[44] that it is best, from the point of view of both its intrinsic and its extrinsic values, that science be amoral or value-indifferent in the sense that ethical and axiological considerations shall not influence its conclusions or form part of its evidence. Even pragmatists believe this in an important sense. It is true, as I have indicated, that some of us seem to doubt this, and complain of the effects of the value-indifferent pursuit of science — for example, as this manifests itself in Germany today. But those who do appear on analysis to mean either that science must be supplemented by religion or philosophy, not that it should take on a bias, or that its pursuit and application should be

[38] Cf. Perry, *Shall Not Perish from the Earth,* Chap. IV.
[39] Cf. Becker, *op. cit.* (see note 25), pp. 96–123.
[40] Cf. Rader, *op. cit.,* Chap. V; Morrow, *op cit., passim;* Perry, *Our Side Is Right,* Chap. III.
[41] Stace, *op. cit.,* pp. 161–166.
[42] *Adventures of Ideas* (1933), p. 105.
[43] Cf. Perry, *loc. cit.; On All Fronts,* Chaps. III–IV; B. F. Pittenger, *Indoctrination for American Democracy* (1941).
[44] Russell, B., *Mysticism and Logic* (1917), pp. 29–31.

dominated by a moral purpose, not that it should allow ethical considerations to constitute part of its evidence for its conclusions.

These are our beliefs and disbeliefs in reason. In what we disbelieve we are opposed to the Age of Reason, and so far we are irrationalists; in what we believe we are opposed to Nazi-Fascist irrationalism, and so far we are rationalists — and the Age of Reason is not over. Of course, a faith in reason, especially one as qualified as ours, does not guarantee that a reign of reason is forthcoming, but it does hold out a promise of one, since a faith without works is dead.

University of Michigan

SOCIOLOGY

THE COMMUNITY BASIS OF PEACE

CHARLES R. HOFFER

WITHIN the last quarter of a century the concept "community" has gained widespread acceptance in the social sciences. It has been defined in various ways, but the central idea is that a community is a group of people inhabiting a definite area, whose members have common interests and activities as a consequence of its habitation. Whenever a group has interests and activities originating in this manner, a community comes into existence. As the number of these bonds within a particular area increases, the importance of a community for that area is enhanced. Thus communities may be small, like a crossroads store in a rural area, or large multiple-service centers such as a metropolitan city. The territorial expanse of a community depends upon the distance existing between persons who influence one another and its importance upon the number of interests and activities which are effective in an area.

It follows from the foregoing remarks that communities will vary in number and size as means of communication and transportation reduce or increase their spatial dimension and the manner by which people can influence one another. Thus, in America the original rural neighborhood has been superseded to a considerable extent by the town-country community of which it is a part. Automobiles and the electric train contributed to the growth of the suburban community. Continuous improvements in transportation and scientific discoveries of various kinds have made the nation more than a political entity. In many ways it has become an actual community.

Indeed, these advancements have developed so far that in some respects virtually the entire world is a vast community. If anyone is inclined to doubt the reality of such a community, the statistics regarding international trade will show the degree of interdependence among nations which has gradually developed. In the United States, for example, imports for consumption purposes, including duties paid, amounted to 2,251 million dollars in 1938.[1] In 1929 the League of

[1] United States Department of Agriculture, *Farmers in a Changing World, Yearbook of Agriculture*, 1940, p. 354. Data from *United States Department of Commerce Survey of Current Business*, Sept., 1939, p. 11.

Nations issued a directory of international organizations then active. The number listed totaled 470. They were distributed in the following manner:

	Number of organizations
Pacifism	13
Law and administration	34
Labor	56
Education	28
Feminism	30
Humanitarianism, religion, and morals	80
Economics and finance	9
Agriculture	16
Trade and industry	26
Communication and transit	33
Arts and sciences	73
Medicine and hygiene	31
Miscellaneous (including international languages, protection of nature, politics, library, etc.)	41 [2]

Another directory issued in 1942 contains a list of approximately five hundred organizations concerned with international affairs.[3] "There are at present about twenty-one commercial organizations in the United States actively promoting trade relations between the United States and the country represented."[4] These organizations try to foster commercial relations and assist in every way possible to encourage business and friendly relationships with our nation. Further proof of the basis of an international community is manifested in the timetables of steamship companies and airplane routes.

It is a characteristic of community life that common interests and activities tend to create institutions, mores, folkways, and other means of social control that maintain order in the community. Each community acquires these agencies and, if it be small, the elements of a true *Gemeinschaft* develop.[5] In most communities laws are largely observed or enforced to the extent that other community interests become effective and sanction them.[6] When these controls are present,

[2] League of Nations, *Handbook of International Organizations* (Geneva, 1929), pp. 295–312.

[3] Savord, Ruth, *American Agencies Interested in International Affairs.* Council on Foreign Relations, New York, 1942.

[4] *Ibid.*, p. 164.

[5] Herberle, Rudolph, "Fundamental Concepts in Rural Community Studies," *Rural Sociology*, 6 (1941) : 203–210.

[6] Hoffer, Charles R., "The Local Community and Social Control," *Rural Sociology*, 7 (1942) : 81–84.

a community can maintain a tolerable degree of harmony and peace. Gradually the principle of coöperation becomes dominant in community organization, and people are expected to observe accepted modes of behavior. If some members of the community do not, force may be used to make them conform.

When a similar analysis is applied to international relations, it becomes evident that a new community is emerging, albeit at a tremendous expense and loss of human life. The most obvious evidence of such a community is in the realm of economic relationships. Trade among nations has developed to a point where national prosperity during peacetime without it seems well-nigh impossible. But regardless of the basic importance of this interest in international community life, it has not contributed to harmony and good will. The process of competition has prevailed in this area, and consequently there has been a tendency for each nation to strive for immediate maximum economic advantage. Motives of competition are powerful. Consequently some nations have pushed trade relations and economic activities generally to a point where the processes of exploitation begin to operate with the result that conflict, rather than coöperation, emerges. Several decades ago Herbert Spencer advanced the idea that industrialism and militarism within a society were antagonistic to each other, for he reasoned that industrial development is dependent upon trade, and trade can occur best when peaceful relationships prevail.[7] But other factors, notably imperialistic nationalism, tend to disturb this relationship by causing military power to be perfected and used for the protection and expansion of economic development. Under such conditions conflict is inevitable.

In fully developed communities exaggerated emphasis on any particular activity is held in check by the various social controls which are operative therein. Besides police power, there are institutions, mores, customs, and public opinion which exert influence in favor of order and coöperative activities. In the international community these controls are nonexistent or not sufficiently numerous and potent to hold in check the behavior of a nation which is detrimental to the well-being of the community. Under such conditions war emerges as a logical, even unavoidable, consequence. The paucity of social controls on an international basis in comparison with local and national communities is indicated by the lists in Chart I. There

[7] Spencer, Herbert, *Principles of Sociology,* II, Part 3 (1897): 366–367.

CHART I

ORGANIZATIONS PROMOTING IMPORTANT INTERESTS OR ACTIVITIES IN LOCAL, NATIONAL, AND INTERNATIONAL COMMUNITIES

Interest or activity	Local community	National community	International community
Police power . . .	Police officers	Army and Navy law enforcement agencies (Federal Bureau of Investigation)	None, except by specific agreement
Economic life . . .	Local industries and agriculture	National economy (tariff, economic planning, etc.)	International trade
Education	Public schools	Nationally supported educational systems	Voluntary coöperation only
Government	Local organization; laws and ordinances	National legislation	International law
Religion	Local churches	State-supported church systems; denominational organization	Voluntary world organization partially achieved

is nothing in the international community comparable to police power in a local community or naval and military organization at the national level. The nearest approach to an international police force is the voluntary agreement among two or more nations to marshal their forces against a common enemy. In economic activities the local community has agriculture and industrial enterprises that are of general concern to most of the inhabitants. The national community has tariff programs and economic planning. In the international community there is only such international trade as can be developed within the framework of a group of states, each seeking maximum advantage. In matters involving legal controls the international community has only the pattern of international law, the enforcement of which depends not upon a common police power, as in the local or national community, but upon the decision of separate nations to observe the various regulations. Education to promote an understanding of its nature and problems exists only on a voluntary basis in the international community, whereas in local communities and in the nation taxation and legislation promote this interest. World-wide organization has been only partially achieved in religious affairs, but in local communities and in national life churches have made religion an effective influence in favor of justice and good will.

It is not surprising that with such a paucity of social controls in the international community conflict, that is, war, occurs with relentless regularity, even though the peoples in a large portion of the earth do not want war. They cannot prevent it, however, because at its present state of development the international community lacks the means to do so. As previously explained, the principal activity in community life at this level is economic. It is based on the process of competition, which in turn is fostered by nationalistic ideologies of an extreme nature. One factor tends to supplement the other. Extreme nationalism favors competition, and competition thus promoted tends to produce conflict, especially when other interests and controls are comparatively ineffective.

The preceding paragraphs show that the maintenance of peace depends upon the development of interests and controls in the international community which will bring about two conditions: (*a*) a recognition of mutual interests actually existing among peoples of different nations at the international level so that they may choose to coöperate rather than compete with one another; and (*b*) the development of international controls which will hold in check any unwarranted tendency to supplant coöperation with conflict among nations. It is evident that, viewed in this way, the attainment and maintenance of international peace is a tremendous task far surpassing in its purview a majority of programs which have been contemplated to accomplish this objective. Yet it is an approach which is indicated by a sociological analysis of the problem. The failure of peace movements as they have emerged and receded during the last century may be attributed, one is inclined to conclude, not to any fault in the purposes they tried to accomplish, for they were truly laudable, but, rather, to an absence of a sociological analysis and approach to the problem. A plan to bring about peace which deals with the results rather than the roots of a problem is certain to be ineffective. Programs designed to promote peace must be grafted on existing patterns of social organization, so that each nation will act as a member of the international community when matters concerning this community are involved. Participation in the activities of the international community need not weaken or endanger the existence of a nation. Indeed, it might prove to be the best safeguard the national community could have.

Even though motivated by the highest ideals known to mankind, war does not necessarily produce a durable peace. The outbreak of a war is incontestable evidence that the conditions before it occurred were intolerable, and a return to those conditions paves the way for a recurrence of international conflict. When such a policy is adopted, peace declarations gradually become ineffective, no matter how carefully they have been prepared. Peace is maintained not only by formal statements but by effecting adjustments in national and international relationships which will remove the occasion for war.

What, then, does a program designed to create conditions favorable to an enduring peace involve? It seems necessary, first of all, that some form of political organization be established in the international community and vested with power to hold in check any unwarranted use of military activity by a nation for exploitative purposes. It is too soon, perhaps, to determine in detail the exact nature of such an organization, though numerous proposals have been made. One of the most comprehensive is contained in the recommendation by a study conference on a just and durable peace called by the Federal Council of Churches. It states "that certain powers now exercised by national government must, therefore, be delegated to international government, organized and acting in accordance with a world system of law. Among the powers so delegated must be final judgment in controversies between nations, the maintenance and use of armed forces except for the preservation of domestic order and the regulation of international trade and population movements among nations." [8] The novel element in this proposal is the use of power by an international authority, but it is not inconsistent with successful social organization. Power is a great ally of man when regulated and used to attain conditions favorable to community life, but, uncontrolled, it is a destructive force without an equal. Vesting a supranational body with military resources to curb nations with

[8] Commission to Study the Bases of a Just and Durable Peace, *A Message from the National Study Conference on the Churches and a Just and Durable Peace*, p. 19. Sponsored by the Federal Council of the Churches of Christ in America, New York City, 1942.

imperialistic or exploitative ambitions constitutes a necessary, and moral, use of power in the international community, just as a police force is considered an essential agency in a local community.

In conjunction with the development of legal and political machinery to produce effective controls in the international community it is necessary to revise economic relationships so that contacts of nations in this realm of activity may not lead to exploitation and discontent which foment war. The philosophy of protectionism among nations that has been in existence for many decades and seemingly reached a climax during the period from the Versailles Treaty to the outbreak of World War II [9] must be modified to harmonize with the responsibility of a nation as a member of the international community. Social trends affecting international relations now make such a course necessary. Dominance of the pattern of coöperation is essential in community life, and the international community is no exception to this principle. A gradual shift toward freer exchange of goods among nations is indicated. As a matter of fact, an international community in the economic sphere is already in existence, although in a chaotic form, and is a greater social force than economic imperialism on a national basis. Time may yet show that Spencer's idea is a sound one.

Besides the adoption of a policy facilitating the exchange of goods among nations, there is need for further democratization of the benefits of economic activity so that people may have a greater degree of security. This need is well recognized since specific mention was made of it in the Atlantic Charter. The point to emphasize here is that economic insecurity, especially if engendered by exploitation, furnishes a basis of social unrest which is favorable to war. "The foundations of democracy can be rendered safe only when people everywhere have an opportunity to work and buy and sell with reasonable assurance that they will be able to enjoy the fruits of their work." [10]

In thus outlining political and economic organization which is conducive to peace there is no intention to overlook the fact that such measures cannot be effective unless various social organizations and

[9] Heyman, Hans, *Plan for Permanent Peace* (New York: Harper and Brothers, 1942), p. 43.

[10] Wallace, H. A., "Foundations of the Peace," *America Organizes to Win the War: A Handbook on the American War Effort* (Harcourt, Brace and Co., 1942), p. 366.

agencies seek to make them so. Without such support, proposals of this kind remain as abstract statements apart from the cultural values of the people. An international community in the realm of religion, education, and art is a necessary and logical development to give strength and meaning to economic and political organization, and the mechanical means for such a program are at hand.

It is a responsibility of Christianity, or of any other religion, for that matter, to emphasize continually throughout the world the idea that human life is sacred and that men are brothers more than they are enemies. This may be done in part by oral or written declaration, but such assertions will not have their maximum effect unless the ideas involved are actually demonstrated in local communities throughout the land. This implies continued and further coöperation among church denominations, support of programs designed to dissolve race prejudices, and encouragement of democratic programs and policies in economic affairs. A suggested list of guiding principles for churches in the postwar period states succinctly the goals for endeavor in this connection:

1. "There is a moral order which is fundamental and eternal, and which is relevant to the corporate life of men and the ordering of human society."

2. "Penitence is demanded of us as individuals and nations. 'All share in responsibility for the present evils'."

3. "There is no place for the spirit of revenge and retaliation."

4. A. "The principle of cooperation and mutual concern, implicit in the moral order and essential to a just and durable peace, calls for a true community of nations."

B. "The interdependent life of nations must be ordered by agencies having the duty and the power to promote and safeguard the general welfare of all peoples."

5. "Economic security is essential to a just and durable peace."[11]

If it should appear that these are ideas of idealistically inclined church leaders, suffice it to say that every statement is supported by quotations from national leaders in political and military affairs in the United States, England, China, and other countries favoring the cause of the United Nations.

[11] Federal Council of the Churches of Christ in America, *Post-War*, I, No. 2 (January, 1943): 1–2.

To the school and other agencies in the field of education belongs the task of informing people about the life and problems of mankind in different countries. Studies in geography need to be paralleled by a sociological description of the social organization and culture existing in the different regions, at both the high school and college levels, to the end that understanding and tolerance may supplant suspicion and distrust. Sociology and anthropology have amply demonstrated that the great variety of customs and mores regulating behavior have relevance in the particular environment where they originate. But these sciences have also demonstrated that amid these diversities there are values which appear repeatedly in a great majority of cultures throughout the world. Such values are desire for freedom, love of home and family, economic security, regard for truth, and appreciation of art symbols.[12] These constitute the psychosocial basis of the international community. A lasting peace will prevail when mankind becomes convinced that such values cannot be denied to one part of the international community and be enjoyed by another. Science has made it inevitable that all peoples in the world must have freedom commensurate with their ability to exercise it or else none can have it. The general acceptance of this idea and the inauguration of programs and policies among nations to make it a reality may yet be the great achievement of the twentieth century.

Michigan State College
East Lansing, Michigan

[12] Hocking, William Ernest, "The Cultural and Religious Organization of the Future," in Robinson, Howard, Ed., *Toward International Organization*, (New York: Harper and Brothers, 1942), pp. 170–172.

IS SOCIOLOGICAL METHODOLOGY STERILE?

ALFRED McCLUNG LEE

TO AVOID possible misunderstandings an elaboration of my title is desirable at the outset. It might be stated at more length thus: "Is sociological methodology relatively unproductive of advances in our knowledge of the structure and functioning of society?" Worded in this manner, my answer is: "Yes." But in view of the enthusiasm with which many academic sociologists devote their better efforts to the elaboration of this Neo-Scholastic pursuit, I know that I shall have considerable difficulty to convince even a few that my conclusion has merit.

Methodology is defined quite satisfactorily, from my standpoint, as "a branch of logic dealing with principles of procedure."[1] It thus includes among its "principles of procedure" those of mathematics, with its techniques for measuring, sampling, summarizing, and comparing that are labeled otherwise as statistics; those of semantics, with its emphasis upon *a priori* definitions and their delusory "clarifications"; and, I assume, those associated with such other research tools as the English and other languages, with their grammar, vocabulary, and idioms; writing techniques, photography, sound-recording machinery, and printing; and an array of other mechanical gadgets and logical devices, to be drawn upon as needed for current scientific purposes.

From this definition of methodology one would gather that the term labels a storehouse of useful tools, a great warehouse in which an experienced scientist can find the implements of his trade regardless of his field. And beyond this, one would understand that the methodologists have worked out principles concerning the proper use of such tools and procedures, principles that might help any scientist attain a maximum of productivity.

Methodology, including semantics and statistics and whatnot, would have an untarnished record of aid to science if my definition to this point gave a wholly accurate impression, an impression that

[1] *Webster's New International Dictionary*, 1935.

goes beyond superficialities. It does not. Methodology started out as a modest — albeit metaphysical — handmaiden of scientists, but she could not permanently accept such a lowly status. Under another label, "Logic," she had been no less than the household mistress for the philosophers through the ages, and she craved a return to the "good old days." What though philosophers had lost their vigor; Logic — garbed as Methodology — could gain new prestige in the household of a new and more vigorous breed, the scientists.

To change the metaphor and restrict the field, the devotees of sociological methodology sought to magnify the significance of what they did in their comfortable armchairs. They liked the way in which they could remain far removed from the fleshpots and cesspools of human society, out of range of such nasty conceptions as the struggle for existence and antagonistic coöperation, and yet call themselves sociologists, authorities somehow on the world of men. The fact that they did not know enough about human problems to be called in by the New Deal in peace or war annoyed them even after they were given sop assignments in opinion-sampling orgies, those paradises of the methodologists where a mathematically correct sample overcomes any possible absurdities along psychological or sociological lines. Well, as I was pointing out, these authorities on the world of men liked the way in which they could discourse learnedly in an abstract jargon flatteringly reminiscent of the terminology promulgated by the lordly physicists. Under such leadership the sociological dictum — "You must be scientific" — became the sociological dogma, "To be scientific, one must utilize the Scientific Method," or, stated otherwise, "To be scientific, one must utilize the latest advances in Methodology." Other criteria for scientific contributions might occur to a Darwin or a Freud or a Sumner, but the methodologists are intimate with the secrets of semantics and statistics, and they naturally have found themselves the keepers of that magical and alchemical principle, the Scientific Method. Methodology thus becomes not a tool but an end in itself.

Let us permit this thread of discussion to dangle a bit while I introduce another one: What does a scientist do? But, of course, some will say: "Everyone knows what a scientist does. This is a scientific civilization." I should like to take issue with those statements and especially with the latter before I attempt to answer my question. This is *not* a scientific civilization. This is a civilization in which the hard-wrought conclusions of scientists are seized upon, applied to bring forth the miracles of our gadget civilization, and treated as dogmas that are drilled into the minds of a part of our school children. Few children learn how scientists discover principles; even few men and women learn of the actual significance of the struggles some men make to broaden human knowledge. On the contrary, our science courses are based upon cookbooks in the physical sciences, prescribed routines through which students go to learn laboratory method, and upon manuals of accepted dogmas with suitable illustrations in the social sciences. The latter attach more significance to grasping a "minimum list of concepts" than to learning how to observe and understand the structure and functioning of human society!

In short, this is a civilization in which a few of our young people learn how to become scientists *in spite of* the combined efforts of our public, private, and parochial school systems in which formulas and method rather than observation and curiosity are glorified.

What does a scientist do? In answering this question I do not pose as being privy to the confidences of some great master. Neither am I going to explore the cloying verbosities with which speculative philosophers try to pull down and fondle the incontestable leaders in scientific development. I shall merely summarize what I have been able to learn of the procedures and fertile curiosities of Darwin, Freud, Sumner, and a few others whose impact upon their fields has become unmistakable, regardless of the antagonism with which their contributions were uniformly received.

What does a scientist do? Let us see. He is an observer, a *curious* observer of some area who painstakingly records, verifies, and analyzes his observations and who thus occasionally discovers some new characterizations of reality not hitherto included in our collection of acceptable bits of knowledge. This statement will amuse some readers no end because now they can shout: "But what is the point of the futile accumulation of facts? Is there some great merit in stamp collecting or in the sort of thing done by historical anthropologists as they follow subtle variations in the back-scratcher complex from tribe to tribe?" To this I would retort, for whatever a retort is worth, that a useful social scientist is curious about *significant* aspects of human society, aspects associated with powerful influences upon the lives of people.

But let me give some examples that demonstrate why I share with W. H. Hudson [2] the view that "an observer is a rarer thing than a genius":

Logical inferences revealed to man that disease is caused by sin. Observers discovered that disease is caused by microscopic "plants and animals . . . propagated according to strict laws of heredity."

Logical inferences fixed the notion in man's mind that the lights "in the sky are personalities which are busy with determining human destiny," but astronomical observers have concluded that all the motions of the stars might be accounted for by supposing such things as that the earth revolves around the sun.

Logical inferences led men to think that fossils "were made by Satan to deceive men and cause them to lose their souls by reasoning irreligiously," but scientists have demonstrated to the satisfaction of many that fossils "are remnants of ancient life; by careful study of them we can read the history of the earth." [3]

Logical inferences gave men the impression some time back in our culture that there is an imperative of human progress. We have progressed, they said, and we shall progress. Scientists, with less ethnocentric criteria for the status and workability of cultures and with a knowledge of more cultures than have most men, came to the conception of cultural adaptation, a value-free conception, and left the term "progress" for the philosophers, politicians, and other propagandists to juggle.

Logical inferences gave men the conviction that there is but one true and proper code of ethics or morality. Hardy fact gatherers furnished the evidence that has been summarized in the conception of "cultural relativity" or "moretic relativity."

I want to point to two outstanding characteristics of these examples: (1) In each, curious-minded fellows exploded preëxisting theories with unexplained but verified observations, and (2) In each, a wealth of such verified observations were summarized into a new theory that had to be accepted as the best theory available at the time by all interested in accurate conceptions of reality. And perhaps some additional notations are also in order: The observers, not being hard-breathing and precocious geniuses, got along quite well without pretentious frames of reference and other methodological

[2] Quoted by C. H. Ward in his *Builders of Delusion* (Indianapolis: Bobbs, Merrill, 1931), p. 223.

[3] Ward, *op. cit.*, p. 42.

gimcracks; they were seeking new knowledge; they were not preparing to write academically respectable papers; and they held no definitions of terms, no fine-wrought concepts, and no frames of reference in superstitious awe; they knew that all these things have temporary merit but that they can also strangle intellectual innovation.

The scientist is thus a curious-minded fellow who is not satisfied with what the "authorities" have reported as suitable theories concerning butterflies, fossils, stars, minds, or society and who sets about finding out whether or not the "authorities" were as accurate and as inclusive as possible. If he permits himself to be hog-tied with the *a priori* definitions and theories and frames of reference, he may not realize that existing methodological tools are not helpfully *methodical* in a scientific sense; in other words, he may not see that existing words and definitions do not fit his evolving conceptions or — even worse — that he cannot push the conceptions beyond the *a priori* definitions. Think of how many sociologists have been unable to get beyond the conception of evolution as a progressive process formulated by Spencer, despite the wealth of facts to the contrary. The scientist regards his definitions and conclusions as tentative, subject to modification in the light of increased experiential knowledge.

Thus, when his problem in hand suggests it, the scientific worker goes to the storehouse of techniques available to him in our and other societies and selects what he needs, but he tries to steer clear of Methodology, with its Neo-Scholastic bickerings and its other evidences of futilitarianism. He is *methodical,* not methodological. If he is a Darwin, he collects all the facts on organic change that can come to his hand from the worlds of fossils, breeders, amateur and professional naturalists, and others. If he is a Freud, he makes his first-hand observations as a laboratory experimenter, a clinician, and a private practitioner and avails himself of the experience of as many others as possible. If he is a Sumner, he makes his first-hand observations in his travels, in the city council of New Haven, and in state and national political arenas, and he avails himself of the experience of as many others in as many languages as his reading time will permit.

These men were methodical, not methodological, I want to repeat. Their published works have been attacked with great erudition by hordes of methodologists who can see their methodological weaknesses, but who cannot make comparable contributions to human knowledge. These scientists did not develop some special language to suit their purposes; they merely supplemented German, English, and other languages with only very necessary additional terms; they tried to write, as nearly as practicable, in a reasonably grammatical and lucid version of their mother tongues. Similarly, they did not develop the equivalent of such things as sociological statistics or sociological logic or sociological semantics or sociological thingamajigs[4] of one sort or another. They merely used whatever mathematics, statistics, typewriters, logic, and the like were needed as they were needed. They were interested in the experience that others had had with such tools, but they were distrustful of "principles of procedure" worked out by systematizers.

And now I should like to try to bring together the two threads of the discussion. In view of what scientists do, if I am reasonably accurate in my portrayal of their activities, is sociological methodology sterile? In such a brief paper it is difficult to demonstrate much, but it is my contention that sociological methodology is worse than sterile. It is strangling. It is a flight from the central business of those who want to know more about society. In its Never-Never Land of multiple coefficients of correlation and sociometric diagrams one is constrained to say with the late great Russell Gordon Smith[5] such things as he did when he wrote as follows:

> When I took my oral examination for that Teutonic atavism called Ph.D., I had acquired laryngeal dexterity with the fifty-seven systems of sociology and the hundred and fifty-seven theories of society then in vogue. . . . Then I stepped out, so to speak, in quest of this thing I was supposed to understand — this thing named society, human society.
>
> . . . it came to me with a shock that human society, as described by the classical sociologists, was a Platonic idea, a conceptual idealization, having about as much correspondence to the fire and sparkle, the dirt and tears and blood of real life, as Kant's *Ding an sich* has to a pregnant skunk.
>
> . . . To teach such formulae, based for the most part on pure deduction and applicable only in an intellectual fourth dimension, to unspoiled men on the threshold of life, was a form of intellectual dishonesty which I found peculiarly distasteful.

WAYNE UNIVERSITY
DETROIT, MICHIGAN

[4] Littell, Robert, "Sociological Thingamajigs," *Today*, 5, No. 16 (Feb. 8, 1936): 17.

[5] *Fugitive Papers* (New York: Columbia University Press, 1930), pp. 26–27.

INDEX OF AUTHORS, WITH SUBJECTS

PAPERS OF THE MICHIGAN ACADEMY OF SCIENCE, ARTS, AND LETTERS

Orders should be directed to the University of Michigan Press.

VOL. I (1921). Pp. xi + 424. $2.00.

VOL. II (1922). Pp. xi + 226. $2.00. Bound in paper, $1.50.

VOL. III (1923). Pp. xii + 473. $3.00. Bound in paper, $2.25.

VOL. IV (1924), PART I. Pp. xii + 631. $3.00. Bound in paper, $2.25.

VOL. IV (1924), PART II. A KEY TO THE SNAKES OF THE UNITED STATES, CANADA AND LOWER CALIFORNIA. By Frank N. Blanchard. Reprint in paper cover, $.80.

VOL. V (1925). Pp. xii + 479. $3.00. Bound in paper, $2.25.

VOL. VI (1926). (Botany only.) Pp. xii + 406. $3.00. Bound in paper, $2.25.

VOL. VII (1926). (No botany.) Pp. xii + 435. $3.00. Bound in paper, $2.25.

VOL. VIII (1927). Pp. xiv + 456. $3.00. Bound in paper, $2.25.

VOL. IX (1928). (Botany and forestry only.) Pp. xiii + 597. $4.00. Bound in paper, $2.25.

VOL. X (1928). (No botany or forestry.) Pp. xvii + 620. $4.00. Bound in paper, $2.25.

VOL. XI (1929). (Botany and zoology only.) Pp. xii + 493. $3.50. Bound in paper, $2.25.

VOL. XII (1929). (No botany or zoology.) Pp. xii + 348. $3.00. Bound in paper, $2.25.

VOL. XIII (1930). (Botany and zoology only.) Pp. xii + 603. $4.00. Bound in paper, $2.25.

VOL. XIV (1930). (No botany or zoology.) Pp. xv + 650. $4.00. Bound in paper, $2.25.

VOL. XV (1931). (Botany, forestry, and zoology only.) Pp. x + 511. $3.50. Bound in paper, $2.25.

VOL. XVI (1931). (No botany, forestry, or zoology.) Pp. x + 521. $3.50. Bound in paper, $2.25.

VOL. XVII (1932). (Botany, forestry, and zoology only.) Pp. x + 738. $4.00. Bound in paper, $2.25.

VOL. XVIII (1932). (No botany, forestry, or zoology.) Pp. xiii + 623. $4.00. Bound in paper, $2.25.

VOL. XIX (1933). Pp. xii + 662. $4.00. Bound in paper, $2.25.

VOL. XX (1934). Pp. xv + 755. $4.00. Bound in paper, $2.25.

VOL. XXI (1935). Pp. xv + 757. $4.00. Bound in paper, $2.25.

VOL. XXII (1936). Pp. xiv + 717. $4.00. Bound in paper, $2.25.

VOL. XXIII (1937). Pp. xiv + 671. $4.00. Bound in paper, $2.25.

VOL. XXIV (1938). Complete volume. Cloth. $5.00.

Separate Parts in Paper Covers

Part I. Botany and Forestry. 208 p. $1.75.

Part II. Zoology. 221 p. $1.75.

Part III. Geography. 134 p. $1.50.

Part IV. General Section. 152 p. $1.50.

Vol. XXV (1939). Pp. xiii + 643. Cloth. $5.00.

Separate Parts in Paper Covers

Part I. Botany and Forestry. Pp. 1–185. $1.75.

Part II. Zoology. Pp. 187–370. $1.75.

Part III. Geography and Geology. Pp. 371–496. $1.50.

Part IV. General Section. Pp. 497–640. $1.50.

Vol. XXVI (1940). Pp. xiii + 586. Cloth. $5.00.

Separate Parts in Paper Covers

Part I. Botany and Forestry. Pp. 1–158. $1.75.

Part II. Zoology. Pp. 159–342. $1.75.

Part III. Geography and Geology. Pp. 343–448. $1.50.

Part IV. General Section. Pp. 449–583. $1.50.

Vol. XXVII (1941). Pp. xii + 713. Cloth. $5.00.

Separate Parts in Paper Covers

Part I. Botany and Forestry. Pp. 1–180. $1.75.

Part II. Zoology. Pp. 181–334. $1.75.

Part III. Geography and Geology. Pp. 335–478. $1.50.

Part IV. General Section. Pp. 481–710. $1.75.

Vol. XXVIII (1942). Pp. xiii + 701. Cloth. $5.00.

Separate Parts in Paper Covers

Part I. Botany and Forestry. Pp. 1–277. $2.00.

Part II. Zoology. Pp. 279–431. $1.75.

Part III. Geography and Geology. Pp. 433–502. $1.00.

Part IV. General Section. Pp. 503–698. $1.75.

Vol. XXIX (1943). Pp. xiii + 606. Cloth. $5.00.

Separate Parts in Paper Covers

Part I. Botany and Forestry. Pp. 1–112. $1.25.

Part II. Zoology. Pp. 113–312. $1.75.

Part III. Geography and Geology. Pp. 313–354. $.50.

Part IV. General Section. Pp. 355–604. $2.00.

Printed and bound by CPI Group (UK) Ltd, Croydon, CR0 4YY
16/09/2026
14963492-0008